Modeling Solar Radiation at the Earth's Surface

Viorel Badescu (Ed.)

Modeling Solar Radiation at the Earth's Surface

Recent Advances

 Springer

Professor Viorel Badescu
Candida Oancea Institute
Polytechnic University of Bucharest
Spl. Independentei 313
Bucharest 060042
Romania
badescu@theta.termo.pub.ro

ISBN: 978-3-540-77454-9   e-ISBN: 978-3-540-77455-6

Library of Congress Control Number: 2007942168

Cover design: Erich Kirchner, Heidelberg, Germany

Printed on acid-free paper

9 8 7 6 5 4 3 2 1

springer.com

Foreword

Reading the twenty chapters of this book caused me mixed reactions, though all were positive. My responses were shaped by several factors. Although I have maintained a "watching brief" on the relevant literature, my last substantive writing on these topics was a series of papers published in 1993 in *Renewable Energy*. I am impressed to see how far the field has progressed in just over a decade – and not only in selected areas, but across all dimensions of solar radiation theory, measurement, modelling and application.

I was also pleased to be reminded of the solid knowledge-base that was generated by the research community of the 1980s and early 1990s – both the theoreticians and those of us with a more applied focus. The work reported in *Modeling Solar Radiation at the Earth's Surface* suggests incremental rather than revolutionary changes in our knowledge and understanding.

Another factor which influenced my response to the individual chapters, and collectively to the book, is the fact that solar radiation is now a mainstream source of energy that is making significant contributions to meeting the diverse and growing needs for "clean" energy. The joint attributes of being renewable and low carbon gives solar energy a status that now places it centre stage in discussions of energy futures. But a comprehensive and integrated understanding of the spatial, temporal, spectral and directional attributes of the resource, at relevant scales, is required if this energy source is to be utilized to its fullest potential. *Modeling Solar Radiation at the Earth's Surface* demonstrates unequivocally that the necessary capabilities now exist, and that they are mature and ready to be applied in the practical world. Of course many already have.

This leads me to wonder what my reactions will be when I (hopefully) reflect on progress a decade or so from now. How close to zero carbon will we be? What role will solar be playing in the energy mix? But of more importance in the context of *Modeling Solar Radiation at the Earth's Surface*, will I be noting further substantial progress in modelling insolation at the Earth's surface, or congratulating the authors of the present volume for the lasting relevance of their current endeavours?

Regardless, *Modeling Solar Radiation at the Earth's Surface* represents a significant milestone in documenting our knowledge of solar radiation theory, measurement, modelling and application. The subject has surely come of age. Dr. Viorel Badescu is to be congratulated on bringing together a team of authors who are acknowledged leaders in these areas of study. There is excellent representation from both the developing and developed worlds, and an appropriate balance reporting both theoretical and applied studies. The authors document our current knowledge and technical capacities, and how these have evolved to date. They also highlight shortcomings in our understanding and capabilities, making insightful suggestions how these might best be addressed through future research, monitoring and modeling.

John E. Hay

Institute for Global Change Adaptation Science, Ibaraki University, Mito City, Japan

Preface

Solar radiation data at ground level are important for a wide range of applications in meteorology, engineering, agricultural sciences (particularly for soil physics, agricultural hydrology, crop modeling and estimating crop evapo-transpiration), as well as in the health sector and in research in many fields of the natural sciences. A few examples showing the diversity of applications may include: architecture and building design (e.g. air conditioning and cooling systems); solar heating system design and use; solar power generation and solar powered car races; weather and climate prediction models; evaporation and irrigation; calculation of water requirements for crops; monitoring plant growth and disease control and skin cancer research.

The solar radiation reaching the Earth upper atmosphere is a quantity rather constant in time. But the radiation reaching some point on Earth surface is random in nature, due to the gases, clouds and dust within the atmosphere, which absorb and/or scatter radiation at different wavelengths. Obtaining reliable radiation data at ground level requires systematic measurements. However, in most countries the spatial density of actinometric stations is inadequate. For example, the ratio of weather stations collecting solar radiation data relative to those collecting temperature data in the USA is approximately 1:100 and worldwide the estimate is approximately 1:500. Even in the developed countries there is a dearth of measured long-term solar radiation and daylight data. This situation prompted the development of calculation procedures to provide radiation estimates for places where measurements are not carried out and for places where there are gaps in the measurement records. Also, the utility of existing weather data sets is greatly expanded by including information on solar radiation. Radiation estimates for historical weather can be obtained by predicting it using either a site-specific radiation model or a mechanistic prediction model. A site-specific model relies on empirical relationships of solar radiation with commonly recorded weather station variables. Although a site-specific equation requires a data set with actual solar radiation data for determining appropriate coefficients, this approach is frequently simpler to compute and may be more accurate than complicated mechanistic models. These simple, site-specific equations, therefore, may be very useful to those interested in sites near to where these models are developed.

The need for solar radiation data became more and more important mainly as a result of the increasing number of solar energy applications. A large number of solar radiation computation models were developed, ranging from very complicated computer codes to empirical relations. Choosing among these models usually takes into account two features: (1) the availability of meteorological and other kind of data used as input by the model and (2) the model accuracy. For most practical purposes and users the first criterion renders the sophisticated programs based on the solution of the radiative transfer equation unusable. As a consequence, the other models were widely tested.

The kind of solar radiation data required depends on application and user. For example, *monthly or daily averaged* data are required for climatologic studies or to conduct feasibility studies for solar energy systems. Data for *hourly* (or shorter) periods are needed to simulate the performance of solar devices or during collector testing and other activities. With the proliferation of cheap, high performance desktop computers, there is a growing need in various branches of science and engineering for detailed (*hourly or sub-hourly*) solar radiation data to be used for process simulation or design and optimum device sizing. As a best example, the past decade has seen a boom in the construction of energy efficient buildings which use solar architectural features to maximize the exploitation of daylight.

There is a need for a review of the existing "simple" methods of estimating solar radiation on horizontal and inclined surfaces. The goal of this book is to gather together a number of existing, as well as new models to compute solar radiation. Our objective is to classify various computing methods and models, to review statistical performance criteria and to recommend data sets suitable for validation. The book covers most aspects of solar radiation broadband computing models. Both statistical and deterministic methods are envisaged. Also, systematic information on the accuracy of each method is included. This information allows the user to choose the best available estimating model for his/her application when considering available data and demands for accuracy. The reader is implicitly provided with a solid understanding of the main mechanisms which determine the behavior of solar radiation on Earth surface and how solar radiation is estimated, measured and interpreted in an applied world.

The book is structured along logical lines of progressive thought. After an introductory Chapter 1 presenting the progress in solar radiation measurements, a group of two chapters (2 and 3) refer to fractal and statistical techniques used to quantify the properties of global irradiance. Chapter 4 is devoted to computation of solar radiation during clear days. The next four chapters present surveys of mature methods to compute solar radiation. The first three chapters in this series (5, 6 and 7) are related to correlations between solar irradiation and relative sunshine, cloud cover and air temperature, respectively. The series continues with Chapter 8, where the models to compute the diffuse solar fraction are presented. The methods of solar radiation estimation based on Artificial Neural Networks are discussed in Chapter 9. The next part of the book is oriented toward time-series procedures. In Chapter 10 the dynamic behavior of the solar radiation is discussed while chapter 11 shows how ARMA models are applied to solar radiation time series. Chapters 12 and 13 are devoted to new models used to generate series of actinometric data. The next chapters (14 and 15) present the details of MRM and METEONORM, respectively, which are very useful tools for solar radiation estimation. Chapters 16 and 17 refer to computation of UV solar radiation and sky diffuse radiance, respectively. These are important quantities used in the health sector and building design, respectively. Modern methods able to provide solar radiation information derived from Satellite Images are described in Chapters 18 and 19. A wide variety of techniques had been used during the years to describe model performance. This makes difficult comparison of models developed by different authors. Therefore, to meet the original objective would necessitate initially that validation method be classified, that models be catalogued uniformly, that statistical performance criteria are reviewed and that data sets suitable for validation be compiled. Chapter 20 deals with some of these aspects of model validation.

More details about the twenty chapters of the book are given below.

Chapter 1 by Christian Gueymard and Daryl Myers is designed to be an introduction to both solar radiation measurements and the concepts of solar radiation model validation. The authors discuss solar radiation fundamentals, components of solar radiation in the atmosphere, and instrumentation used to measure these components. Accuracy of solar measurements depends upon instrumentation performance, the reference scale and calibration techniques used. The physical principles of solar radiometer measurements, the World Radiometric Reference (WRR) reference scale, calibration and characterization techniques and the basic measurement uncertainty to be expected in measured data are described and commented. A brief discussion of measurement networks and data quality is presented. Very general and basic concepts behind solar radiation model types and their validation are outlined, leading to the detailed modeling and validation concepts that appear in the following chapters, especially chapter 20 which discusses model performance and validation in greater detail. The basic uncertainties in the *best* practical solar radiation data available today are on the order of 3% in direct beam, 5% in total global horizontal, $3\% +/- 2$ Watt in diffuse horizontal irradiance (measured with a black and white or corrected all-black pyranometer), 15% to 20% in diffuse radiation measured with uncorrected all black pyranometers behind a shadow band, and perhaps 5% to 20% in sunshine duration, for digital (including pyrheliometer) and analog (burning) sunshine recorders, respectively.

Chapter 2 by Samia Harrouni deals with fractal classification of daily solar irradiances according to different weather classes. The aim of this new approach is to estimate the fractal dimensions in order to perform daily solar irradiances classification. Indeed, for daily solar irradiances, the fractal dimension (D) ranges from 1 to 2. D close to 1 describes a clear sky state without clouds while a value of D close to 2 reveals a perturbed sky state with clouds. In fact, a straight line has a fractal dimension of one, just like its Euclidean dimension, once the line shows curls, its fractal dimension increases. The curling line will fill the plane more and more, and once the plane is filled up, it has a fractal dimension of two. Thus, the fractal dimension of a temporal signal has a fractal dimension between 1 and 2. To measure the fractal dimension of time series, several methods and algorithms based on various coverings have been elaborated. In order to improve the complexity and the precision of the fractal dimension estimation of the discreet time series, the author developed a simple method called "Rectangular covering method" based on a multi-scale covering using the rectangle as a structuring element of covering. An optimization technique has been associated to this method in order to determine the optimal time interval through which the line log-log is fitted whose slope represents the fractal dimension. This optimization technique permitted to improve the precision and to decrease the computing time of the method. In order to measure the performance and the robustness of the proposed method, one applied it to fractal parametric signals whose theoretical fractal dimension is known, namely: the Weierstrass function and the fractional Brownian motion. Experimental results show that the proposed method presents a good precision since the estimation error averaged over 180 tests for the two types of signals is 3.7%. The "Rectangular covering method" is then applied to estimate the fractal dimension of solar irradiances of five sites of different climates: Boulder and Golden located at Colorado, Tahifet and Imehrou situated in the Algerian south and Palo Alto in California. Then the author proposed a classification method of irradiances using the estimated fractal dimension. The method which defines fractal dimensions thresholds leads to classify the days of the five sites into three classes: clear sky day, partially clouded sky day and clouded sky day. Annually and monthly analysis of the obtained classes demonstrate that this classification method may be used to construct three typical days from global solar irradiances, which allows to reduce a long time series of several variables into typical days.

Chapter 3 by Joaquin Tovar-Pescador starts with fundamental ideas about statistical research techniques and their applications to solar radiation. An exhaustive revision of the research for modelling the statistic behaviour of solar radiation, from first works by Ångström until now, by means of normalised indices k_t, k_b and k_d, and in several temporal intervals (daily, hourly and instantaneous distributions), is made. Finally, the work is focused on the analysis of instantaneous distributions of global, direct and diffuse components of solar radiation, for different values of optical mass and different intervals of temporal integration (conditional distributions). The author proposes a method to model this behaviour by using a special type of functions, based on Boltzmann statistics, whose parameters are related to sky conditions.

Chapter 4 by Amiran Ianetz and Avraham Kudish starts with the common observation that the terrestrial solar irradiation is a function of solar altitude, site altitude, albedo, atmospheric transparency and cloudiness. The atmospheric transparency is a function of aerosol concentration, water vapor as well as other factors. The solar global radiation on a clear day is a function of all the abovementioned parameters with the exception of the degree of cloudiness. The analysis of the relative magnitudes of the measured solar global irradiation and the solar global irradiation on a clear day, as determined by a suitable model, provides a platform for studying the influence of cloudiness on solar global irradiation. Also, the magnitude of the solar global irradiation on clear day provides an estimate of the maximum solar energy available for conversion on a particular day. This chapter deals with the classification of the clear days and investigates a number of models for determining the solar global irradiation on clear sky, recommending the most suitable model. A clear day global index is defined and argued to be a better indicator of the degree of cloudiness than the widely reported clearness index. A clear day horizontal diffuse index and horizontal beam index are also defined and the correlation between them is studied. In addition, the analysis of frequency distribution types with regard to solar irradiation are discussed and defined on the basis of the skewness and kurtosis of the database. The preferred types of frequency distribution, viz., most suitable for solar energy conversion systems, are ranked and the reasoning behind the order of preference of the distribution types is explained. The average solar irradiation at a site, either global and/or beam, is of the utmost importance when designing a solar conversion system but the frequency distribution of the irradiation intensity is also a critical parameter.

Chapter 5 by Bulent Akinoglu shows that the relation between bright sunshine hours and solar radiation has quite a long history, which started at the beginning of the last century. Angström (1924) proposed a linear relation and since then many other different forms appeared in the literature. In this chapter, the physical base of this relation is explained and basic approaches of modeling are summarized. Among these approaches, two recent models are chosen and described in details. One of them is a hybrid model which uses wide band spectral information but also preserves a simple form. In this approach some wide band spectral data is used to calculate the ratio of actual global solar radiation to the clear sky value and this ratio is latter on correlated with the fractional bright sunshine. The second type of correlation discussed is the quadratic type expressions. In this modeling, a quadratic relation between bright sunshine hours and the global solar radiation has emerged within a physical formalism in which the ground reflected radiation is included. Also, the relation between the two Angström coefficients is discussed, which indicates that a quadratic relation should exits between the ratio of global solar radiation to the extraterrestrial radiation and fractional bright sunshine hours. A method is described to obtain a quadratic correlation using the relation between the Angström coefficients. The reasons why these models give relatively better estimates are discussed. A conclusion and some future prospects are also given in the chapter.

Chapter 6 by Ahmet Duran Şahin and Zekai Şen presents several new methods applied to the Angström equation and proposes a new alternative methodology to describe the dynamic behavior of this equation and solar irradiation variables. A dynamic model estimation procedure (denoted SSM) is proposed, which leads to a sequence of parameters. This makes possible to look at the frequency distribution function (probability distribution function) of model parameters. This allows deciding whether the arithmetic average of the parameters or the mode (the most frequently occurring parameter value) should be used in further solar irradiation estimations. In addition, it is easy and practical to do statistical analysis of Angström equation parameters and variables with SSM. Apart from the dynamic model parameter estimation procedure, an unrestricted solar irradiation parameter estimation procedure (UM) is presented, which considers only the conservation of the arithmetic mean and standard deviation of model's input and output variables, without the use of least squares technique. Around the average values, solar irradiation and sunshine duration values are close to each other for Angström and UM models, however, the UM approach alleviates these biased-estimation situations. Lastly, an alternative formulation to Angström equation is proposed for sunshine duration and solar irradiation variables estimation. This formulation and Angström equation procedure are compared and it is proven that there are some physical problems with the classical Angström approach.

Chapter 7 by Marius Paulescu refers to simple formulae that can be used to calculate daily global solar irradiation from air temperature data. These models either using air temperature as additional parameter to cloudiness or using only air temperature, are equally viable alternatives to the classical equations based on sunshine duration. Consequently, these models may be useful in many locations where sunshine duration measurements are missing but air temperature measurements are available in many-year database. A distinct case is the model built inside fuzzy logic, which may exhibit the flexibility needed in solar energy forecast. A C program included on the CD-ROM, which enable fuzzy calculation for daily global solar irradiation is presented. The arguments itemized are leading to the conclusion that air temperature, an all-important parameter worldwide recorded, can be used with success in the estimation of the available solar energy.

Chapter 8 by John Boland and Barbara Ridley refers to models of diffuse solar fraction. The authors previously developed a validated model for Australian conditions, using a logistic function instead of piecewise linear or simple nonlinear functions. Recently, it was proved that this model performs well for locations in Cyprus and that the form of the proposed relationship corresponds well to a logistic function. In this chapter the authors made significant advances in both the physically based and mathematical justification of the use of the logistic function. The theoretical development of the model utilises advanced non-parametric statistical methods. One has also constructed a method of identifying values that are likely to be erroneous. Using quadratic programming, one can eliminate outliers in diffuse radiation values, the data most prone to errors in measurement. Additionally, this is a first step in identifying the means for developing a generic model for estimating diffuse from global radiation values and other predictors. The more recent investigations focus on examining the effects of adding extra explanatory variables to enhance the predictability of the model. Examples for Australian and other locations are presented.

Chapter 9 by Filippos Tymvios, Silas Michaelides and Chara Skouteli discusses the applicability of Artificial Neural Networks (ANN) as a modern tool for the retrieval of surface solar radiation. It also comprises a survey of published research in this field, focusing on the neural methodology that was adopted, the database that was employed and the validation that was subsequently performed. Overall, this chapter aims at assisting the reader to obtain a good understanding of the capabilities and the applicability of the use of ANN in estimating solar radiation, to provide the basic theoretical background material for the issues discussed and to present some software tools that may offer the reader the assistance to build these models. Hopefully, the information presented in this chapter will trigger further research in more diverse areas related to solar radiation and renewable energy issues.

Chapter 10 by Teolan Tomson, Viivi Russak and Ain Kallis shows that solar radiation on the infinitely (in practice – sufficiently) long time axis is a stationary ergodic process that includes both periodical and stochastic components. Still, solar radiation could be a non-stationary process during some shorter time interval, intended for practical problem-solving. Different approaches have to be used for the analysis of the dynamical behavior of solar radiation. These approaches are explained in the chapter as well the technology of the analysis. The chapter is addressed mainly to engineers working on utilization of solar energy converted from the global radiation. Authors expect that the reader is acquainted with fundamentals and terminology of solar engineering.

Chapter 11 by John Boland starts by reminding that measurements of the components of solar radiation - global, diffuse and direct – traditionally have been made at only a limited number of sites. In recent times, for various reasons including the increased use of satellite images, this coverage has decreased further. To use simulation models to predict output from systems under the influence of solar radiation, hourly data values are usually needed. There are various approaches to generating synthetic sequences of solar radiation, using alternatively Markov models, state space models, neural networks or Box and Jenkins methods. The author uses the latter, which he also denotes as classical time series modeling structures. There is a specific reason for choosing this methodology. It is the approach that gives the most knowledge of the underlying physical nature of the phenomenon. The author describes the behaviour of global solar radiation on both daily and hourly time scales. In so doing, one identifies the various components inherent in the time series, seasonality, autoregressive structure, and the statistical properties of the white noise. Subsequently, procedures for generating synthetic sequences are presented, as well as procedures for generating sequences on a sub-diurnal time scale when only daily values (or inferred daily values) are available.

Chapter 12 by Llanos Mora-Lopez presents a model to generate synthetic series of hourly exposure of global radiation. This model has been constructed using a machine learning approach. The model is based on a subclass of probabilistic finite automata which can be used for variable order Markov processes. This model allows to describing the different relationships and the representative information observed in the hourly series of global radiation; the variable order Markov process can be used as a natural way to represent different types of days, and to take into account the "variable memory" of cloudiness. A method to generate new series of hourly global radiation, which incorporates the randomness observed in recorded series, has been also proposed. This method only uses, as input data, the mean monthly value of the daily solar global radiation and the probabilistic finite automata constructed.

Chapter 13 by Viorel Badescu proposes a new kind of solar radiation computing model. The novelty is that the approach uses two parameters to describe the state of the sky. The parameters are the common total cloud amount and a new two-value parameter - the sunshine number - stating whether the sun is covered or uncovered by clouds. Regression formulas to compute instantaneous cloudy sky global and diffuse irradiance on a horizontal surface are proposed. Fitting these relationships to Romanian data shows low bias errors for global radiation and larger errors for diffuse radiation. The physical meaning of the regression coefficients is explained. The model's accuracy is significantly higher than that based on total cloud amount alone. The model is applied to synthesize time-series solar radiation data. A first approximation relationship neglecting auto-correlation of the sunshine number is used in computations. Visual inspection as well as a statistical analysis shows a reasonably good similarity between the sequential features of measured and synthetic data. When the time interval is in the range of a few minutes, the sequential features of the generated time-series change significantly.

Chapter 14 by Harry Kambezidis and Basil Psiloglou describes the history of the various versions of the Meteorological Radiation Model (MRM), a computer code which estimates (broadband) solar irradiance values on a horizontal surface, using as input information only widely available meteorological parameters, viz. air temperature, relative humidity, barometric pressure and sunshine duration. Description of the various versions of MRM is given in detail together with their drawbacks, which triggered the development of the next version. Now MRM is at its version 5, especially developed by the Atmospheric Research Team (ART) at the National Observatory of Athens (NOA) for the purpose of this book. The performance of the recent MRM code is examined by comparing its results against solar radiation data from different locations in the Mediterranean area. Though MRM seems to work very well on cloudless days, an algorithm for calculating the solar radiation components on cloudy days has been added. This part of the code is based on the detailed sunshine duration information provided. If this consists of hourly values, the accuracy of the MRM is better than having just the daily sunshine duration instead. Nevertheless, the MRM results are on the same time steps as the input data, i.e. instant, half-hourly, hourly or daily values.

Chapter 15 by Jan Remund describes the formulation of a chain of algorithms for computing shortwave radiation used in Meteonorm Version 6 (Edition 2007). The basic inputs into the chain are monthly mean values of the Linke turbidity factor and global radiation. The outputs of the chain are time series of hourly values of global shortwave radiation on inclined planes. They correspond to typical years. This is achieved via stochastic generation of daily and hourly values of global radiation, splitting the global into beam and diffuse radiation and finally calculating the radiation on inclined planes. The short validation shows that the quality is good for the focused user group. The root mean square error of yearly means of computed beam radiation comes to 7%, that of the computed radiation on inclined plane is 6%.

Chapter 16 by John Davies and Jacqueline Binyamin refers to a climatological model for calculating spectral solar irradiances in the ultra-violet B waveband. The model is described and evaluated with Brewer spectrophotometer measurements at Resolute, Churchill, Winnipeg and Toronto in Canada. The model linearly combines cloudless and overcast irradiance components calculated with either the delta-Eddington or the discrete ordinates methods. It uses daily measurements of atmospheric ozone depth from the Brewer instrument, hourly observations of total cloud amount and standard climatological vertical profiles of temperature, pressure, humidity, ozone and aerosol properties for midlatitude and subarctic conditions. Cloud optical depth was calculated by iteration using single scattering albedo and asymmetry factor calculated by the Mie theory. Surface albedo varies linearly from 0.05 for snow-free ground to 0.75 for a complete snow cover.

Chapter 17 by José Luis Torres and Luis Miguel Torres carries on a revision of different models proposed for determining the angular distribution of diffuse radiance in the sky vault. Among all the models which can be found in literature, the ones that claim to be valid for all kinds of sky have been selected. Moreover, in many cases, given the similar origin of radiance and luminance, the models can be used for determining both quantities by means of the right set of coefficients. Although most of the models are empirical, as they are widely used, both a semiempirical model and another model that takes into consideration the stochastic nature of radiance in the sky vault are considered. In order to explain the procedure that is to be followed, an example of application of one of the newer empirical models is included. This model requires starting data usually available in conventional meteorological stations and uses an easy procedure for selecting the kind of sky. Knowledge of the angular distribution of diffuse radiance in the sky vault allows for a more precise calculation of the incident irradiance on a sloped surface. In this way, some of the simplifications used in many current models can be overcome. This new approach is especially interesting in urban environments and in terrains of complex orography where the incidence of obstacles can be very important. Finally, a section is devoted to show some of the measure equipments of radiance/luminance in the sky vault. Data registered by them has allowed for the elaboration of the empirical models and the evaluation of the different proposals. As these equipments became more widely used, they will be able to calibrate the models in places different from the ones where they were obtained. Among these equipments are considered both those in which the sensor that measures radiance/luminance moves sweeping different areas of the sky vault and those where the radiance/luminance is measured simultaneously in every area.

Chapter 18 by Jesús Polo, Luis Zarzalejo and Lourdes Ramírez deals with methods to derive solar radiation from satellite images. This approach has become an increasingly important and effective way of developing site-time specific solar resource assessments over large areas. Geostationary satellites observe the earth-atmosphere system from a fixed point offering continuous information for very large areas at temporal resolution of up to 15 minutes and spatial resolution of up to 1 km. Several methods and models have been developed during the last twenty years for estimating the solar radiation from satellite images. Most of them rely on transforming the radiance (the physical magnitude actually measured by the satellite sensor) into the cloud index, which is a relative measure of the cloud cover. Finally, the cloud index is related with the solar irradiation at the earth surface. A review of the most currently used models is made throughout this chapter, after describing the fundamentals concerning meteorological satellites observing the earth-atmosphere system and the cloud index concept. Finally, the degree of maturity in this technology has resulted in a number of web services that provide solar radiation data from satellite information. The different web services are briefly described at the end of this chapter.

Chapter 19 by Serm Janjai refers to the usage of satellite data to generate solar radiation at ground level. Solar radiation maps of Lao People's Democratic Republic (Lao PDR) have been generated by using 12-year period (1995–2006) of geostationary satellite data. To generate the maps, a physical model relating incident solar radiation on the ground with satellite-derived reflectivity and scattering and absorption due to various atmospheric constituents was developed. The satellite data provided cloud information for the model. The absorption of solar radiation due to water vapour was computed from precipitable water obtained from ambient relative humidity and temperature at 17 meteorological stations. The ozone data from the TOMS/EP satellite were used to compute the solar radiation absorption by ozone. The depletion of radiation due to aerosols was estimated from visibility data and the 5S radiative transfer model. Pyranometer stations were established in 5 locations in Lao PDR: Vientiane, Luangprabang, Xamnua, Thakhak and Pakxe. Global radiation measured at these stations was used to validate the model. The validation was also performed by using existing solar radiation collected at 3 stations in Thailand: Nongkhai, Nakhon Panom and Ubon Ratchathani. The solar radiation calculated from the model was in good agreement with that obtained from the measurements, with a root mean square difference of 7.2%. After the validation, the model was used to calculate the monthly average daily global solar radiation for the entire country. The results were displayed as monthly radiation maps and a yearly map. The monthly maps revealed the seasonal variation of solar radiation affected by the monsoons and local geography and solar radiation is highest in April for most parts of the country. From the yearly map, it was observed that the western parts of the country received high solar radiation with the values of 17–$18\,\mathrm{MJ/m^2}$-day. The areas which receive the highest solar radiation are in the south with the values of 18–$19\,\mathrm{MJ/m^2}$-day. The yearly average of solar radiation for the entire country was found to be $15.8\,\mathrm{MJ/m^2}$-day. This solar radiation data revealed that Lao PDR has relatively high solar energy potentials which can be utilized for various solar energy applications.

Chapter 20 by Christian Gueymard and Daryl Myers refers to different types of models which have been developed to provide the community with predictions of solar radiation when or where it is not measured appropriately or at all. An accepted typology of solar radiation models does not currently exist. Thus technical approaches to the evaluation and validation of solar radiation models vary widely. This chapter discusses classification of models by methodology, input criteria, spatial, temporal and spectral resolution, and discusses issues regarding the testing and validation of solar radiation models in general. The importance, and examples of, establishing sensitivity to model input data errors are described. Methods and examples of qualitative and quantitative quality assessment of input data, validation data, and model output data are discussed. The methods discussed include the need for totally independent validation data sets, evaluation of scatter plots, various statistical tests, and the principle of radiative closure. Performance assessment results, including comparison of model performance, for fifteen popular models are presented. Several approaches to evaluating the relative ranking of collections of models are presented, and relative ranking of the 15 example models using the various techniques are shown. This chapter emphasizes to the newcomer as well as the experienced solar radiation model developer, tester, and user, the nuances of model validation and performance evaluation. Section 2 addressed seven criteria describing typical solar radiation model approaches or types. Sections 3 and 4 described the principles of model validation and uncertainty analysis required for both measured validation data and uncertainties in model estimates. Sections 5.1 and 5.2 addressed qualitative and quantitative measured data quality and model performance. Section 5.3 emphasized seven constituent elements of model validation that must be addressed in any evaluation, including validation and input data quality, independence, and uncertainty to consistency of temporal and spatial extent, and validation limits. Section 6 discussed evolution and validation of model component parts (Sect. 6.1), the importance of, and difficulties associated with, interpreting independent model validation (Sect. 6.2), as well as demonstrated the practice (and difficulties) of comparing the performance of many models (Sect. 6.3).

The book facilitates the calculation of solar radiation required by engineers, designers and scientists and, as a result, increases the access to needed solar radiation data. To help the user of solar radiation computing models, a CD-ROM with computer programs and other useful information is attached to the book.

The Editor

Acknowledgments

A critical part of writing any book is the reviewing process, and the authors and editor are very much obliged to the following researchers who patiently helped them read through subsequent chapters and who made valuable suggestions: Dr. Ricardo Aguiar (INETI, Lisboa, Portugal), Prof. Bulent Akinoglu (Middle East Technical University, Turkey), Dr. Bulent Aksoy (Turkish State Meteorological Service, Ankara, Turkey), Prof. Adolfo De Francisco (University Polytechnic of Madrid, Spain), Dr. Saturnino de la Plaza Pérez (University Polytechnic of Madrid, Spain), Prof. Ibrahim Dincer (University of Ontario, Institute of Technology, Canada), Prof Jongjit Hirunlabh (King Mongkut University of Technology, Bankok, Thailand), Prof. Kostis P. Iakovides (University of Athens, Greece), Prof. Mossad El-Metwally (Faculty of Education at Port Said, Egypt), Prof. John E. Frederick (University of Chicago, USA), Dr. Chris Gueymard (Solar Consulting Services, Colebrook, USA), Prof. Detlev Heinemann (Oldenburg University, Germany), Prof. Pierre Ineichen (University of Geneva, Switzerland), Dr. C.G. Justus (Georgia Institute of Technology, USA), Dr. Harry D. Kambezidis (National Observatory of Athens, Greece), Dr. Gabriel Lopez Rodriguez (Universidad de Huelva, Spain), Dr. Michael Mack (Solar Engineering Decker & Mack GmbH, Hannover, Germany), Dr. Sasha Madronich (National Center for Atmospheric Research, Boulder, USA), Dr. James Mubiru (Makerere University, Uganda), Prof. Tariq Muneer (Napier University, Edinburgh, UK), Dr. Costas Neocleous (Higher Technological Institute of Cyprus, Nicosia), Prof. Atsumu Ohmura (Institute for Atmospheric and Climate Science, Zürich, Switzerland), Dr. Costas Pattichis (University of Cyprus, Nicosia), Dr. Marius Paulescu (West University of Timisoara, Romania), Prof. Richard Perez (University at Albany, New York, USA), Dr. Lourdes Ramírez Santigosa (CIEMAT, Madrid, Spain), Dr. Christoph Schillings (German Aerospace Center, Stuttgart, Germany), Dr. Mariano Sidrach-de-Cardona (University of Malaga, Spain), Dr. Thomas Stoffel (National Renewable Energy Laboratory, USA), Dr. Didier Thevenard (Numerical Logics Inc., Waterloo, Canada), Prof. Chigueru Tiba (Universidade Federal de Pernambuco, Brazil), Prof. Teolan Tomson (Tallinn University of Technology, Estonia), Dr. Joaquin Tovar-Pescador (University of Jaén, Spain), Professor John Twidell (AMSET Centre, Horninghold, UK), Prof. T. Nejat Veziroglu (University of Miami,

USA), Prof. Frank Vignola (University of Oregon, Eugene, USA), Dr. Kun Yang (University of Tokyo, Japan), Prof. M. M. Abdel Wahab (Cairo University, Egypt), Dr. Stephen Wilcox (National Renewable Energy Laboratory, USA).

In preparing this volume the editor has been assisted by Dr. Thomas Ditzinger, Monika Riepl and Heather King (Springer NL), to whom thanks are kindly addressed. The editor is particularly indebt to Dr. Chris Gueymard for continuous help with the preparation of the book. The editor, furthermore, owes a debt of gratitude to all authors. Collaborating with these stimulating colleagues has been a privilege and a very satisfying experience.

Contents

List of Contributors

Bulent G. Akinoglu
Middle East Technical University, Fizik Bol. 06531 Ankara, Turkey,
e-mail: bulent@newton.physics.metu.edu.tr

Viorel Badescu
Candida Oancea Institute, Polytechnic University of Bucharest, Spl. Independentei
313, Bucharest 060042, Romania, e-mail: badescu@theta.termo.pub.ro

Jacqueline Binyamin
Department of Geography, University of Winnipeg, Canada,
e-mail: binyamin@winnipeg.ca

John Boland
School of Mathematics and Statistics, Institute of Sustainable Systems and
Technologies, University of South Australia, Mawson Lakes, SA, Australia,
e-mail: john.boland@unisa.edu.au

John Davies
School of Geography and Earth Sciences, McMaster University, Hamilton, Ontario,
Canada, e-mail: davies.j@sympatico.ca

Christian A. Gueymard
Solar Consulting Services, P.O. Box 392, Colebrook, NH 03576, USA,
e-mail: chris@solarconsultingservices.com

Samia Harrouni
Solar Instrumentation & Modeling Group / LINS – Faculty of Electronics and
Computer, University of Science and Technology H. Boumediene, P.O Box 32,
El-Alia, 16111, Algiers, Algeria, e-mail: sharrouni@yahoo.fr

Amiran Ianetz
Ben-Gurion University of the Negev, P.O. Box 653, Beer Sheva 84105, Israel,
e-mail: amirani@zahav.net.il

Serm Janjai
Department of Physics, Faculty of Science, Silpakorn University, Muang Nakhon
Pathom District, Nakhon Pathom 73000, Thailand, e-mail: serm@su.ac.th

Ain Kallis
Marine Systems Institute, Tallinn University of Technology, Ehitajate tee 5, 19086,
Tallinn, Estonia, e-mail: kallis@aai.ee

Harry D. Kambezidis
Atmospheric Research Team, Institute of Environmental Research & Sustainable
Development, National Observatory of Athens, P.O. Box 20048, GR-11810 Athens,
Greece, e-mail: harry@meteo.noa.gr

Avraham Kudish
Ben-Gurion University of the Negev, P.O. Box 653, Beer Sheva 84105, Israel,
e-mail: akudish@bgumail.bgu.ac.il

Silas Chr. Michaelides
Meteorological Service, P. O. Box 43059, Nicosa, CY-6650, Cyprus, e-mail:
silas@ucy.ac.cy

Llanos Mora-López
Dpto Lenguajes y Ciencias de la Computación, E.T. S. I. Informática, Universidad
de Málaga, Campus Teatinos, 29071 Málaga, Spain, e-mail: llanos@lcc.uma.es

Daryl R. Myers
National Renewable Energy Laboratory, 1617 Cole Blvd. MS 3411, Golden CO
80401, USA, e-mail: daryl_myers@nrel.gov

Marius Paulescu
Physics Department, West University of Timisoara, Street V Parvan 4, 300223
Timisoara, Romania, e-mail: marius@physics.uvt.ro

Jesús Polo
Energy Department, CIEMAT, Solar Platform of Almería, Spain,
e-mail: jesus.polo@ciemat.es

Basil E. Psiloglou
Atmospheric Research Team, Institute of Environmental Research & Sustainable
Development, National Observatory of Athens, P.O. Box 20048, GR-11810 Athens,
Greece, e-mail: bill@meteo.noa.gr

Lourdes Ramírez
Energy Department, CIEMAT, Solar Platform of Almería, Spain,
e-mail: lourdes.ramirez@ciemat.es

Jan Remund
Meteotest, Fabrikstrasse 14, CH-3012 Bern, Switzerland,
e-mail: remund@meteotest.ch

Barbara Ridley
School of Mathematics and Statistics, Institute of Sustainable Systems and
Technologies, University of South Australia, Mawson Lakes, SA, Australia,
e-mail: barbara.ridley@unisa.edu.au

Viivi Russak
Tartu Observatory, 61602, Tõravere, Estonia, e-mail: russak@aai.ee

Chara S. Skouteli
Department of Computer Science, University of Cyprus, P.O. Box 20537, Nicosia
1678, Cyprus, e-mail: chara@ucy.ac.cy

Ahmet Duran Şahin
Istanbul Technical University, Aeronautic and Astronautic Faculty, Meteorology
Department, Maslak 34469 Istanbul, Turkey, e-mail: sahind@itu.edu.tr

Zekai Şen
Istanbul Technical University, Civil Engineering Faculty, Civil Engineering
Department, Hydraulic Division, Maslak 34469, Turkey, e-mail: zsen@itu.edu.tr

Teolan Tomson
Department of Materials Science, Tallinn University of Technology. Ehitajate tee 5,
19086, Tallinn, Estonia, e-mail: teolan@staff.ttu.ee

José Luis Torres
Department of Projects and Rural Engineering, Public University of Navarre,
Campus de Arrosadía, 31006 Pamplona, Spain, e-mail: jlte@unavarra.es

Luis Miguel Torres
Department of Projects and Rural Engineering, Public University of Navarre,
Campus de Arrosadía, 31006 Pamplona, Spain, e-mail: lmtorresgarcia@gmail.com

Joaquin Tovar-Pescador
Department of Physics, University of Jaén, Campus Las Lagunillas, s.n., 23071,
Jaén, España, e-mail: jtovar@ujaen.es

Filippos S. Tymvios
Meteorological Service, P. O. Box 43059, Nicosa, CY-6650, Cyprus, e-mail:
ftymvios@spidernet.net

Luis F. Zarzalejo
Energy Department, CIEMAT, Solar Platform of Almería, Spain,
e-mail: lf.zarzalejo@ciemat.es

Chapter 1
Solar Radiation Measurement: Progress in Radiometry for Improved Modeling

Christian A. Gueymard and Daryl R. Myers

1 Introduction

This chapter is designed to be a concise introduction to modern solar radiometry and how solar radiation measurements can be properly used for optimal solar radiation model development and validation. We discuss solar radiation fundamentals, components of solar radiation in the atmosphere, instrumentation used to measure these components, and accuracy of these measurements depending upon instrumentation performance, reference scale, calibration techniques, and quality control. The sources of measurement error are explained in detail, and a comprehensive assessment of the recent advances in radiometry is presented. Existing measurement networks and data sources are described. Various data quality assessment methods are also discussed. The concepts presented here can be successfully applied toward a more informed use of solar radiation data in a large range of applications.

2 Solar Radiation Measurement Fundamentals

Solar radiation consists of electromagnetic radiation emitted by the Sun in spectral regions ranging from X-rays to radio waves. Terrestrial applications of renewable energy utilizing solar radiation generally rely on radiation, or photons, referred to as "optical radiation", with a spectral range of about 300–4000 nm. Broadband measurements in this range are the most common and are described further in the next sections. Figure 1.1 shows the extraterrestrial solar spectrum (ETS) at the mean Sun-Earth distance, with extra detail for the ultraviolet (UV), i.e., below 400 nm, where a lot of spectral structure is obvious.

Christian A. Gueymard
Solar Consulting Services, Colebrook NH, USA, e-mail: chris@solarconsultingservices.com

Daryl R. Myers
National Renewable Energy Laboratory, Golden CO, USA, e-mail: daryl_myers@nrel.gov

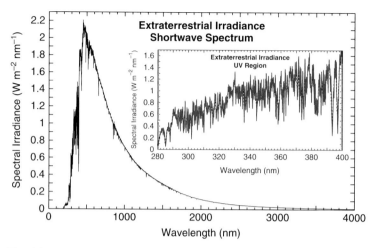

Fig. 1.1 Extraterrestrial solar spectrum in the shortwave at low resolution (0.5 to 5 nm). The highly-structured UV part of this spectrum is detailed in the inset at low (0.5 nm; thick black dots) and high (0.05 nm; thin gray line) resolution

The determination of ETS has evolved over time (Gueymard 2006), based on measurements from terrestrial observatories and spaceborne instruments, model calculations, or their combination. The low-resolution ETS in Fig. 1.1 is sufficiently detailed for most solar energy applications. It is a composite spectrum that uses all the types of data sources just mentioned, with proper weighting (Gueymard 2004). This dataset is provided in the file 'Gueymard_spectrum_2003.txt' on the accompanying CD. [It is also available, along with other similar datasets, from http://rredc.nrel.gov/solar/spectra/am0/].

The spectral integration of the ETS over all possible wavelengths (0 to infinity) is usually referred to as the "solar constant" or "air mass zero" (AM0) spectrum. In recent years, a more proper name, Total Solar Irradiance (TSI), has been introduced, since the Sun's output is not constant but varies slightly over short (daily) to long (decadal or more) periods (Fröhlich 1998). These variations have been monitored from space since 1978 with various broadband instruments, called absolute cavity radiometers (ACR). ACR uncertainty is about an order of magnitude lower than that of instruments used to measure the spectral distribution of the ETS, therefore TSI is more precisely known than its spectral details, shown in Fig. 1.1. Over a typical 11-year Sun cycle, there is a variation about $\pm 1\,\mathrm{W\,m}^{-2}$ around the solar constant. Short-term variations of about $\pm 4\,\mathrm{W\,m}^{-2}$ due to sunspots, solar flares, and other phenomena have been observed. The current best estimate of the average TSI based on 25 years of data is $1366.1\,\mathrm{W\,m}^{-2}$ (ASTM 2000; Gueymard 2004). However, recent measurements using a different type of instrument from the Solar Radiation and Climate Experiment (SORCE) satellite indicate a systematically lower value, of $\approx 1361\,\mathrm{W\,m}^{-2}$ (Rottman 2005). Work is underway to understand and resolve this discrepancy. It is highly probable that the revised value of the solar constant to be proposed in the near future will be somewhere between 1361 and $1366\,\mathrm{W\,m}^{-2}$. The daily or yearly excursions in TSI (± 0.1 to 0.2%) are small compared to all the other uncertainties involved in measuring or modeling solar radiation, and hence are usually not considered in terrestrial applications. In what follows, only the solar constant value matters, along with its predictable daily variation induced by the Sun-Earth distance, as described in Sect. 3.

The spectral distribution shown in Fig. 1.1 is modified and segregated into various component elements by the passage of the radiation through various layers of the Earth's atmosphere. A discussion of these spectral features or their measurement is beyond the scope of this chapter. Further information may be obtained elsewhere (e.g., Gueymard and Kambezidis 2004).

The science behind the measurement of electromagnetic radiation is called *radiometry*. Historically, simple instrumentation has been long used to evaluate the duration of bright sunshine in relation to day length. Radiometers of various designs have then been perfected to measure the energy in specific "components" of terrestrial solar radiation, as will be defined in Sect. 3. The interested reader should consult other textbooks (e.g., Coulson 1975; Iqbal 1983) for historical and technical details about common instruments used in solar radiometry.

Radiometers are constituted of different parts, mainly a casing or body, a radiation detector, and some electronics, including electrical circuits. For the instruments under scrutiny here, whose main purpose is to measure shortwave radiation (as opposed to UV or thermal radiation), detectors can be of three main types: thermopile, blackbody cavity, and solid state (semiconductor). The detector has a known spectral response to incident radiation. It is generally protected from the environment with some type of optical window, which can be transparent (e.g., glass or quartz), colored (e.g., interference filter), or translucent (e.g., white diffuser). The window optical transmission further limits the spectral range of the radiation actually measured.

In the following sections, we discuss solar radiation components, the measurement scale and reference against which solar instrumentation are calibrated, measurement principles used for the instrumentation, and—of greatest importance for the modeler of solar radiation—the uncertainty or accuracy to be expected from typical instrumentation. Recent advances in radiometric techniques are explained in detail. We begin with a description of the components of solar radiation that are created by the interaction of extraterrestrial solar radiation with various extinction processes within the Earth's atmosphere.

3 Components of Solar Radiation in the Atmosphere

From the Earth, the solar disk subtends a solid angle of about $0.5°$ on average. Due to the eccentricity of the Earth's elliptical orbit (0.0167), the distance from the Earth to the Sun varies throughout the year by $\pm 1.7\%$, resulting in a $\pm 3.4\%$ variation in the intensity of the solar radiation at the top of the atmosphere. The Sun thus acts as a quasi point source, illuminating the Earth with very nearly parallel rays of radiation. This quasi-collimated beam is the extraterrestrial direct beam, or extraterrestrial radiation, referred to as ETR.

As the ETR beam traverses the atmosphere, interaction between the photons in the beam and the atmosphere result in scattering and absorption of photons out of the beam into random paths in the atmosphere. Scattered photons (mostly at short wavelengths) produce the diffuse sky radiation, which we will denote by D. The remaining unabsorbed and unscattered photons, still nearly collimated, constitute the direct beam radiation, responsible for the casting of shadows, which we will denote as B. The total radiation flux on a horizontal surface in the presence of diffuse and beam radiation is often called "total" or "global" radiation. We will denote this global solar radiation on a horizontal surface as G. The term "global" refers to the concept that the radiation on a horizontal surface is received from the entire 2π solid angle of the sky dome. The difference between G at ground level and its corresponding value at the top of the atmosphere is what has been absorbed or reflected away by the atmosphere. On average, the Earth reflects about 29% of the incident solar irradiance back to space.

The total solar radiation received by a tilted (non-horizontal) surface is a combination of direct beam, diffuse sky, and additional radiation reflected from the ground (which we will denote as R), and should be referred to as total hemispherical radiation on a tilted surface. However, it is most often described by the simpler term "global tilted" radiation. Figure 1.2 illustrates the various components of solar radiation on intercepting surfaces.

The nearly collimated rays of the solar direct beam, in combination with the constantly changing altitude and azimuth of the Sun throughout the day, produces a constantly changing angle of incidence of the direct beam on a horizontal or tilted surface. Lambert's cosine law states that the flux on a plane surface produced by a collimated beam is proportional to the cosine of the incidence angle of the beam with the surface.

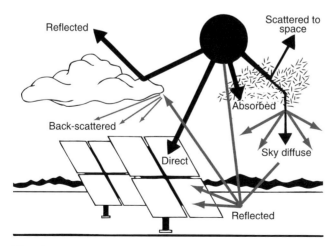

Fig. 1.2 Solar radiation components segregated by the atmosphere and surface

The incidence angle (i) of the solar beam upon a horizontal surface is equal to the solar zenith angle (z), i.e., the complement of the solar elevation (h). Thus the basic relation between the total global horizontal radiation G, direct beam radiation B at normal incidence, and diffuse radiation D on a horizontal surface can be described by Eq. (1.1):

$$G = B\cos(z) + D = B\sin(h) + D. \tag{1.1}$$

Equation (1.1) is fundamental to the calibration of solar instrumentation. For tilted surfaces, Eq. (1.1) needs to be rewritten as:

$$G = B\cos(\theta) + R_d D + R \tag{1.2}$$

where θ is the incidence angle with respect to the normal of the tilted surface, and R_d is a conversion factor that accounts for the reduction of the sky view factor and anisotropic scattering, and R is radiation reflected from the ground that is intercepted by the tilted surface (Iqbal 1983). Modeling each of the components of Eqs. (1.1) or (1.2) is the objective of many investigations and of other chapters in this book.

4 Instrumentation: Solar Radiometer

A pyrheliometer measures B, the direct beam radiation. Pyrheliometers have a narrow aperture (generally between 5° and 6° total solid angle), admitting only beam radiation with some inadvertent circumsolar contribution from the Sun's aureole within the field of view of the instrument, but still excluding all diffuse radiation from the rest of the sky (WMO 1983). Pyrheliometers must be pointed at, and track the Sun throughout the day. Their sensor is always normal to the direct beam, so that B is often called "direct normal irradiance" (DNI).

A pyranometer measures G, the global total hemispherical, or D, the diffuse sky hemispherical radiation. Pyranometers have a 180° (2π steradian) field of view. The measurement of D is accomplished by blocking out the beam radiation with a disk or ball placed over the instrument and in the path of the direct beam that subtends a solid angle matching the pyrheliometer field of view. This requires tracking the Sun with the blocking device through the day. A lower-cost alternative is to use a fixed band or ring of opaque material placed to shadow the pyranometer throughout the day. The shading band/ring approach introduces errors into the measurement of D, since part of the sky radiation is blocked by the shading device. This blocking effect varies with the shading device's geometry, time, and atmospheric conditions. An attempt to compensate for this is usually done by applying a geometric or empirical correction function to the data (e.g., Drummond 1956; Siren 1987), but this is far from perfect. For this reason, only the tracking-shade method is used at research-class sites.

Figure 1.3 portrays typical instruments used to measure G, B, and D in the field. These instruments all have thermopile detectors, except as noted. The thermopile-based detectors are sensitive to the whole shortwave spectrum, in contrast with

Fig. 1.3 Typical instruments for measuring solar radiation components. Pyranometers (top left), pyrheliometers (top right), shaded pyranometers (bottom). The five pyrheliometers shown use (left to right) silicon photodiode (triangular flange), thermopile (circular flanges), and cavity (oval cap) detectors

solid-state detectors, discussed further below. Note, however, that nearly all radiometers are protected from the elements by a window. This limits the spectral sensitivity of thermopile-based instruments to either 290–2800 nm for glass domes (used in most pyranometers) or 290–4000 nm for quartz plane windows (used in pyrheliometers).

The blackened absorbing surface of a thermopile is heated by the incident solar radiation. A number of thermojunctions between dissimilar metals (typically "type T" thermocouple junctions made of copper and constantan) are in contact with the absorbing surface. Thermal flux upon the junctions produces a voltage proportional to the difference in temperature between the heated junctions, and a similar set of "cold junctions" in series with the hot junctions. The output of thermocouples is slightly nonlinear, resulting in some curvature in the relationship between signal and temperature (NBS 1974).

Typically, a thermopile is made of approximately 20 to 40 junctions, and temperature differences between hot and cold junctions are $5°$ C for a $1000\,\mathrm{W\,m^{-2}}$ optical input, resulting in a $4\,\mathrm{mV}$ to $8\,\mathrm{mV}$ signal. In addition to the nonlinearities in the thermal response of the thermocouples described above, the absorbing surfaces of the detectors are not perfect isotropic (or "Lambertian") surfaces; and finally there are exchanges of infrared radiation between the radiometers/detectors and the (usually much colder) sky, all of which contribute to the uncertainty in calibrations and measurements using these detectors (Haeffelin et al. 2001).

Solid-state silicon photodiodes, mounted beneath diffusers, respond to incident radiation by generating a photocurrent, which is proportional to the incident flux. However, these devices have narrow spectral response ranges (e.g., about 350–1000 nm for crystalline silicon) and do not produce a signal proportional to the entire optical radiation spectrum. Since the path length of solar radiation through the atmosphere varies, and the atmosphere is not stable in composition throughout the day, changing infrared spectral content of the solar radiation is not captured by photodiode radiometers. As a result, solid-state detectors are less accurate than most thermopile radiometers discussed above.

"Burning" sunshine recorders were first developed by John Francis Campbell in 1853 and later modified in 1879 by Sir George Gabriel Stokes. The original instrument was based upon glass spheres filled with water, and later solid glass spheres. The latter device, which is known as the Campbell-Stokes (CS) recorder (Fig. 1.4), is still manufactured and used today, and constitutes the oldest solar radiation instrument still in service.

Modern instrumentation may be used to determine percent sunshine as well, by comparing the amount of time a pyrheliometer signal is above the bright sunshine threshold of $120\,\mathrm{W\,m^{-2}}$ with the day length (WMO 1996). Specially designed electronic sunshine recorders (using photodiodes) detect when the beam is above the

Fig. 1.4 Campbell-Stokes (left) and electronic sunshine recorder (right). The glass sphere on the left focuses solar beam radiation on a special paper, which is burned with a trace proportional in length to the time the beam is present

threshold (Fig. 1.4). These modern, automated devices have a much finer time reso-lution, a far more precise threshold, and eliminate the daily burden of replacing the special card used by CS instruments and of manually analyzing the burnt trace to estimate the daily hours of sunshine. These advantages considerably improve the re-liability, value, and accuracy of this measurement. Side-by-side experimental com-parisons, however, have demonstrated that there are significant and non-systematic differences between the crude CS sunshine data and the more refined electronic sunshine data. This prevents the replacement of older instrumentation at many sites with long records, due to the unwanted discontinuity in climatological sunshine trends that such a change produces. Considering the limited value of sunshine data compared to irradiance data, the former type of measurement is now considered ob-solete and has essentially lost its role in atmospheric research. Consequently, some countries (such as the USA) have already stopped its routine measurement.

5 Radiometric Reference and Calibration Methods

In this section, we discuss both calibration and characterization of solar radiome-ters. According to he United States National Institute of Standards and Technology (NIST) Engineering Statistics Handbook (available at http://www.nist.gov/div898/handbook):

> Calibration is a measurement process that assigns value to the property of an artifact or to the response of an instrument relative to reference standards or a designated measurement process. The purpose of calibration is to eliminate or reduce bias in the user's measurement system relative to the reference base.

Note that *characterization* is distinct from *calibration*, as defined in the same NIST handbook:

> The purpose of characterization is to develop an understanding of the sources of error in the measurement process and how they affect specific measurement results.

Once calibration has been accomplished, characterization can further address specific sources of error to produce results with improved accuracy. Also impor-tant is the concept of *traceability* of measurements. In the International Standards Organization Vocabulary of International Metrology (VIM), (ISO 1996), definition 6.10, "traceability" is defined:

> The property of the result of a measurement or the value of a standard whereby it can be related to stated references, usually national or international standards, through an unbroken chain of comparisons all having stated uncertainties.

For a measurement program to claim traceability the provider of a measurement must document the measurement process or system used to establish the claim and provide a description of the *chain of comparisons* that establish a *connection to a particular* **stated reference**.

5.1 *The Solar Radiation Measurement Reference: WRR*

There is no laboratory artifact for direct calibration of broadband solar radiometers. Instead, the Sun itself is used as a source. A group of specialized ACR instruments (first mentioned in Sect. 2) defines the reference, and the solar radiation scale. These radiometers are a sophisticated type of pyrheliometer. They match the temperature rise induced by absorption of sunlight by a cavity with a precision aperture to the temperature rise induced by an electrical current through the cavity walls while the cavity is blocked from the Sun. The precision aperture is normally *not* protected by a window, and therefore is sensitive to all wavelengths of the incident spectrum. (The WRR ACRs are only operated under favorable sky conditions; they are stored in an adjacent room at all other times.) The temperature rise in each case is measured as the voltage output of thermocouples in thermal contact with the cavity. The area of the precision aperture, heating current, and thermocouple voltages are "absolute" measurements used to compute the equivalent electrical and solar optical power density, thus the adjective "absolute" in the reference radiometer name (Kendall 1970; Willson 1973).

The World Radiometric Reference (WRR) is the measurement reference standard of irradiance for solar radiometry. The WRR was introduced to ensure world-wide homogeneity of solar radiation measurements. The WRR was originally determined from the weighted mean of the measurements of a group of 15 ACRs which were fully characterized. It has an estimated precision of 0.1%, accuracy of 0.3%, and stability of better than 0.01% per year (Fröhlich 1991). This determination establishes an experimental radiometric scale, which has been verified to closely correspond to the absolute definition of the International System of Units (SI) irradiance scale (Romero et al. 1991, 1995).

The World Meteorological Organization (WMO) introduced the mandatory use of WRR in its status in 1979, as a replacement for the older International Pyrheliometric Scale of 1956 (IPS56). This scale change also meant that older irradiance data had to be increased by 2.2% for consistency. The WRR is now realized by a group of ACRs called the World Standard Group (WSG). At the moment, the WSG is composed of 6 reference instruments: PMO-2, PMO-5, CROM-2L, PACRAD-3, TMI-67814 and HF-18748, operated by the World Radiation Center at the Physikalish-Meteorologisches Observatorium Davos (PMOD/WRC) in Davos, Switzerland (WMO 1983, 1996). This setup is shown in Fig. 1.5.

Every five years, an International Pyrheliometer Comparison (IPC) is held at the PMOD/WRC to transfer the WRR to the participating national reference instruments. The IPC is intended for the calibration of the ACRs from the Regional Radiation Centers of the six WMO regions. Thus the WSG/WRR is the "stated reference" that is the basis for traceability of solar measurements, through comparisons to the WSG. The procedures for the implementation of the transfer of WRR from the WSG are described elsewhere (Reda et al. 1996).

A slight degradation of uncertainty (from 0.3 to 0.35%) results from the transfer of WRR to regional ACRs during an IPC. More generally, the 0.3% uncertainty in the WRR is the highest definitive accuracy that can be achieved for a direct beam

Fig. 1.5 The WRR group of reference radiometers in normal operation at the World Radiation Center in Davos, Switzerland. (Photo courtesy PMOD/WRC)

measurement alone. The uncertainty in any other solar radiation measurement system other than the WSG must be greater. There are no internationally accepted references for the total global or diffuse sky radiation components independent of the WRR.

5.2 Calibration of Solar Radiometers

Since the WRR/WSG is the only reference for solar radiometer calibrations, pyrheliometers are calibrated by direct comparison with an ACR traceable to the WRR. The calibration of working pyrheliometers is similar to the transfer of the WRR to working reference absolute cavity radiometers (ASTM 2005a).

The ratio of the voltage signal of a test pyrheliometer to the ACR beam irradiance determines responsivity, R_s, in units of $\mu V W^{-1} m^2$. Many data samples through a large range of irradiance levels and solar geometry are collected, and an average responsivity calculated. When such calibrations are performed, the R_s values derived are often not flat or uniform, and exhibit some biases, which are discussed below. Some types of pyrheliometer are more sensitive to environmental conditions (mainly temperature and wind) than others. Most often, the average responsivity is divided into the field pyrheliometer voltage signal to produce measured irradiance. Characterization of the pyrheliometers is needed to produce data with lower uncertainty.

Pyranometer calibrations are accomplished by applying Eq. (1.1) (ISO 1990; ASTM 2005b). The reference ACR measures the direct beam, just as for the pyrheliometer calibrations. Exposed to the total hemispherical irradiance, G, a pyranometer output signal V_g is generated. If the pyranometer is then shaded by a device subtending the same solid angle as the field of view of the pyrheliometer, the pyranometer responds only to the diffuse sky irradiance, D, generating an output voltage V_d. Equation (1.1) implies that the vertical component of the direct beam is equal to the difference between the total and diffuse sky radiation:

$$B\cos(z) = G - D. \tag{1.3}$$

If the voltages V_g and V_d are used for G and D, the responsivity of the pyranometer, R_S can be computed from the vertical component of the direct beam:

$$R_S = (V_g - V_d)/[B\cos(z)]. \tag{1.4}$$

It is observed that R_S varies during the day, mainly as a function of z. A convenient single value of R_S is obtained for $z = 45°$, but a more refined method is desirable (see Sect. 6.1). The procedure described above is referred to as "shade-unshade" calibration, and is suitable for calibrating small numbers of pyranometers at a time. Once a "reference" pyranometer is calibrated in this manner, comparison (ratios) of test and reference pyranometer output voltages can be used to calibrate other pyranometers. However, this simple one-to-one comparison method is not recommended because it results in *larger* uncertainties (since the characteristic response curves of both pyranometers are most likely different) than either a direct shade-unshade calibration or the "component-summation" technique described next.

In the component-summation calibration method, a shade-unshade calibrated pyranometer monitors the diffuse "reference" irradiance, D_R. The direct beam is measured with an ACR. The responsivity of pyranometer i, R_{Si}, exposed to global irradiance with voltage output V_{gi} is computed from:

$$R_{Si} = V_{gi}/[B\cos(z) + D_R]. \tag{1.5}$$

Responsivities as a function of zenith angle for sample pyrheliometers and pyranometers are shown in Figs. 1.6 and 1.7, respectively. Note that these curves are not representative of radiometer make or model, as every instrument has a different individual response curve. Characterization of each pyranometer is needed to produce data with lower uncertainty. We discuss the sources of uncertainty and characterization of solar radiometers in the next section.

The detailed procedures to correctly calibrate solar radiometers and transfer these calibrations to other instruments are described in national and international standards such as those developed by the American Society for Testing and Materials (ASTM) and the International Standards Organization (ISO) as referenced above.

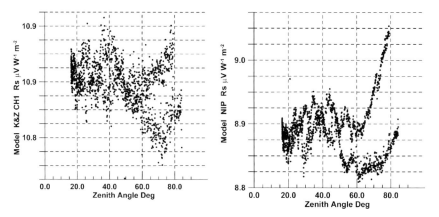

Fig. 1.6 Response curves for sample Kipp & Zonen CH1 and Eppley NIP pyrheliometers as a function of zenith angle, resulting from calibration against an ACR

Examples of response-curve characterizations of different pyrheliometers and pyranometers are included on the CD-ROM. These examples include the reports from the Broadband Outdoor Calibration (BORCAL) conducted in 2006 at NREL ('2006-02_NREL_SRRL_BMS.pdf') and at the Southern Great Plains Atmospheric Radiation Measurement site ('2006-02_ARM_SGP_Full.pdf').

6 Uncertainty and Characterization of Solar Radiometers

Every measurement only approximates the quantity being measured, and is incomplete without a quantitative uncertainty. ISO defines uncertainty as:

> A parameter, associated with the result of a measurement that characterizes the dispersion of the values that could reasonably be attributed to the measurand [the measured quantity].

Every element of a measurement system contributes elements of uncertainty. The Guide to Measurement Uncertainty (GUM) of the International Bureau of Weights and Measures is presently the accepted guide for measurement uncertainty (BIPM 1995). The GUM defines Type-A uncertainty as derived from statistical methods, and Type-B sources as evaluated by "other means", such as scientific judgment, experience, specifications, comparisons, or calibration data. These components of uncertainty are identified and combined in a rigorous manner to produce a standard, and eventually an expanded uncertainty reflecting the best estimate of uncertainty in a measurement parameter.

Various sources of uncertainty are reviewed in the next subsections. These in-depth developments largely result from recent investigations.

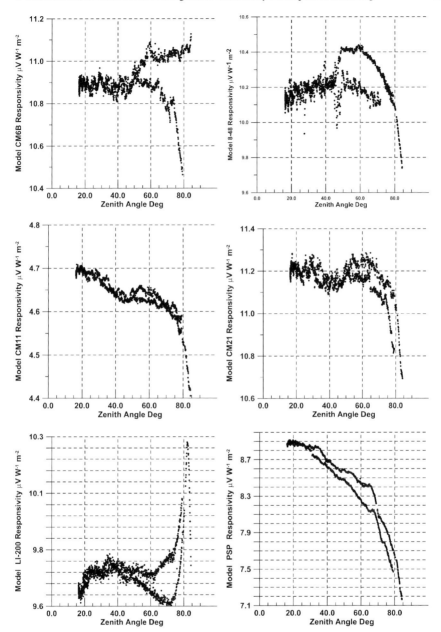

Fig. 1.7 Response versus zenith angle for sample pyranometers. Top left to lower right, Kipp and Zonen (KZ) CM 6b, Eppley 8-48, KZ CM 11, KZ CM 21, Li-Cor silicon-cell pyranometer, and Eppley PSP

6.1 Radiometer Uncertainty Sources

Sensitivity Functions

Combined uncertainties depend on the product of the sensitivity functions (partial derivatives of the response with respect to the measurement equation variables) and error source magnitudes, e_i. The largest contributions to shade-unshade uncertainty are from the e_{Vg} and e_{Vd} *which must include estimates of the thermal offset (10–70 μV) described below*, and data acquisition measurement uncertainty (typically $<10\,\mu V$). For example, for an all-black sensor pyranometer with a responsivity of $7.0\,\mathrm{mV}$ per $1000\,\mathrm{W\,m^{-2}}$, a 70-μV offset corresponds to an irradiance error of $-10\,\mathrm{W\,m^{-2}}$.

For the shade-unshade technique, with fixed direct beam error $e_B \approx 4.0\,\mathrm{W\,m^{-2}}$, zenith angle error $e_z \approx 0.06°$, and diffuse irradiance error $e_D \approx 2.0\,\mathrm{W\,m^{-2}}$ (black-and-white sensor), the uncertainty in R_S for a pyranometer is a function of: zenith angle, uncertainty in the pyranometer voltages, and magnitude of the beam; it ranges from about 1.0% at small zenith angles to >10% for zenith angles greater than 85°. The component-summation technique normally has lower total uncertainties, since the uncertainty in the direct beam is essentially the same as in the shade-unshade calibration, the uncertainty contribution from the diffuse measurement is rather low, and only one voltage measurement is involved for the instruments under test, as opposed to two voltage measurements in the shade-unshade technique (Myers et al. 2004).

Thermal Offsets

Climate research studies of solar radiation instrumentation, such as those made by the Baseline Surface Radiation Network (BSRN) participants, have characterized thermal offsets in thermopile pyranometers with all-black sensors measuring diffuse or global radiation (see http://www.gewex.org/bsrn.html). Thermal offsets produce negative data at night, and lower clear-sky diffuse or global irradiances during daytime.

This systematic negative bias explains in great part the discrepancy found between measurements and predictions from sophisticated radiative transfer models (e.g., Arking 1996; Philipona 2002). Similarly, thermal offsets explain why diffuse irradiance under very clean conditions has been reported lower than what pure Rayleigh-scattering theory (with no additional atmospheric constituents) predicts (Kato 1999; Cess et al. 2000). Other investigations confirmed the importance of thermal offsets, and offered correction methods (e.g., Dutton et al. 2001) as well as improved techniques for optimal pyranometry (Michalsky et al. 1999), which are summarized in Sect. 7. Thermal offsets produce absolute errors of typically -5 to $-20\,\mathrm{W\,m^{-2}}$ in clear-sky diffuse or global irradiance with all-black thermopile pyranometers, and are dependent on instrument installation (e.g., use of ventilators or heaters, etc.), design, deployment site, and atmospheric conditions. Current calibration methods cannot compensate directly for these errors.

For black-and-white sensors, the reference and absorbing thermopile junctions are in a similar thermal environment. These radiometers have lower (typically ± 0–$2\,\mathrm{W\,m^{-2}}$) offsets and normally produce more accurate diffuse sky measurements than all-black sensors without appropriate post-measurement corrections, or special considerations in their construction, such as compensating thermopiles.

Other Spectral Effects

Diffuse sky radiation has little energy in the shortwave near-infrared region 1000–2800 nm, while the direct beam has significant energy in that region. Therefore, nothing affecting the direct beam total irradiance between 1000 and 2800 nm, such as atmospheric water vapor, affects a shaded pyranometer signal. Consequently, for several different water vapor concentrations, and direct normal irradiances, the same shaded signal is possible from the pyranometer. By varying total precipitable water vapor from 0.5 to 3.5 cm, this "spectral mismatch" effect can be shown to result in differences of about 0.5% in R_s (Myers et al. 2004).

Geometric, Environmental and Equipment Uncertainty

Additional contributors to the uncertainty of all radiometers include: temperature coefficients, linearity, thermal electromotive forces, and electromagnetic interference. Moreover, the field of view of many pyrheliometers differs from that of the reference ACR, which results in slight differences in the circumsolar radiation sensed during calibration. For pyranometers, inaccuracies in the zenith angle calculation and in the sensor's cosine response must be considered. The latter issue is usually significant, particularly when measuring global irradiance under clear skies, because of the predominance of the direct beam. This issue is further discussed below and in Sect. 7. Finally, the specifications and performance of the data logging equipment (resolution, precision, and accuracy) must also be considered.

A pyranometer's departure from perfect Lambertian response is often called "cosine error". This has been documented in various publications (e.g., Michalsky et al. 1995; Wardle et al. 1996), with the result that improved calibration techniques using variable responsivity coefficients (rather than the conventional fixed single calibration number) became the recommended procedure (Lester and Myers 2006). What follows is an overview of the most advanced method currently used to calibrate pyranometers at research-class sites.

This method must be performed during a whole clear summer day, with z reaching values as close to 0 as possible. The responsivity for each zenith angle, $R_s(z)$, is calculated as before. The calibration data for the morning and afternoon are separately segregated into a number of zenith angle intervals. These data points are then fitted to a high-order polynomial in the form of Eq. (1.6):

$$R_S(z)_{AM/PM} = \sum_{i=0}^{n} a_i \cos^i(z) \tag{1.6}$$

where the a_i are $n+1$ coefficients for each morning and afternoon set of z. Thus there are two n-degree polynomials in $\cos(z)$ mapping the responsivity curve of each pyranometer. This method can be used with various z intervals. The original version used 5-degree intervals (Reda 1998). The current version uses 2-degree intervals, with $n = 48$.

With this sophisticated approach, uncertainties of no more than $\pm 2.1\%$ in measured pyranometer data can be achieved. This is a significant improvement over the conventional method of using a single value, $R_s(z_0)$, which may induce errors of up to $\pm 10\%$ at zenith angles largely separated from z_0.

Example data for various calibration results for a single pyranometer, reporting responsivity as a function of zenith angle in bins of $2°$, and $9°$, as well as derived coefficients for a fit to Eq. (1.6), may be found on the CD-ROM, in the folder CM22_all_Rs_NREL2006_02. In the folder NREL2006_02COEFF are the results of coefficient fits to several models of pyranometer (Kipp and Zonen CM-22 and CM6b, Eppley PSP, and Li-Cor LI200SB). Also included in that folder is a spreadsheet file, 'RCC_Function_RsCalculator.xls', which implements calculation of responsivities as a function of zenith angle using the coefficient files. The calculated uncertainties published in the '2006-02_NREL_SRRL _BMS.pdf' and '2006-02_ARM_SGP_Full.pdf' reports on the CD-ROM are based upon the techniques specified in the GUM and current knowledge of the sources of uncertainty, and their estimated magnitudes, during outdoor calibrations.

Basic calibration uncertainties of about 2.1% for pyranometers, and 1.8% for pyrheliometers, at "full scale" (i.e., $1000 \, \mathrm{W \, m^{-2}}$ or "1 sun") are the very best that can be expected with present instrumentation. This is equivalent to an uncertainty of $21 \, \mathrm{W \, m^{-2}}$ for global solar radiation and $18 \, \mathrm{W \, m^{-2}}$ for direct normal radiation.

When radiometers are deployed to the field, further sources of uncertainty arise, such as differing (usually lower resolution) data logging, cleanliness, and even atmospheric conditions, which must be considered in addition to the basic calibration uncertainty. Field measurements under varying, sometimes harsh, environmental conditions can easily double or triple the basic uncertainties (Myers 2005; Myers et al. 2004).

Solar radiation model developers must be aware that random and bias errors in models represent how well the model reproduces the measured data, and not necessarily the absolute accuracy of the radiation component. The particular case of *empirical* radiation models is worth discussing further. Such models are not based on algorithms that attempt to describe the physics of the various extinction processes in the atmosphere, but on simple relationships using some correlations between different phenomena and the observed irradiance. For instance, it has long been known that monthly-average global irradiation was roughly linearly correlated with sunshine duration. Because such models use irradiance observations for their development, any systematic or random error in irradiance measurement is embedded in the model. If the model is based on irradiance data that have been measured at site X with one set of instruments, and is used to compare its predictions to irradiance measured at site Y with a different set of instruments, a part of the apparent prediction errors at site Y will be due to a mismatch between the instrument characteristics. This insidious problem is usually overlooked, but is one possible reason why such models are rarely found of "universal" applicability.

7 Optimal Radiometry and Correction Techniques

Previous sections have described the sources of errors that affect solar radiation measurements. During the last decade, the results from high-quality research in radiometry have considerably improved our understanding of these errors and have provided ways to correct them or avoid them altogether. For instance, Eq. (1.6) above provides an effective method to correct a specific pyranometer for its cosine response error (but not necessarily thermal offset errors).

Another way of looking at the cosine error problem has resulted from the consideration that it is mostly caused by direct irradiance, the main component under clear skies. As mentioned above, direct irradiance measurements have a lower uncertainty than global measurements. Hence the development of a method derived from the component-summation technique described in Sect. 5. In such a setup, the global irradiance is *calculated* as the sum of the measured diffuse and direct irradiances according to Eq. (1.1). An unshaded pyranometer is still useful, first to obtain an independent measurement of global irradiance (which might be needed in case of tracking problems, etc.), and second, for quality assurance (by experimentally verifying that Eq. (1.1) is respected). The gain in accuracy is significant, assuming high-quality instrumentation and maintenance, typically $\approx 15\,\mathrm{W\,m^{-2}}$ (Michalsky et al. 1999).

This technique requires that diffuse irradiance be measured properly. This implies a ventilated instrument to homogenize temperatures and avoid condensation or frost on the dome. Furthermore, the thermal offset must be minimized (less than $\approx 2\,\mathrm{W\,m^{-2}}$), by using a pyranometer with either a black-and-white sensor or an all-black sensor with proper correction. Some correction techniques have been proposed (Bush 2000; Dutton et al. 2001; Haeffelin et al. 2001; Philipona 2002), based on the observed relationship between thermal offset and the nighttime net infrared (IR) radiation balance between the instrument and the sky. Accurate diffuse measurements with an all-black sensor require monitoring upwelling and downwelling IR by a colocated pyrgeometer. The comparison of the performance of various types (all-black and black-and-white) of shaded pyranometers has been the subject of a series of experiments (Michalsky et al. 2003, 2005, 2007; Reda et al. 2005). This research has resulted into a proposed working standard for the measurement of diffuse irradiance.

What is the practical significance of all these recent changes in measurement procedures? Only modest improvements in direct irradiance result from the adoption of windowed ACRs or pyrheliometers with low environmental influences on their signals. Conversely, large improvements are obtained if a pyrheliometer is used in lieu of the common indirect method where DNI is obtained by computation from global and diffuse data, through application of Eq. (1.1). This is particularly the case if diffuse irradiance is measured with a shadow band, and if the pyranometers are not corrected for cosine errors and thermal offset (Gueymard and Myers 2007). The implementation of optimal techniques for the measurement of diffuse and global irradiance also results in significant improvements, particularly under clear skies in winter (Gueymard and Myers 2007). Therefore, the development of empirical solar radiation models, or the validation of any type of radiation model, should only be based on optimal data.

All the modifications described above to the conventional measurements of dif-fuse and global irradiance (which were the norm only ten years ago) have induced noticeable effects on the operation of research sites. More sophisticated equipment is required, with redundancy and higher measurement frequency (e.g., 1 minute), and increasing cost. Calibration and maintenance are more stringent, requiring more skilled personnel. Finally, efficient techniques for quality assessment and dissemina-tion of the measured data need to be established. These new constraints require sig-nificant resources, limiting the commissioning of these high-quality radiation sites to only a few in the world. These research-class, high-end sites have been made possible because of their key role in the current climate change context, in which the radiative forcing of the climate must be precisely understood and predicted. The next section gives an overview of the conventional and high-end networks that exist in the world.

8 Measurement Networks and Data Quality Assessment

Finding measured solar radiation data involves locating measurement networks and experimental stations. Most countries have established and maintained national net-works, but the number and quality of these networks is constantly evolving, usually due to financial problems or changes in scientific priorities. Measurement stations being too scarce compared to the need for data in all applications, it is generally nec-essary to complement purely measured (or "primary") data by partly- or completely-modeled data (referred to here as "secondary data").

8.1 Large Networks and Primary Data Sources

Member countries of WMO contribute measured data to the World Radiation Data Center (WRDC), located at the Main Geophysical Observatory in St. Petersburg, Russia. The WRDC serves as a central depository for solar radiation data col-lected at over 1000 measurement sites throughout the world. The WRDC was es-tablished in accordance with a resolution of WMO in 1964. This data set is highly summarized, and difficult to keep up to date, as massive contributions are made on a monthly basis, and intensive quality control measures are implemented before the data becomes available. The majority of data available from 1964 to 1993 is accessible at http://wrdc-mgo.nrel.gov, and data from 1994 to present is available at http://wrdc.mgo.rssi.ru. (Unfortunately, these sites present data in dissimilar formats.) The WRDC archive contains mainly global solar radiation and sunshine duration data. Rarely, other data such as direct solar radiation or net total radiation are recorded. Not all observations are made at all sites. Data are usually available in the form of daily sums of global radiation. Monthly-mean sunshine data are also available at many sites. The map in Fig. 1.8 shows those stations from which at least some solar radiation data has been collected and reported. Similar data, but limited to monthly-average global irradiation, are also available from the International Solar Irradiation Database (http://energy.caeds.eng.uml.edu).

Other WMO-supported networks include BSRN (see Sect. 6.1) and the Global Atmospheric Watch (GAW; http://www.wmo.ch/web/arep/gaw/ gaw_home.html). A map showing the current and projected BSRN sites appears in Fig. 1.9. As mentioned earlier, BSRN has been a leader in developing high-quality radiation data. It adds its own stringent quality-control process to that from the member organizations collecting the data. Some publications detailing these quality-control procedures and other experimental issues are available from http://bsrn.ethz.ch.

The International Daylight Measurement Programme (IDMP; http://idmp.entpe. fr) consists of a specialized international network that is worth mentioning because it combines photometers to measure illuminance and radiometers for irradiance, with a unified quality-control procedure. Measuring stations are scattered around 22 countries, but many are not operational anymore.

The U.S. National Oceanic and Atmospheric Administration (NOAA) Surface Radiation, or SURFRAD (http://www.srrb.noaa.gov/surfrad/ sitepage.html) and U.S. Department of Energy Atmospheric Radiation Measurement (ARM) program (http:// www.arm.gov) have also established national research networks, whose sites are

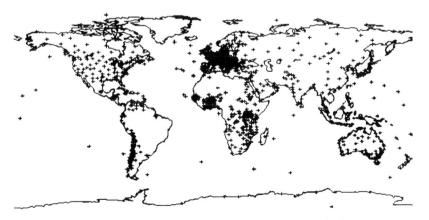

Fig. 1.8 Map of stations contributing at least some data to the WRDC database

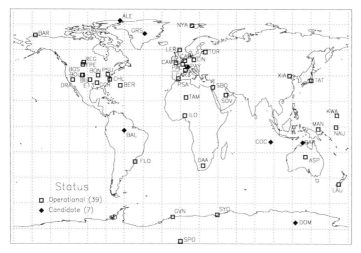

Fig. 1.9 Stations of the BSRN network

shown in Fig. 1.10. (Note that most of these sites are also part of BSRN.) Other important U.S. sources of quality data are grouped into the CONFRRM network http:// rredc.nrel.gov/solar/new_data/confrrm. Older data are accessible from http:// rredc.nrel.gov/solar/#archived.

Other nations with large solar radiation measurement networks include Brazil (Fig. 1.11), India, and Australia. More national networks do exist, but their datasets are sometimes difficult (or expensive) to obtain, and/or difficult to quality assess. Many times, the full complement of solar components (direct beam, total

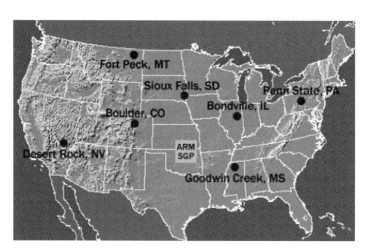

Fig. 1.10 United States SURFRAD National Oceanic and Atmospheric Administration and Department of Energy ARM Southern Great Plains (SGP) measurement networks

Fig. 1.11 Brazilian solar radiation measurement network

horizontal and diffuse horizontal) are not measured so quality checks on the balance of the three components are not available. In addition, calibration and measurement schedules may be irregular and intermittent.

8.2 Secondary Data Sources

Appropriate models are now widely used to fill gaps in measured data, expand their measurement period (to, e.g., 30 years), and estimate solar radiation either at specific sites where only other meteorological data are available, or at the world or continent scale using gridded satellite data. Although such datasets may contain some measured radiation data, the bulk of their content is always modeled values. However, they are usually merged with other meteorological data, which is highly useful in most applications. Examples of such datasets are given below.

Europe has been very active about collecting measured radiation data, supplementing them with models, and developing solar maps since the late 1970s. Successive editions of the European Solar Radiation Atlas have appeared. Information about the latest edition can be found from HelioClim (http://www.helioclim.org).

Other European-based data sources (for Europe or the world) include SoDa (http://www.soda-is.com), Satel-Light (http://www.satel-light.com), DLR-ISIS (http://www.pa.op.dlr.de/ISIS), SOLEMI (http://www.solemi.com), and Meteonorm (http://www.meteonorm.com). (Note that some of them are of a commercial nature.) Radiation datasets and maps for other continents have been recently developed through the international SWERA project (http://swera.unep.net). Daily solar radiation maps are also available for Australia (http://www.bom.gov.au/ reguser/by_prod/radiation) and Scandinavia (http://produkter.smhi.se/strang).

In North America, the most important sources of data and maps are from NREL (http://rredc.nrel.gov/solar/old_data/nsrdb;   http://www.nrel.gov/gis/solar.html), NASA (http://earth-www.larc.nasa.gov/solar; http://power.larc.nasa.gov; http://data. giss.nasa.gov/seawifs; http://eosweb.larc. nasa.gov/PRODOCS/srb/table_srb.html; http://flashflux.larc.nasa.gov), and Environment Canada (http:// www. ec.gc.ca).

Finally, convenient sites centralize links to various sources of world weather data for building applications (http://www.eere.energy.gov/buildings/energyplus/cfm/weather_data.cfm) or photovoltaic applications (http://re.jrc.cec.eu.int/pvgis/solrad/index.htm).

8.3 Data Quality Assessment

The simplest approach to quality assessment of solar radiation data is comparison with physical limits. Are the data within "reasonable" bounds? Another basis for data quality assessment is Eq. (1.1), the "closure" equation between the solar components. However, given the *instrumentation* issues addressed above, it is clear that the closure can occur even if large errors are present in any one or all of the three components. This problem is also traceable to the component-summation and even shade-unshade calibration approaches, where unknown bias (type B) errors are embedded in the test instrumentation. Another approach to identifying the type-B errors and characterizing them under clear-sky conditions has been comparison with modeled clear-sky data. Physical limits, closure, and model comparison approaches will briefly be described in the next subsections.

Quality Assessment Based Upon Physical Limits

As the name implies, the physical limits approach to solar data quality assessment compares measured data with estimated or defined limits. For instance, is the radiation component within the range of zero to the maximum possible expected value? Is the direct normal irradiance greater than zero and less than the extraterrestrial value? Is the global horizontal no greater than the vertical component of the extraterrestrial beam? Is the diffuse irradiance more than the expected Rayleigh diffuse sky? Note that while somewhat crude, and allowing the possibility of one or more components to pass such tests even when bad (such as direct beam of zero on a clear day, because the tracker was not pointing correctly, or an attenuated DNI due to a dirty pyrheliometer window), the last example above did lead to the identification of the thermal offset problem described in Sect. 5.

However, for the most part, the physical limits tests cannot provide the level of accuracy needed to assure that measurement instrumentation is indeed functioning properly, unless used with intensive human interaction. This interaction includes regular cleaning of windows, checking of tracker and shading disk alignment, etc. Other intensive human operations involve daily inspection of diurnal irradiance profiles, combined with knowledge of meteorological conditions on the day being inspected. Clearly, these are difficult procedures to automate and implement. They also involve a good deal of qualitative, rather than quantitative evaluation. Thus, the quality assessment may be more subjective than objective.

Quality Assessment Based Upon Closure

An alternative to the physical limits approach is to rely on Eq. (1.1), the theoretical relation between the three components. This approach has important advantages: it is objective and can be performed a posteriori. It can also be implemented directly, or more simply using irradiance values normalized to extraterrestrial beam (I_o) values. These normalized values are referred to as clearness indices, $K_n = B/I_o$ for direct beam, $K_t = G/[I_o \cos(z)]$ for global total hemispherical, and $K_d = D/[I_o \cos(z)]$ for diffuse sky clearness. The closure equation, Eq. (1.1), then becomes

$$K_t = K_n + K_d. \tag{1.7}$$

With a large collection of historical data, the site-specific relation between any two of the clearness indices with any other can be developed, and the physical limits boundaries greatly reduced to an envelope of acceptable data, bounded by limiting curves, rather than zero and some upper limit. Figure 1.12 shows a schematic relationship between K_t and K_n for a site, with analytical boundaries defined by double-exponential Gompertz functions (Maxwell 1993; Younes et al. 2005).

The equation of the Gompertz curves (Parton and Innes 1972) is:

$$Y = A \cdot W^{C \cdot W^{D \cdot X}} \tag{1.8}$$

where choice of A, W, C, and D, along with judicious "shifting" left and right along the X-axis, result in the proper "S" shaped boundaries around the data. Acceptable data then falls 'within' the analytic boundary curves. A library of curves can be build up for sites, times, and air mass conditions. An important point to keep in mind regarding either a direct computation of the closure condition or the clearness index approach is that with the known uncertainties in measured data, a tolerance, or acceptable deviation from perfect closure is needed. Typically, with measurement data uncertainties of 3% to 5% in total global and direct beam data, tolerances

Fig. 1.12 Clearness index
relation between K_t and K_n
showing schematic envelope
of analytical (Gompertz)
functions (solid curved lines)
which can be assigned for
acceptable data

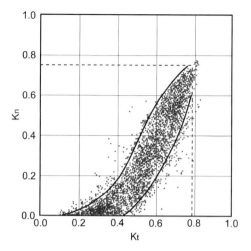

of ±5% in the balance are allowed. This means tolerances of about 0.02 to 0.03 in
the clearness-index approach.

Note again that if calibrations and measurements are performed with bias (type
B) errors inherent in each of the instruments for measuring different components, the
closure test can be "passed" even though the measurements themselves still contain
errors, perhaps several times as large as the tolerances.

Other approaches to establishing envelopes around physical relations between
measured data components (such as diffuse and direct to total global ratio) have
also been developed (Younes et al. 2005).

Quality Assessment Based Upon Comparison with Models

Many models based on the physics of radiation transfer through the clear atmosphere have been developed (Lacis and Hansen 1974; Atwater and Ball 1978;
Hoyt 1978; Bird and Hulstrom 1981a, 1981b; Davies, McKay 1982; Gueymard
1993, 2008). These models can be compared with clear-sky measured data to determine if measured data deviate significantly from "expected" data.

Satellite-based remote sensors and the development of algorithms for estimating
the solar flux at the surface has also lead to the possibility of using satellite-based
estimates to evaluate the performance of ground-based sensors (Perez et al. 1997,
2002; Myers 2005). An example of a web-based means of evaluating ground-based
measurements with respect to estimates derived from satellite data has been described (Geiger et al. 2002).

Of course, in all cases, model inputs or flux estimation algorithms have to be
relatively accurate, and there must still be tolerances on measured data to account for
possible instrumentation bias errors, as well as an idea of the additional sources of
uncertainty in the models and algorithms themselves (Gueymard and Myers 2007).

9 Conclusions

We urge the reader to remember, as he or she studies the rest of this book, that there are no perfect measurements. However, the field of solar radiometry has critically progressed in the last few years, resulting in significantly improved measurement quality at research-class sites. Much more work yet needs to be done to obtain better instruments, reference scales, calibrations, characterizations, and corrections for measuring solar radiation accurately in the field, and improve the data quality at the vast majority of sites that still rely on suboptimal experimental techniques.

The basic uncertainties in the *best practical* solar radiation data available today are still on the order of 3% in direct beam, 5% in total global horizontal, 3% $\pm 2\,\mathrm{W\,m^{-2}}$ in diffuse horizontal irradiance (measured with a black-and-white or corrected all-black pyranometer), 15% to 20% in diffuse radiation measured with uncorrected all-black pyranometers behind a shadow band, and perhaps 5 to 20% in sunshine duration, for digital (including pyrheliometer) and analog (burning) sunshine recorders, respectively. For the future, we can only hope for better models through better instrumentation and improved measurement techniques.

References

Arking A (1996) Absorption of solar energy in the atmosphere: Discrepancy between model and observations. Science 273: 779–782

ASTM (2000). Standard solar constant and zero air mass solar spectral irradiance tables. Standard E490-00, American Society for Testing and Materials, West Conshohocken, PA

ASTM (2005a). Standard test method for calibration of pyrheliometers by comparison to reference pyrheliometers. Standard ASTM E816-05, American Society for Testing and Materials, West Conshohocken, PA

ASTM (2005b). Standard test method for calibration of a pyranometer using a pyrheliometer. Standard ASTM G167-05, American Society for Testing and Materials, West Conshohocken, PA

Atwater MA, Ball JT (1978). A numerical solar radiation model based on standard meteorological observations. Solar Energy 21: 163–170

BIPM (1995). Guide to the expression of uncertainty in measurement. Rep. ISBN 92-67-10188-9, International Bureau of Weights and Measures (BIPM) by International Standards Organization (ISO), Geneva, Switzerland

Bird RE, Hulstrom RL (1981a) A simplified clear sky model for direct and diffuse insolation on horizontal surfaces. Tech. Rep. SERI TR-642-761. Golden, CO, Solar Energy Research Institute (on line at http://rredc.nrel.gov/solar/models/clearsky)

Bird RE, Hulstrom RL (1981b) Review, evaluation, and improvement of direct irradiance models. Trans. ASME J. Solar Energy Engineering 103: 182–192

Bush BC, Valero FPJ, Simpson AS, Bignone L (2000) Characterization of thermal effects in pyranometers: a data correction algorithm for improved measurement of surface insolation. J. Atmos. Ocean. Technol. 17: 165–175

Cess RD, Qian T, Sun M (2000) Consistency tests applied to the measurement of total, direct, and diffuse shortwave radiation at the surface. J. Geophys. Res. 105D: 24881–24887

Coulson KL (1975) Solar and terrestrial radiation. Academic Press

Davies JA, McKay DC (1982). Estimating solar irradiance and components. Solar Energy 29: 55–64

Drummond AJ (1956) On the measurement of sky radiation. Arch. Met. Geophys. Biokl. A7: 413–437

Dutton EG, Michalsky JJ, Stoffel T, Forgan BW, Hickey J, Nelson DW, Alberta TL, Reda I (2001) Measurement of broadband diffuse solar irradiance using current commercial instrumentation with a correction for thermal offset errors. J. Atmos. Ocean. Technol. 18: 297–314

Fröhlich C (1991) History of solar radiometry and the World Radiometric Reference. Metrologia 28: 111–115

Geiger M, Diabate L, Menard L, Wald L (2002) A web service for controlling the quality of measurements of global solar radiation. Solar Energy 73: 475–580

Gueymard CA (1993) Critical analysis and performance assessment of clear sky solar irradiance models using theoretical and measured data. Solar Energy 51: 121–138

Gueymard CA (2004) The sun's total and spectral irradiance for solar energy applications and solar radiation models. Solar Energy 76: 423–453

Gueymard CA (2006) Reference solar spectra: their evolution, standardization and comparison to recent measurements. Adv. Space Res. 37: 323–340

Gueymard, CA (2008) REST2: High performance solar radiation model for cloudless-sky irradiance, illuminance and photosynthetically active radiation—Validation with a benchmark dataset. Solar Energy, 82, doi:10.1016/j.solener.2007.04.008

Gueymard CA, Kambezidis HD (2004) Solar spectral radiation. In: Muneer T (ed) Solar radiation and daylight models. Elsevier, pp 221–301

Gueymard CA, Myers DR (2007) Performance assessment of routine solar radiation measurements for improved solar resource and radiative modeling. In: Campbell-Howell B (ed) Solar 2007 Conference, Cleveland, OH, American Solar Energy Society

Haeffelin M, Kato S, Smith M, Rutledge CK, Charlock TP, Mahan JR (2001) Determination of the thermal offset of the Eppley precision spectral pyranometer. Appl. Opt. 40: 472–484

Hoyt DV (1978) A model for the calculation of solar global insolation. Solar Energy 21: 27–35

Iqbal M (1983) An introduction to solar radiation, Academic Press

ISO (1990) Standard 9050: Solar energy—Calibration of field pyrheliometers by comparison to a reference pyrheliometer. American National Standards Institute, New York NY

ISO (1996) ISO Guide 99: International vocabulary of basic and general terms in metrology (VIM). ISO Secretariat, Geneva, Switzerland

Kato S (1999) A comparison of modeled and measured surface shortwave irradiance for a molecular atmosphere. J. Quant. Spectrosc. Radiat. Transf. 61: 493–502

Kendall JM, Berdahl CM (1970) Two blackbody radiometers of high accuracy. Appl. Opt. 9: 1082–1091

Lacis AL, Hansen JE (1974) A parameterization of absorption of solar Radiation in the Earth's atmosphere. J. Atmos. Sci. 31: 118–133

Lester A, Myers DR (2006) A method for improving global pyranometer measurements by modeling responsivity functions. Solar Energy 80: 322–331

Maxwell E (1993) Users manual for SERI QC software—Assessing the quality of solar radiation data. Solar Energy Research Institute, Golden, CO (available at http://www.osti.gov/bridge)

Michalsky JJ, Harrison LC, Berkheiser WE (1995) Cosine response characteristics of some radiometric and photometric sensors. Solar Energy 54: 397–402

Michalsky J, Dutton E, Rubes M, Nelson D, Stoffel T, Wesley M, Splitt M, DeLuisi J (1999) Optimal measurement of surface shortwave irradiance using current instrumentation. J. Atmos. Ocean. Technol. 16: 55–69

Michalsky JJ, Dolce R, Dutton EG, Haeffelin M, Major G, Schlemmer JA, Slater DW, Hickey J, Jeffries WQ, Los A, Mathias D, McArthur LJB, Philipona R, Reda I, Stoffel T (2003) Results from the first ARM diffuse horizontal shortwave irradiance comparison. J. Geophys. Res. 108(D3), 4108, doi:10.1029/2002JD002825

Michalsky JJ, Dolce R, Dutton EG, Haeffeln M, Jeffries W, Stoffel T, Hickey J, Los A, Mathias D, McArthur LJB, Nelson D, Philipona R, Reda I, Rutledge K, Zerlaut G, Forgan B, Kiedron P, Long C, Gueymard C (2005) Toward the development of a diffuse horizontal shortwave irradiance working standard. J. Geophys. Res. 110, D06107, doi:10.1029/2004JD005265

Michalsky JJ, Gueymard C, Kiedron P, McArthur LJB, Philipona R, Stoffel T (2007) A proposed working standard for the measurement of diffuse horizontal shortwave irradiance. J. Geophys. Res. 112D, doi:10.1029/2007JD008651

Myers D (2005) Solar radiation modeling and measurements for renewable energy applications: data and model quality. Energy 30: 1517–1531

Myers D, Reda I, Wilcox, S, Stoffel T (2004) Uncertainty analysis for broadband solar radiometric instrumentation calibrations and measurements: an update. In: Sayigh AAM (ed) World Renewable Energy Congress 2004, Denver, CO. Elsevier

Myers DR, Wilcox S, Marion W, George R, Anderberg M (2005) Broadband model performance for an updated national solar radiation Database in the United States of America. In: Goswami DY, Vijayaraghaven S, Campbell-Howell B (eds) Proc. 2005 Solar World Congress. Orlando, FL, International Solar Energy Society

NBS (1974) Thermocouple reference tables based on the IPTS-68. NBS Monograph 125, National Bureau of Standards, Washington, DC

Parton WJ, Innes GS (1972) Some graphs and their functional forms. Tech. Report No. 153, National Resource Ecology Laboratory, Colorado State University, Ft. Collins, CO

Perez R, Ineichen P, Moore K, Kmiecik M, Chain C, George R, Vignola F (2002) A new operational model for satellite derived irradiances, description and validation. Solar Energy 73: 307–317

Perez R, Seals R, Zelenka A (1997) Comparing satellite remote sensing and ground network measurements for the production of site/time specific irradiance data. Solar Energy 60: 89–96

Philipona R (2002) Underestimation of solar global and diffuse radiation measured at Earth's surface. J. Geophys. Res. 107(D22), 4654, doi:10.1029/2002JD002396

Reda I (1998) Improving the accuracy of using pyranometers to measure the clear sky global solar irradiance. Tech. Rep. NREL TP-560-24833, National Renewable Energy Laboratory, Golden, CO

Reda I, Hickey J, Long C, Myers D, Stoffel T, Wilcox S, Michalsky JJ, Dutton EG, Nelson D (2005) Using a blackbody to calculate net-longwave responsivity of shortwave solar pyranometers to correct for their thermal offset error during outdoor calibration using the component sum method. J. Atmos. Ocean. Technol. 22: 1531–1540

Reda I, Stoffel T, Myers D (1996) Calibration of a solar absolute cavity radiometer with traceability to the world radiometric reference. Tech. Rep. NREL TP-463-20619 National Renewable Energy Laboratory, Golden, CO

Romero J, Fox NP, Fröhlich C (1991) First comparison of the solar and SI radiometric scales. Metrologia 28: 125–128

Romero J, Fox NP, Fröhlich C (1995) Improved intercomparison of the world radiometric reference and the SI scale. Metrologia 32: 523–524

Rottman G (2005) The SORCE mission. Solar Physics 230: 7–25

Siren KE (1987) The shadow band correction for diffuse irradiation based on a two-component sky radiance model. Solar Energy 39: 433–438

Wardle DI, Dahlgren L, Dehne K, Liedquist JL, McArthur LJB, Miyake Y, Motshka O, Velds CA, Wells CV (1996) Improved measurement of solar irradiance by means of detailed pyranometer characterisation, International Energy Agency, Tech. Rep. SHCP Task 9C-2

Willson, RC (1973) Active cavity radiometer. Appl. Opt. 12: 810–817

WMO (1983) OMM No. 8, Guide to meteorological instruments and methods of observation; 5th edn, Secretariat of the World Meteorological Organization, Geneva, Switzerland

WMO (1996) OMM No. 8, Guide to meteorological instruments and methods of observation; 6th edn, Part I, sec. 8.1.1. Secretariat of the World Meteorological Organization, Geneva, Switzerland

Younes SR, Claywell R, Muneer T (2005) Quality control of solar radiation data: present status and new approaches. Energy 30: 1533–1549

Chapter 2
Fractal Classification of Typical Meteorological Days from Global Solar Irradiance: Application to Five Sites of Different Climates

Samia Harrouni

1 Introduction

To electrify remote areas, the use of solar energy is the best economical and techno-logical solution. The choice of the sites for the installation of photovoltaic systems and the analysis of their performances require the knowledge of the solar irradiation data. To meet these requirements, we have to classify the days into typical cases for a given site.

Many studies have investigated the problem of typical day's classification. These studies differ by the parameters used as criterion for the classification. This chapter presents a classification method of daily solar irradiances which is mainly based on fractals.

Fractals are objects presenting high degree of geometrical complexity, their description and modeling is carried out using a powerful index called fractal dimension. This later contains information about geometrical irregularities of fractal objects over multiple scales. The fractal dimension of a curve, for instance, will lie between 1 and 2, depending on how much area it fills. The fractal dimension can then be used to compare the complexity of two curves (Dubuc et al. 1989). In solar field, the fractal dimension is directly related to the temporal fluctuation of the irradiance signals. We can then quantify the solar irradiance fluctuations in order to establish a classification according to the atmospheric state (Maafi and Harrouni 2000, 2003; Harrouni and Guessoum 2003; Harrouni and Maafi 2002).

Our classification method defines two thresholds of the fractal dimensions using first a heuristic method then a statistical one. This allows determining three classes of days: clear sky day, partially clouded sky day and clouded sky day.

Samia Harrouni
Solar Instrumentation & Modeling Group/LIWS - Faculty of Electronics and Computer, University of Science and Technology H. Boumediene, Algiers, Algeria,
e-mail: sharrouni@yahoo.fr

This chapter is devoted to the fractal classification of typical meteorological days from global solar irradiances. We start in Section 2 with generalities on the solar radiation especially the most commonly used models to estimate the amount of radiation falling on a tilted plane. Then, we deal in the Section 3 with the problem of the fractal dimension estimation giving a short survey of existing methods. In Section 4, we present a new method to evaluate the fractal dimension of discrete temporal signals or curves with an optimization technique: the "Rectangular covering method". To evaluate its accuracy, the proposed method is applied to fractal signals whose theoretical fractal dimensions are known: Weierstrass function (WF) and fractional Brownian motion (FBM). Section 5 focuses on the classification of irradiances into typical days. This section begins with a survey of existing methods, and then the "Rectangular covering methods" is presented. Thereafter, we will be interested in the application of this method to five sites of different climates. Finally, in Section 6, we give a conclusion and discuss experimental results.

2 Solar Radiation

This section reviews the properties of solar radiation on Earth and summaries well-known models which are used to estimate the amount of radiation falling on a tilted plane.

Extraterrestrial solar radiation falling on a surface normal to the sun's rays at the mean sun earth distance is given by *solar constant* (I_{sc}). The current accepted value of I_{sc} is 1367 W/m^2.

When solar radiation enters the Earth's atmosphere, a part of the incident energy is removed by scattering or absorption by air molecules, clouds and particulate matter usually referred to as aerosols. The radiation that is not reflected or scattered and reaches the surface straight forwardly from the solar disk is called direct or beam radiation. The scattered radiation which reaches the ground is called diffuse radiation. Some of the radiation may reach a panel after reflection from the ground, and is called the ground reflected irradiation. In the Liu and Jordon approach the diffuse and ground reflected radiations are assumed to be isotropic. The total radiation consisting of these three components is called global or total radiation as shown in Fig. 2.1.

In many cases it is necessary to know the amount of energy incident on tilted surface, as shown in Fig. 2.1. However, measured total and diffuse radiation on horizontal surface are given in most available solar radiation databases. There are many models to estimate the average global radiation on tilted surfaces.

In this section we present the isotropic model developed by Liu and Jordan (Liu and Jordan 1963) which also estimates the average hourly radiation from the average daily radiation on a tilted surface.

The daily total radiation incident on a tilted surface H_T can be written as

$$H_T = H_{b,T} + H_{d,T} + H_{r,T} \qquad (2.1)$$

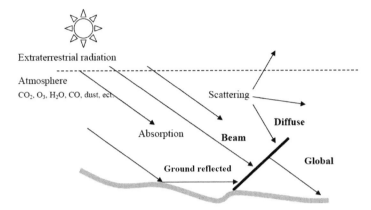

Extraterrestrial radiation

Atmosphere
CO₂, O₃, H₂O, CO, dust, ect

Scattering

Diffuse

Absorption   Beam

Global

Ground reflected

Fig. 2.1 Solar radiation components

where H_T, $H_{b,T}$, $H_{d,T}$ and $H_{r,T}$ are daily total, beam, diffuse and ground reflected radiation, respectively, on the tilted surface.

In this model, (Liu and Jordan 1963) assumed that the intensity of diffuse radiation is uniform over the sky dome. Also, the reflected radiation is diffuse and assumed to be isotropic. Consequently, the daily total radiation on a tilted surface is given by

$$H_T = H_b R_b + H_d \left(\frac{1 + \cos \beta}{2} \right) + H \rho \left(\frac{1 - \cos \beta}{2} \right) \tag{2.2}$$

where H_b, H_d and H are daily beam, diffuse, total radiation, respectively, on a horizontal surface. β represents a tilt angle, ρ the ground albedo and R_b the ratio of the daily beam radiation incident on an inclined plane to that on horizontal plane. For the northern hemisphere and south facing surfaces R_b is given by

$$R_b = \frac{\cos(\phi - \beta) \cos \delta \sin \omega_s' + \omega_s' \sin(\phi - \beta) \sin \delta}{\cos \phi \cos \delta \sin \omega_s + \omega_s \sin \phi \sin \delta} \tag{2.3}$$

where ϕ, δ and ω_s are the latitude, the declination and the sunset hour angle for the horizontal surface, respectively. ω_s is given by

$$\omega_s = \cos^{-1}(-\tan \phi \tan \delta) \tag{2.4}$$

ω_s' is the sunset hour angle for the tilted surface; it is given by

$$\omega_s' = \min \left\{ \cos^{-1}(-\tan \phi \tan \delta), \cos^{-1}(-\tan(\phi - \beta) \tan \delta) \right\} \tag{2.5}$$

In the relation (2.3) ω_s and ω_s' are given in radian.

The daily clearness index K_T is defined as the ratio of the daily global radiation on a horizontal surface to the daily extraterrestrial radiation on a horizontal surface. Therefore,

$$K_T = \frac{H}{H_0} \tag{2.6}$$

where H_0 is the daily extraterrestrial radiation on a horizontal surface. H_0 is given by (Sayigh 1977; Kolhe et al. 2003)

$$H_0 = \frac{24}{\pi} H_{sc} \left[1 + 0.033 \cos \left(\frac{2\pi j_d}{365} \right) \right] (\cos \phi \cos \delta \sin \omega_s + \omega_s \sin \phi \sin \delta) \quad (2.7)$$

where j_d is the Julian day of the year.

Outside the atmosphere there is neither diffuse radiation nor ground albedo. H_0 is then assumed to be composed only of the beam radiation. Similarly, for tilted surfaces, the daily extraterrestrial radiation above the location of interest H_{T0} is constituted only of direct component. Then, according to the relation

$$H_{bT} = H_b R_b \quad (2.8)$$

H_{T0} can be computed as follows

$$H_{T0} = H_0 R_b \quad (2.9)$$

3 Fractal Dimension Estimation

3.1 Preliminaries

Mathematically, any metric space has a characteristic number associated with it called *dimension*, the most frequently used is the so-called *topological* or *Euclidean* dimension. The usual geometrical figures have integer Euclidean dimensions. Thus, points, segments, surfaces and volumes have dimensions 0, 1, 2 and 3, respectively.

But what for the fractals objects, it is more complicated. For an example, the coastline is an extremely irregular line in such way that it would seem to have a surface, it is thus not really a line with a dimension 1, nor completely a surface with dimension 2 but, an object whose dimension is between 1 and 2. In the same way, we can meet fractals whose dimension ranges between 0 and 1 (Like the Cantor set which will be seen later) and between 2 and 3 (surface which tends to fill out a volume), etc. So, fractals have dimensions which are not integer but fractional numbers, called *fractal dimension*.

In the classical geometry, an important characteristic of objects whose dimensions are integer is that any curve generated by these elements contours has finite length. Indeed, if we have to measure a straight line of 1 m long with a rule of 20 cm, the number of times that one can apply the rule to the line is 5. If a rule of 10 cm is used, the number of application of the rule will be 10 times, for a rule of 5 cm, the number will be 20 times and so on. If we multiply the rule length used by the number of its utilization we will find the value 1 m for any rule used.

This result if it is true for the traditional geometry objects, it is not valid for the fractals objects. Indeed, let us use the same way to measure a fractal curve, with a rule of 20 cm, the measured length will be underestimated but with a rule of 10 cm, the result will be more exact. More the rule used is short more the measure will be precise. Thus, the length of a fractal curve depends on the rule used for the measurement: the smaller it is, the more large length is found.

It is the conclusion reached by Mandelbrot when he tried to measure the length of the coast of Britain (Mandelbrot 1967). He found that the measured length depends on the scale of measurement: the smaller the increment of measurement, the longer the measured length becomes.

Thus, fractal shapes cannot be measured with a single characteristic length, because of the repeated pattern we continuously discover at different scale levels.

This growth of the length follows a power law found empirically by Richardson and quoted by Benoît Mandelbrot in his 1967 paper (Richardson 1961)

$$L(\eta) \propto \eta^{-\alpha} \tag{2.10}$$

where L is the length of the coast, η is the length of the step used, the exponent α represents the fractal dimension of the coast.

Other main property of fractals is the self-similarity. This characteristic means that an object is composed of sub-units and sub-sub-units on multiple levels that resemble the structure of the whole object. So fractal shapes do not change even when observed under different scale, this nature is also called scale-invariance. Mathematically, this property should hold on all scales. However, in the real world the self-similarity is only observed over some scales the objects are then statistically self-similar or self-affine.

3.2 Experimental Determination of the Fractal Dimension of Natural Objects

Fractal dimension being a measurement in the way in which the fractal occupies space, to determine it we have to draw up the relationship between this way of occupation of space and its variation of scale. If a linear object of size L is measured with a self-similar object of size l, then number of self similar objects within the original object $N(l)$ is related to L/l as

$$N = \left(\frac{L}{l}\right)^D \tag{2.11}$$

where D is the fractal dimension. From where

$$D = \frac{\ln(N)}{\ln\left(\frac{L}{l}\right)} \tag{2.12}$$

For the self-similar fractals, L/l represents the magnification factor and l/L the reduction factor. Nevertheless, when one tries to determine fractal dimension of natural objects, one is often confronted with the fact that the direct application of Eq. (2.12) is ineffective. In fact, the majority of the natural fractal objects existing in our real world are not self-similar but rather self-affine. The magnification factor and the reduction factor are thus difficult to obtain since there is not an exact self-similarity. Other methods are then necessary to estimate the fractal dimensions of these objects.

In practice, to measure a fractal dimension, several methods exist, some of which are general, whereas others are applicable only to special classes of fractals. This section, focuses on the more commonly used methods namely, Box-counting dimension and Minkowski–Bouligand dimension which are based on the great works of Minkowski and Bouligand (Minkowski 1901; Bouligand 1928) and from which derive several other algorithms.

Box–counting dimension: This method is based upon a quantization of the space in which the object is imbedded by a grid of squares of side ε. The number $N(\varepsilon)$ of squares that intersect the fractal object is then counted. The Box-counting dimension is then defined by

$$D_B = \lim_{\varepsilon \to 0} \frac{\ln[N(\varepsilon)]}{\ln\left(\frac{1}{\varepsilon}\right)} \tag{2.13}$$

If one plots $\ln(N(\varepsilon))$ versus $\ln(1/\varepsilon)$, the slope of the straight line gives the estimate of the fractal dimension D_B in the box-counting method.

Figure 2.2 gives an example illustrating this method. The object E (a curve) is covered by a grid of squares of side $\varepsilon_1 = 1/20$, and for this value of ε total number of squares contained in the grid is $20^2 = 400$ and the number of squares intersecting the curve E is 84 (Fig. 2.2a). In Fig. 2.2b, which is obtained using different values of ε, the slope of the straight line fitted by a linear regression constitutes the fractal dimension of the curve E.

Minkowski–Bouligand dimension: This method is based on Minkowski's idea of dilating the object which one wants to calculate the fractal dimension with disks of radius ε and centered at all points of E. The union of these disks thus creates a *Minkowski cover*.

Let $S(\varepsilon)$ be the surface of the object dilated or covered and D_M the Minkowski–Bouligand dimension. Bouligand defined the dimension D_M as follows

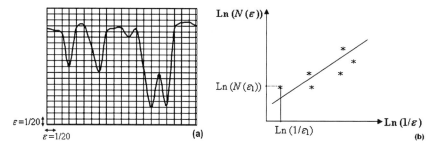

Fig. 2.2 Example illustrating the Box -counting method **a)** Covering the curve by a grid of squares **b)** The log-log plots

$$D_M = 2 - \lambda(S) \tag{2.14}$$

where $\lambda(S)$ is the similarity factor and it represents the infinitesimal order of $S(\varepsilon)$. It is defined by

$$\lambda(S) = \lim_{\varepsilon \to 0} \frac{\ln[S(\varepsilon)]}{\ln(\varepsilon)} \tag{2.15}$$

Inserting Eq. (2.15) in Eq. (2.14) we obtain

$$D = \lim_{\varepsilon \to 0} \left[\frac{2 - \ln[S(\varepsilon)]}{\ln(\varepsilon)} \right] \tag{2.16}$$

The properties of the logarithm permit us to rewrite the relation (2.16) in the following form

$$D = \lim_{\varepsilon \to 0} \frac{\ln\left[\frac{S(\varepsilon)}{\varepsilon}\right]}{\ln\left(\frac{1}{\varepsilon}\right)} \tag{2.17}$$

or, rearranged

$$\ln\left(\frac{S(\varepsilon)}{\varepsilon}\right) = D\ln\left(\frac{1}{\varepsilon}\right) + \text{constant, as } \varepsilon \to 0 \tag{2.18}$$

The fractal dimension can then be estimated by the slope of the log–log plot: $\ln(S(\varepsilon)/\varepsilon) = f(\ln(1/\varepsilon))$ fitted by the least squares method. Figure 2.3a shows the Minkowski covering $E(\varepsilon)$ composed of the union of disks of radius ε.

3.3 Discussion of the Two Methods

According to the analysis of Dubuc et al. (Dubuc et al. 1989), the Box–counting dimension and the Minkowski–Bouligand dimension are mathematically equivalent

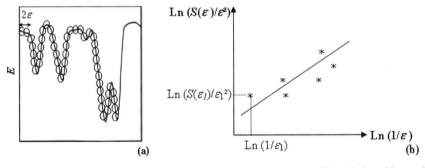

Fig. 2.3 Example illustrating the Minkowski–Bouligand method **a)** The Minkowski covering **b)** The log–log plots

in limit thus, $D_B = D_M$. However, they are completely different in practice because of the way that limits are taken, and the manner in which they approach zero.

Experimental results published in the literature (Dubuc et al. 1989; Maragos and Sun 1993; Zeng et al. 2001) showed that these two methods suffer from inaccuracy and uncertainty. Indeed, according to Zeng et al. (2001) the precision of these estimators are mainly related to the following aspects:

– **Real Value of the Fractal Dimension D:** With big values of D, the estimation error is always very high. This can be explained by the effect of resolution (Huang et al. 1994). When the value of D increases, its estimates can not reflect the roughness of the object and higher resolution is then needed.
– **Resolution:** In the case of the temporal curves, the resolution consists of observation size of the curve (minute, hour, day...). According to Tricot et al. (1988) estimated fractal dimension decreases with the step of observation. This is due to the fact that a curve tends to become a horizontal line segment and appears more regular.
– **Effect of Theoretical Approximations:** Imprecision of the Box-counting and the Minkowski-Bouligand methods is also related to constraints occurring in theoretical approximations of these estimators. For example, the Box-counting dimension causes jumps on the log-log plots (Dubuc et al. 1989) which generate dispersion of the points of the log-log plots with respect to the straight line obtained by linear regression. Moreover, the value of $N(\varepsilon)$ must be integer in this method. The inaccuracy of the method of Minkowski-Bouligand is due to the fact that the Minkowski covering is too thick.
– **Choice of the Interval $[\varepsilon_0, \varepsilon_{max}]$:** The precision of the estimators is influenced much by the choice of the interval $[\varepsilon_0, \varepsilon_{max}]$ through which the line of the log-log plots is adjusted. ε_o is the minimum value that can be assigned to the step. When ε_0 is too large, the curve is covered per few elements (limp or balls). Conversely, when the value ε_{max} is too small, the number of elements which cover the curve is too large and each element covers few points or pixels. Some researchers tried to choose this "optimal" interval in order to minimize the error in estimation (Dubuc et al. 1989; Huang et al. 1994). For example, Liebovitch and Toth (1989) proposed a method for determining this interval, Maragos and Sun (1993) used an empirical rule to determine ε_{max} for temporal signals. In practice, these optimal intervals improve considerably the precision of the fractal dimension estimate for special cases but not in all cases.

4 Measuring the Fractal Dimension of Signals

4.1 A Survey of Existing Methods

Many natural processes described by time series (e.g., noises, economical and demographic data, electric signals... etc.) are also fractals in the sense that their graph is a fractal set (Maragos and Sun 1993). Thus, modeling fractal signals is of great interest in signal processing.

Considering the importance of this index and the impact of its use in practice, the precision of its estimate is necessary. Methods of Box–counting and of Minkowski–Bouligand prove then ineffective due to the fact that they suffer from inaccuracy as we already mentioned. Inspired by the Minkowski–Bouligand method, a class of approaches to compute the fractal dimension of signal curves or one-dimensional profiles called "covering methods" is then proposed by several researchers.

These methods consist in creating multiscale covers around the signal's graph. Indeed, each covering is formed by the union of specified structuring elements. In the method of Box–counting, the structuring element used is the square or limp, that of Minkowski–Bouligand uses the disk.

Dubuc et al. (1989 and Tricot et al. (1988) proposed a new method called "Variation method". This one criticizes the standard methods of fractal dimension estimation namely: Box–counting and Minkowski–Bouligand. Indeed, "Variation method" applied to various fractal curves showed a high degree of accuracy and robustness.

Maragos and Sun (1993) generalized the method of Minkowski–Bouligand by proposing the "Morphological covering method" which uses multiscale morphological operations with varying structuring elements. Thus, this method unifies and improves other covering methods. Experimentally, "Morphological covering method" demonstrated a good performance, since it has experimentally been found to yield average estimation errors of about 2%–4% or less for discrete fractal signals whose fractal dimension is theoretically known (Maragos and Sun 1993). For deterministic fractal signals (these signals will be detailed further in this chapter) Maragos and Sun developed an optimization method which showed an excellent performance, since the estimation error was found between 0 % and 0.07 %.

4.2 New Method for Estimating the Fractal Dimension of Discrete Temporal Signals

In order to contribute in improving the accuracy of fractal dimension estimation of the discrete temporal signals we developed a simple method based on a covering by rectangles called Rectangular Covering Method.

Presentation of the Method

The method based on Minkowski–Bouligand approach consists in covering the curve for which we want to estimate fractal dimension by rectangles. The choice of this type of structuring element is due to the discrete character of the studied signals.

From the mathematical point of view, the use of the rectangle as structuring element for the covering is justified. Indeed, Bouligand (1928) showed that D_M

(Minkowski–Bouligand dimension) can be obtained by also replacing the disks in the previous covers with any other arbitrarily shaped compact sets that posses a nonzero minimum and maximum distance from their center to their boundary.

Thus, as shown in Fig. 2.4, for different time intervals $\Delta\tau$, the area $S(\Delta\tau)$ of this covered curve is calculated by using the following relation

$$S(\Delta\tau) = \sum_{n=0}^{N-1} \Delta\tau \, |f(t_n + \Delta\tau) - f(t_n)| \tag{2.19}$$

where N denotes the signal length, $f(t_n)$ is the value of the function representing the signal at the time t_n and $|f(t_n + \Delta\tau) - f(t_n)|$ is the function variation related to the interval $\Delta\tau$. The fractal dimension is then deduced from Eq. (2.20) where ε is replaced by the time interval $\Delta\tau$. Hence

$$\ln\left(\frac{S(\Delta\tau)}{\Delta\tau}\right) = D\ln\left(\frac{1}{\Delta\tau}\right) + \text{constant}, \text{ as } \Delta\tau \to 0 \tag{2.20}$$

Thus, to determine the fractal dimension D which represents the slope of the straight line of Eq. (2.20), it is necessary to use various time scales $\Delta\tau$ and to measure the corresponding area $S(\Delta\tau)$. We then obtain several points $(\Delta\tau_i, S(\Delta\tau_i))$ constituting the line.

A good estimation of the fractal dimension D requires a good fitting of the log-log plot defined by Eq. (2.20). Therefore, the number of points constituting the plot is important. This number is fixed by $\Delta\tau_{max}$ which is the maximum interval through which the line of the log-log plots is fitted.

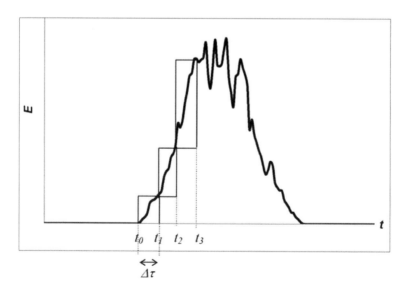

Fig. 2.4 An example of temporal curve covered by rectangles

As mentioned above, to estimate the fractal dimension most of methods determine $\Delta\tau_{max}$ experimentally. This procedure requires much time and suffers from precision. Also, we developed an optimization technique to estimate $\Delta\tau_{max}$.

Optimization Technique

Experience shows that $\Delta\tau_{max}$ required for a good estimation of D depends on several parameters, especially the time length of the signal N. $\Delta\tau$ should not be too weak, in order not to skew the fitting of the line, and it must not exceed $N/2$. $\Delta\tau$ must also satisfy the condition of linearity of the line.

Our optimization technique (Harrouni et al. 2002) consists first in taking a $\Delta\tau_{max}$ initial about 10, because the number of points constituting the plot should not be very small as signaled above; then, $\Delta\tau_{max}$ is incremented by step of 1 until $N/2$. We hence obtain several straight log-log lines which are fitted using the least squares estimation. The $\Delta\tau_{max}$ optimal is the one corresponding to the log-log straight line with the minimum least square error. This later is defined by the following formula

$$E_{quad} = \frac{\sum\limits_{i=1}^{n} d_i}{n} \qquad (2.21)$$

In this relation n denotes the number of points used for the straight log–log line fitting, d_i represents the distance between the points $\left(\ln\left(1/\Delta\tau\right), \ln\left(S\left(\Delta\tau\right)/\Delta\tau^2\right)\right)$ and the fitted straight log–log line.

Validation of the Method

In order to test the validity and the accuracy of the rectangular covering method, we applied it to two different types of parametric fractal signals whose theoretical fractal dimension is known, these test signals are the Weirstrass function (WF) which is a deterministic signal and the random signal of the fractional Brownian motion (FBM). These fractal signals that will be briefly defined below are most commonly used in various applications.

The Weierstrass Function (FW): It is defined as (Hardy 1916; Mandelbrot 1982; Berry and Lewis 1980)

$$W_H(t) = \sum_{k=0}^{\infty} y^{-kH} \cos\left(2\pi y^k t\right), \text{ as } 0 < H < 1 \qquad (2.22)$$

This function is continuous but nowhere differentiable; γ is an integer such as $\gamma > 1$. This parameter is fixed by the experimenter so that he can choose the shape of the signal, the fractal dimension of this function is $D = 2 - H$. In our experiments, we synthesized discrete time signals from WF's by sampling $t \in [0,1]$ at $N+1$ equidistant points, using $\gamma = 2.1$ and truncating the infinite series so that the summation is done only for $0 \leq k \leq k_{max}$. The k_{max} is determined by the inequality $2\pi\gamma k \leq 10^{12}$ established by Maragos and Sun (1993).

Fractional Brownian Motion (FBM): It is one of the most mathematical models used to describe self-affine fractals existing in nature. Mandelbrot and Wallis proposed an extension of this motion: the fractional Brownian motion. The function of the Brownian fractional motion $B_H(t)$ with parameter $0 < H < 1$ is a time varying random function with stationary, Gaussian distributed, and statically self-affine increments. So

$$\left\langle [B_H(t) - B_H(t_0)]^2 \right\rangle = 2D(t - t_0)^{2H} \text{, as } 0 < H < 1 \qquad (2.23)$$

The fractal dimension D of $B_H(t)$ is $D = 2 - H$. To synthesize FBM signals, several methods exist (Mandelbrot and Wallis 1969; Voss 1988; Lundahl and all. 1986) the most known are: Choleski decomposition method, Durbin–Levinson algorithm, FFT method and circulant matrix method. In our experiments we synthesized FBM signals via the Durbin–Levinson method.

To validate our rectangular covering method we applied it to these synthesized test signals. For this purpose, the error between the theoretical fractal dimension and the estimated one is used. The experimental results indicate that, for the two fractal signals WF and FBM, the rectangular covering method performs well in estimating dimensions $D \in [1.1, 1.9]$, since the estimation error is less than or equal 6 % for the WF signals and 7 % for FBM signals (Harrouni and Guessoum 2006). By varying the signals' length $N \in [100, 1000]$ with a step of 100 we have also observed similar performance of this method. Over 99 different combinations of (D, N) the average estimation error of the rectangular covering method was 4 % for both WF's and FBM's.

5 Classification of the Solar Irradiances to Typical Days

5.1 A Survey of Existing Methods

Modeling random fluctuations of the solar irradiance has already been the object of several studies published in the literature. These are based mostly on the random processes. The Markovian approaches in particular, contributed extensively to this modeling. One can see for example, the works of Brinkworth (1977), Bartoli et al. (1981), Lestienne et al. (1979), Aguiar et al. (1988) and Maafi (1991). This last reference treated the problem of the classification of the insolation and the daily irradiation indirectly by joining them to the states of the sky: clear sky, covered sky, etc. (Maafi 1991, 1998).

Other statistical methods were used for classification of typical meteorological days such as automatic classification (Bouroubi 1998), the analysis of the correlations (Louche and al. 1991) and the Ward's method (Muselli et al. 1991).

More recent studies are interested by the modeling of the random character of the solar radiation using neural networks (Guessoum et al. 1998; Sfetsos and Coonick 2000). In addition to the originality of these new approaches, these studies aim to value the contribution of their formalisms in the description of the solar radiation fluctuating character.

However, very few works treating the classification of the solar radiation signals using the fractal analysis were published (Maafi and Harrouni 2000, 2003; Harrouni and Guessoum 2003; Harrouni and Maafi 2002; Louche et al. 1991). In this section the contribution of the fractal analysis to the classification of the solar irradiance signals is given. This examination leads to the determination of different sky types in a given time interval as: clear sky, partially covered sky, covered sky etc. which is useful for planning and analyzing solar energy systems. Hence, a classification method is proposed which allows the categorization of the solar radiation fluctuations based on the fractal dimension (Harrouni et al. 2005).

5.2 Fractal Classification of Solar Irradiance

Methodology

Our method classification uses the fractal dimension as a basic criterion to achieve the classification of the solar irradiance and to yield different types of days, i.e., clear sky day, covered sky day, a cloudy day, etc. Our research reveals that some daily solar irradiance signals have the same fractal dimension but corresponding to days with different weather conditions. Indeed, a uniformly cloudy day and a sunny one have regular irradiance shapes and practically the same value for D but have daily different clearness indexes. That is why the daily clearness index K_T is calculated along with D as a second criterion in the categorization algorithm which allows sorting daily irradiances into three classes according to the following classification:

Class I: Clear sky day
$1 \leq D \leq D_I$ and $K_T \geq (K_T)_I$
Class II: Partially cloudy sky
$D_I < D \leq D_{II}$ and $K_T \geq (K_T)_I$
Class III: Completely cloudy sky
$D > D_{II}$ or $D \leq D_{II}$ and $K_T < (K_T)_I$

D_I, D_{II}, are the thresholds for D and $(K_T)_I$ is the one for K_T for the different classes.

The thresholds for D and K_T are new parameters to be determined in order to achieve the classification of the irradiances. The value 0.5 is chosen for $(K_T)_I$; this value permits to distinguish the covered sky day class from the one of clear sky day. Indeed, experimental results reveals that for some days of class III (covered sky day), the fractal dimension D is closer to 1, this is due to the fact that these days are so covered that the corresponding irradiance curve is regular but the clearness index is very low (lower than 0.5).

To determine the thresholds of the fractal dimension D_I and D_{II} we first used a heuristic approach then a statistical one. The heuristic approach consists in analyzing all daily solar irradiances shapes and their corresponding fractal dimension. For each day of the year the histograms of the irradiance signals are constructed. These histograms are built by class of $100\,W/m^2$. By observing their various forms, i.e. preponderance of low or high frequencies, we noted that there were three kinds of histograms (Maafi and Harrouni 2003):

- Histograms in the shape of J
- Histograms in the shape of U
- Histograms in the shape of L

By identifying the relations of classification established above with these three types of histograms, we can determine the D thresholds correspondent to the three classes. (An example of these histograms is given for Tahifet on the accompanying CD).

The statistical method is based on the cumulative distribution function (CDF) $F_X(x)$. This latter describes the probability distribution of a real-valued random variable X. For every real number x, the CDF of X is the probability that the random variable X takes on a value less than or equal to x. Thus, the two thresholds of D correspond respectively to the fractal dimension whose the cumulative distribution function $F_X(x)$ are:

$$F_x(x) = \frac{\max(F_x(x)) - \min(F_x(x))}{3} \text{ and} \qquad (2.24)$$

$$F_x(x) = \frac{2(\max(F_x(x)) - \min(F_x(x)))}{3}$$

Data Bank

The experimental database contains global irradiance data measured at five sites of different climates. Two south Algerian sites: Tahifet (Tamanrasset) and Imehrou (Illizi), two sites of Colorado: Golden and Boulder and the last site is Palo Alto located in California.The geographical coordinates of these sites are given in Table 2.1.

Algerian sites data are recorded from the operation of two stand-alone photovoltaic power installations during 1992-year on a $10°$-tilted surface with a time step of 10 minutes. These systems have been installed by the National Company from

Table 2.1 Geographical coordinates of the studied sites

Site	Latitude	Longitude	Altitude (m)
Tahifet	22°53′N	06°00′E	1400
Imehrou	26°00′N	08°50′E	600
Golden	39°74′N	105°18′W	1829
Boulder	39°91′N	105°25′W	1855
Palo alto	37°42′N	122°.15′W	12.192

Electricity and Gaz (SONELGAZ). For Colorado sites the irradiance data have been collected during the year 2003 on a horizontal surface. These data are provided by MIDC (Measurement and Instrumentation Data Center) (MIDC, 2007). Data at Palo Alto have been recorded from the operation of one grid-connected system during 2003-year on a 30°-tilted surface with a time step of 15 minutes. This PV system was installed in May 2000 by CPAU (City of Palo Alto utilities) (CPAU, 2007).

By integrating the measured irradiances we determined the daily irradiation. Then, we calculated the daily clearness index K_T using Eq. (2.6). The measured daily global and extraterrestrial irradiation together with daily clearness index for all studied sites are included on the accompanying CD.

H_0 is calculated by using Eq. (2.7) for irradiations received on horizontal surface (Boulder and Golden) and Eq. (2.9) for irradiations on tilted plane (Tahifet, Imehrou and Palo Alto).

Fractal Treatment of Solar Irradiances

Figure 2.5 presents two examples of the log–log lines permitting the estimation of the fractal dimension of irradiance curves. This figure shows that the log–log points are grouped around the fitting line which demonstrates the self affinity of the studied solar irradiances.

The fractal dimensions obtained from the slopes of the log–log lines for all sites are given in the accompanying CD. Figure 2.6 gives representative examples for the daily irradiation values corresponding to different fractal dimensions from three classes. As can be observed there is good correspondences between the shapes of the signals and the corresponding fractal dimensions.

Figure 2.7 gives the annual evolution of the monthly average of D for the studied sites. This figure shows clearly that D fluctuates.

In order to quantify this fluctuation we calculated the annual average $<D>$ of the fractal dimension and the corresponding standard deviation σ which are tabulated in Table 2.2. These values suggest that the solar irradiances of Tahifet and Boulder exhibit the similar fluctuations ($<D> = 1.16$ for Tahifet and 1.13 for Imehrou).

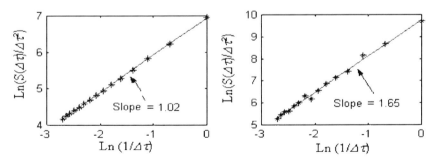

Fig. 2.5 Two examples (Golden site) of log–log plots fitted by the least-squares estimation with their slopes which represent the estimated fractal dimension

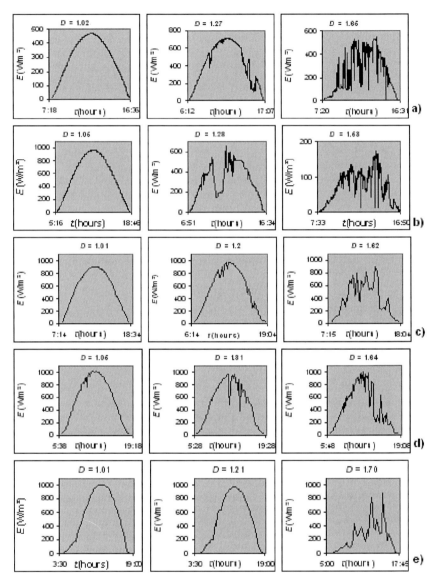

Fig. 2.6 Typical daily irradiance values for the three classifications and fractal dimensions for the sites under consideration **a)** Golden, **b)** Boulder, **c)** Tahifet, **d)** Imehrou, **e)** Palo Alto

This is also observed for Golden and Boulder ($<D> = 1.38$ for Golden and 1.39 for Boulder). To compare the degree of fluctuation of the solar irradiances of the different sites we can refer to the values of $<D>$. Hence, the two sites of Colorado are fluctuating, those of the Algerian sites fluctuate slightly, they are practically regular, and in Palo Alto irradiances are fairly fluctuating. However, the analysis of D month by month permits the detection of the months where the fluctuations of the irradiances are most intense – June and December for Tahifet, March and June for

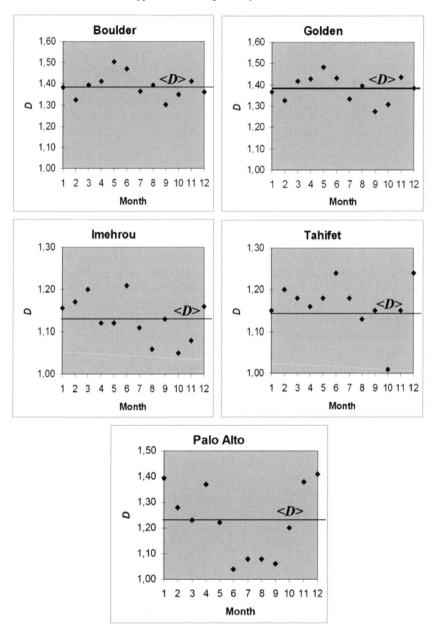

Fig. 2.7 Annual variation of the monthly means of the estimated fractal dimension D, the straight solid line represents the annual mean of the fractal dimension $<D>$

Table 2.2 Annual averages $<D>$ and standard deviations σ of the estimated fractal dimensions

Site	$<D>$	$\sigma(\%)$
Tahifet	1.16	19
Imehrou	1.13	17
Golden	1.38	18
Boulder	1.39	18
Palo alto	1.23	20

Imehrou, May and June for Golden and Boulder, January and December for Palo Alto – and those where these irradiances are very regular – October for the two sites Tahifet and Imehrou, September for Golden and Boulder, June for Palo Alto. These informations are very useful to refine the sizing of photovoltaic systems. Indeed, the anomalies in the operating of the photovoltaic systems installed in these sites appear during these months. There is for example for Tahifet excess of energy in October and storage is requested in June and December much more than in other months.

5.2.1 Annually and Monthly Classification Analysis

The thresholds D_I and D_{II} have first been determined for the sites of Tahifet and Imehrou. For this purpose, the heuristic method and the statistical one has been used, Table 2.3 gathers the thresholds obtained with the two methods. We notice that the empirical and statistical thresholds are very close. Since the empirical approach is very expensive in time to build histograms and to carry out their meticulous examination, we chose the statistical thresholds to classify the days of the studied sites. The obtained thresholds for all sites are illustrated by Table 2.4.

Table 2.5 gives the distribution of the probability of occurrence of daily solar irradiances for each class obtained from our classification. For Tahifet and Imehrou daily irradiances of class I have the largest probability of occurrence as compared to irradiances of the two other classes. These results confirm the pre-eminence of days with clear sky for the two sites; this is due to the climate of the south Algerian which is characterized by irradiances rarely fluctuated. However Class III (completely covered sky) is preponderant for the Californian sites. Class I is also important, whereas class II has less frequency of occurrence. These results demonstrate that the two

Table 2.3 Fractal dimension thresholds obtained with the two methods: heuristic and statistic for Tahifet and Imehrou sites

Site	D_I (heuristic)	D_I (statistic)	D_{II} (heuristic)	D_{II} (statistic)
Tahifet	1.14	1.10	1.34	1.25
Imehrou	1.12	1.10	1.27	1.25

Table 2.4 Statistical Fractal dimension thresholds for all studied sites

Site	D_I	D_{II}
Tahifet	1.14	1.34
Imehrou	1.12	1.27
Golden	1.35	1.49
Boulder	1.35	1.50
Palo Alto	1.19	1.37

Table 2.5 Probability of occurrence of daily solar irradiance shapes of each class

Site	Class I (%)	Class II (%)	Class III(%)
Tahifet	58	16	26
Imehrou	62	16	21
Golden	24	22	53
Boulder	26	22	52
Palo Alto	49	17	34

studied sites are characterized by disturbed climate since the overcast sky days are preponderant at the two sites. At Palo Alto, classes I and III are pre-eminent which demonstrate that this site has a climate fairly disturbed.

On the accompanying CD, Tables of day's class are included for each studied site.

To validate the classification results, the average of the fractal dimension $<D>$, of clearness index $<K_T>$ and their standard deviations $\sigma(D)$ and $\sigma(K_T)$ have been computed for each class. They are summarized by Table 2.6.

These statistical properties show that our classification method leads to homogeneous groupings of the studied days since the standard deviations of D and K_T are weak compared to their averages. Indeed, in all the sites $\sigma(K_T)$ is lower than 10% for all classes and except for Golden and Boulder we note the same thing for $\sigma(D)$ but only for classes I and II. The more important value of this standard deviation for class III (upper than 10%) is due to the fact that this class contains rainy days whose irradiance signals have a regular form thus a fractal dimension near to 1 like already explained. For example, the shape of solar daily irradiance of class III

Table 2.6 Mean value and standard deviation of D and K_T for the different classes of days

	Site	Golden			Boulder			Tahifet			Imehrou			Palo Alto		
	Class	I	II	III	I	II	III	I	II	III	I	II	III	I	II	III
Average	$<D>$	1.15	1.43	1.47	1.17	1.43	1.48	1.03	1.24	1.42	1.02	1.19	1.40	1.06	1.27	1.46
	$<K_T>$	0.70	0.63	0.46	0.69	0.64	0.47	0.66	0.60	0.45	0.69	0.62	0.50	0.70	0.61	0.33
Standard	$\sigma(D)$	0.12	0.03	0.14	0.12	0.04	0.13	0.04	0.05	0.13	0.03	0.04	0.14	0.06	0.05	0.15
deviation	$\sigma(K_T)$	0.07	0.07	0.18	0.08	0.08	0.12	0.04	0.04	0.12	0.04	0.04	0.13	0.07	0.05	0.17

(see Fig. 2.8) corresponds to a rainy day in Golden. Its fractal dimension is equal to 1.11 and its related K_T is 0.47. Using D, this daily irradiance should be classified in class II. But, when using D and K_T together it is categorized as class III. The fairly high values of $\sigma(D)$ for the class I in the two sites of Colorado is explained by the high value of D_I due to the irradiances character of these sites which is very fluctuating.

In order to better characterize the three classes obtained our statistical analysis was refined by carrying out it on a monthly scale. In Table 2.7 monthly results of the frequency of each class, averages and standard deviations of the two parameters: D and K_T are presented. Table 2.7 shows that the distribution of the classes differs from a site to another.

As it can be observed from Table 2.7, Class III days have high frequency of occurrence for the sites Golden and Boulder, reaching a maximum in May and June. Only for the month September for Golden and February for Boulder class I have higher frequency of occurrences which are 51.6% and 39.3%, respectively. However, in Tahifet and Imehrou class I has higher frequency of occurrences for all the months, reaching maximum values in October and minimum in May and June.

In Palo Alto on the other hand we notice a seasonal distribution of the days. Indeed, class I presents high values in winter (January, February, November and December) where the maximum is detected in December and class III high values in summer (June–September).

These results are confirmed by the transition probabilities between two consecutive days having the same or different classes. For the two sites of Algeria, while transition probabilities from class I to the same class were quite high (65% and 40%), all other transitions were low. However for Golden and Boulder all transition probabilities are quite close in the ranges of 5 to 20%.

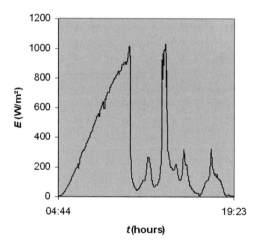

Fig. 2.8 An example of a rainy day with an enough regular shape, $D = 1.11$ and $K_T = 0.47$

Table 2.7 Monthly characteristics of each class obtained for the various sites

Site		Golden			Boulder			Tahifet			Imehrou			Palo Alto		
	Class	I	II	III	I	II	III	I	II	III	I	II	III	I	II	III
January	Freq (%)	19.4	25.8	54.8	19.4	22.6	58.0	64.5	12.9	22.6	58.1	12.9	29.0	12.9	22.6	64.5
	$\langle D \rangle$	1.20	1.44	1.39	1.16	1.45	1.43	1.02	1.25	1.47	1.02	1.17	1.42	1.18	1.25	1.49
	$\sigma(D)$	0.15	0.03	0.16	0.10	0.05	0.14	0.23	0.06	0.10	0.03	0.02	0.12	0.01	0.05	0.14
	$\langle K_T \rangle$	0.60	0.58	0.36	0.65	0.62	0.40	0.69	0.60	0.34	0.66	0.60	0.39	0.59	0.59	0.28
	$\sigma(K_T)$	0.05	0.05	0.14	0.07	0.04	0.16	0.15	0.05	0.15	0.03	0.05	0.15	0.06	0.03	0.13
February	Freq (%)	21.4	14.3	64.3	39.3	25.0	35.7	48.3	17.2	34.5	44.8	34.5	24.1	25.9	25.9	48.1
	$\langle D \rangle$	1.13	1.43	1.37	1.21	1.43	1.37	1.03	1.26	1.40	1.04	1.17	1.41	1.16	1.27	1.35
	$\sigma(D)$	0.16	0.04	0.13	0.13	0.04	0.16	0.23	0.05	0.07	0.04	0.03	0.19	0.01	0.05	0.10
	$\langle K_T \rangle$	0.66	0.65	0.41	0.68	0.66	0.42	0.69	0.62	0.49	0.66	0.64	0.52	0.63	0.60	0.25
	$\sigma(K_T)$	0.07	0.06	0.14	0.12	0.11	0.18	0.15	0.02	0.13	0.03	0.05	0.09	0.03	0.05	0.14
March	Freq (%)	12.9	35.5	51.6	19.4	29.0	51.6	54.8	12.9	32.3	45.2	12.9	41.9	45.2	32.3	22.6
	$\langle D \rangle$	1.13	1.43	1.48	1.19	1.42	1.45	1.01	1.24	1.43	1.04	1.19	1.39	1.07	1.29	1.47
	$\sigma(D)$	0.10	0.03	0.16	0.11	0.04	0.19	0.01	0.07	0.18	0.04	0.06	0.15	0.05	0.06	0.15
	$\langle K_T \rangle$	0.69	0.66	0.43	0.72	0.70	0.50	0.65	0.61	0.39	0.67	0.57	0.41	0.68	0.62	0.38
	$\sigma(K_T)$	0.02	0.04	0.23	0.09	0.08	0.21	0.04	0.03	0.11	0.04	0.04	0.12	0.04	0.04	0.12
April	Freq (%)	13.3	30.0	56.7	16.7	30.0	53.3	60.0	16.7	23.3	76.7	0.00	23.3	20.0	13.3	66.7
	$\langle D \rangle$	1.19	1.42	1.48	1.17	1.41	1.49	1.03	1.19	1.46	1.02		1.44	1.08	1.28	1.48
	$\sigma(D)$	0.18	0.02	0.10	0.15	0.04	0.11	0.04	0.04	0.08	0.03		0.11	0.06	0.02	0.14
	$\langle K_T \rangle$	0.76	0.64	0.49	0.74	0.66	0.51	0.67	0.63	0.51	0.70		0.53	0.68	0.65	0.46
	$\sigma(K_T)$	0.07	0.08	0.19	0.10	0.09	0.18	0.03	0.01	0.10	0.04		0.17	0.07	0.09	0.13

Table 2.7 (continued)

Site		Golden			Boulder			Tahifet			Imehrou			Palo Alto		
	Class	I	II	III	I	II	III	I	II	III	I	II	III	I	II	III
May	Freq (%)	3.23	19.3	77.4	6.50	12.9	80.6	41.9	38.7	19.4	51.6	29.0	19.4	54.8	16.1	29.0
	$\langle D\rangle$	1.34	1.43	1.50	1.29	1.42	1.54	1.04	1.21	1.39	1.02	1.17	1.32	1.03	1.30	1.52
	$\sigma(D)$	0.00	0.04	0.13	0.07	0.06	0.09	0.04	0.05	0.15	0.02	0.04	0.14	0.04	0.02	0.10
	$\langle K_T\rangle$	0.73	0.66	0.49	0.68	0.67	0.50	0.65	0.60	0.53	0.68	0.64	0.60	0.76	0.65	0.52
	$\sigma(K_T)$	0.00	0.10	0.17	0.03	0.08	0.17	0.03	0.04	0.07	0.05	0.04	0.09	0.06	0.10	0.15
June	Freq (%)	16.7	13.3	70.0	13.3	23.3	63.3	36.7	26.7	36.7	36.7	40.0	23.3	96.7	3.30	0.00
	$\langle D\rangle$	1.22	1.46	1.48	1.21	1.44	1.53	1.05	1.25	1.42	1.04	1.21	1.46	1.04	1.21	
	$\sigma(D)$	0.11	0.02	0.14	0.09	0.04	0.10	0.04	0.04	0.09	0.04	0.05	0.13	0.04	0.00	
	$\langle K_T\rangle$	0.60	0.59	0.49	0.65	0.61	0.47	0.59	0.57	0.52	0.68	0.63	0.57	0.77	0.56	
	$\sigma(K_T)$	0.08	0.06	0.20	0.10	0.09	0.16	0.03	0.05	0.07	0.02	0.04	0.06	0.03	0.00	
July	Freq.(%)	35.5	22.6	41.9	25.8	22.6	51.6	48.4	3.20	48.4	71.0	9.70	19.4	80.6	9.70	9.70
	$\langle D\rangle$	1.16	1.41	1.43	1.14	1.44	1.44	1.04	1.19	1.33	1.03	1.19	1.34	1.02	1.28	1.42
	$\sigma(D)$	0.12	0.04	0.16	0.13	0.06	0.11	0.05	0.00	0.11	0.04	0.05	0.19	0.03	0.07	0.04
	$\langle K_T\rangle$	0.69	0.61	0.54	0.67	0.62	0.51	0.60	0.52	0.42	0.69	0.64	0.48	0.77	0.66	0.49
	$\sigma(K_T)$	0.08	0.07	0.12	0.09	0.07	0.11	0.03	0.00	0.09	0.03	0.02	0.14	0.03	0.07	0.19
August	Freq.(%)	22.6	29.0	48.4	29.0	22.6	48.4	58.1	29.0	12.9	83.9	12.9	3.20	80.6	12.9	6.50
	$\langle D\rangle$	1.13	1.45	1.49	1.23	1.43	1.48	1.02	1.25	1.40	1.02	1.21	1.30	1.02	1.26	1.49
	$\sigma(D)$	0.12	0.03	0.10	0.11	0.03	0.11	0.04	0.06	0.13	0.04	0.02	0.00	0.03	0.08	0.07
	$\langle K_T\rangle$	0.70	0.61	0.49	0.68	0.61	0.48	0.67	0.59	0.47	0.71	0.64	0.68	0.72	0.65	0.55
	$\sigma(K_T)$	0.07	0.08	0.11	0.08	0.06	0.11	0.02	0.05	0.09	0.02	0.03	0.00	0.03	0.08	0.04

Table 2.7 (continued)

Site		Golden			Boulder			Tahifet			Imehrou			Palo Alto		
	Class	I	II	III	I	II	III	I	II	III	I	II	III	I	II	III
September	Freq.(%)	50.0	16.7	33.3	46.7	16.7	36.7	66.7	10.0	23.3	60.0	20.0	20.0	90.0	0.00	10.0
	$\langle D\rangle$	1.09	1.43	1.48	1.10	1.46	1.49	1.04	1.27	1.41	1.03	1.23	1.36	1.04		1.30
	$\sigma(D)$	0.09	0.04	0.17	0.11	0.06	0.12	0.05	0.05	0.09	0.03	0.03	0.08	0.04		0.25
	$\langle K_T\rangle$	0.74	0.68		0.74	0.65	0.47	0.67	0.57	0.55	0.70	0.63	0.58	0.65		0.45
	$\sigma(K_T)$	0.02	0.07	0.20	0.03	0.07	0.14	0.04	0.05	0.11	0.01	0.03	0.09	0.05		0.19
October	Freq.(%)	51.6	9.70	38.7	38.7	16.1	45.2	100	0.00	0.00	87.1	3.20	9.70	67.7	19.4	12.9
	$\langle D\rangle$	1.15	1.41	1.50	1.14	1.42	1.50	1.01			1.01	1.16	1.43	1.12	1.29	1.46
	$\sigma(D)$	0.13	0.04	0.13	0.11	0.06	0.11	0.03			0.02	0.00	0.13	0.05	0.03	0.06
	$\langle K_T\rangle$	0.73	0.64	0.47	0.71	0.68	0.48	0.68			0.71	0.63	0.61	0.63	0.58	0.46
	$\sigma(K_T)$	0.04	0.07	0.21	0.05	0.03	0.20	0.02			0.02	0.00	0.07	0.02	0.02	0.14
November	Freq.(%)	20.0	20.0	60.0	23.3	20.0	56.7	63.3	16.7	20.0	80.0	3.30	16.7	0.00	40.0	60.0
	$\langle D\rangle$	1.13	1.44	1.53	1.16	1.45	1.50	1.02	1.24	1.47	1.01	1.13	1.42		1.26	1.46
	$\sigma(D)$	0.14	0.04	0.11	0.12	0.05	0.13	0.03	0.03	0.09	0.02	0.00	0.10		0.03	0.16
	$\langle K_T\rangle$	0.70	0.63	0.46	0.68	0.65	0.41	0.68	0.61	0.42	0.69	0.63	0.48		0.59	0.28
	$\sigma(K_T)$	0.07	0.08	0.22	0.07	0.08	0.18	0.02	0.04	0.13	0.02	0.00	0.11		0.02	0.14
December	Freq.(%)	25.8	29.0	45.2	32.3	22.6	45.2	51.6	6.50	41.9	54.8	19.4	25.8	9.70	12.9	77.4
	$\langle D\rangle$	1.16	1.42	1.50	1.16	1.43	1.48	1.04	1.28	1.49	1.03	1.19	1.44	1.13	1.28	1.46
	$\sigma(D)$	0.12	0.03	1.50	0.11	0.04	0.14	0.05	0.11	0.17	0.03	0.06	0.12	0.04	0.06	0.17
	$\langle K_T\rangle$	0.68	0.62	0.46	0.67	0.61	0.45	0.66	0.42	0.42	0.65	0.59	0.51	0.58	0.58	0.21
	$\sigma(K_T)$	0.08	0.05	0.46	0.06	0.05	0.17	0.02	0.09	0.10	0.04	0.04	0.09	0.06	0.02	0.12

6 Conclusions

In this chapter, a classification procedure for solar irradiances is presented and discussed for five locations. This procedure uses fractal dimension analysis. A new method of estimating fractal dimensions is utilized which gives satisfactory results. This method based on covering multi scale, using rectangles as the structuring element. The method is tested for two well-known functions and an average error of 3.7% is obtained for over 180 tests.

The validation of the classification method is carried out by annual and monthly analysis using the fractal dimension and the clearness index of the daily irradiances. Three different classes of the days are determined to be a reasonable classification. Results for the sites with similar climates give the same type of classifications of the days as it is observed from their annual and monthly average classification parameters. Observed standard deviations of the monthly parameters from an annual mean value are relatively small.

Classification of the daily solar irradiance is important in design and installation of solar energy systems, especially PV arrays. Trends in the patterns of daily solar irradiance became significant information due to the recent interests in renewable technologies. This interest is essentially due to global warming and other negative effects to our environment. Such analyses presented in this chapter are of great interest as they reduce the initial costs by appropriate design and construction of solar energy systems suitable to the climate of the site of interest.

References

Aguiar RJ, Collores-Pereira M, Conde JP (1988) Simple procedure for generating sequences of daily radiation values using a library of Markov transition Matrices. Solar Energy 40:269–279

Bartoli B, Coluzzi B, Cuomo V, Francesca M, Serio S (1981) Autocorrelation of daily global solar radiation. II Nuovo Cimento, 4C:113–122

Berry MV, Lewis ZV (1980). On the Weierstrass–Mandelbrot fractal function. Proc roy Soc Ser a 370:459–484

Bouligand G (1928) Ensembles impropres et nombre dimensionnel. Bull Sci Math II-52: 320–344, 361–376

Bouroubi MY (1998) Modélisation de l'irradiation solaire à l'échelle journalière et horaire, pour l'Algérie. Master thesis, USTHB University

Brinkworth BJ (1977) Autocorrelation and stochastic modelling of insolation sequences. Solar Energy 19:343–347

CPAU, 2007, http://www.cpau.com/programs/pv-partners/pvdata.html

Dubuc B, Quiniou F, Roques-Carmes C, Tricot C, Zucker SW (1989) Evaluating the fractal dimension of profiles. Phys Rev A 39:1500–1512

Guessoum AS, Boubekeur A, Maafi A (1998) Global irradiation model using radial basis function neural networks. In Proceedings of World Renewable Energy Congress V (WREC'98), Florence (Italy), pp 169–178

Hardy GH (1916) Weierstrass's nondifferentiable function. Trans amer Math Soc 17:322–323

Harrouni S, Guessoum A (2003) Fractal Classification of solar irradiances into typical days using a cumulative distribution function. ICREPQ'03, Vigo (Spain)

Harrouni S, Guessoum A (2006) New method for estimating the time series fractal dimension: Application to solar irradiances signals. In: Solar Energy: New research Book. Nova Science Publishers, New York, pp 277–307

Harrouni S, Guessoum A, Maafi A (2002) Optimisation de la mesure de la dimension fractale des signaux: Application aux éclairements solaires. SNAS'02, Université d'Annaba

Harrouni S, Guessoum A, Maafi A (2005) Classification of daily solar irradiation by fractional analysis of 10–min–means of solar irradiance. Theoretical and Applied Climatology 80: 27–36

Harrouni S, Maafi A (2002) Classification des éclairements solaires à l'aide de l'analyse fractale. Revue Internationale des énergies renouvelables (CDER) 5:107–122

Huang Q, Lorch JR, Dubes RC (1994) Can the fractal dimension of images be measured?. Pattern Recognition 27:339–349

Kolhe M, Agbossou K, Hamelin J, Bose TK (2003) Analytical model for predicting the performance of photovoltaic array coupled with a wind turbine in a stand-alone renewable energy system based on hydrogen. Renewable Energy 28:727–742

Lestienne R, Bois Ph, Obled Ch (1979) Analyse temporelle et cartographie de la matrice stochastique pour le modèle Markovien dans le midi de la France. La Météorologie 17:83–122

Liebovitch LS, Toth A (1989) Fast algorithm to determine fractal dimension by box counting. Phys Lett A 14:386–390

Liu BYH, Jordan RC (1963) The long-term average performance of flat-plate solar-energy collectors: With design data for the US its outlying possessions and Canada. Solar Energy 7:53–74

Louche A, Notton G, Poggi P, Simonot G (1991) Classification of direct irradiation days in view of energetic applications. Solar Energy 46:255–259

Lundahl T, Ohley WJ, Kay SM, Siffert R (1986) Fractional Brownian motion: A maximum likelihood estimator and its application to image texture. IEEE Trans Med Imaging MI-5:152–160

Maafi A (1991) Mise en évidence d'aspects physiques du modèle Markovien du premier ordre à deux états en météorologie solaire: Application à la conversion photovoltaïque. Doctorat thesis, USTHB University

Maafi A (1998) Markov-Models in discrete time for solar radiation. In Proceedings of Multiconference on Computational Engineering in Systems Applications (IMACS–IEEE), Nabeul-Hammamet (Tunisia) pp 319–322

Maafi A, Harrouni S (2000) Measuring the fractal dimension of solar irradiance in view of PV systems performance analysis. Proc 6th W REC, Brighton (UK), pp 2032–2035

Maafi A, Harrouni S (2003) Preliminary results of fractal classification of daily solar irradiance. Solar Energy 75:53–61

Mandelbrot B (1967) How Long Is the Coast of Britain? Statistical Self-Similarity and Fractional Dimension. Science, New Series 156:636–638

Mandelbrot BB (1982/1983) The fractal geometry of nature, Freeman, New York

Mandelbrot BB, Wallis JR (1969) Computer experiments with fractional Brownian motion. Water Resources Res 5:228–267

Maragos P, Sun FK (1993) Measuring the fractal dimension of signals: morphological covers and iterative optimization. IEEE Transaction on Signal Processing 41:108–121

MIDC, 2007, http://www.nrel.gov/midc/

Minkowski H (1901) Uber die Bgriffe Lange, Oberflache und Volumen, Jahresber, Deutch. Mathematikerverein 9:115–121

Muselli M, Poggi P, Notton G, Louche A (2000) Classification of typical meteorological days from global irradiation records and comparison between two Mediterranean coastal sites in Corsica Island. Energy Conversion and Management 41:1043–1063

Richardson LF (1961) The problem of contiguity: an appendix of statistics of deadly quarrels. Genmal Systems Yearbook 6:139–187

Sayigh AAM (1977) Solar energy engineering chap 4. Academic Press, New York

Sfetsos A, Coonick AH (2000) Univariate and multivariate forecasting of hourly solar radiation with artificial intelligence techniques. Solar Energy 68:169–178

Tricot C, Quiniou JF, Wehbi D, Roques-Carmes C, Dubuc B (1988) Evaluation de la dimension fractale d'un graphe. Rev Phys 23:111–124

Voss RF (1988) Fractals in nature: From characterization to simulation. In The science of fractal Images, HO Peitgen and D Saupe. Springer-Verlag, New York

Zeng X, Koehl L, Vasseur C (2001) Design and implementation of an estimator of fractal dimension using fuzzy techniques. Pattern Recognition 34:151–169

Chapter 3
Modelling the Statistical Properties of Solar Radiation and Proposal of a Technique Based on Boltzmann Statistics

Joaquin Tovar-Pescador

1 Introduction

Solar radiation affects all the Earth's processes related to the environment and plays a fundamental role in the development of human activities. Among these processes, solar radiation influences water evaporation into the atmosphere and, consequently, also humidity of ground and air. Therefore, solar radiation strongly affects the agricultural and ecological processes. The knowledge of solar radiation is also important for solar energy conversion systems, such as photovoltaic, thermal and thermosyphon applications. Finally, solar radiation determines the Earth's energy balance and, therefore even, it is a key parameter for the understanding of the climatic change.

Solar radiation has been measured for a long time, but even today there are many unknown characteristics of its behaviour for remote areas with no direct measurement.

Along the last century, and particularly in its second half, a notably theoretical and experimental research effort has been conducted to develop solar energy conversion devices. These studies have contributed to great technological know-how in the use of solar energy and, nowadays, thermoelectric and photovoltaic solar energy production facilities are found in many countries of the world. The need of use of renewable energies, particularly in recent years, has contributed to the use of solar energy, too. The latter represents a small amount in relation to other type of energies, but a significant increase of solar facilities both thermal and photovoltaic is foreseen for the next years. Data given by the Official Energy Statistics of the U.S. Government in February 2007 [Report DOE/EIA-0383(2007) table 16] foresee a great increase in the renewable energy sector during the next years. Particularly, photovoltaic solar energy will undergo the most important increases between 2005 and 2030.

Joaquin Tovar-Pescador
University of Jaén, Spain, e-mail: jtovar@ujaen.es

Despite the long period in solar radiation research, the most important advances took place in the last two decades. These advances triggered important improvements in the efficiency of the conversion methods. The solar energy conversion systems have a fast and non-linear response to incident radiation (Suehrcke and McCormick 1989). Therefore, the knowledge of temporal variability of solar radiation is important for the study of these systems based on thermal or photovoltaic principles, and the fluctuating nature of solar radiation has to be taken into account (Gansler et al. 1995).

2 Physical and Statistical Modelling of Solar Radiation

In order to understand and to model the behaviour of solar radiation, two approaches can be used (Festa and Ratto, 1993).

a) The first one is called "physical modelling", and studies the physical processes occurring in the atmosphere and influencing solar radiation.

In the upper atmosphere, the incoming solar radiation is affected by atmospheric components, such as molecular gases, aerosols, water vapour or clouds. Part of this radiation is backscattered to space, another part is absorbed and the rest falls into the Earth's surface. This latter component interacts with the surface, part is absorbed and the rest is reflected back to space. Therefore, the diffuse radiation is composed by the radiation backscattered by the atmosphere before reaching the ground and by the component reflected by the Earth's surface. Finally, the radiation on the surface depends on the absorption and scattering processes in the atmosphere.

The physical method is exclusively based on physical considerations, allowing that the radiant energy exchanges take place within the Earth-atmosphere system. This approach dictates models that account for the estimated solar irradiation at the ground in terms of a certain number of physical parameters (water vapour content, dust, aerosols, clouds and cloud types, etc.).

b) The second approach, which could be called "statistical solar climatology", arose mainly as a tool to reach immediate goals in solar energy conversion, rapidly becoming an autonomous field of solar energy research. This methodology can ideally be subdivided into the following topics:

- descriptive statistical analysis, for each place and period of the year, of the main quantities of interest (such as hourly or daily global, diffuse or beam solar irradiation) and statistical modelling of the observed empirical frequency distributions;
- investigation on the statistical relationship among the main solar radiation components on the one hand (for instance, diffuse versus global irradiation) and the spatial correlation between simultaneous solar data at different places on the other;
- research on the statistical interrelationship between the main solar irradiation components and other available meteorological parameters such as sunshine duration, cloudiness, temperature, etc;

- forecasting of solar radiation values at a given place or time based on historical data. The statistical forecasting models often constitute a method used in climate prediction. It is also an appropriate methodology to estimate the probabilistic future behaviour of a system based on its historical behaviour.

The application of the statistical methods to solar radiation research involves a wide range of studies:

- characterisation of numerical data to describe concisely the measurements and to aid to understand the behaviour of a system or process;
- to aid in the estimation of the uncertainties involved in observational data and those related to subsequent calculations based on observational data;
- characterisation of numerical outputs from physical models to understand the model behaviour and to assess the model ability to simulate important features of the natural system (model validation). Feeding this information back into the model enhances the performance;
- estimation of probabilistic future behaviour of a system based on historical information;
- spatial and temporal extrapolation or interpolation of data based on a mathematical fitting method;
- estimation of input parameters for more complex physical models;
- estimation of the frequency spectra of observations and model outputs.

The main advantage offered by the physical methods, in comparison to the statistical ones, is their spatial independence. In addition, they do not require solar radiation data measured at the Earth's surface. However, the physical method needs complementary meteorological data to characterise the interactions of solar radiation with the atmosphere.

The physical and statistical methods are related to each other. On the one hand, the parameters which govern a physical model take values, which fluctuate according to the changes in the meteorological conditions. Thus, if we are interested in using a physical model in order to estimate data in a determined site, statistics must be introduced at the level of the model parameters. On the other hand, any statistical analysis, which does not carefully choose the "right" quantities by taking into account their fundamental physical and meteorological relationships, is condemned to give trivial and/or useless results.

3 Stochastic Processes. Stationarity

Stochastic processes concern sequences of events governed by probabilistic laws (Karlin and Taylor 1975). A stochastic process $X=\{X(t), t \in T\}$ is a collection of random variables. That is, for each "t" in the domain T, $X(t)$ is a random variable. We often interpret "t" as time and call $X(t)$ the state of the process at time "t". The domain T can be a discrete stochastic process, or a continuous one. Any set of X values is a sample.

We assume that the climate at a given site is fully described by a set of stochastic processes $X_1(t)$, $X_2(t)$, $X_3(t)\ldots$, each of them representing the stochastic time evolution of a specific climatological quantity (solar irradiance, wind speed, wind direction, temperature, humidity, pressure, etc).

According to the goal, the continuous stochastic processes $X_1(t)$, $X_2(t)$, $X_3(t)\ldots$ can be substituted by stochastic sequences (i.e. stochastic processes with discrete time parameter t), obtained by averaging, integrating or sampling them on an appropriate time basis (for instance, hour or day).

Atmospheric observations separated by relatively short time intervals tend to correlate (Wilks 2006). The analysis of the nature of these correlations can be useful for both understanding atmospheric processes and forecasting future atmospheric events.

We do not expect the future values of a data series to be identical to past values of existing observations. However, in many cases, it may be very reasonable to assume that their statistical properties will be similar. The idea that past and future values of a time series will be similar in the statistical sense is an informal expression of what is called stationarity. Usually, this term refers to what is considered "weak" stationarity. In this sense, stationarity implies that the mean and autocorrelation function of the data series do not change in time. Different ensembles of a stationary time series can be regarded as having the same mean and variance.

Most methods of time series analysis assume stationarity of the data. However, many atmospheric processes are distinctly not stationary. Obvious examples of non-stationary atmospheric variables are those exhibiting annual or daily cycles. For instance, solar radiation exhibits inter-annual cycles.

When studying solar radiation, we expect stationarity conditions to be met only in its annual or diurnal cycles. If separated sub-series are compared each other for a period equal to a cycle we expect them to be consistent. Thus, for example, the daily global irradiation can be represented by a cycle-stationary stochastic sequence with a time step of one day and period of one year. Even with this simplification, a considerable number of years with data are needed in order to obtain a satisfactory statistical knowledge of the behaviour the parameters in concern. Since no secular changes in climatic conditions are considered, the set of processes corresponding to solar radiation is stochastically periodic (with period equal to one year), i.e. with probabilistic parameters periodically varying in time.

There are two approaches to deal with non-stationary variables. Both aim at processing the data in a way that will subsequently lead to stationarity. The first approach is to transform the non-stationary data to, approximately, stationary. For example, by subtracting a periodic function from the data subject to an annual cycle can derive a transformed data series with constant mean. In order to produce a series with both constant mean and variance, it might be necessary to parameterize these anomalies.

The alternative way is to stratify the data. That is, to conduct separate analyses for different subset of the data that are short enough to be regarded as nearly stationary, for instance monthly subsets of daily solar irradiation values (Wilks 2006).

4 The "k" Indices

The seasonal and diurnal variations of solar irradiance are described by well-established astronomical relationships. At the short-term, the behaviour of solar radiation is mainly ruled by the stochastic parameters: frequency and height of the clouds and their optical properties, atmospheric aerosols, ground albedo, water vapour and atmospheric turbidity (Woyte et al. 2007). As a consequence, the actual solar irradiance can be considered as the sum of two components: deterministic and stochastic. We can isolate the stochastic component by defining the instantaneous clearness index as:

$$k_t = \frac{G}{I_{0h}} = \frac{G}{I_{0n}\cos\theta_z} = \frac{G}{I_{SC}E_0\cos\theta_z},\qquad(3.1)$$

where G is the horizontal global irradiance at ground; I_{0h}, the extraterrestrial horizontal solar irradiance; $I_{SC} = 1367\,W/m^2$, the solar constant; E_0, the eccentricity correction factor and θ_z, the zenith angle. The sub-indices denote: h, horizontal; n, normal and 0, extraterrestrial.

E_0 and θ_z depend on astronomical relationships only and can analytically be determined for each time instant (Iqbal 1983). The instantaneous clearness index k_t accounts for all meteorological, thus stochastic, influences. Therefore, clearness index is the quantity needed to focus on the analysis of fluctuations in solar irradiance. It gives the ratio of the actual energy on the ground to that initially available at the top of the atmosphere accounting, therefore, for the transparency of the atmosphere.

Similarly, we can define the k_b and k_d indices for the diffuse and direct radiation components, respectively.

$$k_d = \frac{D}{I_{0h}} = \frac{D}{I_{0n}\cos\theta_z},\qquad(3.2)$$

is called diffuse fraction (in some literature, diffuse coefficient) and is defined as the ratio of the diffuse irradiance on the ground to the extraterrestrial global horizontal one.

$$k_b = \frac{I_n\cos\theta_z}{I_{0h}} = \frac{I_n}{I_{0n}},\qquad(3.3)$$

is called direct fraction and is defined as the ratio of the horizontal direct irradiance on the ground to the extraterrestrial global horizontal one.

From the well-known expression, $G = I_n\cos\theta_z + D$, it is evident that:

$$k_t = k_b + k_d.\qquad(3.4)$$

These indices can also be defined for the irradiation by integrating the irradiance values over a given time interval Δt. The clearness index will be then denoted by $k_t^{\Delta t}$ and defined as the relation between the horizontal global irradiation on the ground and the extraterrestrial global irradiation over the same time interval Δt:

$$k_t^{\Delta t} = \frac{\int_{\Delta t} G dt}{\int_{\Delta t} I_{0h} dt} = \frac{H}{H_0}. \tag{3.5}$$

The most usual integration periods are the day and the hour, although other periods, as the month, can also be used. When Δt is less than 5–10 minutes, the clearness index is said to be instantaneous. Therefore, according to the integration period, we deal with monthly (k_t^M), daily (k_t^D), hourly (k_t^H) or instantaneous (k_t) clearness index. Similarly, there can be defined the diffuse index:

$$k_d^{\Delta t} = \frac{\int_{\Delta t} D dt}{\int_{\Delta t} I_{0h} dt}, \tag{3.6}$$

and, then, the monthly (k_d^M), daily (k_d^D), hourly (k_d^H) or instantaneous (k_d) diffuse fraction. Finally, we can define the direct index:

$$k_b^{\Delta t} = \frac{\int_{\Delta t} I_n \cos \theta_z dt}{\int_{\Delta t} I_{0h} dt}, \tag{3.7}$$

and, then, the monthly (k_b^M), daily (k_b^D), hourly (k_b^H) or instantaneous (k_b) direct fraction.

The ensemble study of the k_t, k_d and k_b indices provides an adequate information to characterise the actual state of the atmosphere and to know the solar energy availability at a given place.

5 Density and Cumulative Distribution Functions

In mathematical sense a histogram is simply a mapping m_i that counts the number of observations (frequencies) that fall into various disjoint categories (known as bins or intervals). The histograms also are called frequency distributions. If we let n be the total number of observations and k the total number of bins, the histogram meets the following condition:

$$n = \sum_{i=1}^{k} m_i$$

A cumulative histogram is a mapping that counts the cumulative number of observations in all of the bins up to the specified bin. That is, cumulative histogram M_i of a histogram m_i is defined as

$$M_i = \sum_{j=1}^{i} m_j$$

Conversationally, the probability density function (PDF) is the curve that adjusts the histogram, and the cumulative distribution function (CDF) the curve that adjusts the cumulative histogram and completely describes the probability distribution of a real random variable.

Usually, the statistical behaviour of the random variables such as k_t, k_b and k_d is carried out using the cumulative distribution function, which represents the probability that the event $x(t)$, at the time instant t, be less than a given value x:

$$F(x,t) = P(x(t) \le x). \tag{3.8}$$

For stochastic variables, this quantity also represents the fraction of time that the stochastic variable is below a given value (fractional time). This second interpretation is more appropriate in certain cases.

The minimum number of intervals to be chosen in order to correctly draw the frequency histogram depends on the number of available data. As we will see later, different authors have used different number of intervals within the range of variation of these indices (from 0 to 1). We will use $(x_0|\Delta x|x_f)$ to denote the first value of the interval (x_0), its width (Δx) and the last value (x_f). Therefore, for example, $(0|0.02|1)$ represents a distribution with 0 as the first value, 0.02 as the interval width and 1 as the last value. This implies a total number of 50 intervals.

The statistical behaviour can also be characterised by the probability density function $f(x,t)$ defined as:

$$f(x,t) = \frac{\partial F(x,t)}{\partial x}. \tag{3.9}$$

The functions are normalised in a way that the area under the $f(x,t)$ curve is equal to unity. That is:

$$\int_{-\infty}^{\infty} f(x,t)dx = 1. \tag{3.10}$$

In case of a finite range of variation, the integration limits in Eq. (3.10) only are extended to this range, since $f(x,t) = 0$ outside the range of variation. Particularly, in the study of k_t, k_d and k_b, the normalised functions will verify that:

$$\int_{-\infty}^{\infty} f(x,t)dx = \int_{0}^{1} f(x,t)dx = 1. \tag{3.11}$$

Hereinafter, the parameter "t" will be omitted for the sake of clarity in the expressions.

The distributions of k_t, k_b and k_d provide statistical information about the absolute frequency of these values. However, frequently it is more interesting to analyse the probability distribution of these indices under certain conditions. This is known as "conditional probability". The density function is written as $f(x|y)$, and is the distribution function of "x" when "y" fulfils a particular condition. It provides more accurate information on the index behaviour under the given conditions. Particularly, because of the interest of these distributions to estimate the performance solar conversion systems, the conditional probability distributions of k_t, k_b and k_d are expressed in terms of the optical air mass, $f(k_t|m_a)$, or in terms of the mean value in a determined period, for example $f(k_t|\bar{k}_t^H)$. We will refer to the cumulative conditional probability distributions as $F(x|y)$.

6 A Research Survey on the Statistical Behaviour of Solar Radiation Components

We will shortly review here the most important studies on the statistical solar radiation behaviour, with special focus on those works that have produced important advances on solar radiation modelisation. Until relatively recently, most of the studies were focused on the daily distributions. Nevertheless, in the last years, the analysis of the instantaneous behaviour has given rise to a special consideration. This is because the need to use the instantaneous values in specific applications, as photovoltaic devices and estimation of erythemal doses, and the appreciable differences observed with regards to the monthly, daily and hourly distributions. Such differences must be specially taken in account in energy and biological applications, where the systems have a non-linear response and are very sensitive to the instantaneous values.

Several studies on these topics have been carried out. The first results go back to the famous works of Ångström (1924, 1956), who derived regression expressions for the different components of the daily solar radiation based on the sunshine duration. Other works, which are worth to point out are those of Black et al. (1954) and Glover and McCulloch (1958). These works revealed certain bimodality or, at least, a strong asymmetry in respect to the mean, as confirmed afterwards by several authors: Bennet (1965, 1967), Klink (1974a), Andretta et al. (1982), Barbaro et al. (1983).

Among the solar irradiation measurements, the daily global values have been, for a long period, the most frequently studied. Nevertheless, they have not adequately been analysed using statistical or even graphical methods. A previously suggested normalisation by the corresponding extraterrestrial irradiation values has recently become a customary practice. The monthly frequency distributions for many U.S. stations have been presented by Bennet (1967), showing a relative bimodality and skewness in the distributions. The monthly frequency distributions have also been studied by Klink (1974b) and Baker and Klink (1975), showing a negative skewness (or to the left), platycurtic and a tendency to bimodality.

6.1 Daily Distributions of Global Radiation

The most important, probably, study on the daily global irradiation distributions, cited in most works, is that by Liu and Jordan (1960). Based on this study, several works have been carried out using different methodologies as well as different types of equations to model the solar radiation variability.

For instance, the daily global irradiation scaled to the mean daily global irradiation over a month (Liu and Jordan 1960), and the daily global irradiation scaled to the clear-sky daily global irradiation (Bois et al 1977; Exell 1981), have been studied. Nevertheless, most of the studies deal with the daily global clearness index, as those by Whillier (1956), Liu and Jordan (1960), Bendt et al. (1981),

Hollands and Huget (1983), Olseth and Skarveit (1984), Saunier et al. (1987), Graham et al. (1988), Gordon and Reddy (1988), Feuillard and Abillon (1989), Rönnelid and Karlsson (1997) or Babu and Satyamurty (2001).

The earlier work by Liu and Jordan (1960) examines the daily clearness index distribution for certain monthly mean values of the clearness index, $\bar{k}_t^M$. They used 5 years of daily data from 27 locations in the United States with latitudes from $19°$ to $55°$ North. The authors pointed out that the curves of CDFs of k_t^D do not significantly change with the month and location, but they rather depend on the monthly average, $\bar{k}_t^M$, of daily values, for each considered month. The cumulative distributions, $F = F(k_t^D \,|\, \bar{k}_t^M)$, were generated for the monthly averages clearness index values, $\bar{k}_t^M = (0.3|0.1|0.7)$. The authors made the hypothesis of universal validity of the CDF curves although they did not provide any fit function for the distributions (Fig. 3.1).

Bendt et al. (1981) studied the different frequency distributions from which purely random sequences of daily clearness index, k_t^D, can be generated, with the restriction of $\bar{k}_t^M$ to be bound to a specified value. They proposed the following expression for the density function:

$$f(x \,|\, \bar{x}) = \frac{\gamma e^{\gamma x}}{e^{\gamma x_{max}} - e^{\gamma x_{min}}}, \qquad (3.12)$$

being $x = k_t^D$ and $\bar{x} = \bar{k}_t^M$, with $x_{min} \leq x \leq x_{max}$ and where $x_{min} = 0.05$ and x_{max} conveniently selected for each month. The parameter γ can be calculated from the following equation:

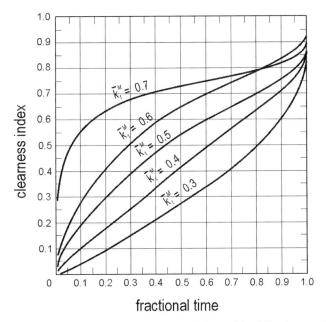

Fig. 3.1 The CDFs of the monthly distributions of the daily clearness index adapted from Liu and Jordan (1960)

$$\bar{x} = \frac{\left(x_{min} - \frac{1}{\gamma}\right) e^{\gamma x_{min}} - \left(x_{max} - \frac{1}{\gamma}\right) e^{\gamma x_{max}}}{e^{\gamma x_{min}} - e^{\gamma x_{max}}}. \tag{3.13}$$

The corresponding CDF is expressed as:

$$F(x|\bar{x}) = \frac{e^{\gamma x_{min}} - e^{\gamma x}}{e^{\gamma x_{min}} - e^{\gamma x_{max}}}. \tag{3.14}$$

Figure 3.2 shows a plot of Eq. (3.14) for different values of $\bar{x}$. The authors also found that the distributions depend on the season (Fig. 3.3).

Because of the difficulty in obtaining the parameter γ from Eq. (3.13), Suehrcke and McCormick (1987) proposed the following simplified expression:

$$\gamma = A \cdot tg\left(\pi \frac{\bar{x} - (x_{max} - x_{min})/2}{x_{max} - x_{min}}\right), \tag{3.15}$$

where $x_{min} = 0.05$ and $A = 15.51 - 20.63x_{max} + 9.0x_{max}^2$.

Based on this distribution, Reddy et al. (1985) suggested that the maximum value of the Bendt's distribution can be yielded from the linear expression:

$$x_{max} = 0.362 + 0.597\bar{x}. \tag{3.16}$$

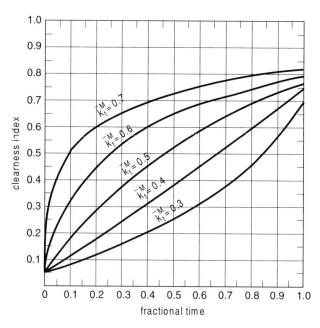

Fig. 3.2 The CDFs based on Bend's model. The curves are similar to that of Liu and Jordan (1960), but exhibit a different behaviour around the unity value of fractional time. Adapted from Bendt et al. (1981)

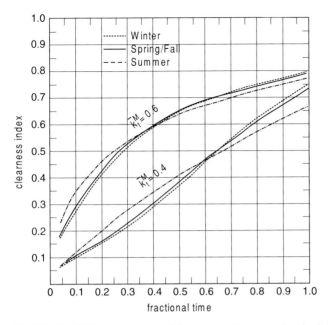

Fig. 3.3 The CDFs as a function of the season. The curves for the 0.4 and 0.6 average monthly clearness index are shown as an example. Adapted from Bendt et al. (1981)

Hollands and Huget (1983) proposed the use of a modified gamma PDF such as:

$$f(x|\bar{x}) = C\frac{(x_{max} - x)}{x_{max}}e^{\lambda x},$$
(3.17)

with $x = k_t^D$, $\bar{x} = \bar{k}_t^M$, and $0 \leq x \leq x_{max}$. The parameters C and λ depend on x_{max} and $\bar{x}$ and C is yielded by:

$$C = \frac{\lambda^2 x_{max}}{e^{\lambda x_{max}} - 1 - \lambda x_{max}}.$$
(3.18)

The relation between x_{max} and λ is given by:

$$\bar{x} = \frac{\left[\left(\frac{2}{\lambda} + x_{max}\right)(1 - e^{\lambda x_{max}}) + 2x_{max}e^{\lambda x_{max}}\right]}{e^{\lambda x_{max}} - 1 - \lambda x_{max}}.$$
(3.19)

The CDF is then:

$$F(x|\bar{x}) = \frac{C(1 + \lambda x_{max})}{\lambda^2 x_{max}}\left[e^{\lambda x_{max}}\left(1 - \frac{\lambda x}{(1 + \lambda x_{max})}\right) - 1\right].$$
(3.20)

Figure 3.4 shows the gamma functions based on Eq. (3.17). Notice the unimodal character of the distributions. Figure 3.5, obtained based on Eq. (3.20), shows the

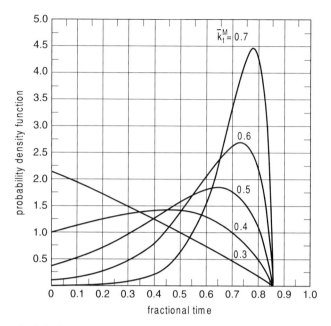

Fig. 3.4 The density distribution functions proposed by Hollands and Huget. All the curves are unimodal. Adapted from Hollands and Huget (1983)

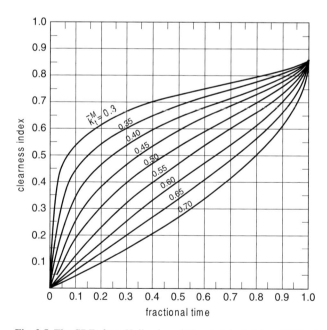

Fig. 3.5 The CDFs from Hollands and Huget. Adapted from Hollands and Huget (1983)

CDFs for different monthly mean values of $\bar{k}_t^M$. It can be observed the similarity with those generated from the equations suggested by Bendt.

Olseth and Skarveit (1984) used another type of normalisation based on an index. This index, ϕ, depends on the maximum and minimum irradiation values of the considered location on the Earth, defined as:

$$\phi = \frac{x - x_{min}}{x_{max} - x_{min}}, \tag{3.21}$$

with partitions $\phi = (0.0|0.1|1.0)$ and being $x = k_t^D$. Then, they fitted the curves using two modified gamma distributions in such a way that the bimodality of the frequency distributions was captured. That is,

$$f(\phi|\bar{\phi}) = \omega G(\phi, \lambda_1) + (1 - \omega)G(\phi, \lambda_2), \tag{3.22}$$

being:

$$\lambda_1 = -6.0 + 21.3\bar{\phi}, \tag{3.23}$$

$$\lambda_2 = 3.7 + 35e^{-5.3\bar{\phi}}. \tag{3.24}$$

These authors introduced for the first time a bimodal distribution with a clear-sky mode at high ϕ values and an overcast mode at low ϕ values. By means of this new methodology, they achieved to accurately reproduce the distributions for high latitudes. The results were compared with those predicted by the Hollands-Huget model showing significant differences and yielding a better fitting to the data. Figure 3.6 shows the observed differences between the fitting of the Olseth-Skarveit model and that of the Hollands-Huget model.

Most of these research studies on solar radiation have been carried out with the main aim of predicting the long-term average energy delivered to solar collectors. The daily distributions have been studied by Hansen (1999) for 10 locations in the United States, with the aim to be used in biological applications. This author describes three alternative models for the distributions and emphasises the strong non-normality of them. Wang et al. (2002) also analysed the behaviour of these distributions and the fluctuations introduced by the topography, with the aim of using the results for terrestrial ecosystem studies. All the distributions showed to be asymmetrical. The asymmetry shown by the annual irradiation distributions at high latitudes was studied by Rönnelid (2000). The analysis conducted by Ibáñez et al. (2003) for 50 locations in the USA aimed to test the modality of the daily clearness index distributions. This study concluded that 60% of the distributions showed a bimodal behaviour. Given the predominance of the bimodal shape of the probability density distributions, Ibáñez et al. (2002) proposed a bi-exponential probability density function. This function used the mean monthly clearness index and the mean monthly solar altitude at noon to fit the observed behaviour of the daily clearness indices.

Tiba et al. (2006) analysed the CDFs for 23 sites located in the Southern hemisphere. These authors concluded that the Liu and Jordan CDFs do not have a universal character as previously stated by other authors, such as Saunier et al. (1987).

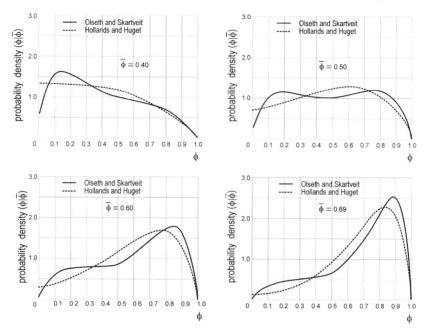

Fig. 3.6 Differences between the fitting of the Olseth-Skarveit model and that of the Hollands-Huget model. The Olseth-Skarveit PDF is bimodal. Adapted from Olseth and Skarveit (1984)

Gueymard (1999) presented two new models to predict the monthly-average hourly global irradiation distributions from its daily counterpart; whereas Mefti et al. (2003) used the monthly mean sunshine duration to estimate the probability density functions of hourly clearness index for inclined surfaces in Algeria.

6.2 Hourly Distributions of Global Radiation

The number of studies about the hourly irradiance is less than for longer time scales. Some authors, as Engels et al. (1981) or Olseth and Skartveit (1987, 1993) emphasise that the hourly distributions are similar to the daily ones and they even use the same fitting procedures. Ettoumi et al. (2002) used Beta distributions to model the behaviour of the global solar irradiation in Algeria. Only few authors are pointing out an increase in the bimodality with regard to the daily distributions.

6.3 Instantaneous Distributions of Global Radiation

The distributions that can be considered as instantaneous (less than 10-minutes) show a different shape, since the transient effects caused by clouds are now evident and contribute to the increase of the bimodality. The first authors who proved the strong bimodality of these clearness index distributions (with 1-minute basis data) were Suehrcke and McCormick (1988a). They analysed a year of irradiance data collected in Perth (Australia) for different values of optical air mass and proposed a fitting function. Tovar et al. (1998a) confirmed the hypothesis of Suehrcke and McCormick using 1-minute data collected during almost three years in Armilla, near Granada, (Southern Spain). They observed the strong bimodality of the distribution conditioned by the optical air mass, increasing as the air mass increases. These authors proposed a model based on the Boltzmann statistic. Later, they studied the variability of the probability density based on the daily irradiation (Tovar et al. 2001). These last distributions presented unimodal features, unlike the distributions conditioned by the optical air mass. Nevertheless, the same type of fitting function can be used (Tovar et al. 2001). The model of Tovar et al. (1998a) has been used by other authors, like Varo et al. (2006), who evaluated the model with data collected in Córdoba (Southern Spain). They achieved reasonable performance using different fitting parameters according to the local climatic conditions. Vijayakumar et al. (2006) analysed the instantaneous distributions, with the aim of exploring the differences between hourly and instantaneous distributions. They concluded that the variations in solar radiation within an hour cannot be considered negligible when conducting performance analyses of solar energy systems. Depending on the critical level, location and month, an analysis using hourly data rather than short-term data can underestimate the performance between 5% and 50%. Tomson and Tamm (2006) analysed the distribution functions of the increments of solar radiation mean values over a period of time, classifying the solar "climate" in stable and highly variable. They found that the distributions functions can be explained by the superposition of two exponential functions with different exponents. The study of Woyte et al. (2007) introduced the wavelet techniques to analyse the cumulative frequency distributions of the instantaneous clearness index for four datasets from three different locations. The analysis resulted in the known bimodal pattern of the distributions. The wavelet technique allows the identification of fluctuations of the instantaneous clearness index and their specific behaviour in the time dimension.

6.4 Distributions of Diffuse and Direct Components of Global Radiation

Research about the direct and diffuse components of the solar radiation is less prolific, mainly because of the scarce availability of such data. Moreover, most of the direct component analyses involve the use of data obtained from the differences between measured global and diffuse radiations. Lestienne's works (1978, 1979) suggested the use of two different types of exponential functions for the daily behaviour: one for cloudy days and the other for clear days. Stuart and Hollands (1988) analysed the shape of the hourly direct component suggesting a polynomial fitting for the cumulative distribution function. Later, Callegari et al. (1992) developed a dynamical statistical analysis to reproduce daily direct solar component.

Skartveit and Olseth (1992) compared the global and direct instantaneous distributions, proposing some models for the PDFs of short-term (5-minutes) irradiances. These distributions are not unique functions of the hourly mean, but they depend on the averaging time and also on the inter-hourly variability among 3-hourly averages, namely, the hour of the study, the preceding and the following. Tovar et al. (1998b) continued their previous work and proposed some functions based on the Boltzmann statistic to explain the clearness index distributions. Regarding the diffuse component, the work is worthy of being mentioned by Suehrcke and McCormick (1988b). The results of this study have been used by Tovar et al. (1998b).

So far, the analyses of the solar radiation variability followed very different approaches. Nevertheless, one shared pitfall for most of them is that they were carried out using local data bases. This implies that the proposed models are site dependent and new evaluations must be performed when using data sets from other locations. Finally, it is worth to point out that the analysis devoted to the behaviour of the direct and diffuse components are still scarce. In the last years, these components are being investigated for some special spectral regions from the statistics point of view. For instance, among other works, the photosynthetically active region (PAR, 400–700 nm) have been studied by Ross et al. (1998) and Tovar-Pescador et al (2004), and the ultraviolet region (UV, 290–385 nm) was studied by Varo et al. (2005).

7 Modelling the Instantaneous Distributions Conditioned by the Optical Air Mass

There are two types of approach to obtain the statistical behaviour of solar radiation: the study of the distributions conditioned by the optical air mass and the study of the distributions conditioned by the mean clearness index. For the case of the daily distributions, the distributions conditioned by the monthly clearness index have been profusely studied.

7.1 Statistical Investigation of the Clearness Index

7.1.1 Bimodal Character of the Probability Density Functions

The measurements of instantaneous solar radiation values allow to considerer the effect of optical air mass. As a consequence, the distributions conditioned by the optical air mass not only describe how the instantaneous radiation values are distributed for a given mean value, but also how the instantaneous solar radiation varies with air mass and the time of the day (Suehrcke and McCormick 1988a).

Prior to studying the 1-minute distributions, the influence of the temporal integration interval on the shape of the density distribution function has been analysed. Notice that the size of the integration interval has an important influence on the bimodal character of the distributions. Figure 3.7 shows the comparison between 1-minute frequency distribution $f(k_t|m_a)$ and the hourly frequency distributions $f(k_t^H|m_a)$ for the same value of the optical air mass ($m_a = 3.0$).

The distribution curve corresponding to 1-minute k_t values presents a more marked bimodality than the corresponding to the hourly clearness index values. This behaviour was first noticed by Suehrcke and McCormick (1988a), who analysed the effect of internal averaging on the CDFs (Fig. 3.8) and confirmed later by Jurado et al. (1995) and by Gansler et al. (1995). Suehrcke and McCormick (1988a) suggested that, for averaging periods longer than 60 minutes, there is no evidence of bimodality in the k_t distributions. Thus, they obtained similar distributions to those derived by Bendt et al. (1981) for daily averages. Nevertheless, some degree of bimodality persists for the hourly distribution corresponding to higher values of optical air mass (Fig. 3.7).

Figure 3.9 shows some degree of bimodality in all the distributions. This feature increases with increasing optical air mass. The present finding is in line with the results reported in Suehrcke and McCormick (1988a), Skartveit and Olseth (1992) and Jurado et al. (1995). However, Gansler et al. (1995) found a different behaviour in three U.S. locations.

The distribution densities in Fig. 3.9 show that the probability for values of clearness index in the intermediate range is low. The low probability associated with intermediate k_t values indicates that it is possible to relate the curves of the distribution with two levels of irradiation in the atmosphere, for each optical air mass. The major peak in the density function corresponds to high values of k_t, associated with cloudless conditions, and the secondary maximum corresponds to low values of k_t, associated to cloudy conditions. An increase in optical air mass implies a decrease in the intensity of the first maximum and a subsequent increase in the secondary maximum.

Furthermore, by increasing optical air mass, the principal maximum is shifted towards lower k_t values. The decrease in probability density for the principal maximum implies an increase in the probability density of k_t in the lower range

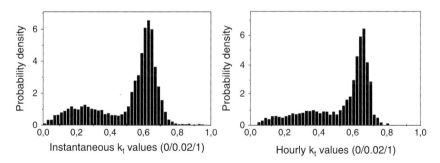

Fig. 3.7 Comparison between 1-minute (instantaneous) frequency distribution and hourly frequency distributions for the same value of the optical air mass ($m_a = 3.0$). The distribution corresponding to 1-minute k_t values presents a more marked bimodality than the corresponding to the hourly clearness index values

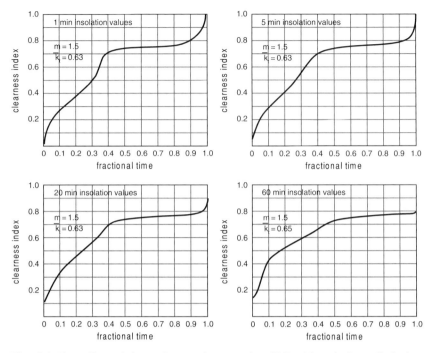

Fig. 3.8 The effect of internal averaging on the CDF. Adapted from Suehrcke and McCormick (1988a)

that leads to the enhancement of the second maximum in order for the area under the curve to remain constant. When the optical air mass tends to higher values, there is an increase in the probability density for the lower k_t range. This result can be associated with the fact that, for small zenith angles, the clouds shade a smaller Earth surface area than for larger angles (Fig. 3.11). On the other hand, for horizontal layers of clouds, the effective thickness of the clouds is also larger for high zenith angles; hence, the clouds are less transparent. Thus, for higher values of optical air mass, the effect of the clouds is stronger, and the bimodality suffers an increase. On the other hand, the probability density of the intermediate states of k_t does not vary considerably with the optical air mass. The increase of the bimodal character when the optical air mass increases is explained by the fact that the largest k_t values and their frequency tends to decrease, therefore increasing the lowest partitions.

In Fig. 3.9 also we can appreciate that k_t reach values close to unity, specially at low optical air mass due to multiple cloud reflections of solar radiation (Fig. 3.10).

Suehrcke and McCormick (1988a) proposed a model, based on the Boltzmann statistics, to explain the bimodal character of the distributions by using three functions associated with three different irradiation levels. Particularly, two of the functions were associated with the extreme conditions related to the cloudless and cloudy conditions, i.e. higher and lower k_t values, and a third function was associated with the intermediate k_t range. Figure 3.12 shows the clearness index distributions for Armilla (Granada, Spain) and Perth (Australia) in Suehrcke and McCormick's work.

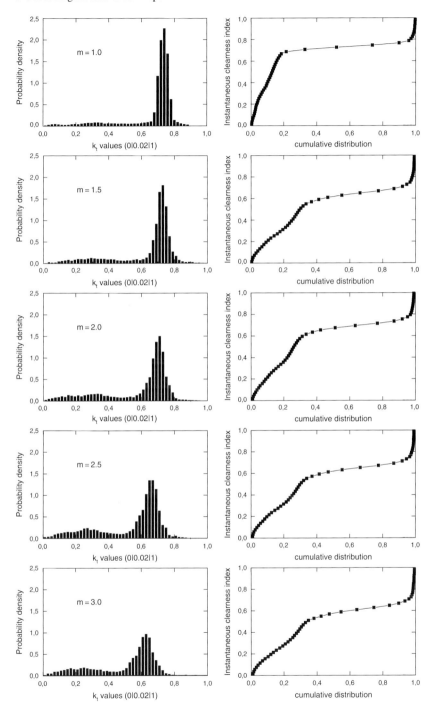

Fig. 3.9 Density distributions of instantaneous clearness index measured in Armilla (Spain), for different optical air masses (left), and their respective CDFs (right)

Fig. 3.10 The clearness index, k_t, at point P can reach values close to unity, especially, at low optical air mass. These high k_t values are associated with an enhancement due to cloud reflections (Suehrcke and McCormick 1998a)

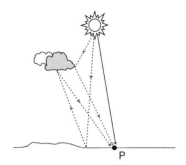

A similarity in the shape of the distributions can be observed. The figure also depicts the adjusted curve proposed by the authors from their data.

There are several other approaches to model bimodality. Many of them use the sum of two functions, each of them describing the behaviour around one of the two maxima. Intermediate values can be obtained as the sum of the "tails" of the distributions between the maxima. For instance, Jurado et al. (1995) proposed the use of two Gaussian distributions.

In order to properly describe the shape of the distribution we should use functions, which meet certain criteria:

- The bimodal character may be expressed as the sum of two functions corresponding to two discernible atmospheric conditions: clear and overcast skies.
- The function must be as simple as possible, such as we do not have to implicitly assess any parameter.
- The parameters governing the function should be interpreted in terms of the climatic and atmospheric variables involve in the process: average clearness index, optical mass, climatology, etc.
- The function should be versatile enough to adapt to any kind of distribution.
- The function may be used to model the direct and diffuse radiation distributions.

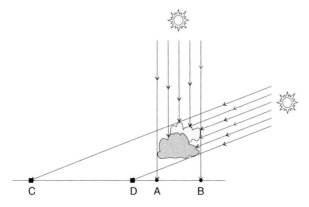

Fig. 3.11 For small zenith angles the clouds shade a smaller area on the surface of the Earth than for larger angles

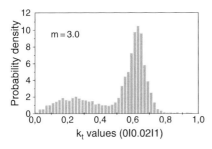 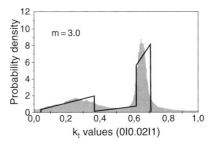

Fig. 3.12 The PDFs of the k_t values for Armilla (left) and Perth (right). The Suehrcke and McCormick adjusted function for data Perth (Australia), is also shown (solid line). Adapted from the authors

- All the function parameters may be formulated by means of the optical air mass, the determinant variable in the process.

With the aim of achieving these goals, several types of functions have been analysed: Beta, Gamma, Gompert, Gauss, Lorentz and Boltzmann distributions. Among them, the latter density function has been selected since it verifies the above requirements:

$$f(x) = A \frac{\lambda e^{(x-x_0)\lambda}}{\left[1 + e^{(x-x_0)\lambda}\right]^2}. \tag{3.25}$$

This function is symmetrical, centred at x_0 and its width is determined by the parameter λ. The introduction of a parameter β in the denominator of the exponential allows to get asymmetrical distributions, as Fig. 3.13 reveals.

$$f(x) = A \frac{\lambda e^{(x-x_0)\lambda}}{\left[1 + e^{(x-x_0)(\lambda+\beta)}\right]^2}. \tag{3.26}$$

7.1.2 Modelling with Boltzmann Distribution

As mentioned early, the experimental distributions are described by the sum of two functions:

$$f(k_t \,|\, m_a) = f_1(k_t) + f_2(k_t), \tag{3.27}$$

subject to the normalisation condition:

$$\int_0^1 f(k_t \,|\, m_a) \; dk_t = 1. \tag{3.28}$$

The f_1 and f_2 functions are obtained from the Boltzmann statistic:

$$f_i(k_t) = A_i \frac{\lambda_i \; e^{(k_t - k_{t0i})\lambda_i}}{\left[1 + e^{(k_t - k_{t0i})\lambda_i}\right]^2} \quad i = 1, 2. \tag{3.29}$$

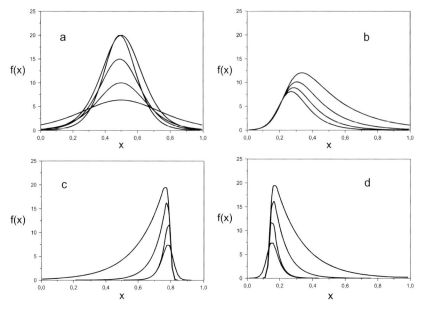

Fig. 3.13 The behaviour of the modified Boltzmann distribution through the β parameter. a) $\beta = 0$ and various values of λ: all the curves are symmetrical, b) curves for a constant $\beta < 0$ and various values of λ, c) curves for various $\beta > 0$, varying simultaneously with other parameters. d) Curves for various $\beta < 0$, varying simultaneously with other parameters

This function yields unimodal symmetrical curves around k_{t0i}, where the function reaches its maximum. A_i determines the function height and λ_i is related to the width of the distribution function. This can be integrated to:

$$F_i(k_t) = A_i \left[1 - \frac{1}{1 + e^{(k_t - k_{t0i})\lambda_i}} \right] \quad i = 1, 2 \tag{3.30}$$

and this can be analytically inverted:

$$k_t = k_{t0i} + \frac{1}{\lambda_i} \ln \frac{F_i(k_t)}{A_i - F_i(k_t)} \quad i = 1, 2. \tag{3.31}$$

These characteristics allow the generation of synthetic data of instantaneous values from the CDFs by methods of inferential statistics. Also it is possible to obtain explicitly the k_t coefficients.

The coefficients A_1 and A_2 must satisfy the normalisation condition:

$$\int_0^1 f(k_t)dk_t = A_1 \int_0^1 \frac{\lambda_1 e^{(k_t - k_{t01})\lambda_1}}{\left[1 + e^{(k_t - k_{t01})\lambda_1}\right]^2}dk_t + A_2 \int_0^1 \frac{\lambda_2 e^{(k_t - k_{t02})\lambda_2}}{\left[1 + e^{(k_t - k_{t02})\lambda_2}\right]^2}dk_t = 1. \tag{3.32}$$

After fitting the various density distributions, corresponding to each optical air mass, we obtain the parameters A_1, A_2, k_{t01}, k_{t02}, λ_1, λ_2, that can be adjusted by functions depending on m_a.

When modelling this dependence for the data of Armilla (Granada, Spain), we found for the maxima distributions, k_{t01} and k_{t02}, the following expressions:

$$k_{t01} = 0.763 - 0.0152m_a - 0.012m_a^2, \text{ with } R^2 = 0.996, \tag{3.33}$$

$$k_{t02} = 0.469 - 0.0954m_a + 0.01m_a^2, \text{ with } R^2 = 0.992, \tag{3.34}$$

where R^2 is understood as the proportion of response variation "explained" by the parameters in the model.

The position of the principal maximum, k_{t01}, shifts towards lower values as the optical air mass increases. The same trend occurs for the value of k_{t02}, corresponding to the second maximum of the distribution. However, the shift is smaller than that associated with the principal maximum k_{t01}, as it can be concluded by comparing the coefficients of the optical air mass terms in each equation. This implies that, when the optical air mass increases, the two maxima tend to be closer.

The values of the width parameters, λ_1 and λ_2, can also be expressed in terms of the optical air mass:

$$\lambda_1 = 91.375 - 40.092m_a + 6.489m_a^2, \text{ with } R^2 = 0.999, \tag{3.35}$$

$$\lambda_2 = 6.737 + 1.248m_a + 0.4246m_a^2, \text{ with } R^2 = 0.975. \tag{3.36}$$

The coefficient A_1 has been fitted using the following expression:

$$A_1 = 0.699 + 0.1217m_a^{-2.1416}, \text{ with } R^2 = 0.994. \tag{3.37}$$

Considering the A_1 and A_2 dependence ($A_1 + A_2 = 1$, because of the normalisation condition), it is obvious that while A_1 decreases with air mass, A_2 shows the opposite trend. The ratio between the intensity of the two peaks depends on m_a. This ratio decreases when m_a increases, that is a decrease in m_a implies an enhancement of the first maximum relative to the second one.

Figure 3.14 shows the fitting curves using both, the Suehrcke and McCormick's model and the Tovar's model based on the Boltzmann statistics. Figure 3.14a shows the case of the best adjustment provided by the Suehrcke-McCormick model for data collected in Armilla (Granada), adapting conveniently the parameters to fit the maxima of the distribution. Figure 3.14b shows the Tovar model adjustment for the same set. Figure 3.14c shows the results by applying the Tovar model to the

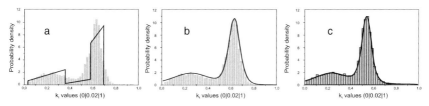

Fig. 3.14 1-minute kt values fitted distribution functions: a) Suehrcke and McCormick model to data of Armilla, b) Tovar model to data of Armilla, c) Tovar model to data of Córdoba (from Varo et al. 2006)

data collected in Córdoba. It can be observed that the Boltzmann model provides a reasonable adjustment (Tovar et al. 1998a; Varo et al. 2006). Nevertheless, the maxima of the bimodal distribution depend on the location and its climatic features. Therefore, the fitting parameters in Eqs. (3.33–3.37) will also depend on the location and its climatic features. However, there are some common characteristics for all the functions used to fit the distributions.

7.1.3 Parameter's Dependence on the Local Climatology

As we have shown, the distributions and models proposed depend on the local climate. In the case of the Tovar's model the parameters A_1, A_2, k_{t01}, k_{t02}, λ_1, λ_2 should be analysed for other locations. To this end, appropriate data bases for different latitudes are needed. However, based on the previous results, it can be concluded that:

- The A_1 and A_2 values depend on sky conditions (clear, completely overcast and partially-cloudy conditions). As observed in Eq. (3.32), an increase in the number of clear-sky and overcast-sky events implies a decrease in the partially-cloudy conditions. Additionally, it should be taken in account that A_1 and A_2 are also influenced by the optical air mass and the latitude.
- The places, which exhibit climatology with predominance of clear-sky conditions experience an increase of A_1 and, thus a decrease of A_2. The opposite occurs in those places with a dominance of overcast conditions.
- The parameters k_{t01} and k_{t02} provide information about the position of the distribution's maxima. Obviously, these positions fundamentally depend on the optical air mass in the case of clear skies, and also on the climatology for overcast conditions, the latter being more important.
- The parameters λ_1 and λ_2 are related to the width of the maxima. For clear skies, (λ_1), and given the optical air mass, the width of the density distribution is associated with the particles in suspension in the atmosphere, which modifies the transmittance of the atmosphere. For overcast conditions, (λ_2), the main factors influencing the width are the amount and type of clouds, which can dramatically change the atmospheric transmitivity conditions.

In short, the affirmation supported by several authors that the distribution functions depend on local conditions and are not universal seems suitable. They could be adjusted for a certain region, but the local climatic conditions must be taken in account in order to get a proper fit of the statistic behaviour of the local clearness index. Certainly, all the parameters of the model depend on the optical air mass, the latitude, the climatology and even on the atmospheric turbidity conditions. Hence, the adjustment shown for the parameters A_1, A_2, k_{t01}, k_{t02}, λ_1, λ_2 should be reviewed in accordance to these conclusions.

We want to underline that this experimental features can be applied to the analysis of the solar radiation components. Additionally, we think important to provide a function, which accurately describes all the local statistic behaviour.

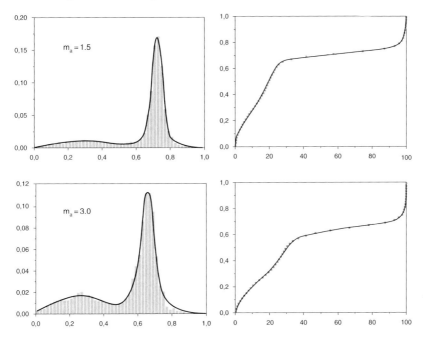

Fig. 3.15 Adjustment of the Boltzmann function to the 1-minute k_t values distributions (left), and their respective CDFs (right), obtained for Armilla and for 1.5 and 3.0 values of the optical mass

Finally, we would like to highlight that the time interval considered strongly influences the distributions. The instantaneous values clearly provide bimodal distributions, while data for a greater time average tend to make this bimodal character disappear. Figure 3.15 shows the adjustment with the Tovar's model for the data of Armilla.

7.2 Distributions of 1-minute k_b Values Conditioned by Optical Air Mass

Figure 3.16 shows the k_b frequency distributions for different optical air masses for Armilla (Spain). All the curves present a bimodal appearance, with two well-defined maxima. The first maximum, located in the interval (0.00, 0.02), corresponds to conditions associated with direct irradiance close to zero. These conditions correspond to overcast or partially-cloudy skies, more frequent for high optical air mass. For greater zenith angles, scattered clouds hide proportionally greater areas on the surface of the Earth and horizontal cloud layers have large effective thickness than for lower zenith angles. Therefore, for higher values of optical air mass, the blocking effect of clouds is more efficient. The probability of that solar global irradiance

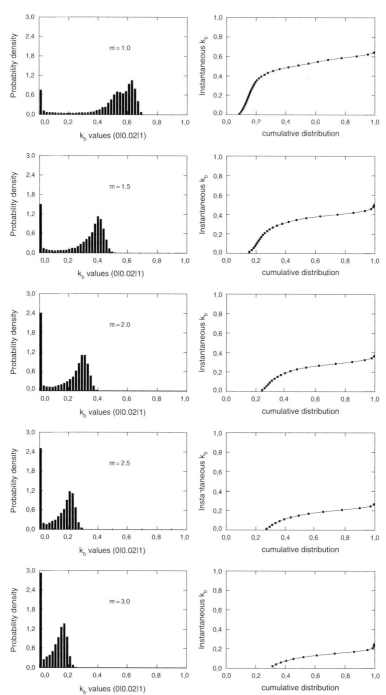

Fig. 3.16 Density distributions of the instantaneous k_b index (left) and their respective CDFs (right) for Armilla (Spain) for different optical air mass

mainly consists of diffuse irradiance increases with the optical air mass. Skartveit and Olseth (1992) have found similar results. The experimental probability for the $(0.00, 0.02)$ interval (Figs. 3.16, 3.18) can be represented by an exponential function, which depends on the relative optical air mass:

$$f_1(m_a) = 17.34 - 36.75 \cdot \exp(-0.975 m_a), \text{ with } R^2 = 0.98. \tag{3.38}$$

The second maximum of the distribution, located on the right, covers a range between 0.7 for m=1.0 and 0.1 for m=3.0. Intermediate values between both maxima present a low probability. When optical air mass increases, the second maximum shifts towards lower values of k_b. This is a result of the enhancement of direct irradiance extinction. The shape of the distribution around this second maximum presents a marked asymmetry towards the left side of the distribution.

In order to model the bimodality that characterises the distribution of the data, the sum of two functions can be used (Fig. 3.17). The first corresponds to the first k_b interval, expressed as a Dirac delta multiplied by a factor depending on the optical air mass. The second corresponds to the remaining intervals, and can be adjusted by means of a function that reproduces the observed asymmetry. To this end, the same kind of functions used in a previous work (Tovar et al. 1998) to model the frequency distribution functions of 1-minute k_t values, can be used. I have modified this function including an additional parameter β, that accounts for the asymmetry of the function. This modified equation is:

$$f_2(k_b) = A \frac{\lambda e^{(k_b - k_{b0})\lambda}}{\left[1 + e^{(k_b - k_{b0})(\lambda + \beta)}\right]^2}, \tag{3.39}$$

and satisfies the normalisation condition:

$$\int_0^1 f(k_b)\, dk_b = \int_0^1 f_1(k_b)\, dk_b + \int_0^1 f_2(k_b)\, dk_b = 1. \tag{3.40}$$

The degree of asymmetry depends on the ratio of the parameters β and λ. The sign of β determines whether the asymmetry goes towards the right or the left side.

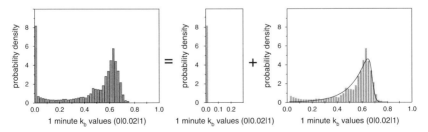

Fig. 3.17 The instantaneous k_b distributions can be divided in two functions. The first corresponding to the $(0.00, 0.02)$ interval is modelled by a delta function, and the second by a modified Boltzmann function

Fig. 3.18 The amplitudes of instantaneous k_b distribution in the interval (0.00, 0.02) and its fitting using Eq. (3.38)

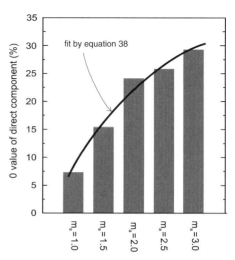

The k_{b0} parameter is related to the position of the maximum in the PDF. The product $A \cdot \lambda$ depends on the size of the frequency distribution maximum.

The use of this equation to describe the statistic behaviour of k_b is interesting since, in this way, a formal coherence with the functions used for the k_t distributions can be maintained (Tovar et al. 1998). The equation modified by the parameter β accounts properly for the experimental values.

The parameter k_{b0} locates the maximum of the distribution function. Note that k_{b0} shifts towards lower k_b values as the optical air mass increases. The tendency to a decrease of the asymmetry with the optical air mass is modelled by the decrement of the ratio β to λ. Figure 3.16 shows the PDFs and their respectives CDFs. Figure 3.17 shows the division of the PDF into two functions: the first corresponding to the (0.00, 0.02) interval and the second corresponding to the rest of the intervals. We can appreciate in the CDFs of the Fig. 3.16 that the initial value for these curves as the optical air mass varies. For the lowest optical mass, the interval (0.02, 1.00) includes 72% of the cases. This percentage diminishes with the optical air mass, mainly due to the increase in the direct beam extinction for greater optical air masses.

The parameter that rules the distributions presents a dependence on the optical air mass. After a linear multiple regression adjust, we obtain for the Armilla data the following results:

$$A = 0.9984 - 0.01686m_a + 0.00171m_a^2, \text{with } R^2 = 0.986, \qquad (3.41)$$

$$k_{b0} = 0.7862 - 0.0546m_a - 0.00543m_a^2, \text{with } R^2 = 0.99, \qquad (3.42)$$

$$\lambda = 8.855 - 0.2666m_a - 0.3229m_a^2, \text{with } R^2 = 0.935, \qquad (3.43)$$

$$\beta = 83.74 - 34.059m_a - 6.1846m_a^2, \text{with } R^2 = 0.975. \qquad (3.44)$$

The R-squared between the experimental values and the modelled ones each air mass is close to 0.98; similar results are obtained for other optical air masses.

7.3 One-Minute k_d Values Distributions Conditioned by the Optical Air Mass

The analysis of the experimental distributions of 1-minute horizontal diffuse irradiation has been carried out following the same approach as for k_t and k_b. The diffuse component has been analysed by means of the coefficient k_d. Figure 3.19 shows the 1-minute k_d distributions for every optical air mass ($k_d|m_a$) using the data of Armilla. The curves are unimodal with a maximum value similar for all the distributions. Nevertheless, there is a slight shift towards higher k_d values when the optical air mass increases. This shift is minor than that encountered in k_t (Tovar et al. 1998) and k_b analyses. This index ranges from 0 to about 0.5. The maximum of the distribution, for all the optical air masses considered, is located between 0.05 and 0.15, presenting a slight displacement toward higher values as the optical air mass increases. However, the optical air mass influence on the shape of the distribution is minor. The CDFs are presented in Fig. 3.19; note their similarity. The greatest differences between the curves correspond to the k_d values in the range 0.1 to 0.2.

The experimental distribution has been fitted using the same modified Boltzmann functions previously used:

$$f(k_d) = A \frac{\lambda e^{(k_d - kd_{d0})\lambda}}{\left[1 + e^{(k_d - k_{d0})(\lambda + \beta)}\right]^2}. \tag{3.45}$$

The parameters present a dependence on optical air mass that could be adjusted by multiple linear regression; we have obtained the following results:

$$A = 0.07062 - 0.07609m_a + 0.02989m_a^2, \text{ with } R^2 = 0.98, \tag{3.46}$$

$$k_{d0} = 0.0248 + 0.0222m_a, \text{ with } R^2 = 0.98, \tag{3.47}$$

$$\lambda = 538.69 - 152.34m_a, \text{ with } R^2 = 0.996, \tag{3.48}$$

$$\beta = -305.27 + 122.154m_a - 11.468m_a^2, \text{ with } R^2 = 0.998. \tag{3.49}$$

It must be pointed out that all the distributions show similar k_{d0} values. Additionally, the parameter A also presents similar values for all distributions. The similarity of both parameters, governing the position and amplitude of the maximum of the distribution, is related to the evident similarity of the experimental distributions. These results could be explained as follows. The scattering process increases with the optical air mass, leading to an enhancement of solar diffuse irradiance. On the other hand, the extinction process by both scattering and absorption increases as the optical air mass increases, thus diminishing the solar radiation reaching the Earth's surface. In this way, regarding the slight difference among the k_d distribution associated with different optical air mass, it seems that there is a compensation of these two effects. Therefore, there is a relative independence of the k_d distribution function with the optical air mass.

However, λ and β present a more marked difference for the different distributions. Note that λ decreases as the optical air mass increases, evidencing an increase of the

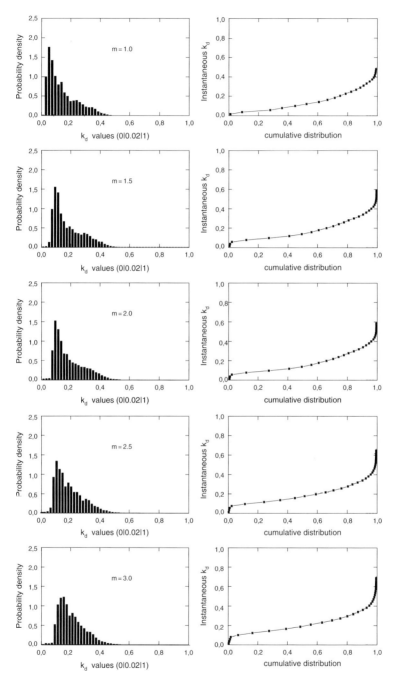

Fig. 3.19 Density distributions functions of the instantaneous k_d index (left), at different optical air mass, and their respective CDFs (right) for Armilla (Spain)

dispersion in the experimental distributions. Nevertheless, the ratio β to λ slightly changes with the optical air mass, reflecting the fact that the asymmetry is rather similar for all distributions.

8 Conditioned Distributions $\bar{k}_t^H$

We have also analysed the PDFs and the CDFs of the k_t conditioned by the k_t hourly average value represented, respectively, by $f(k_t|\bar{k}_t^H)$ and $F(k_t|\bar{k}_t^H)$ expressions.

In order to obtain the 1-minute conditional probability distributions of k_t, we have computed the hourly average values of solar global irradiance corresponding to the three-year period of data available at Armilla (Granada, Spain). Particularly, we have classified the data in intervals of 0.01, centered at $\bar{k}_t^H = (0.3, 0.35, 0.4, 0.45, 0.5, 0.55, 0.6, 0.65, 0.7, 0.75)$. The 1-minute data have been classified according to these criteria. These intervals are grouped into two teams. The first includes the intervals with hourly average values centered at 0.3, 0.4, 0.5, 0.6 and 0.7, including 91575 values. The second corresponds to distribution of values around 0.35, 0.45, 0.55, 0.65 and 0.75, and includes 98113 data points. The second group has been reserved for validation purposes.

Figure 3.20 shows the density probability distributions of k_t for given $\bar{k}_t^H$ values (0.30–0.7) and their respective CDFs. These distributions show a marked unimodality that contrasts with the bimodality that characterises the distributions conditioned by the optical air mass. This fact can be explained in terms of the reduced range of the k_t values associated with a given $\bar{k}_t^H$. On the contrary, when the intervals are defined as a function of the optical air mass $f(k_t|m_a)$ the distributions tend to be bimodal.

The distributions present a marked symmetry around a central value that is close to the corresponding $\bar{k}_t^H$ values. This feature is more marked for $\bar{k}_t^H$ in the range 0.45–0.65, while the distributions corresponding to $\bar{k}_t^H$ out of this range show a slight asymmetry. For values of $\bar{k}_t^H$ below 0.45, there is an asymmetry toward higher values, indicating that $\bar{k}_t^H$ in this range can be the result of a combination of very low and very high instantaneous values of k_t. This can be related to transient conditions under partial cloud cover, with clouds close to the Sun position that, in a short time, can block the Sun or enhance the Sun direct beam due to reflections from the edges of the clouds.

Another relevant feature is the range of k_t instantaneous values associated with a given $\bar{k}_t^H$. Excluding the higher values of $\bar{k}_t^H$, the range of k_t instantaneous values is rather wide. This indicates that, for these categories, we included partially-covered skies characterised by a great variability of instantaneous k_t values, especially if the clouds are close to the Sun. For higher $\bar{k}_t^H$, the range of the k_t instantaneous values is reduced, indicating that these higher hourly values are associated with cloudless-sky conditions. A rather narrow range of k_t values characterises these distributions.

For the distributions corresponding to intermediate values of k_t, associated with partially-cloudy skies, we observe the highest k_t values. This is a result of multiple

86   J. Tovar-Pescador;

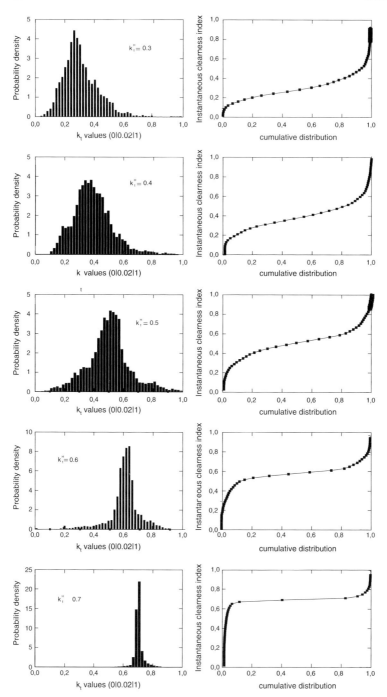

Fig. 3.20 Density distributions of instantaneous k_d index (left) for Armilla (Spain) at different optical air mass and their respective CDFs (right)

reflections from the clouds located close to the position of Sun. Under these conditions, the reflections from the cloud edges lead to an increment of global irradiance due to the enhancement of the diffuse component.

Considering the shape of the curves, we have approximated the probability distributions using a function based on Boltzmann's statistics. This function has been used previously for modelling 1-minute distributions conditioned by the air mass.

To account for the asymmetry of the analysed distributions, we also use the parameter (β) in the above function. The modified equation reads as follows:

$$f(k_t|\bar{k}_t^H) = \frac{A\,\lambda\,e^{(k_t-k_{t0})\lambda}}{\left[1+e^{(k_t-k_{t0})(\lambda+\beta)}\right]^2}, \tag{3.50}$$

that satisfies the normalisation condition:

$$\int_0^1 f(k_t|\bar{k}_t^H)dk_t = A \int_0^1 \frac{\lambda\,e^{(k_t-k_{t0})\lambda}}{\left[1+e^{(k_t-k_{t0})(\lambda+\beta)}\right]^2}dk_t = 1. \tag{3.51}$$

This function provides also reasonable fits, even for distributions that exhibit a high degree of asymmetry.

For this kind of distributions, the dependence of the coefficients k_{t0}, λ and β on $\bar{k}_t^H$ may be formulated by means of polynomial functions:

$$k_{t0} = -0,006 + 1,010\,\bar{k}_t^H, \text{with } R^2 = 0.999, \tag{3.52}$$

$$\lambda = 11,284 + 1150,37\,(\bar{k}_t^H)^{7.205}, \text{with } R^2 = 0.935, \tag{3.53}$$

$$\beta = 0,293 + 6,093\,\bar{k}_t^H - 15,643\,(\bar{k}_t^H)^2, \text{with } R^2 = 0.984, \tag{3.54}$$

The performance of these functions depends on $\bar{k}_t^H$ and the ratio β/λ, that provide information about the distribution asymmetry. Note that this ratio presents a sign change about $\bar{k}_t^H = 0.45$. The values of the parameter A must satisfy the normalisation condition and the computation of the integrand:

$$\int_0^1 k_t\,f(k_t|\bar{k}_t^H)dk_t$$

must return the corresponding value of $\bar{k}_t^H$. The A parameter can be fitted by a polynomial function:

$$A = 2.81307 - 237859\,\bar{k}_t^H + 143.72282\,(\bar{k}_t^H)^2 - 456.57093\,(\bar{k}_t^H)^3 +$$
$$+771.8157\,(\bar{k}_t^H)^4 - 656.1663\,(\bar{k}_t^H)^5 + 220.62361\,(\bar{k}_t^H)^6 \tag{3.55}$$

for the range 0 to 1; although the experimental values that we found for $\bar{k}_t^H$ are in the range 0.3 to 0.8. The associated R^2 value is about 0.999 when considering the range $0.05 \le \bar{k}_t^H \le 0.95$.

9 Conclusions and Future Work

We would like to relate some of the main results of this work and to discuss the future research in the field of the modelling of the statistical properties of the instantaneous values of solar radiation.

The distribution of the solar radiation components at the Earth's surface can be analysed based on the k_t, k_b y k_d indices. This analysis is widely used. Most of the bibliography is related to the analysis of clearness index, while still scarce works have dealt with the direct and diffuse components of the solar radiation.

In most cases, the distributions are analysed conditionally by the optical air mass or based on time-averaged intervals. In this latter case, the distribution of the daily index k_t for a given monthly-average value is the most widely used.

In many cases, although the bibliography shows a wide range of results, the distributions show a marked bimodality. For daily distributions, this bimodality seems related to climatic features with relatively high cloudiness, while this bimodality is not observed for climates with relatively low cloudiness.

The hypothesis by Liu and Jordan (1960) of a universal character of the CDFs is not supported, as also deduced by works of Bendt et al. (1981), Reddy et al. (1985), Olseth and Skarveit (1984), Saunier et al. (1987) and Rönnelid (2000).

Additionally, the attempts of different authors to provide general models do not seem feasible. On the opposite, the bibliography is plenty of works proposing models developed for specific regions, due to the non-adjust performance of these general models.

The instantaneous distribution shows a marked different behaviour from those of the hourly, daily and monthly distributions. Particularly, for the distributions conditioned by the optical air mass, an increment in the bimodal character is observed.

We have proposed a model for representing distributions of data collected in a wide range of climatic conditions. Particularly, the proposed distributions allow easily calibrate the parameter of these distributions based on the climatology of the study area. This makes the model, in some sense, general. Additionally, based on the proposed distribution functions it is easy to obtain the CDFs. and, also, the generation of synthetic solar radiation time series. The CD attached to this book contains the programs and the code source used to fit the density distribution functions with the models proposed by the author.

We would like to highlight that, in our opinion, any attempt of solar radiation modelisation should take into account the climatic characteristics of the study area, since these characteristics strongly influence the solar radiation values and the associated statistics.

Some authors have used normalisation processes in their statistical analysis, trying to overcome the site dependences of the models. We think, on the opposite, that these dependences should be taken into account in the modelling process, through adequate parameter tuning.

We would also like to point out that there are still scarce works dealing with the statistical analysis of spectral solar radiation values. On the other hand, there are some preliminary works dealing with solar radiation instantaneous values collected at intervals less than 1-minute. Finally, in the last years, the wavelet spectral techniques have begun to be used in the analysis of solar radiation data. This technique allows to identify the most important modes for the variability of the solar radiation time series and to allocate time flags when these modes take place.

References

Andretta A, Bartoli B, Colucci B, Cuomo V, Francesca M, Serio C (1982) Global solar radiation estimation from relative sunshine hours in Italy. Journal of Applied Meteorology 21:1377–1384

Ångström A (1924) Solar and terrestrial radiation. Q J Roy Met Soc 50:121–126

Ångström A (1956) On the computation of global radiation from records of sunshine. Arkiv fur Geofisik 5:41–49

Babu KS, Satyamurty VV (2001) Frequency distribution of daily clearness index through generalized parameters. Solar Energy 70:35–43

Baker DG, Klink JC (1975) Solar radiation reception, probabilities, and a real distribution in the north-central region. Agricultural experiment Station, University of Minnesota, p 225

Barbaro S, Cannata G, Coppolino S (1983) Monthly reference distribution of daily relative sunshine values. Solar Energy 31:63–67

Bendt P, Collares-Pereira M, Rabl A (1981) The frequency distribution of daily insolation values. Solar Energy, 27:1–5

Bennet I (1965) Monthly maps of mean daily insolation for the United States. Solar Energy, 9:145–158

Bennet I (1967) Frequency of daily insolation in Anglo North America during June and December. Solar Energy 11:41–55

Black JN, Bonython CW, Prescott JM (1954) Solar Radiation and the duration of sunshine. Q J Roy Met Soc 80:231–235

Bois P, Goussebaile J, Mejon MJ, Vachaud G (1977) Analyse de donnees d'energie solaire en vue des applications. Solar Energy Conversion and Application, Chassagne et al., editors, pp 171–191

Callegari M, Festa R, Ratto CF (1992) Stochastic modelling of daily beam irradiation. Renewable Energy. 6:611–624

Engels JD, Pollock SM, Clark JA (1981) Observation on the statistical nature of terrestrial irradiation. Solar Energy, 26:91–92

Ettoumi FY, Mefti A, Adane A, Bouroubi MY (2002) Statistical analysis of solar measurements in Algeria using beta distributions. Renewable Energy, 26:47–67

Exell RHB (1981) A mathematical model for solar radiation in South-East Asia (Thailand). Solar Energy, 26:161–168

Festa R, Ratto CF (1993) International Energy Agency. Solar heating and Cooling Programme Report-IEA-SCHP-9E-4

Feuillard T, Abillon JM (1989) The probability density function of the clearness index: a new approach. Solar Energy, 43:363:372

Gansler RA, Klein SA, Beckman WA (1995) Investigation of minute solar radiation data. Solar Energy, 55:21–27

Glover J, McCulloch JSG (1958) The empirical relation between solar radiation and hours of sunshine. Q J Roy Met Soc 84:172–175

Gordon JM, Reddy TA (1988) Time series analysis of daily horizontal solar radiation. Solar Energy, 41:215–226

Graham VA, Hollands KGT, Unny TE (1988) A time series model for kt, with applications to global synthetic weather generation. Solar Energy, 40:83–92

Gueymard C (1999) Prediction and performance assessment of mean hourly global radiation. Solar Energy, 68:285–303

Hansen JW (1999) Stochastic daily solar irradiance for biological modeling applications. Agricultural and Forest Meteorology, 94:53–63

Hollands KGT, Huget RG (1983) A probability density function for the clearness index, with applications. Solar Energy, 30:195–209

Ibáñez M, Beckman WA, Klein SA (2002) Frequency distributions for hourly and daily clearness index. J. Solar Energy Eng, 124:28–33

Ibáñez M, Rosell JI, Beckman WA (2003) A bi-variable probability density function for the daily clearness index. Solar Energy, 75:73–80

Iqbal, M (1983) An Introduction to Solar Radiation. Academic Press. New York

Jurado M, Caridad JM, Ruiz V (1995) Statistical distribution of the clearness index with radiation data integrated over five minutes intervals. Solar Energy 55:469–473

Karlin & Taylor (1975) A First Course in Stochastic Processes, Academic Press, New York

Klink SA (1974a) Calculation of monthly average insolation on tilted surfaces. Solar Energy 19:325–329

Klink JC (1974b) A global solar radiation climatology for Minneapolis_St. Paul. PhD Tesis

Mefti A, Bouroubi MY, Adane A (2003) Generation of hourly solar radiation for inclined surfaces using monthly mean sunshine duration in Algeria. Energy Conversion and Management, 44:3125–3141

Lestienne R (1978) Modele markovien simplifie de meteorologie a deux etats. L'exemple d'Odeillo. La Meteorologie, 12:53–64

Lestienne R (1979) Application du modele markovien simplifie a l'etude du comportement du stockage d'une centrale solaire. Revue de Physique Appliquee, 14:139–144

Liu BYH, Jordan RC (1960) The interrelationship and the characteristics distribution of direct, diffuse and total solar radiation. Solar Energy 4:1–19

Olseth JA, Skartveit A (1984) A probability density function for daily insolation within the temperate storm belts. Solar Energy, 33:533–542

Olseth JA, Skartveit A (1987) A probability density model for hourly total and beam irradiance on arbitrarily orientated planes. Solar Energy, 39:343–351

Olseth JA, Skartveit A (1993) Characteristics of hourly global irradiance modelled from cloud data. Solar Energy, 51:197–204

Reddy TA, Kumar S, Gaunier GY (1985) Review of solar radiation analysis techniques for prediction long-term thermal collector performance. Renewable Energy Review J, 7, 56

Rönnelid M (2000) The origin of the asymmetric annual irradiation distribution at high latitudes. Renewable Energy, 19:345–358

Rönnelid M, Karlsson B (1987) Irradiation distribution diagrams and their use for estimating collectable energy. Solar Energy 61:191–201

Ross J, Sukev M, Saarelaid P (1998) Statistical treatment of the PAR variability and its application to willow coppice. Agricultural and Forest Meteorology, 91:1–21

Saunier GY, Reddy TA, Kumar S (1987) A monthly probability distribution function of daily global irradiation values appropriate for both tropical and temperate locations. Solar Energy, 38:169–177

Skartveit A, Olseth JA (1992) The probability density and autocorrelation of short-term global and beam irradiance. Solar Energy, 49:477–487

Stuart RW, Hollands KTG (1988) A probability density function for the beam transmittance. Solar Energy, 49:463–467

Suerhrcke, H. y McCormick, P.G. (1987) An aproximation for γ of the fractional time distribution of daily clearness index. Solar Energy, 39: 369–370

Suehrcke H, McCormick PG (1988a) The frequency distribution of instantaneous insolation values. Solar Energy, 40:413–422

Suehrcke H, McCormick PG (1988b) The diffuse fraction of instantaneous solar radiation. Solar Energy, 40:423–430

Suehrcke H, McCormick PG (1989) Solar radiation utilizability. Solar Energy 43:339–345

Tiba C, Siqueira AN, Fraidenraich N (2007) Cumulative distribution curves of daily clearness index in a southern tropical climate. Renewable Energy, in press doi:10.1016/j.renene.2006.11.014

Tomson T, Tamm G (2006) Short-term variability of solar radiation. Solar Energy, 80:600–606

Tovar J, Olmo FJ, Alados-Arboledas L (1998a) One-minute global irradiance probability density distributions conditioned to the optical air mass. Solar Energy, 62:387–393

Tovar J, Olmo FJ, Batlles FJ, Alados-Arboledas L (1998b) One-minute kb and kd probability density distributions conditioned to the optical air mass. Solar Energy, 65:297–304

Tovar J, Olmo FJ, Batlles FJ, Alados-Arboledas L (2001) Dependence of one-minute global irradiance probability density distributions on hourly irradiation. Energy, 26:659–668

Tovar-Pescador J, Pozo-Vazquez D, Batlles J, López G, Muñoz-Vicente D (2004) Proposal of a function for modeling the hourly frequency distributions of photosynthetically active radiation. Theoretical and Applied Climatology, 79:71–79

Varo M, Pedrós G, Martínez-Jimenez P (2005) Modelling of broad band ultraviolet clearness index distributions for Cordoba, Spain. Agricultural and Forest Meteorology, 135:346–351

Varo M, Pedrós G, Martínez-Jiménez P, Aguilera MJ (2006) Global solar irradiance in Cordoba: Clearness index distributions conditioned to the optical air mass. Renewable Energy, 31:1321–1332

Vijayakumar G, Kummert M, Klein SA, Beckman WA (2005) Analysis of short-term solar radiation data. Solar Energy, 79:495–504

Wang S, Chen W, Cihlar J (2002) New calculation methods of diurnal distribution of solar radiation and its interception by canopy over complex terrain. Ecological modelling 00:1–14

Whillier A (1956) The determination of hourly values of total solar radiation from daily summation. Archiv fur Meteorologie Geophysik und Bioklimatologie, B7:197

Wilks DS (2006) Statistical methods in the Atmospheric Sciences, International Geophysics Series, Academic Press, USA, Chapter 6.

Woyte A, Belmans R, Nijs J (2007) Fluctuations in instantaneous clearness index: Analysis and statistics. Solar Energy, 81:195–206

Chapter 4
A Method for Determining the Solar Global and Defining the Diffuse and Beam Irradiation on a Clear Day

Amiran Ianetz and Avraham Kudish

1 Introduction

The terrestrial solar irradiation is a function of solar altitude, site altitude, albedo, atmospheric transparency and cloudiness. The atmospheric transparency is a function of aerosol concentration, water vapor as well as other factors. The presence of aerosols in the atmosphere attenuates the beam component, whereas it increases the diffuse component of the solar global irradiation. In essence, the beam component is converted to diffuse irradiation. Consequently, it may have a relatively small effect on the total solar global irradiation. Water vapor, on the other hand, attenuates both the beam and diffuse components and, thereby, decreases the total solar global irradiation.

The determination of the magnitude of the solar irradiation on a clear day is contingent on the criteria used to define a clear day. A priori a clear day is characterized by a perfectly cloudless sky assuming an average transparency state of the atmosphere (Sivkov 1971). The degree of cloudiness can be quantified by human observation of cloud cover and/or sunshine duration measurements. It should be noted that (a) cloud cover observations are usually made only intermittently, i.e., varies between hourly or a number of times per day, and (b) there is an inherent uncertainty in utilizing sunshine duration measurements to define a clear day, viz., the existence of clouds in the sky that are not in the optical path between the sunshine duration measuring device and the sun are not observed by the instrument. The atmospheric transparency can be quantified by determining either a turbidity coefficient or aerosol optical thickness (AOT). It is also reasonable to expect that a clear day will be associated with a measured maximum of the solar global irradiation intensity, i.e., relative to some time interval, e.g., a month. Nevertheless, it

Amiran Ianetz
Ben-Gurion University of the Negev, Beer Sheva, Israel, e-mail: amirani@zahav.net.il

Avraham Kudish
Ben-Gurion University of the Negev, Beer Sheva, Israel, e-mail: akudish@bgumail.bgu.ac.il

is conceivable that such a relative maximum may be observed under special cloud cover conditions, viz., the existence of clouds in the vicinity of the sun, but not directly blocking the sun, which create a funneling-effect on the solar irradiation. In practice, the criteria used to define a clear day at the particular site under consideration will be contingent on the database of measured parameters available. Databases consisting of measured solar global irradiation on a horizontal surface are available from most meteorological stations, whereas the existence of cloud cover and/or sunshine duration measurements concurrent with solar global irradiation measurements is much less common. Atmospheric transparency measurement databases are also quite rare. Consequently, the following analysis to determine the solar global irradiation on a clear day will be limited to the criterion based upon the most readily available database, viz., that consisting of solar global irradiation.

2 Solar Global Irradiation on a Clear Day

It is of interest to study the solar global irradiation on a clear day, which is a function of all the abovementioned parameters with the exception of degree of cloudiness, i.e., by definition there is a total absence of clouds. The analysis of the relative magnitudes of the measured solar global irradiation and the solar global irradiation on a clear day, as determined by a suitable model, provides a platform for studying the influence of cloudiness on solar global irradiation. Also, the magnitude of the solar global irradiation on clear day provides an estimate of the maximum solar energy available for conversion on a particular day.

2.1 Classification of Clear Days

There exist two, generally accepted, methods for classifying the day type as clear, partially cloudy or cloudy with regard to solar global irradiation. Barbaro et al. (1981) suggested that the classification of day type be based upon the degree of cloudiness. They defined day type as a function of degree of cloudiness, both in octas and tenths, as reported in Table 4.1.

Iqbal (1983) proposed utilizing the magnitude of the daily clearness index K_T (the ratio of the solar global to the extraterrestrial solar irradiation) to define sky conditions, cf., Table 4.2.

Table 4.1 Classification of days according to cloud cover (Barbaro et al. 1981)

Day type	Octas	Tenths
Clear	0 – 2	0 – 3
Partially cloudy	3 – 5	4 – 7
Cloudy	6 – 8	8 – 10

Table 4.2 Classification of days according to clearness index (Iqbal 1983)

Day type	K_T
Clear	$0.7 \leq K_T < 0.9$
Partially cloudy	$0.3 \leq K_T < 0.7$
Cloudy	$0.0 \leq K_T < 0.3$

These two methods are essentially data filters that classify the day type on the basis of the magnitude of a related parameter. The authors have previously applied the latter definition to two sites in Israel, viz., Beer Sheva (Kudish and Ianetz 1996) and Jerusalem (Ianetz 2002). It is important to note that the monthly average daily solar global irradiation values on a clear day in the case of Beer Sheva reported in Kudish and Ianetz (1996) are very close to those reported in the present analysis despite the somewhat different K_T criteria and different database, cf., section 2.3, Table 4.4. viz., the 1996 publication defined a clear day as that for which $K_T > 0.65$ and the database consisted of the years 1982–1993, whereas the present analysis utilizes for most of the months a $K_T > 0.7$ to define a clear day and the database consisted of the years 1991–2004. The close agreement between the two sets of clear day solar global irradiation values testifies to the stability of the Iqbal filters.

The advantage of the latter method lies in the fact that only solar global irradiation measurements are required, whereas the former method necessitates the availability of a concurrent cloudiness observation database. In addition, as mentioned previously, the latter are usually made only intermittently, i.e., usually three times per day, whereas solar global irradiation data are monitored continuously and usually reported on an hourly basis. The criterion based upon sunshine duration also has not been considered due to the limited availability of such data measured concurrently with solar global irradiation and the abovementioned inherent error involved in such measurements.

It should be noted that two manuscripts treating the subject of clear sky solar global irradiation and its classification have been published recently. Lopez et al. (2007) presented a new model to estimate horizontal solar global irradiation under cloudless sky conditions, i.e., clear day, which requires the following input parameters: latitude, day of year, air temperature, relative humidity, Ångström turbidity coefficient, ground albedo and site elevation along with solar elevation at sunrise or sunset if the site has horizon obstructions. Younes and Muneer (2007) have proposed a clear day identification based upon the clearness index, diffuse ratio, turbidity and cloud cover limits. Once again, the method proposed in the following discussion has the distinct advantage that it requires a database consisting of a single parameter, viz., the solar global irradiation.

2.2 Models for Determining the Global Irradiation on Clear Day

The clear day solar global irradiation intensity at a particular site is also an inherent function of the day for which it is determined, since the extraterrestrial irradiation varies from day to day. The latter parameter sets an upper, although unattainable, limit on the magnitude of the solar global irradiation.

A number of models, essentially empirical correlations, have been developed and reported in the literature that calculate the clear sky solar global irradiation, G_c, based exclusively on site location and astronomical parameters, i.e., the solar zenith angle θ_z. A priori, it is to be expected that such simple empirical correlations will be best suited to sites having similar meteorological parameters. A listing of some of these previously reported empirical clear sky regression equations, where the clear sky global solar irradiation is given in units of W/m^2, includes the following:

Haurwitz (1945, 1946)

$$G_c = 1098[\cos\theta_z \exp(-0.057/\cos\theta_z)], \tag{4.1}$$

Daneshyar-Paltridge-Proctor (Daneshyar (1978); Paltridge and Proctor (1976); Gueymard (2007))

$$G_c = 950.0\{1 - \exp[-0.075(90° - \theta_z)]\} + 2.534 + 3.475(90° - \theta_z), \tag{4.2}$$

Berger-Duffie (1979)
$$G_c = 1350[0.70\cos\theta_z], \tag{4.3}$$

Adnot–Bourges-Campana-Gicquel (1979)

$$G_c = 951.39\cos^{1.15}(\theta_z), \tag{4.4}$$

Kasten-Czeplak (1980)

$$G_c = 910\cos\theta_z - 30, \tag{4.5}$$

Robledo-Soler (2000)

$$G_c = 1159.24(\cos\theta_z)^{1.179} \exp[-0.0019(\pi/2 - \theta_z)]. \tag{4.6}$$

Badescu (1997) tested these empirical clear sky regression equations, viz., Eqs. (4.1–4.5), for the climate and latitude of Romania. He found that Eq. (4.4), based upon measurements made in Western Europe, best modeled clear sky global irradiation in Romania.

Lingamgunta and Veziroglu (2004) have proposed a universal relationship for estimating daily clear sky irradiation using a dimensionless daily clear sky global irradiation, H_{dclv}, as a function of the day of year, n, latitude, ϕ, a dimensionless altitude, A (which is local altitude divided by 452 m - height of Petronas Towers) and hemisphere indicator, i (which is i = 1 for northern hemisphere and i = 2 for southern hemisphere). They defined the dimensionless daily clear sky global irradiation as $H_{dclv} = H_{dc}/(24 \cdot 3600 \cdot G_{sc})$, where H_{dc} is the clear sky global irradiation (Wh/m^2) at the site and G_{sc} is the solar constant, $1367\,W/m^2$. Their universal relationship is given as

$$H_{\text{dclv}} = \{[0.123 + 0.016(-1)^i](\phi/90)^{1.5} + (0.305 + 0.051[(90/\phi)^2 - 1]^{1.5}) \quad (4.7)$$
$$\cos\phi - 0.1(1+A)^{-0.1}\}\{1 + [0.975 + 0.075(-1)^i]$$
$$\sin((72/73)(n-81))\tan(3\phi/4).$$

It should be noted that the solar zenith angle is not a parameter in Eq. (4.7), as opposed to the other models under discussion.

2.3 Berlynd Model for Determining Global Irradiation on Clear Days

In 1956 Berlynd proposed a model, which is a function of astronomical parameters, albedo and meteorological parameters, that is reported in Kondratyev (1969) and is given as

$$G_c = G_{\text{sc}} \cos\theta_z / [1 + f\sec\theta_z], \quad (4.8)$$

where the coefficient f is a function of albedo, atmosphere optical thickness in the zenith direction and parameters that characterize the diffuse portion of the global irradiation. The value for f is determined by adjusting the calculated values for G_c to the values measured on a clear day at the site under consideration. It should be emphasized that the Berlynd model is the only one of the models discussed that also takes into consideration meteorological parameters, i.e., the magnitude of the coefficient f.

The authors have recently reported (Ianetz et al. 2007) upon the application of Berlynd and the above listed models to the solar global irradiation being monitored at three meteorological stations located in the semi-arid Negev region of Israel, viz., Beer Sheva, Sde Boker and Arad. The following discussion will be limited to the results for Beer Sheva, which are similar to those corresponding to the other two sites.

The global irradiation is measured utilizing an Eppley PSP pyranometer and a Campbell Scientific Instruments datalogger monitors and stores the data at 10 minute intervals (i.e., the meters are scanned at 10 s intervals and average values at 10 minute intervals are calculated and stored). The database utilized to test the clear sky correlations consisted of measurements made from January 1991 through December 2004. In addition, a much larger database for Beer Sheva, January 1982 through December 2004, was utilized to perform an in-depth analysis of this site. The Beer Sheva meteorological station is part of the national network and the instrument calibration constant is checked periodically. Only those days for which all hourly values were recorded were included in the analysis. The validity of the individual measured hourly values was checked in accordance with WMO recommendations (WMO 1983). Those values that did not comply with the WMO recommendations were considered erroneous and rejected (i.e., the corresponding daily value was rejected).

The Iqbal filter was applied to the database to determine monthly average clear sky global irradiation. The threshold value for a clear sky was set as $K_T \geq 0.7$ but it was observed that there were a number of months that had either a very small number or zero days that met this criterion. Consequently, the threshold was lowered to $K_T \geq 0.67$ for January, February and October and further reduced to $K_T \geq 0.65$ for November and December.

It was observed from a preliminary screening of the simple clear sky correlations, viz., Eqs. (4.1–4.6), that Eqs. (4.2, 4.5 and 4.6) significantly underestimated the clear sky global irradiation as determined by applying the Iqbal filter to the site databases. The same result was observed when Eq. (4.7) was applied to these sites. Consequently, only those models corresponding to Eqs. (4.1, 4.3 and 4.4), referred to as the H, BD and ABCG models, respectively, were tested.

The Berlynd model was also applied to the database. There exist tabulated monthly values of Berlynd coefficient f corresponding to moderate latitudes, cf., Kondratyev (1969), but a set of monthly values for Beer Sheva was determined as was done previously by Kuusk (1978) for his site in Estonia. As mentioned above, the f values were determined by adjusting the calculated values for G_c to those values measured on a clear day at the site, i.e., Beer Sheva. It was observed that the f values exhibit a maximum in June (0.320) and a minimum in January (0.170), cf., Table 4.3.

An inter-comparison between the monthly average clear day solar global irradiation as defined by the Iqbal filter and that determined by the application of the four models to the Beer Sheva database is reported in Table 4.4. The deviation (%) of the monthly average clear day solar global irradiation is on the average smallest for the case of the Berlynd model. Its maximum deviation is 5.06% (October) and its average monthly deviation is 1.95%. It should be emphasized that the Berlynd model is fitted to the site conditions via the coefficient f, whereas the other models are regression equations expressed as a function of the zenith angle.

The solar global irradiation database consisting of the years 1982 to 2004 (23 years) was utilized to perform an in-depth analysis for Beer Sheva. The monthly average daily extraterrestrial solar irradiation, the absolute monthly maximum measured daily global irradiation (i.e., a single value for each month), the monthly average of the maximum daily global irradiation measured each year (i.e., average of 23 values for each month), monthly average clear day global irradiation using the Iqbal filter, monthly average measured daily global irradiation, the monthly average of the minimum daily global irradiation measured each year (i.e., average of 23 values for each month) and the absolute monthly minimum measured daily global irradiation (i.e., a single value for each month) during the years 1982 to 2004 is shown in Fig. 4.1. It should be noted that both the maximum and minimum solar global

Table 4.3 Monthly 'f' values for Berlynd model

Month	J	F	M	A	M	J	J	A	S	O	N	D
Beer Sheva	0.170	0.190	0.230	0.275	0.310	0.320	0.312	0.300	0.280	0.210	0.190	0.176

Table 4.4 Inter-comparison of the clear day solar global irradiation as determined by the Iqbal filter and the models (kWh/m^2)

Month (days)	Iqbal	Berlynd	B-D	A-B-C-G	H
January (54[a])	4.06	**4.02**[c]	*3.81*[d]	*3.43*	3.87
February (104[a])	4.87	5.02	4.70	*4.30*	**4.87**
March (62)	4.34	**4.30**	*5.98*	*5.62*	4.28
April (96)	7.39	**7.34**	7.09	*4.76*	7.53
May (147)	8.02	**7.93**	7.87	*7.53*	8.34
June (278)	8.24	**8.18**	**8.18**	7.84	*8.68*
July (198)	8.09	**8.10**	8.04	7.70	*8.54*
August (91)	7.52	7.69	**7.44**	*7.13*	7.89
September (41)	4.62	**4.51**	4.41	*4.08*	4.75
October (60[a])	5.34	*5.61*	5.13	*4.73*	**5.34**
November (32[b])	4.14	4.27	**4.10**	*3.70*	4.21
December (28[b])	3.56	3.70	**3.58**	*3.19*	3.60

[a] $K_T \geq 0.67$.
[b] $K_T \geq 0.65$.
[c] Bold type- model result with smallest deviation from Iqbal.
[d] Italic type- model result with > 5% deviation from Iqbal.

irradiation values for a particular month for each year are independent and can be applied to determine the standard error of the average. The standard error of the average has been found previously (Kudish et al. 2005) to be less than the inherent measurement error of the instrument.

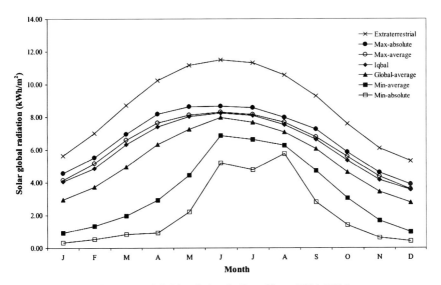

Fig. 4.1 Analysis of the solar global irradiation for Beer Sheva (1984–2004)

It is observed from Fig. 4.1, that the magnitudes of the monthly average clear day global irradiation values based upon the Iqbal filter are very close to the corresponding average of the maximum daily global irradiation measured each year. Also, it is observed that the magnitudes of the monthly average daily global irradiation values approach that of the corresponding Iqbal filter values during the summer months. This testifies to the prevalence of clear sky conditions during the summer months in Beer Sheva, cf., Ianetz et al. (2000).

2.4 Correlation Between Clear Day Global Index K_C and K_T

It is of interest to utilize the clear day solar global irradiation to determine a 'daily clear day index', K_c, which we define as

$$K_c = H/H_c, \tag{4.9}$$

viz., the ratio of the daily solar global irradiation, H, to the daily clear day solar global irradiation, H_c. The daily clear sky solar global irradiation was calculated using the Berlynd model, since it gave the values that most closely agreed to those determined using the Iqbal filter, cf., previous section. Pöldmaa (1975) has applied a similar analysis to investigate the effect that the cloud cupola exerts on the global irradiation in Estonia, viz., he normalized the measured solar global irradiation by the virtual solar global irradiation under a cloudless sky. He calculated the latter by applying the Berlynd model.

The monthly average daily K_c values together with the corresponding median and coefficient of variation (C_v) were determined. The coefficient of variation is defined as the ratio of the standard deviation to the average value and is a measure of the degree of scatter of the data around the mean value. The results of such an analysis are reported in Table 4.5.

Table 4.5 Monthly average K_c, median and coefficient of variation (%) for Beer Sheva

Month	Average	Median	C_v
January	0.728	0.792	29.6
February	0.748	0.795	27.8
March	0.783	0.839	23.2
April	0.842	0.889	18.6
May	0.899	0.945	13.7
June	0.958	0.972	5.0
July	0.945	0.956	4.9
August	0.935	0.943	4.2
September	0.925	0.939	4.3
October	0.852	0.876	11.0
November	0.774	0.837	20.4
December	0.731	0.808	27.8

It is observed from Table 4.5 that the monthly average K_c values range from a maximum of 0.958 for June to a minimum of 0.728 for January and that the magnitude of the median exceeds that of the average for all months, i.e., more than half of the data values exceed the average value. In addition, the magnitude of the coefficient of variation, C_V, is lowest for the months June through September (values range from 4.2 to 4.3%) and highest for the months January through March, November and December (values range from 20.4 to 29.6%). This, once again, testifies to the prevalence of clear sky conditions during the summer months as reported for Beer Sheva and as expected for the Negev region, cf., Ianetz et al. 2000.

The daily clearness index has been utilized, in the past, by many researchers as an indication of the degree of cloudiness at a particular site, i.e., $1 - K_T$. Intuitively, it appears that the daily clear day index, K_c, is better suited for this task. The magnitude of K_T is a measure of both atmospheric transparency and cloudiness. Consequently, the degree of cloudiness as determined on the basis of K_T also includes the other parameters contributing to the sky transparency. The daily clear sky index, K_c, is defined such that the effect of cloudiness is not a parameter. Viz., the daily clear day solar global irradiation, H_c, is determined under clear sky conditions and only those parameters contributing to sky transparency are involved. Consequently, the term $1 - K_c$ is a much better indication of the degree of cloudiness at a particular site, viz., $(1 - K_c) \to 0$ for a clear day and $(1 - K_c) \to 1$ for a cloudy day. We believe that this is sufficient justification for determining such an index.

A linear regression analysis was performed on the individual monthly databases to determine the correlation between the K_c and K_T, viz.,

$$K_c = aK_T. \tag{4.10}$$

A priori, it is assumed that the intercept of the linear regression curve should be at the origin of the axes. The results of this analysis, i.e., the slope 'a' of the monthly linear regression curve and the corresponding correlation coefficient (R^2) are reported in Table 4.4. In Figs. 4.2 (a) and (b) the data and linear regression curves for January and July, which are representative of those obtained for all months, are shown.

It is observed from Table 4.6 and Fig. 4.2 that K_c and K_T are highly correlated. A Fisher's statistic (F) analysis performed on the monthly databases showed that the regression equations explain almost 100% of the data variance. Consequently, it can be assumed that these correlations will be applicable to future measurements, since the data are representative (each individual monthly database consists of a minimum of 250 measurements).

It is apparent from Table 4.6 that there is a very small variation in the slopes of the individual monthly linear regression equations, i.e., they vary between 1.3666 and 1.4461. Thus an average annual correlation between K_c and K_T has been determined based upon a database consisting of all 12 months of data, viz., 4087 data pairs. The data and linear regression curve are shown in Fig. 4.3. The coefficient of correlation

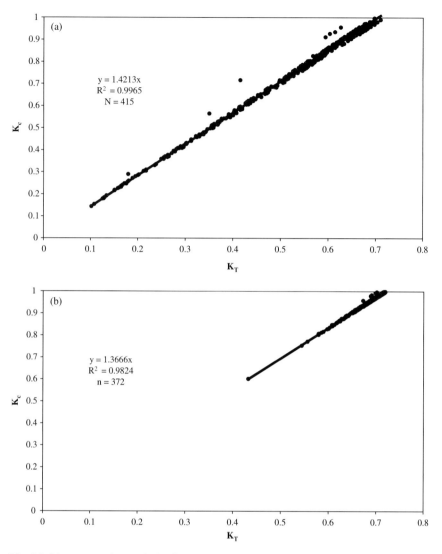

Fig. 4.2 Linear regression analysis of K_c as a function of K_T for (a) January, (b) July

of the annual linear regression curve is only minimally smaller than those obtained for the individual monthly databases. A statistical analysis was performed on the clear day global index as calculated utilizing the annual linear regression curve, $K_{c,calc}$, i.e., inter-comparison with that determined using the Berlynd model K_c. The Mean Bias Error (MBE) and Root Mean Square Error (RMSE) were determined to be -0.023 and 0.019, respectively, whereas the value of the annual average K_c is 0.83_5. It can be concluded that from this analysis that an annual correlation may be sufficient in the case of Beer Sheva.

Table 4.6 Monthly slope and correlation coefficient for $K_c = aK_T$

Month (data)	a	R^2
January (415)	1.4213	0.9997
February (373)	1.3996	0.9995
March (403)	1.3937	0.9991
April (363)	1.3981	0.9997
May (354)	1.4067	0.9999
June (263)	1.4013	0.99995
July (373)	1.3666	0.9824
August (250)	1.3915	0.99998
September (318)	1.4192	0.99998
October (265)	1.3833	0.9996
November (326)	1.4267	0.99998
December (384)	1.4461	0.9995

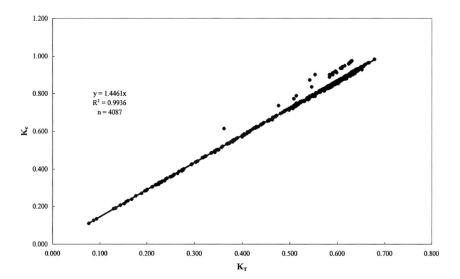

Fig. 4.3 Linear regression analysis of K_c as a function of K_T for all months

3 Solar Horizontal Diffuse and Beam Irradiation on Clear Days

There exist a number of models to determine the solar horizontal diffuse irradiation on a clear day (Kondratyev 1969) but they are complex and have very stringent conditions. Similarly, there also exist models to determine solar horizontal beam irradiation on a clear day (Kondratyev 1969; Sivkov 1971 and Ohvril et al. 1999) but they are dependent on the Forbes effect (i.e., the atmospheric transparency is a function of the air density) and quite complex. Consequently, the clear day solar horizontal diffuse and beam irradiation is defined in terms of the clearness index K_T.

3.1 Clear Day Horizontal Diffuse Irradiation

The clear day horizontal diffuse irradiation is defined as that incident on a horizontal surface on a clear day as defined by Iqbal, viz., when $K_T \geq 0.7$. The clear day horizontal diffuse index $K_{d,c}$ is thus defined in terms of the extraterrestrial solar irradiation H_o as

$$K_{d,c} = H_d/H_o \text{ for } K_T > 0.7, \tag{4.11}$$

where H_d is the horizontal diffuse irradiation measured on a clear day. If there is a particular month with either a very small number or zero days that meet the $K_T > 0.7$ threshold, then it may be lowered by small deficits as was the case for Beer Sheva, cf., Table 4.4.

The solar horizontal diffuse irradiation database consisting of the years 1991 to 2006 (16 years) was utilized to perform an in-depth analysis for Beer Sheva. The monthly average daily extraterrestrial solar irradiation (Extraterrestrial) , the absolute monthly maximum measured daily horizontal diffuse irradiation (i.e., a single value for each month) (Max-absolute), the monthly average of the maximum daily horizontal diffuse irradiation measured each year (i.e., average of 16 values for each month) (Max-average), monthly average clear day horizontal diffuse irradiation using the Iqbal filter (Iqbal), monthly average measured daily horizontal diffuse irradiation (Diffuse-average), the monthly average of the minimum daily horizontal diffuse irradiation measured each year (i.e., average of 16 values for each month) (Min-average) and the absolute monthly minimum measured daily horizontal diffuse irradiation (i.e., a single value for each month) (Min-absolute) during the years 1991 to 2006 is shown in Fig. 4.4.

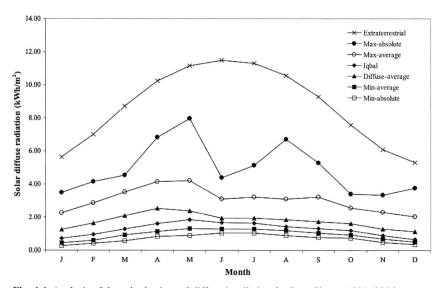

Fig. 4.4 Analysis of the solar horizontal diffuse irradiation for Beer Sheva (1991–2006)

It is observed from Fig. 4.4, that the magnitudes of the monthly average clear day horizontal diffuse irradiation values based upon the Iqbal filter are closest to the corresponding average of the minimum daily horizontal diffuse irradiation measured each year. Also, it is observed that the magnitudes of the monthly average daily horizontal diffuse irradiation values approach that of the corresponding Iqbal filter values during the summer months. This, once again, testifies to the prevalence of clear sky conditions during the summer months in Beer Sheva, cf., Ianetz et al. (2000).

3.2 Clear Day Horizontal Beam Irradiation

The clear day horizontal beam irradiation is also defined as that incident on a horizontal surface on a clear day, viz., $K_T \geq 0.7$. The clear day horizontal beam index $K_{b,c}$ is defined in terms of the extraterrestrial solar irradiation H_o as

$$K_{b,c} = H_b/H_o \text{ for } K_T > 0.7 \tag{4.12}$$

where H_b is the horizontal beam irradiation on a clear day. Once again, if there is a particular month with either a very small number or zero days that meet the $K_T > 0.7$ threshold, then it may be lowered by small deficits.

The solar horizontal beam irradiation database consisting of the years 1991 to 2006 was utilized to perform an in-depth analysis similar to that performed for the horizontal diffuse irradiation and the results are presented in Fig. 4.5.

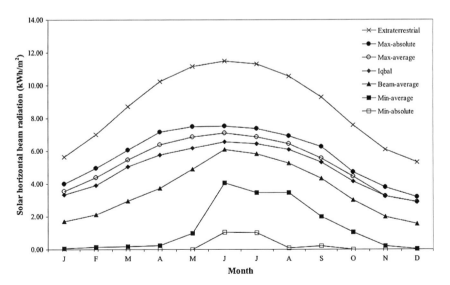

Fig. 4.5 Analysis of the solar horizontal beam irradiation for Beer Sheva (1991–2006)

It is observed from Fig. 4.5, that the magnitudes of the monthly average clear day horizontal beam irradiation values based upon the Iqbal filter are closest to the corresponding average of the maximum daily horizontal beam irradiation measured each year. Also, it is observed that the magnitudes of the monthly average daily horizontal beam irradiation values approach that of the corresponding Iqbal filter values during the summer months. This, once again, testifies to the prevalence of clear sky conditions during the summer months in Beer Sheva, cf., Ianetz et al. (2000).

3.3 Correlation Between Clear Day Horizontal Diffuse and Beam Indices

Ever since the publication of the pioneering work by Liu and Jordan (1960) there have been persistent efforts to develop correlations between global, direct and diffuse irradiation. We have previously developed (Ianetz et al. 2001) empirical regression equations that expressed K_d the ratio of the daily diffuse on a horizontal surface to the daily extraterrestrial irradiation on a horizontal surface, as a function of K_b, the ratio of the daily beam on a horizontal surface to the daily extraterrestrial on a horizontal surface, irrespective of day type, as presented graphically by Liu and Jordan (1960). The regression equations that gave the best fit to the data were found to be non-linear and exponential in form, viz., $K_d = a[\exp(bK_b + cK_b^2)]$. When this analysis was applied to a database consisting of solar horizontal diffuse and beam irradiation on clear days, i.e., the present discussion, it was observed that linear regression equations gave the best fit to the data. It can be concluded that the non-linearity observed in the previous analysis is caused by inclusion of both cloudy and partially cloudy days within the database.

The results of this analysis, i.e., the slope 'a' and intercept 'b' of the monthly linear regression curve and the corresponding correlation coefficient are reported in Table 4.7 for Beer Sheva. In Figs. 4.6 (a) and (b) the data and linear regression curves

Table 4.7 Monthly regression equation coefficients and correlation coefficient for $K_{d,c} = aK_{b,c} + b$

Month (data)	a	b	R^2
January (56)	−1.1787	0.7899	0.9131
February (104)	−0.9128	0.6427	0.9518
March (62)	−0.8712	0.6449	0.9396
April (96)	−0.9128	0.6427	0.9518
May (147)	−0.8324	0.6248	0.9379
June (278)	−0.7860	0.5961	0.9115
July (198)	−0.8730	0.6435	0.9282
August (91)	−0.7772	0.5833	0.9234
September (41)	−0.9292	0.6712	0.9827
October (60)	−0.8254	0.5928	0.8810
November (32)	−0.8132	0.5683	0.8950
December (28)	−0.9301	0.6279	0.9084

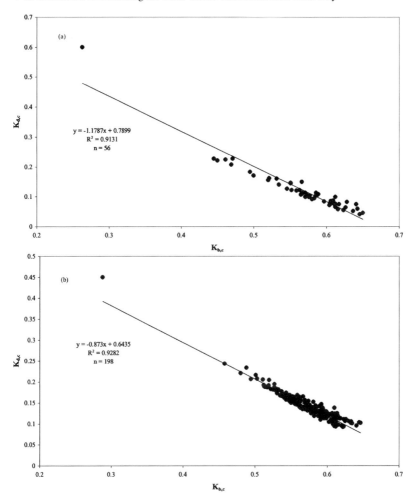

Fig. 4.6 Linear regression analysis of $K_{d,c}$ as a function of $K_{b,c}$ for (a) January, (b) July

for January and July, which are representative of those obtained for all months, are shown.

It is apparent from the magnitude of the coefficients of correlation reported in Table 4.7 for the monthly regression equations that, with the exception of October and November, more than 90% of the variation in $K_{d,c}$ is accounted for by the variation in $K_{b,c}$, i.e., R^2 exceeds 0.90 except for October and November, for which it slightly less than 0.90. It is also observed that the coefficients of the regression equations vary from month-to-month.

In the accompanying CD-ROM the reader will find a ReadMe file containing a step-by-step description of the procedure the authors used to produce the results described in this chapter.

4 Analysis of Solar Irradiation Distribution Types

The analysis of the frequency distribution and distribution type of a particular solar irradiation database are very important parameters in the design of solar irradiation conversion systems. The conversion system size and type (non-concentrating/concentrating) are functions of the magnitude and distribution of the solar irradiation (global/beam). The ability to perform a meaningful economic feasibility analysis of such systems is contingent on the availability of such information.

4.1 Distribution Types

The values for the skewness (As) and kurtosis (K) can be utilized to define the frequency distribution type for a particular database, viz., to describe the breadth of the distribution curve, its degree of asymmetry and its shape relative to that for a normal distribution curve. This following discussion is based upon generally accepted rules of statistical analysis, cf., Brooks and Carruthers 1953. The frequency distribution types as a function of the skewness and kurtosis values are defined in Table 4.8.

The preferred types of distribution, viz., most suitable for solar energy conversion systems, in descending order are as follows: V > IV > I > VI > II and III. The reasoning behind this order of preference of the distribution types is as follows.

- Statistically, a **type V** distribution frequency has a higher occurrence of values greater than the average value, 35-40%, relative to a normal distribution (type I);
- A **type IV** also has a higher occurrence of values greater than the average value, 25%, relative to a normal distribution (type I).
- Consequently, if **types V** and **IV** possess the same average value as a normal distribution, they will both afford a greater number of days with values in excess of the average value relative to that afforded by a normal distribution. **Type V** is

Table 4.8 Definition of frequency distribution types as a function of the range of the kurtosis and skewness values

Distribution Type No.	Distribution Curve	Skewness (As)	Kurtosis (K)
I	Normal	$-0.4 < As < 0.4$	$-0.8 < K < 0.8$
II	almost normal with positive tail	$As \geq 0.4$	$-0.8 < K < 0.8$
III	narrow peak with positive tail	$As \geq 0.4$	$K \leq -0.8$ $K \geq 0.8$
IV	almost normal with negative tail	$As \leq -0.4$	$-0.8 < K < 0.8$
V	narrow peak with negative tail	$As \leq -0.4$	$K \geq 0.8$
VI	bimodal, symmetrical with flat peak	$-0.4 < As < 0.4$	$K \leq -0.8$

preferred because it has a higher occurrence of values greater than the average value;

- **Types II** and **III** are both characterized by a relatively low average value and are, therefore, much less preferred for solar conversion systems. Their preference will be dictated by the relative magnitude of their average value.
- **Type VI** is rated, with regard to preference, between types V/IV and types II/III since it is characterized by a peak value that exceeds the average value.

Obviously, the average solar irradiation intensity at a site, either global and/or beam, is of the utmost importance when designing a solar conversion system but the distribution of the irradiation intensity is also a critical parameter.

4.2 Frequency Distribution and Distribution Types for Clear Day Irradiation Index

The frequency distribution type for each monthly K_c database was determined based upon their respective skewness and kurtosis values. It was observed that January, February, March and December are classified as Type IV distribution, whereas the remaining months are classified as Type V distributions, the most preferred distribution types with regard to solar energy conversion systems. The frequency distribution data for January (type IV) and July (type V) are shown in Figs. 4.7 and 4.8, respectively.

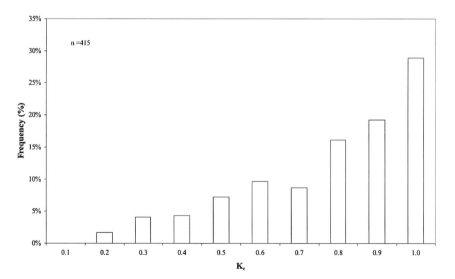

Fig. 4.7 Frequency distribution of K_c for January- type IV

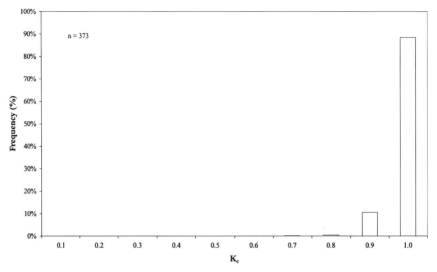

Fig. 4.8 Frequency distribution of K_c for July- type V

4.3 Frequency Distribution and Distribution Types for Clear Day Horizontal Beam Irradiation Index

The frequency distribution of the clear day horizontal beam irradiation index for the database consisting of all 12 months is shown in Fig. 4.9. Since the data are normalized, viz., $K_{b,c} = H_b/H_o$, it is to be expected that the distribution type should be independent of month or season. This was in fact confirmed by comparing the

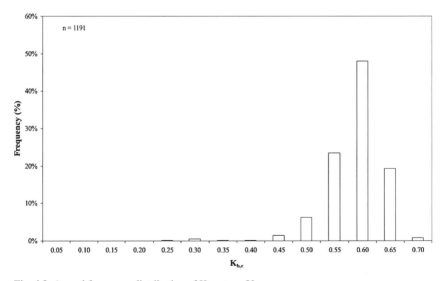

Fig. 4.9 Annual frequency distribution of $K_{b,c}$- type V

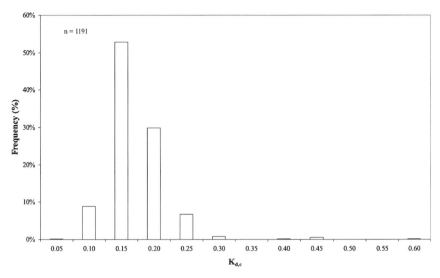

Fig. 4.10 Annual frequency distribution of $K_{d,c}$- type III

distribution types for the four seasons (the individual monthly databases were too narrow for distribution analysis). It is observed from Fig. 4.9 that the distribution of $K_{b,c}$ is a type V, i.e., a narrow peak with a negative tail, which is the most preferred type for solar energy conversion systems.

4.4 Frequency Distribution and Distribution Types for Clear Day Horizontal Diffuse Irradiation Index

The frequency distribution of the clear day horizontal diffuse irradiation index for the database consisting of all 12 months is shown in Fig. 4.10. It is apparent from Fig. 4.10 that the distribution of $K_{d,c}$ is a type III, i.e., a narrow peak with a positive tail. In essence, it is a mirror image of that for $K_{b,c}$, which is to be expected since their respective irradiation intensities constitute the clear day solar global irradiation.

5 Conclusions

The solar global irradiation intensity on a clear day provides both a platform for studying the influence of cloudiness and sets an upper, although unattainable, limit on the magnitude of the solar global irradiation. The Iqbal (1983) classification of days according to clearness index was applied to define a clear day and the Berlynd model for determining the solar global irradiation on a clear day was found to give the best agreement with the clear day solar global irradiation as defined by Iqbal's criteria. Consequently, it is proposed that the Berlynd model be utilized to determine the solar global irradiation on a clear day and that the Berlynd coefficient 'f', cf. Eq. (4.8), be determined for the site being studied.

It is suggested that the daily clear day index K_c, defined as H/H_c, is a better indicator of the degree of cloudiness than the oft used K_T, where the daily clear sky solar global irradiation is calculated utilizing the site-specific Berlynd model. The magnitude of K_T is a measure of both atmospheric transparency and cloudiness. Consequently, the degree of cloudiness as determined on the basis of K_T also includes the other parameters contributing to the sky transparency. The daily clear sky index, K_c, is defined such that the effect of cloudiness is not a parameter. Viz., the daily clear day solar global irradiation, H_c, is determined under clear sky conditions and only those parameters contributing to sky transparency are involved.

The determination of the frequency distribution of the solar global/beam irradiation intensity or the corresponding indices, i.e., K_c and $K_{b,c}$, are very important parameters in the design of solar irradiation conversion systems. They provide another criterion, in addition to the average magnitude of the solar irradiation, for determining the feasibility of conversion system with regard to both size and type (non-concentrating/concentrating). The availability of such information enhances the economic feasibility analysis of such solar energy conversion systems.

The relationship between the ratio of the daily diffuse on a horizontal surface to the daily extraterrestrial irradiation on a horizontal surface, K_d, and the ratio of the daily beam on a horizontal surface to the daily extraterrestrial on a horizontal surface, K_b, irrespective of day type, was found to be non-linear and of exponential form. When the database was limited to clear days only, linear regression equations give the best fit to the data, i.e., $K_{d,c} = aK_{b,c} + b$. It can be concluded that the non-linearity observed in the first case is caused by inclusion of both cloudy and partially cloudy days within the database.

References

Adnot J, Bourges B, Campana D, Gicquel R (1979) Utilisation des courbes de frequence cumulees pour le calcul des installation solaires (in French). In: Lestienne R. (ed), Analise Statistique des Processus Meteorologiques Appliquee al'Energie Solaire. CNRS, Paris, pp 9–40.

Badescu V (1997) Verification of some very simple clear and cloudy sky models to evaluate global solar irradiance. Solar Energy 61:251–264.

Barbaro S, Cannata G, Coppolino S, Leone C, Sinagra E (1981) Correlation between relative sunshine and state of the sky. Solar Energy 26:537–550.

Berger X (1979) Etude du Climat en Region Nicoise en vue d'Applications a l'Habitat Solaire (in French). CNRS, Paris.

Brooks CEP, Carruthers N (eds) (1953) Handbook of Statistical Methods in Meteorology. Her Majesty's Stationary Office, London.

Daneshyar M (1978) Solar radiation statistics for Iran. Solar Energy 21:345–349.

Gueymard CA (2007) Personal communication.

Haurwitz B (1945) Insolation in relation to cloudiness and cloud density. Journal Meteorology 2:154–164.

Haurwitz B (1946) Insolation in relation to cloud type. Journal Meteorology 3:123–124.

Ianetz A, Lyubansky V, Setter I, Evseev EG, Kudish AI (2000) A method for characterization and inter-comparison of sites with regard to solar energy utilization by statistical analysis of their solar radiation data as performed for three sites in the Negev region of Israel. Solar Energy 69:283–293.

Ianetz A, Lyubansky V, Evseev EG, Kudish AI (2001) Regression equations for determining the daily diffuse radiation as a function of daily beam radiation on a horizontal surface in the semi-arid Negev region of Israel. Theoretical and Applied Climatology 69:213–220.

Ianetz A (2002) Clearness index in Jerusalem (in Hebrew). Judea and Samaria Research Studies 11:301–312.

Ianetz A, Lyubansky V, Setter I, Kriheli B, Evseev EG, Kudish AI (2007) Inter-comparison of different models for estimating clear sky solar global radiation for the Negev Region of Israel. Energy Conversion and Management 48:259–268.

Iqbal M (1983) An introduction to solar radiation. Academic Press, Canada.

Kasten F, Czeplak G (1980) Solar and terrestrial radiation dependent on the amount and type of clouds. Solar Energy 24:177–189.

Kondratyev KY (1969) Radiation in the atmosphere. Academic Press, New York, p 463.

Kudish AI, Ianetz A (1996) Analysis of daily clearness-index, global and beam radiation for Beer Sheva, Israel: Partition according to day type and statistical analysis. Energy Conversion and Management 37:405–414.

Kudish AI, Lyubansky V, Evseev EG, Ianetz A (2005) Statistical analysis and inter-comparison of the solar UVB, UVA and global radiation for Beer Sheva and Neve Zohar (Dead Sea), Israel. Theoretical and Applied Climatology 80:1–15.

Kuusk A (1978) Diurnal variation of the total solar radiation of a clear sky. In: Variability of cloudiness and radiation field (in Russian). Academy of Sciences of the Estonian SSR, Tartu, pp. 39–48.

Lingamgunta C, Veziroglu TN (2004) A universal relationship for estimating daily clear sky insolation. Energy Conversion and Management 45:2313–2333.

Liu BYH, Jordan RC, (1960) The interrelationship and characteristic distribution of direct, diffuse and total solar radiation. Solar Energy 4:1–19.

Lopez G, Batlles FJ, Tovar-Pescador J (2007) A new simple parameterization of daily clear-sky global solar radiation including horizon effects. Energy Conversion and Management 48:226–233.

Ohvril H, Okulov O, Teral H Teral K (1999) The atmospheric integral transparency coefficient and the Forbes effect. Solar Energy 66:305–317.

Paltridge GW, Proctor D (1976) Monthly mean solar radiation statistics for Australia. Solar Energy 18:234–243.

Pöldmaa VK, (1975) On methods of normalizing the fluxes of short-wave radiation. In: Cloudiness and radiation (in Russian). Academy of Sciences of the Estonian SSR, Tartu, pp. 89–94.

Robledo L, Soler A (2000) Luminous efficacy of global solar radiation for clear skies. Energy Conversion and Management 41:1769–1779.

Sivkov SI, (1971) Computation of solar radiation characteristics. Israel Program for Scientific Translations, Jerusalem.

World Meteorological Organization, 1983. World Climate Program report, WCP-48.

Younes S, Muneer T (2007) Clear-sky classification procedures and models using a world-wide data-base. Applied Energy 84:623–645.

Chapter 5
Recent Advances in the Relations between Bright Sunshine Hours and Solar Irradiation

Bulent G. Akinoglu

1 Introduction

It seems quite a realistic view to state that the data of bright sunshine hours are the only long term, reliable and readily available measured information that can be used to reach highly accurate estimates of solar irradiation values on the Earth surface. Kimball at 1919 demonstrated for the first time the existence of the relation between the average daily irradiation obtained by means of phyroheliometric and photometric measurements and the duration of sunshine measured by a Marvin sunshine recorder. He presented the relations graphically and included also the relation between the solar irradiation and cloudiness. Using the data of several locations in USA he came to a conclusion that: "In fact, the radiation-ratio and sunshine data plot very nearly on the straight line connecting 100% sunshine and 0% sunshine radiation intensities" (Kimball 1919).

Alternatively, one might exclaim that the history started at 1924, with a simple empirical linear relation proposed by Angström (1924). Since then hundreds of articles appeared in the literature from all over the world which made use of this well-known Angström's linear correlation in the same, similar and/or modified manner. The correlation derived by Angström from measured data of Stockholm, in its original form, was:

$$H = H_c \left(0.25 + 0.75 \frac{n}{N} \right) \tag{5.1}$$

where H and H_c are the total irradiation income on horizontal surface for a day and for a perfectly clear sky, respectively while n/N is the time of sunshine expressed as the fraction of greatest possible time of sunshine. One of the chief results that Angström reached was "A clear conception of the radiation climate ... cannot be obtained without a detailed experience of the amount of energy furnished by the diffuse radiation" (Angström 1924).

Bulent G. Akinoglu
Middle East Technical University, Ankara, Turkey, e-mail: bulent@newton.physics.metu.edu.tr

It is rather hard to define a perfectly clear sky which barely depends on the geographical parameters and the climate of the locality that is mainly the air mass, atmospheric constituents and the cloud amount and type. Another untidiness is the records of sunshine which is not only dependent on the instrument calibration and the burning paper strip type but also on the climate conditions and sun's altitude. Nevertheless, sunshine records evidently possess very valuable surface measured information, are already used in too many applications and will certainly be used in future applications.

In 1940 Prescott replaced the clear sky reference value by a rather more generalized 'Angot's value' that is the radiation on a horizontal surface with a transparent atmosphere. He used the only available measured solar irradiation data in the continent to obtain the regression constants and utilize them to estimate the solar income for Acton, Canberra close to Mount Stromlo. The formula was then named as Angström-Prescott correlation and the correlations and/or models which use the bright sunshine hours to estimate solar irradiation were named as sunshine-based models. The regression coefficients were named as Angström coefficients.

This chapter starts with a brief description of the measuring instruments and some information on the available data. The physical basis of Angström-Prescott relation is discussed especially with reference to the recent advancements on the subject matter. Some of the recent successful models are discussed in details emphasizing especially a broadband hybrid model and the quadratic form. Finally, model comparisons and validation techniques are summarized which is followed by a discussion and conclusion section together with a future prospect.

2 Measurements and Data Availability

In the construction and validation of all types of models and/or correlations accurate long-term and spatially wide range surface data is needed. In addition to this, and may be more important is that the accurate surface data is used to understand our natural environment, especially the atmosphere which we live in. The derived information can be utilized to develop new revenues in order to reduce the human-made hazards given to our Globe (Page 2005).

Early instruments to measure the solar irradiation were quite sophisticated but not accurate enough to derive reliable and universal models/correlations to use in the estimations. However, they were rather accurate enough to understand some very basic physically effective relations between irradiation and other measured meteorological parameters such as the temperature, relative humidity and bright sunshine hours. A detailed historical development in the measuring instruments is given in Duffie and Beckmann (1991) and somewhat extensive discussion can be found in Coulson (1975).

Angström (1924) used a device constructed by him to record the solar irradiation on a photographic film by means of a deflection of a mirror galvanometer. Following scientific works on the relation between the global solar irradiation and bright sunshine hours, especially for the last fifty years used mostly the measurements of two types of instruments for the irradiation: Robitzsch type actinographs and Eppley type black-and white pyranometers, and for the bright sunshine hours: Campbell-Stokes type recorders. Note also that in most of the stations Robitzsch data exist for longer than fifty years while Campbell-Stokes recorders are taking data for more than 100 years. Measurements of the beam and/or diffuse components of the solar irradiation are also carried out regularly in a small number of stations with normal incidence pyrheliometers that follows the sun and/or with a shading ring on the black and white pyranometers (Duffie and Beckman 1991; Coulson 1975).

Robitzsch type pyranometer uses bimetallic strips, one is a white reflector and the other is a black painted absorber exposed to sun, to convert the thermal expansion into the deflection of the pen of the instrument to record irradiation. This instrument needs frequent calibration which is lacking in most stations but may be the more important is the temporal variations in its calibration constants within a year which seemed also instrument dependent (Akinoglu 1992a). Errors in measurements of this instrument may be as high as 30% even for the averages of the readings and it might be far better to use the sunshine based estimation models for global irradiation instead of using the records of this instrument (Akinoglu 1992a). However, the physical relationship between the bright sunshine hours and solar irradiation measured with these instruments might be used in some manner especially in understanding the physical basis of the empirical relations.

Robitzsch type instruments (and also Eppley 180° pyranometers) were replaced with far more accurate Eppley black and white type instruments (or by Kipp and Zonen instruments which use similar principles) for around 40 years in some stations all over the world. But it must be noted that these instruments also need regular calibrations and/or maintenance to produce data less than $\pm 1.5\%$ error (Duffie and Beckman 1991). These instruments basically use a black (hot) and white (cold) regions exposed to sun connected by thermopile junctions producing mV output signal varying with the solar intensity impinging on it. Number of stations with such new radiation measuring instruments is still not enough in most of the countries and also the amount of accumulated data is insufficient for long term investigations of solar radiation modeling. Fully automatic weather stations monitoring all components of the irradiation together with meteorological parameters are also installed at some of the meteorological stations in the last ten years. Robitzsch type pyranographs yet are still recording the solar radiation on the Earth surface in some stations although its data is not accurate enough to drive reliable estimation models.

Controversially, long term records of bright sunshine hours are relatively accurate as the most common instrument used is very simple and quite free of human interfere to run. It is Campbell-Stokes type sunshine recorders. It basically needs the replacement of its recording paper once a day. It contains a spherical glass lens focusing the sun rays on a sensitive record strip placed behind, burning the paper whenever the sun is shining. The trace of the sun collected in the burnt portion of the paper is then read and recorded usually in hourly intervals in hours. The sum for all the hours within a day gives the value of the daily bright sunshine hours. Solar irradiation must exceed a threshold value for the burning of the strip used. As it is clear from the working principle of these recorders, the recorded hours of sunshine carries information about the solar irradiation. However, burning of the strip is accomplished whenever the sun is shinning, provided that irradiation is above the threshold value, regardless of the time within a day. Therefore, it must be noted that the data collected during sunrise and sunset has less effective contribution to the solar radiation amount within a day than the data collected at noon hours.

There are some other problems with the records of Campbell-Stokes instruments which may affect the relation between solar irradiation and the sunshine duration. The irradiance threshold value that can produce a burn can vary from 100 to $300\,W/m^2$ (Painter 1981) or even a wider range of 16–400 W/m2 as determined by Gueymard (1993a). Humidity, frost etc. are some factors that results the loss of records (Aksoy 1999) yet the more important is the extension of the burn during intermittent strong sunshine. This can produce higher outcomes of sunshine durations than the real values (Painter 1981; Gueymard 1993a). Nevertheless, long-term reliable bright sunshine data must have been accumulated as long as 100 years or more in stations from all over the world which is relatively reliable and available to explore, not only for the estimation models but also to use in possible analysis to understand the long-term temporal variations of our atmosphere we live in. Some recent attempts on the subject can be found in references Aksoy (1997) and Chen et al. (2006). Another fact is that the network of sunshine recording sites is denser compared to the sites that records irradiation.

Normal incidence pyrheliometers following the sun are also used to record the bright sunshine hours to be utilized in the models and in the comparisons with the records of Campbell-Stokes recorder. This may aid to determine the errors that would be introduced due to variations of its threshold value with respect to the standard value of $120\,Wm^{-2}$ (Gueymard 1993a), set by the World Meteorological Organization (WMO).

Photovoltaic response of p-n junction is another means of detecting the solar irradiation and various types appeared in rather recent years. The sensitivity however is less than thermopile type pyranometers mainly because of their unsteady spectral response curves which have their maximums usually above the wavelength of $0.6\,\mu m$ (Duffie and Beckman 1991).

Number of surface stations recording all meteorological parameters should certainly increase to understand our environment and the atmosphere as this would help the future of our Globe. These data should be used in the development and validation of the atmospheric models dynamically and to aid the satellite models of radiation estimation which has great potential of producing spatially continuous solar irradiation maps both for the average and instantaneous energy incomes. To take and reduce the data has become very easy and rather cheap with the new computer technologies and developed user-friendly software which are usually supplied free by the companies selling the instruments.

In literature, a number of data, mostly for the monthly average daily values of both the global solar irradiation and bright sunshine hours, appeared, and those published in between 1978–1989 for 100 locations were tabulated by Akinoglu (1991). These data of course averages of some number of years varying for different locations and also the sensitivity of the instruments have certainly different values. Averaging however, suppress the random errors that are introduced due to sensitivities of the measuring instruments and hence the set given by Akinoglu (1991) can be used for model development, validation and comparison.

An important source of data is the internet site of WMO, namely World Radiation Data Center in St. Petersburg, Russia (http://wrdc-mgo.nrel.gov/). It is possible to acquire this data by getting contact with the site and any reliable data can be sent to the internet site to be included in their database.

3 Angström-Prescott Relation and its Physical Significance

Prescott at 1940 used the data recorded in Mount Stramlo Observatory in Canberra-Australia. Instead of perfect clear sky value of Angström, he used the published data of Angot's (as supplied by Brunt (1934)) values which is the solar radiation that would be received if the atmosphere were transparent. He obtained the regression equation:

$$\frac{H}{H_0} = 0.25 + 0.54\frac{n}{N}.$$ (5.2)

In this expression, H is the monthly averages of daily global solar irradiation on horizontal surface, H_0 is the monthly average daily solar radiation on horizontal surface if there were a transparent atmosphere, n is the monthly average of daily bright sunshine hours and N is the maximum possible sunshine on cloudless day. Instead of measured perfectly clear sky value in Eq. (5.1), use of H_0 makes it possible to utilize such correlations to estimate the irradiation values in locations where no radiation data exists. This is because it is possible to calculate H_0 in any time interval using the solar radiation values reaching outside the atmosphere.

The main reason for the large difference between the coefficient of Prescott, 0.54 and that of Angström's, 0.75 is the use of new normalizing value H_0 instead of the a perfectly clear day value of the site of interest. Now we know that even similar calculated values of H_0 of the site are used for normalizing the solar irradiation, such coefficients span a wide range of values varying with the location (mainly latitude), climatic, atmospheric and seasonal meteorological variations in the site under consideration.

Later, Angström at 1956 has written the expression:

$$H = H_c\left[\alpha_1 + (1 - \alpha_2)\frac{n}{N}\right]$$ (5.3)

and stated that with the definition of N, α_1 will be equal to α_2 and the equation takes the form:

$$H = H_c\left[\alpha + (1 - \alpha)\frac{n}{N}\right]$$ (5.4)

Angström stated that for the fractional bright sunshine period $n/N = 0$, $H = \alpha H_c$ while for $n/N = 1$, $H = H_c$; clarifying also his α value of 0.25 for Stockholm. This value 0.25 means that H for an overcast sky has a value 25% of that of a perfectly clear sky.

In his 1956 article, Angström also gave a detailed physical understanding of the coefficient α, stating that Eq. (5.4) is an idealized form and the coefficient depends on many parameters. Some of these parameters are frequency of atmospheric disturbances, cloud amounts and types, month of the year, altitude and ground reflectance. In the work by Martinez-Lozano et al. (1984) α values for a number of locations are presented; they span a range of 0.22 to 0.68. Now it is very clear that all such parameters mentioned above and others are effective on the regression coefficients of the linear sunshine based models, namely the Angström-Prescott relation:

$$\frac{H}{H_0} = a + b \frac{n}{N}. \qquad (5.5)$$

Calculations of H_0 and N values are explained in details in Duffie and Beckmann (1991), which start with the solar constant, $1367\,\mathrm{W/m^2}$. Solar constant is defined as the extraterrestrial solar intensity outside of the atmosphere incident on a perpendicular surface at the mean sun-earth distance. It is reduced to an instantaneous value outside the atmosphere on a horizontal surface for any location, by taking into account the varying sun-earth distance and multiplying by cosine of the zenith angle of the location of interest. For the total values in any time interval of course one should integrate the instantaneous values in the required time interval, mainly hourly and daily. Monthly averages of daily values, H_0 and N, can be obtained simply by taking the averages of daily values or to simplify by using the values at a specific day number of the year which gives directly the monthly average daily value of that specific month, as explained in Duffie and Beckman (1991).

Angström also derived simply the value of α_1 in Eq. (5.3) in terms of Angström coefficients a and b as $\alpha_1 = a/(a+b)$ (Angström 1956) with a definition that his value H_c in Eq. (5.3) is equal to H in Eq. (5.5) for $n = N$. He calculated the values of α_1 using regression constants a and b of different stations and obtained values in quite a large range of from 0.218 to 0.583. He thus concluded that the values of α or the other coefficients depends on different climatic and geographic parameters stated above. Later as stated by Gueymard et al. (1995), for the mean value of irradiation after monthly averaging, it is a question of "superposability" of two extreme cloud states for two limiting idealized days (fully overcast and perfectly cloudless) which is only a simplifying assumption. Essentially, for $n/N = 0$ and for $n/N = 1$ the values of H/H_0 would certainly be different than a and $a + b$, even for different days of the same month.

It is rather easy to attribute rough physical meanings to the coefficients a and b in Eq. (5.5) using the extreme values of n/N. If there is no cloud obscuring the sun within a day, then $n/N = 1$ and $H/H_0 = a + b$ can be interpreted as the monthly average daily value for the transmittance of a clear day. Note that clear day ($n/N = 1$ in this case) does not always mean a perfectly clear day without appearance of any cloud all the day. Even sometimes the presence of clouds that do not obscure the sun may increase the irradiation reaching the site due to high reflections. Another fact is that the days without any cloud may have different solar irradiation reaching the

Earth due to differences in the air mass and also due to some atmospheric conditions such as dense turbidity. For a completely overcast day, $n/N = 0$ and $H/H_0 = a$, which essentially accounts for the diffuse component. It may represent the average daily transmission of an overcast sky of the site under consideration. The range of values obtained for a and b given below however show that they are affected by many geographical and atmospheric parameters.

Angström coefficients a and b in Eq. (5.5) have quite a wide range of different values, a ranging from 0.089 to 0.460 and b from 0.208 to 0.851 as tabulated for 100 locations in the review article by Akinoglu (1991), or from 0.06 to 0.44 for a and from 0.19 to 0.87 for b as given for 101 locations in the paper of Martinez-Lozano et al. (1984). This variation may be considered to recognize the importance of the above mentioned parameters affecting the regression constants of the empirical correlation Eq. (5.5). In addition, one should think that the variation of a with b might be hindering another conceptual information about the relation between global solar irradiation and bright sunshine hours, to be used in developing an estimation model with higher accuracy and better universal applicability. Another important fact that must not be overlooked is the measurement errors both for the irradiation values and bright sunshine hours.

Nevertheless, many researchers expressed Angström coefficients in terms of different geographical and climatic parameters such as the latitude, altitude, sunshine fraction (see for example Akinoglu and Ecevit (1989); Gopinathan (1988); Rietvel (1978); Abdalla and Baghdady (1985)). An overall conclusion that can be derived from all these works might be summarized as: these coefficients depend on all physical, spatial and the dynamic properties of the atmosphere at the region of interest. One may even state that, for a region the coefficients derived from a long term data of some number of years can be different than those obtained by using the data of same length for the same region but for another set of years. This is of course another research of interest which necessitates long-term reliable data with high accuracy from different regions.

One of the most important facts about the wide range of values of the regression constants is hidden in the diffuse component of the energy income, as Angström (1924) concluded. Diffuse component has mainly three different parts as explained in section 4.3, each of which depends on different physical properties of the elements of our environment and the atmosphere. Use of H_c instead of H_0 may reduce the wide range of the values of the regression coefficients since H_c includes the diffuse irradiation characteristics of the atmosphere of the location of interest. Then, of course, H_c must be calculated by some other means for the locations where the long term measured data is missing.

Gueymard (1993b) developed a model to calculate H_c using reasonable estimates of precipitable water w and Angström turbidity coefficient β which makes it possible to use Angström Equation in its original form. As mentioned above H_c intrinsically contains information on the atmospheric characteristics of the site of interest. Hence, he recommended that the researches should be directed toward the determination of α of Angström expression, namely the Eq. (5.4).

Two self-explanatory excel worksheet are included in the CD-ROM supplied with the book, which calculate the daily and monthly mean daily values of H_0 and N, namely 'daily-calculations-Ho.xls' and 'monthly-mean-daily-calculations-Ho.xls'. Calculations only need to input the latitude of the location of interest to the cell B3. Solar constant is taken to be $1367\,\mathrm{W/m^2}$.

4 Physical Modeling and Some Recent Models

4.1 Introduction

A physical model for the transmission of radiation through a semi-transparent matter of finite thickness, like the atmosphere, should start with monochromatic, non-coherent electromagnetic wave of initial intensity $I_0(\lambda)$, where λ is the wavelength of the electromagnetic wave. If such a wave traveling in free space enters a non-dispersive, homogeneous and isotropic layer of matter then the energy conservation reads:

$$I_0(\lambda) = I_\rho(\lambda) + I_\tau(\lambda) + I_\alpha(\lambda) \qquad (5.6)$$

where $I_\rho(\lambda)$, $I_\tau(\lambda)$ and $I_\alpha(\lambda)$ are the reflected, transmitted and absorbed intensities, respectively at wavelength λ, by the layer, atmosphere for our case. In this expression, even if the incident monochromatic energy is specular (directional) –which is the case for extraterrestrial radiation, impinging on the atmosphere- reflected and transmitted radiation will have also a diffuse component due to the scattering by different constituents of the matter. Another physical fact is the irradiative properties of the heated matter, and in our case it is the atmosphere, the temperature of which is increased due to the absorption of electromagnetic wave propagating through it (Coulson 1975). Hence, propagation of the sun rays through the atmosphere is a sophisticated phenomenon possessing many different physical mechanisms of interaction of matter with the electromagnetic waves. Nevertheless, dividing Eq. (5.6) by $I_0(\lambda)$ and calling $\rho(\lambda)$, $\alpha(\lambda)$ and $\tau(\lambda)$ as the spectral reflectance, absorbance and transmittance of the layer at a specific wavelength λ, one can write the expression:

$$1 = \rho(\lambda) + \alpha(\lambda) + \tau(\lambda). \qquad (5.7)$$

For the solar spectrum, most of the emitted radiation is in the range of wavelengths from 0.20 to $4.0\,\mu\mathrm{m}$, typical of the spectrum of a blackbody at around $6000\,\mathrm{K}$. The values of $\rho(\lambda)$, $\alpha(\lambda)$ and $\tau(\lambda)$ are different for each wavelength (or in wavelength intervals) for the atmosphere and one needs to integrate to find the transmitted amount of instantaneous radiation over all wavelengths, knowing the value of the incident radiation $I_0(\lambda)$.

Some models starts from spectral calculations, assigning different values (most of the time as a function air mass) to the transmittance of the atmosphere for different wavelengths bands considering the interaction of electromagnetic radiation with various constituents of our atmosphere, some examples are Leckner (1978); Atwater and Ball (1978) and Bird (1984). These interacting constituents are mainly water vapor, greenhouse gases, ozone, dust and aerosol, some of which are either changed or given by man-made interventions to the atmospheric and climatic cycles of our globe. The important issues of this approach rest on the definition, formulation and/or measurement of the spectral values for the transmittance due to different components of a dynamic atmosphere which bring in quite cumbersome calculations to reach the instantaneous irradiation values on the earth surface. Scattering due to the atmospheric constituents and the reflection from the ground introduce an extra diffuse component to the irradiation values, the transmittance of which must be handled with care and on a different base. Spectral models define these parameters in different wavelength bands and reach instantaneous transmission of the atmosphere for the whole solar spectrum.

Some other types of models are based on a critical assumption that the transmissions of the atmosphere for different components of irradiation (mainly beam and diffuse) can be obtained using some spectrally averaged values of the transmittances and some other optical properties of the atmospheric components. Actually, such properties should be written in an integral form for a spectral average value, for example for the transmittance of the beam component, as:

$$\bar{\tau} = \frac{\int\limits_{\lambda_{\min}}^{\lambda_{\max}} \tau(\lambda)I_0(\lambda)d\lambda}{\int\limits_{\lambda_{\min}}^{\lambda_{\max}} I_0(\lambda)d\lambda}. \tag{5.8}$$

In this expression the angle of incidence for the light rays are not taken into account which of course will introduce extra complexity in the determination of these spectrally averaged properties. $\lambda_{\min}$ and $\lambda_{\max}$ are the minimum and maximum wavelengths of the energy source under consideration which in our case is the solar spectrum. Such spectrally averaged properties can be directly measured and used to construct models without any concern of their spectral variations. Such approaches of course should be supported and validated by measurements of the total, diffuse and beam components of the global solar irradiation on the surface of the Earth. In any case, one needs assumptions such as that the spectral property is constant at least for some wavelength band to allow a numerical integration unless an analytical form for these properties can be given.

A two-band model for example was developed to estimate clear-sky solar irradiance which divided the solar spectrum into UV/VI (0.29–0.7 μm) and IR (0.7–2.7 μm) bands. In this model effect of atmospheric constituents are parameterized using preliminary integrations of spectral transmission functions (Gueymard 1989).

May be the most important dilemma of all these models is the need of a further integration of these properties for a specified time interval which are mostly an hour or a day for our case. Such integration procedures should contain air mass, and since the incidence angle and thus the column of atmosphere traversed is a function of time, this introduces new complications in modeling. If however instantaneous spectrally averaged properties can be defined, then by making analogy that the time integrations will not change the form of the analytically derived models, some inter-relationships between the regression coefficients of the empirical relations and physically defined average properties can be obtained, a typical example is the coefficients of Angström equation. Hence, solar irradiation on the surface of the earth may be directly written in a similar manner as given by Davies and McKay (1982):

$$G = G_0 \sum_{i=1}^{n} \Psi_i f(\rho_c, \rho_g, \rho_a) \qquad (5.9)$$

where G and G_0 are global and cloudless sky irradiance, Ψ_i is ith cloud layer transmittance as defined by Davies and McKay (1982) which indeed is also a function of the reflectance, transmittance and absorbance of the atmospheric constituents. The function $f(\rho_c, \rho_g, \rho_a)$ depends on cloud-base, ground and clear sky atmospheric reflectance which stands to take into account the multiple reflections between the ground and the atmosphere. In this expression G_0 may be the solar radiation above the atmosphere at the location of interest but of course this replacement may introduce modification in the definition of Ψ. In this approach if one starts with the instantaneous values of the solar radiation and spectrally averaged physical properties, then the only requirement will be the time integrations within a preferred time interval, predominantly hourly and daily.

Following sub-sections review two recent models together with their physical basis which also make use of the simple Angström-Prescott approach. The reasons of choosing these two models are as they represent two different approaches in the search for a physical model for the relationship between the solar irradiation and bright sunshine hours and both end up with a simple Angström-Prescott type correlation. I should note that there exist various models and correlations with good performances but it is rather hard to discuss them all within the content of one chapter. I hope also that the models we discuss herein would be enough to comprehend the subject matter of interest, namely exploring the recent advances between the solar irradiation and bright sunshine hours. Another point is that the following models are compared with some other models and showed good performances as appeared in the literature.

4.2 Yang and Co-Workers Hybrid Model

There are five major spectral transmittances for the atmospheric components: Rayleigh scattering, aerosol extinction, ozone absorption, water vapor absorption and permanent gas absorption, according to the work by Leckner (1978). Spectral transmittance of these different components was written both as a function of wavelength and air mass. Yang et al. (2001) suggested a new form so-called hybrid model which considered the spectral and temporal physical processes and still preserved the simplicity of Angström correlation. In their model they used Eq. (5.8) to obtain the spectrally averaged transmittance for each irradiative transfer process. All these broadband transmittance values can be calculated using the measured or calculated values of some atmospheric-climatic and geographic parameters of the site of interest as outlined in Yang et al. (2001). The values for the clear sky beam and diffuse components of the solar irradiation on a horizontal surface are then calculated with the integral forms:

$$H_b = I_o \int \tau_{b,clear} Sinhdt \qquad (5.10a)$$

$$H_d = I_o \int \tau_{d,clear} Sinhdt \qquad (5.10b)$$

with

$$\tau_{b,clear} \approx \max(0, \tau_{oz} \tau_w \tau_g \tau_r \tau_a - 0.013) \qquad (5.11a)$$

$$\tau_{d,clear} \approx \max\{0, [\tau_{oz} \tau_w \tau_g (1 - \tau_r \tau_a) + 0.013]\} \qquad (5.11b)$$

where I_0 is the integral of the solar spectrum for all the wavelengths outside the atmosphere and h is sun altitude. As it is clear, subscripts oz, w, g, r, and a stand to indicate that various transmittances are for ozone absorption, water vapor absorption, permanent gas absorption, Rayleigh scattering and aerosol extinction, respectively.

In the first version of the hybrid model (Yang et al. 2001), the relation between monthly-mean daily solar irradiation and sunshine duration was established as:

$$H = (a + bn/N)H_b + (c + dn/N)H_d \qquad (5.12)$$

in order to find four regression coefficients a, b, c and d, between H and n/N. In this expression, extraterrestrial daily solar irradiation on horizontal surface, namely H_0, above the location of interest outside the atmosphere and the inclusion of the diffuse component of the site are embedded into the delicate calculation of the beam and diffuse components. Yang and co-workers used the measured H and n values together with the calculated H_b and H_d values using the data of 16 stations of Japan in 1995 and obtained the coefficients as: $a = 0.391$, $b = 0.518$, $c = 0.308$, $d = 0.320$ for $n/N > 0$, and $a = 0.222$ and $c = 0.199$ for $n/N = 0$.

In the model of Yang et al. it is possible to obtain the radiation values in any preferred time interval since the time integrals equation (5.10) gave this flexibility. For example, from the hourly values, daily sums can be obtained. They used 14 different stations from Japan to validate their model from different locations at different latitudes and altitudes. They also compared the model with the model proposed by Gopinathan (1988) and concluded that their model showed better performances.

In fact, to start from hourly values might be significant as the bright sunshine periods at different times within a day have different contributions to the amount of daily radiation as mentioned in section 2. Monthly averages of hourly values of bright sunshine and hourly solar irradiation at the same intervals correlate quite well for a one year data set of a specific location (Akinoglu et al. 2000). A further similar analysis can be carried out using a larger and reliable data set from different locations in which the hourly values of these variables at the same average air mass may also be considered.

Gueymard (2003a), compared 21 broadband spectral models including the above model of Yang et al. (2001) for the predictions of direct solar transmittance and irradiance, and as the result of a detailed investigation recommended four of them, one of which is the Yang and co-worker's model. As stated in their recent article of Yang et al. (2006) which presents a similar but modified version of the model, some other investigators (Paulescu and Schlett (2003); Madkour et al. (2006)) also verified the high performance of this broadband model (Yang et al. 2001).

In updated version of the hybrid model, in addition to some corrections of the first version, Yang et al. (2006) showed that the relation between global solar irradiation and sunshine duration for monthly-mean scale can be easily extended to daily scale and even to hourly scale. Also, this version introduced global aerosol and ozone data sets to improve the accuracy of radiation estimation.

In the modified version, global solar irradiation is written as:

$$R = \tau_c \int_{\Delta T} (\tau_{b,clear} + \tau_{d,clear}) I_0 dt \tag{5.13}$$

where ΔT is any preferred time interval. Bright sunshine hour in this new version is now introduced into the formula through $\tau_c = R/R_{clear}$, which was thought to be a function of n/N, that is $\tau_c = f(n/N)$, a newly defined parameter which is the ratio of the surface solar irradiation R to the surface solar irradiation under clear sky, R_{clear}. This parameter is of course similar to the quantity which Angström used in obtaining his simple correlation between solar radiation and bright sunshine hours, H/H_c of Eq. (5.1). τ_c can be obtained by regression analysis using hourly, daily or monthly bright sunshine records. Yang et al. used a two-step procedure to obtain τ_c: in step one, they regressed the daily data of 67 stations at the year 1995, in Japan. In step two using the result of pass one, excluded the data with RMSE>2.0 MJm^{-2} (Yang et al. 2006) (RMSE is the root mean square error which is the square-root of average of the square of calculated minus measured values of a quantity and explained briefly in section 5). Remaining data was then used to obtain the function $f(n/N)$ for daily and monthly mean daily solar irradiation, respectively as:

$$\tau_c = 0.2505 + 1.1468 n/N - 0.3974(n/N)^2 \tag{5.14}$$

$$\tau_c = 0.2777 + 0.8636 n/N - 0.1413(n/N)^2.$$

A limitation for these equations was for $n/N = 1$, $\tau_c = 1$ must be used since the radiation in this case is the clear sky value R_{clear}.

They tested the model for the seven sites in China, seven in USA and twelve in Saudi Arabia comparing the results with the estimations of two different models and with Angström-Prescott type quadratic correlations that they obtained directly by regression analysis for the same data set they used in obtaining Eq. (5.14). Their results showed that the new formalism which started with spectral considerations has better performance and hence universally applicable.

It may be questioned that why the quadratic forms, Eq. (5.14), are chosen in obtaining the function $f(n/N)$. They used this form due to some reasons (Yang 2007): as their experience on humid regions (Japan) showed that the quadratic form gives better correlations and also following the conclusion reached in a recent article (Suehrcke 2000) that those non-linear forms are better than linear. They also followed the work of Iqbal (1979) who proposed quadratic relations between the fractional monthly averages of daily diffuse and beam components and bright sunshine hours. It is rather interesting that Suehrcke (2000) reached a non-linear form using a physical formalism he developed, which starts from bi-modal character of instantaneous or short-term irradiation. This characteristic is due to the fact that the cloud within a time interval may intermittently obscure the sun rays in rather short times so that significantly reduces the irradiation without changing the bright sunshine records of the instrument.

A physical model which starts by defining various spectrally averaged physical properties (reflectance, transmittance etc.) of the atmosphere within the entire solar spectrum, measurable by the instruments such as pyranometers and pyrheliometers, may give quadratic forms rather than linear as will be discussed in the following section. Such instruments measure the total integrated instantaneous solar global and beam irradiation over the whole spectrum.

The FORTRAN code of the model of Yang et al. is included in the CD-ROM. In the readme.doc file within modelYangetal.zip, some explanations of how to use the programs are given. Clear-sky, hourly, daily and monthly models are in the zip file and each model has an application as an example.

4.3 Direct Approach to Physical Modeling

As mentioned in section 4.1, if the spectral averages of some of the physical properties of the atmosphere can be defined instantaneously as given in Eq. (5.8), then the only requirement is to consider the time variation of these quantities. Consequently, an expression of the form as Eq. (5.9) can be utilized. Instead of G_0, extraterrestrial instantaneous solar irradiation outside the atmosphere on a perpendicular surface, I_{0n} can be used. Then, for the direct component of the solar radiation on a horizontal surface at the bottom of the atmosphere can be written as;

$$I = I_{0n}\Psi Cosz. \tag{5.15}$$

In this expression z is the zenith angle and Ψ is a term that accounts various absorption and/or transmission mechanisms. Monteith (1962) for example used $\Psi = (1 - \phi')(1 - \rho')$ for the direct component where ϕ' stand for Rayleigh, Fowle, and dust scattering while ρ' is for the absorption coefficient of water vapor and dust. In his model, Monteith used the fractional cloudiness c of two monthly mean of daily observations.

One can start with the fractional cloud amount c for the cloudy sky as discussed by Davies et al. (1984), and then taking into account the multiple reflection cycles the solar irradiation can be written as:

$$G = G_0(1 - c + \tau_c c)(1 - \alpha\beta)^{-1} \qquad (5.16)$$

where τ_c is the cloud transmittance α is the ground albedo and β is a total atmospheric back-scattering coefficient. G_0 is theoretical cloudless sky global irradiation. In this expression $1-c$ is for the solar irradiation on the surface coming from the cloudless part of the sky while $\tau_c c$ is for the portion coming after transmitted by the cloudy part. Such approach of course introduce a new question on the time integration of Eq. (5.16) as one also needs the time interval that the sky has the fractional cloud amount c.

Instead of c one can also start with bright sunshine fraction within an infinitesimally small time interval n_i (Akinoglu 1992b). In such a consideration, infinitesimally small means a small enough time interval so that numerical integration on time gives acceptable approximate results for the average parameters. Firstly, by considering the direct component I_D which comes during the bright sunshine period within the specified time interval as:

$$I_D = I_0 n_i \tau. \qquad (5.17)$$

I_0 in this equation is the extraterrestrial solar irradiance at the site of interest on a horizontal surface above the atmosphere coming within the specified time interval and τ is transmittance of the atmosphere during the clear-sky period, an average value for all wavelengths. Diffuse component during the same bright sunshine period can be written as:

$$I_{d1} = I_0 n_i (1 - \tau)\beta' \qquad (5.18)$$

where β' is the atmospheric forward scattering coefficient. Finally, diffuse component during cloudy sky period is:

$$I_{d2} = I_0(1 - n_i)\tau\tau_c. \qquad (5.19)$$

In this expression, τ is also included as the sun rays must pass through the whole atmosphere also in the presence of clouds (Akinoglu 1992b). Including the first reflection cycle between the ground and the atmosphere, namely the first term in the binomial series, total global solar irradiation on the surface may be written as:

$$I = I_0[n_i\tau + n_i(1 - \tau)\beta' + (1 - n_i)\tau\tau_c](1 + \alpha\beta) \qquad (5.20)$$

where $\alpha\beta$ is for the irradiation reflected back by the atmosphere due to all its components. Note that this expression now contains three diffuse parts, two coming from Eqs. (5.18) and (5.19) and one coming because the first multiple reflection cycle between the ground and the atmosphere is considered. The third component can be deduced from Eq. (5.20) as:

$$I_{d3} = I_0[n_i\tau + n_i(1-\tau)\beta' + (1-n_i)\tau\tau_c]\alpha\beta. \tag{5.21}$$

As defined above, β is the total atmospheric back-scattering coefficient and can be defined with two components: back-scattering from the clear atmosphere with a coefficient α_R, and from the cloud base with the coefficient α_c. This consideration yields:

$$\beta = \alpha_R n_i + \alpha_c(1-n_i) \tag{5.22}$$

similar to that explained by Davies and McKay (1982), but they used the cloud amount c. An aerosol scattering term may be introduced as written in the article of Davies and McKay (1982), however this would not change the nature of the relationship that will be obtained below, between the global solar irradiation and bright sunshine hours. This form is quadratic in n_i, and given as:

$$I = I_0[n_i\tau + n_i(1-\tau)\beta' + (1-n_i)\tau\tau_c][1 + \alpha(\alpha_R n_i + \alpha_c(1-n_i))] \tag{5.23}$$

Assuming that the form of the equations does not change after daily integrations and monthly averaging, analog equation for the monthly mean daily values can be written. But then of course all the parameters should be replaced by their monthly effective counterparts. Also n_i must be replaced with the monthly average values of n/N. Thus one can obtain a quadratic relation as:

$$\begin{aligned}H = H_0[&(n/N)(\tau_e + (1-\tau_e)\beta_e') + (1-n/N)\tau_e\tau_{ce}]\\ &\times[1 + \alpha_e(\alpha_{Re}n/N + \alpha_{ce}(1-n/N))]\end{aligned} \tag{5.24}$$

where index e stands to indicate that the parameters are monthly effective parameters. Last equation has a quadratic form as:

$$\frac{H}{H_0} = a_0 + a_1\frac{n}{N} + a_2\left(\frac{n}{N}\right)^2 \tag{5.25}$$

where the coefficients a_0, a_1 and a_2 can be written in terms of the effective parameters in Eq. (5.24). As these parameters have some physical interpretations, so we may conclude that a_0, a_1 and a_2 and the quadratic relation (5.25) have physical base as they are written in terms of them. It should be noted that, although these effective parameters are newly defined, approximate values of some of them can be obtained from the literature. Some of them can be left free to be calculated for each month as a_0, a_1 and a_2, by using the measured diffuse and total component of the monthly average daily solar irradiation at any site on the Earth surface (Akinoglu 1992b). Expressions for the three coefficients in terms of the parameters in Eq. (5.24) are:

$$\begin{aligned}a_0 =& \tau_e\tau_{ce}(1 + \alpha_e\alpha_{ce})\\ a_1 =& \tau_e(1 + \alpha_e\alpha_{ce})(1-\beta_e') + \beta_e'(1 + \alpha_e\alpha_{ce}) + \tau_e\tau_{ce}(\alpha_{Re}\alpha_e - 2\alpha_e\alpha_{ce} - 1)\\ a_2 =& \tau_e\alpha_e(\alpha_{Re} - \alpha_{ce})(1-\beta_e') + \beta_e'\alpha_e(\alpha_{Re} - \alpha_{ce}) - \tau_e\tau_{ce}\alpha_e(\alpha_{Re} - \alpha_{ce})\end{aligned} \tag{5.26}$$

For some further description of the nature of a possible quadratic relation between solar irradiation and bright sunshine hours here several results on the use of Eqs. (5.24) and (5.25) will be given. Some of these results already appeared in the literature (Akinoglu 1992b; Akinoglu 1993).

The value of ground albedo α varies from 0.1 for forest and grass up to 0.7 for fresh snow (Davies et al. 1984). However, for semi-urban and cultivation site 0.2 is a rational value as measured by Ineichen et al. (1990). Although a seasonal variation exists, Ineichen et al. concluded that a unique average ground albedo can give satisfactory results in the calculation of the ground reflected radiation. Cloud reflectance depends on the type, height and amount of clouds but an average value can be assigned as proposed by Fritz (1949), the value he obtained was 0.5. Later, Houghton (1954) and Monteith (1962) determined the same value using different approaches. Using the results obtained by Houghton (1954) a value of around 0.25–0.40 can be derived for the average forward scattering coefficient of the clear atmosphere (Houghton 1954). Finally, for α_R, a value of 0.0685 can be used as Davies and McKay (1982) which was proposed by Lacis and Hansen (1974). In the highlight of these values, one can assign and change within an appropriate interval the values of the effective parameters in Eq. (5.24), namely α_e, α_{ec}, α_{Re} and β'_e. The parameters τ_e and τ_{ec} can be left free to be calculated for each month.

Using measured monthly averages of daily global and diffuse irradiation together with monthly average of bright sunshine hours, monthly values of the coefficients a_0, a_1 and a_2 of the quadratic correlation Eq. (5.25) and the monthly values of the parameters τ_e and τ_{ec} are calculated for four stations (Table 5.1) from different latitudes. Table 5.1 also includes the climate types of the locations as given in Trewartha (1968). The data of three of these stations are those used in the work of Jain (1990) who constructed a similar formalism to write the global solar irradiation in the linear form and applied it to find the Angström coefficients a and b of these three locations. This work of Jain is summarized at the end of this section. Linear form of the outlined formalism above (that is, if the first reflection cycle is not accounted) was applied to a location in Turkey before (Akinoglu 1992b) and the results for this station will also be presented here.

Table 5.2 gives the calculated monthly values of the parameters τ_e and τ_{ec} and the coefficients a_0, a_1 and a_2 of the quadratic correlation. In the calculations, 0.2 is used as the ground reflectance, 0.35 is used as the forward scattering coefficient and 0.5 is used for the cloud albedo. The values of the ground albedo varied between the

Table 5.1 Some characteristics of the locations

Country	Station	Latitude	Longitude	Altitude (m)	Climate (Trewartha 1968)
Zimbabwe	Bulawayo	20.15° S	22.86° E	1341	Aw (Tropical)
	Salisbury	17.83° S	31.05° E	1471	Aw (Tropical)
Turkey	Ankara	39.95° N	32.88° E	894	BS (Dry)
Italy	Macerata	43.30° N	13.45° E	338	Csa (Subtropical)

Table 5.2 Values of the parameters τ_e and τ_{ec}, and the coefficients a_0, a_1 and a_2

Month	Bulawayo					Salisbury				
	τ_e	τ_{ec}	a_0	a_1	a_2	τ_e	τ_{ec}	a_0	a_1	a_2
Jan	0.63	0.47	0.322	0.488	−0.040	0.61	0.47	0.312	0.484	−0.040
Feb	0.61	0.48	0.323	0.473	−0.039	0.58	0.50	0.316	0.458	−0.038
Mar	0.63	0.44	0.306	0.506	−0.042	0.58	0.51	0.323	0.452	−0.037
Apr	0.64	0.40	0.277	0.541	−0.044	0.64	0.42	0.297	0.522	−0.043
May	0.68	0.31	0.232	0.624	−0.050	0.70	0.36	0.281	0.583	−0.047
Jun	0.69	0.34	0.260	0.598	−0.048	0.70	0.36	0.279	0.583	−0.047
Jul	0.68	0.32	0.237	0.618	−0.050	0.69	0.40	0.305	0.547	−0.045
Aug	0.65	0.31	0.221	0.615	−0.050	0.69	0.29	0.221	0.641	−0.052
Sep	0.66	0.32	0.231	0.604	−0.049	0.66	0.38	0.272	0.561	−0.046
Oct	0.63	0.37	0.260	0.557	−0.045	0.63	0.39	0.273	0.542	−0.044
Nov	0.63	0.39	0.270	0.547	−0.045	0.71	0.34	0.265	0.609	−0.049
Dec	0.66	0.41	0.294	0.538	−0.044	0.61	0.45	0.302	0.492	−0.040
Mean	0.65	0.38	0.269	0.559	−0.046	0.65	0.41	0.287	0.540	−0.044

Month	Ankara					Macerata				
	τ_e	τ_{ec}	a_0	a_1	a_2	τ_e	τ_{ec}	a_0	a_1	a_2
Jan	0.73	0.22	0.181	0.713	−0.057	0.70	0.40	0.303	0.557	−0.046
Feb	0.63	0.34	0.236	0.585	−0.047	0.64	0.49	0.347	0.471	−0.039
Mar	0.57	0.32	0.199	0.576	−0.046	0.82	0.32	0.286	0.660	−0.054
Apr	0.57	0.28	0.175	0.607	−0.049	0.78	0.33	0.286	0.633	−0.051
May	0.53	0.30	0.172	0.578	−0.046	0.74	0.38	0.313	0.580	−0.047
Jun	0.49	0.50	0.270	0.448	−0.037	0.77	0.39	0.329	0.582	−0.048
Jul	0.57	0.20	0.124	0.657	−0.052	0.69	0.46	0.349	0.498	−0.041
Aug	0.57	0.11	0.070	0.715	−0.057	0.73	0.44	0.354	0.523	−0.043
Sep	0.60	0.17	0.112	0.690	−0.055	0.89	0.40	0.391	0.598	−0.049
Oct	0.56	0.26	0.161	0.609	−0.049	0.77	0.42	0.359	0.549	−0.045
Nov	0.42	0.49	0.229	0.440	−0.036	0.73	0.40	0.323	0.557	−0.046
Dec	0.49	0.34	0.185	0.535	−0.043	0.83	0.32	0.290	0.667	−0.054
Mean	0.56	0.29	0.176	0.596	−0.048	0.76	0.40	0.328	0.573	−0.047

values 0.154 and 0.220, which are reflecting the range for the semi-urban and cultivation sites (Ineichen et al. 1990), to determine the validity of the assumption for the value 0.2 assigned for this parameter. No considerable changes in the calculated monthly values of the parameters and the coefficients were observed with the values 0.154 and 0.220. However use of extreme values for this parameter affects the results especially the second and third coefficient. One can state that inclusion of first multiple reflection cycle effect in the models may be an indication of natural quadratic relation between the solar irradiation and bright sunshine hours or at least this may be one of the reasons of a slight curvature observed by Ogelman et al. (1984). It may be one of the causes for relatively better performances of quadratic correlations (Akinoglu and Ecevit 1990a, 1990b, 1993; Tasdemiroglu and Sever 1989; Badescu 1999) or the correlation of Rietveld who expressed the Angström coefficients as a function of n/N (Rietveld 1978).

In Table 5.2, mean values of the parameters τ_e and τ_{ec} and the coefficients a_0, a_1 and a_2 are also presented in the last rows. The monthly values of the parameters do not have large deviations for a location which may be thought of as the typical effective value, different for the locations with different climate type and latitude as can be observed from Table 5.2. Two locations, Salisbury and Bulawayo, which are having close latitudes and also similar climates seem to give rather close values for the parameters and also for the coefficients. In fact, climate type is a quite valuable starting point for the accurate solar radiation estimations. A recent study of the present author showed that climate type alone can be used to estimate the annual profile of monthly average daily global solar irradiation on horizontal surface with acceptable accuracy, without any measured input data, at least for the locations in USA (Akinoglu 2004).

Variation of the coefficients with respect to n/N is given in Fig. 5.1, for two locations one from south latitudes and the other north and with different climates to scan a wider range of n/N values. The coefficients a_0, a_1 and a_2 seem to span relatively smaller range of values compared to the wide ranges of values of the Angström coefficients. However, the formalism should be applied to different locations from all over the world to reach any further conclusions about the physical base of quadratic correlation and to talk about the universal superiority of it over linear relation.

In fact, a linear form for the relation between the monthly average solar irradiation and bright sunshine hours can be obtained if the multiple reflection effect is not accounted in the above outlined formalism (Akinoglu 1993). Then, the Angström coefficient of the linear form are obtained as: $a = \tau_e \tau_{ec}$, $b = \tau_e(1 - \tau_{ec}) + \beta_e(1 - \tau_e)$ and $a + b = \tau_e + \beta_e(1 - \tau_e)$ which seem indeed appropriate for the physical meanings attributed to these coefficients by many authors.

Jain (1990) used a similar formalism to obtain a linear relation. He has written Angström coefficients as $a = \beta_m$ and $b = (\gamma_m + \alpha_m - \beta_m)$ where subscript m stands to indicate the monthly average, β_m is similar to τ_e defined above and γ_m and α_m are the monthly average transmittance of atmosphere for the diffuse component during

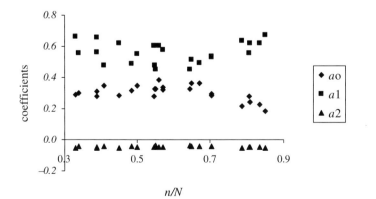

Fig. 5.1 Variations of the coefficients with n/N for Bulawayo and Macereta

cloudy and clear times, respectively. Both formalisms can be used to write various relations between the direct, diffuse and global solar irradiation and the bright sunshine hours. Such relationships were given for his formalism by Jain (1990) and only for the diffuse component by Akinoglu (1993). For example, the coefficients a' and b' in the linear relation between the monthly average diffuse component and bright sunshine hours: $D/H_0 = a' + b'n/N$, can be obtained. Both in the formalisms of Jain and in the above outlined formalism but without multiple reflection effect (Akinoglu 1993; Jain 1990), coefficient a of Angström equation equals to a'. Table 5.3 gives the coefficients a, b and b' obtained by Jain's formalism and for comparison mean values of those derived monthly from the above formalism without multiple reflection effect is also presented.

Ogelman et al. (1984) noticed that quadratic fits to the daily data of two locations in Turkey had better correlation coefficient than the linear fits. They also observed that a single quadratic curve can represent two locations having quite different climatology. They obtained the quadratic fit for the daily data of these two locations as:

$$\frac{H}{H_0} = 0.204 + 0.758\frac{n}{N} - 0.250\left(\frac{n}{N}\right)^2. \tag{5.27}$$

For the monthly average values one needs to take the average of the square of n/N, and the mean of a square of any value is related to its standard deviation σ as $< (n/N)^2 >=< n/N >^2 + \sigma^2$. Then, the quadratic form for the monthly average values can be written as:

$$\frac{H}{H_0} = 0.204 + 0.758\frac{n}{N} + 0.250\left[\left(\frac{n}{N}\right)^2 + \sigma^2\right] \tag{5.28}$$

They obtained an empirical quadratic correlation between σ^2 and n/N using the same daily data set of two locations in Turkey. Then, by inserting this empirical correlation into Eq. (5.28), they have written:

$$\frac{H}{H_0} = 0.195 + 0.676\frac{n}{N} - 0.142\left(\frac{n}{N}\right)^2. \tag{5.29}$$

They proposed that this expression can be used for the estimations of monthly average daily global solar irradiation for the locations without measured data. Later

Table 5.3 Calculated values of the Angström coefficients and a and b and also the coefficient b' by Jain and by the present formalism

Location		a	b	b'
Macereta	Jain	0.290	0.625	−0.121
	Present	0.362	0.480	−0.278
Salisbury	Jain	0.360	0.390	−0.250
	Present	0.333	0.441	−0.212
Bulawayo	Jain	0.345	0.435	−0.230
	Present	0.348	0.433	−0.231

similar procedure as followed by Ogelman et al. (1984) is used in another research for the data of six locations in Turkey and another quadratic form was obtained (Aksoy 1997).

Correlation derived by Ogelman et al. (1984) Eq. (5.29), and the one given in the following section are compared with the estimations of different models and found to be the best (Akinoglu and Ecevit 1990a, 1990b, 1993; Tasdemiroglu and Sever 1989; Badescu 1999).

An excel worksheet is given in the CD which calculates the monthly values of the parameters and the coefficients (Eqs. 5.26) of the model presented in this section, namely 'monthly-mean-daily-quad-parameters-workbook.xls'. Input is monthly mean bright sunshine hours, monthly mean global solar irradiation and monthly mean diffuse irradiation of the location of interest. This input can either be imported to the cells H10-J21 or directly written. The main input of course is the latitude of the location under consideration.

4.4 Quadratic Variation of a with b

Present author and his co-worker (Akinoglu and Ecevit 1990) make use of the variation of the Angström coefficients a with b to find a quadratic form. They only used the published values of a and b of 100 locations all over the World to derive a quadratic form and did not use any measured irradiation and sunshine hours.

Quadratic form dictates the variation of Angström coefficients with respect to n/N values. In other words for a specific value of n/N, the result that would be obtained from $a + b(n/N)$ must be equal to that obtained from the quadratic form. Therefore, slope of the quadratic form is the Angström coefficient b and the intercept of the line having this slope is the Angström coefficient a. These are:

$$b = a_1 + 2a_2(n/N)$$
$$a = a_0 - a_2(n/N)^2 \tag{5.30}$$

which has similar dependence as found by Rietveld, using the coefficients obtained from measured data, for the variation of a and b with respect to n/N (Rietveld 1978). That is, b decreases with n/N while a increases. Note that the third coefficient a_2 is negative and that is why b decreases with n/N while a increases. In fact, the coefficient a_2 always comes with a negative sign, not only in the fits to the real data as in Eq. (5.27) of Ogelman et al. but also in the physical formalism developed and presented in section 4.3 for the monthly values (see Table 5.2 last columns). Note also that monthly a and b values obtained using the formalism presented in section 4.3 but without considering the first reflection cycle (Akinoglu 1993) (which results in a linear relation in n/N) give exactly the similar variations of the monthly values of a and b with n/N as Eq. (5.30).

Another natural outcome of the quadratic relation between H/H_0 and n/N is the quadratic nature of the dependence of a to b, which was validated using the published values of a and b all over the world (Ogelman 1984; Akinoglu and Ecevit 1990). If the quadratic nature of the relation between H/H_0 and n/N has a global implication, then the linear Angström-Prescott forms are the family of straight lines which will be various chords on the quadratic curve, Eq. (5.25).

Consequently, the coefficients of the quadratic form Eq. (5.25) can be obtained in terms of the coefficients of the quadratic relation between a and b. Equating the slopes of the tangents of Eq. (5.25) to b, namely the first expression in Eq. (5.30), the specific value of n/N can be extracted as $n/N = (b - a_1)/2a_2$. At that specific value of n/N, Angström-Prescott equation written with the slope and intercept of the tangents to the quadratic curve, that is $[a + b(b - a_1)/2a_2]$ and the quadratic form itself, that is $\{a_0 + a_1(b - a_1)/2a_2 + a_2[(b - a_1)/2a_2]^2\}$ must give the same results. This equality then gives a relation between a and b of the form:

$$a = \left(a_0 - \frac{a_1^2}{4a_2}\right) + \frac{a_1}{2a_2}b - \frac{1}{4a_2}b^2 \tag{5.31}$$

as given in Akinoglu and Ecevit (1990). Hence, a quadratic fit to the curve a versus b can be used to obtain the coefficients of the quadratic relation, that is a_0, a_1 and a_2 of Eq. (5.25), using this fit and Eq. (5.31). Such a regression fit can be attained using the published Angström coefficients a and b that are obtained by regression analysis between the measured values of global solar irradiation and bright sunshine hours. Therefore, a quadratic relation between H/H_0 and n/N can be obtained without any measured H and n values directly, but by using the relation between a and b that can be found in the literature for different locations, as given in Akinoglu and Ecevit (1990). Such a relation should have global applicability as it will be derived only from the set of a and b values for the locations with different latitudes and climates all over the World. Figure 5.2 shows this relation between a and b values obtained from hundred locations all over the World presented also in Akinoglu and Ecevit (1990), which has the same type of variation as implied by the quadratic relation between H/H_0 and n/N.

Hence, a quadratic curve is fitted to the curve a versus b in Fig. 5.2 using a computer program prepared in Cern Computer Centre (James and Roos 1977). The result obtained was (Akinoglu and Ecevit 1990):

$$a = 0.783 - 1.509b + 0.892b^2. \tag{5.32}$$

The coefficients a_0, a_1 and a_2 of a quadratic expression between H/H_0 and n/N were obtained in terms of the coefficients of Eq. (5.32) using Eq. (5.31) (Akinoglu and Ecevit 1990). The expression was:

$$\frac{H}{H_0} = 0.145 + 0.845\frac{n}{N} - 0.280\left(\frac{n}{N}\right)^2 \tag{5.33}$$

and as mentioned and outlined above it was obtained only by using 100 published values of Angström coefficients a and b, without using any measured data of H and s of any location. Comparisons with 13 other correlations were carried out

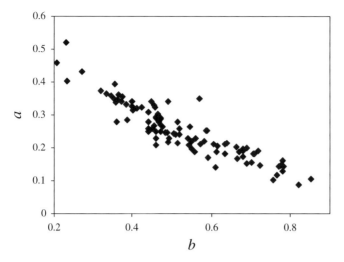

Fig. 5.2 Variation of a with respect to b for 100 locations

and quadratic correlations (Eq. (5.33) and Eq. (5.29)) were observed to have better performances (Akinoglu 1991, 1990a, 1990b, 1993; Tasdemiroglu and Sever 1989; Badescu 1999).

A self-explanatory excel worksheet which calculates the monthly-average daily values using Eq. (5.33) can be found in the CD, namely 'monthly-mean-daily-comparison-workbook.xls'. In this worksheet, calculations for statistical errors MBE and RMSE -which will be explained in the following section-, are also included for comparisons. Of course, any expression (model, correlation etc.) can be written in the cell that calculates H (column J, J10-J21), so that comparisons can be carried out. The data is the measured monthly average bright sunshine hours and monthly average global solar irradiation which must be imported or directly written to the cells H10-I21. The latitude of the location under consideration is of course the main input.

5 Model Validation and Comparison

In this section, some basic discussions about model validation, goodness of the fit and model comparisons are given. Some very fundamental expressions for the model validation and comparisons will be presented but the statistical procedures of calculation and detailed discussions can be found in the literature some of which are given here.

Widely used statistical errors are mean biased error (MBE) and root mean square error (RMSE). Percentage or fractional deviations of the estimated value with respect to the measured value can also be used. First two are defined as:

$$MBE = \left[\sum_{1}^{n} (y_{ci} - y_{mi}) \right] \Big/ n \tag{5.34}$$

and,

$$RMSE = \left\{ \left[\sum_{1}^{n} (y_{ci} - y_{mi})^2 \right] \Big/ n \right\}^{1/2} \tag{5.35}$$

where y_{ci} and y_{mi} are the calculated and measured values of the variable. First one gives the over or under-estimation of a model in the long run while the second may read a high value even if a single measurement has high deviation from that of calculated. They can also used in a fractional form as:

$$MBE = \frac{1}{n} \sum_{1}^{n} \frac{y_{ci} - y_{mi}}{y_{mi}} \tag{5.36}$$

and

$$RMSE = \sqrt{\frac{1}{n} \sum_{1}^{n} \left(\frac{y_{ci} - y_{mi}}{y_{mi}} \right)^2}. \tag{5.37}$$

All the above expressions and also those given in Yorukoglu et al. (2006) can be used in the model validation, goodness of the fit and comparison. The values of the variables y may either be directly irradiation values or the fractional forms normalized by H_0, that is H/H_0 (these values may also be hourly values). The latter should be used essentially for clarifying the goodness of the fit between fractional solar irradiation H/H_0 and fractional bright sunshine hours n/N (Yorukoglu et al. 2006). A work on comparing the two procedures of calculating MBE and $RMSE$ values, namely Eqs. (5.34–5.35) and Eqs. (5.36–5.37), showed that the maximum differences for the statistical indicators are around 3% (Badescu 1988).

Goodness of a fit is the representation quality of an empirical correlation that is obtained by regression analysis or by some statistical means (Yorukoglu et al. 2006) using the measured (or any given data) values between various variables having such relations. It mainly depends on the utilized method (for example the least square method) and the coefficients are only some mathematical constants for calculating one variable in terms of the others. Correlation coefficient R^2, for example, is the most important indicator of the goodness of the fit which is defined as:

$$R^2 = 1 - \frac{RMSE^2}{\sigma^2} \tag{5.38}$$

where σ is the standard deviation. Hence, R^2 can have values between 0 and 1, and closer to 1 means better the regression result.

Model validation should be considered as the justification of a physical or any analytical derivation of the relation between various variables which are believed to have correlations. Hence, the validness of the physical parameters introduced in the development of a physical model is important in the model validation which can either be checked with measurements (if exists any) or with the appropriate limiting values that can be assigned within the physical reasoning of the developed model. Most of the time, pre-given values are used in the models for these parameters but sometimes some of them can be obtained within the calculations of the constructed formalism. Of course, if these parameters can be calculated within the formalisms, their values must be close to those pre-assumed and/or measured values. In the model development for the meteorological variables discussed in this chapter, the physical models usually have both types of such parameters but the good point is that almost all read values within some specified ranges. For example, monthly effective value of the ground albedo might have a value from 0.13 to 0.22 for the semi-urban and cultivation sites as tabulated in Ineichen et al. (1990). In constructing the model essentially a measured data set should be used, but the same data set *can not* be used in justification of its universal applicability and/or in comparison with the estimations of different model approaches.

If some relations exist between variables then it is valuable of course to seek a physical (analytical) means of describing such relations since it highlights the physical details hidden in such correlations. The coefficients then can be written in terms of the physical parameters of the analytical model. In our case, this empirical relation is the Angström-Prescott expression and for a linear form the coefficients are *a* and *b*.

In the solar irradiation and bright sunshine hour relationships, both in the validation and comparison of the models and/or correlations, as mentioned above, measured values from different locations must be used but not those utilized in the construction of the model and correlation. In fact, as outlined in the Handbook of Methods of Estimating Solar Radiation (1984), a data set to be used for the validation and comparison must:

- be randomly selected;
- be independent of models being evaluated;
- span all seasons;
- be selected from various geographical regions;
- be sufficiently large to include a spectrum of weather.

Another point is the uncertainty in the measurements which put limitations to the level of confidence on validation and/or comparisons of the models. These errors of course reduce with increased averaging time interval. Hay and Wardle (1982) showed that the observation error of 5% for an individual observation was appropriate to an hourly time interval and reduced substantially with increased time averaging. Uncertainties that they observed for two locations in Canada had marked seasonal and inter-annual variability and also strong dependence on the observed irradiance (Hay and Wardle 1982).

6 Discussions, Conclusions and Future Prospect

Relation between the solar irradiation and bright sunshine hours is not a static occurrence but essentially a dynamic phenomenon. This is mainly due to our dynamic atmosphere because of its natural short and long term climatic cycles which was heavily affected by human-made hazards in the last 200 years. Hence, researches on solar irradiation and the atmosphere become more and more important not only to understand our environment but also to clarify its natural and man-made alteration upon intervention to its usual cycles.

Discussion of the static physical models as outlined in this chapter is still important as they aid to understand averaged behaviors of our environment, namely the atmospheric interaction of the solar rays in our case. This in turn helps to explain the dynamic activities of our atmosphere which are continuously monitored by the advanced technological satellites at least for the last several years (Mueller et al. 2006). Another important issue of course is the use of such models to estimate solar irradiation for the sites having no long term measurements so that short-term performance and long-term feasibility studies of all types of solar energy applications can be carried out.

Physical modeling of the transfer of solar radiation through the atmosphere is extremely important concept in this sense, and a part of it is the interrelations discussed in this chapter. In the search of these interrelations, physical properties of our atmosphere are mainly utilized and this will introduces new highlights to the interaction of solar rays and the atmosphere.

It might be desirable to construct accurate computations with simple equations; however, as stated by Gueymard (1993), the accuracy and simplicity are inversely proportional. In any case, the models should follow the physical mechanisms of atmospheric interactions of electromagnetic wave with matter as close as possible for a better radiation transfer modeling (Gueymard 1993).

Under the highlight of above considerations, some discussions and conclusions on the relationship between solar irradiation and bright sunshine hours are given in the followings. Discussions seem to start with Kimball's and Angström's pioneering works (Kimball 1919; Angström 1924) and Angström pointed that Angström-Prescott equation between H/H_0 and n/N might depend also on the averaging time intervals. Hence, daily and monthly average daily values should correlate in a different manner as the averaging sweep out some of the information contained in the daily values. After reviewing the literature, a conclusion is reached by Gueymard et al. (1995) that the curvature of the quadratic form observed for daily values does not significantly remain for the monthly averages. In fact, some smoothing can be demonstrated using the second derivatives of the daily quadratic regression correlation, Eq. (5.27) and monthly expression, Eq. (5.29). That is the third coefficient in the monthly form, a_2 have smaller value of 0.142 in magnitude than that of the daily expression, which is 0.25. Gueymard et al. (1995) also noted that the opposite trends in the variations of fractional diffuse and beam components with n/N cancel out to give almost perfectly linear variation between H/H_0 and n/N.

The averaging procedure may indeed partly smooth out the quadratic nature in the daily values, however quadratic forms obtained for monthly average daily values have relatively better estimates of solar radiation as determined by Akinoglu and Ecevit (1990a, 1990b. 1993), Tasdemiroglu and Sever (1989) and Badescu (1999). In any case, non-linear nature of the relation between H/H_0 and n/N for the monthly averages may still need further justifications using accurate and longer data set.

Essentially, it seems that there may be at least three reasons of the non-linearity. First one is the non-linearity in the daily values, basically quadratic form, results in a quadratic form for the monthly averages due to the quadratic dependence of the standard deviation on n/N. Second reason may be the bi-modal characteristics observed by Schuerke (2000) and some other researchers cited by Schuerke, give rise to a non-linear form. Finally, back-scatter effects may lead a non-linear term to the relation between H/H_0 and n/N as discussed in section 4.3. Another fact is the variation of Angström coefficients with n/N imposed by quadratic correlation, namely Eq. (5.30), have similar trends as determined by Rietveld (1978). In addition, variation of a with b seems quadratic in nature which leads a quadratic relation between H/H_0 and n/N as outlined in section 5.4.4. These conclusions however should be verified using accurate and longer data sets and starting from instantaneous considerations researches must be carried out for hourly, daily and monthly average daily values.

In fact, Eq. (5.33) seems to be under-estimating the monthly mean irradiation, at least for two Tibet sites with altitudes 2809 m and 3659 m (Yang 2007). In high altitude regions, due to the fact that the atmospheric parameters are less effective, the value of $a_0 + a_1 + a_2$ for $n/N = 1$ of the quadratic form should be higher than those for the lower altitude regions (Yang 2007). Better estimations of the model of Yang et al. especially for the higher altitudes as presented in Table 6 of Yang et al. (2006) might be due to the fact that their model is taking into account some atmospheric parameters of the site under consideration.

The importance of hourly, daily and monthly average daily considerations are already discussed, yet a recent attempt on correlating the yearly average daily values of H/H_0 and n/N for 38 years data of 51 locations resulted in a linear relation with a regression coefficient of 0.834 (Chen 2006).

Researches in the field of solar radiation should continue, particularly of works that present new and significantly improved ideas or concepts, and which enhance progress toward applications of solar energy (Kasten and Duffie 1993). Gueymard underlined this conclusion supporting the idea that the mere use of Angström's equation to predict global irradiation from local sunshine data would not give significant progress (Gueymard et al. 1995). Following years, works on the simple use of Angström relation appeared rather frequently in the literature but researches on new ideas and concepts were not quite often.

Regional works on the subject may continue to reach local correlations between the solar irradiation and bright sunshine hours using the surface data, however it seems that new prospects of future research is strictly needed. Surface data should unquestionably continue to be collected but it is a must to use new instruments which also need frequent calibrations. Physical models of irradiative transmission through the atmosphere should be incorporated not only with the surface data but also with the data taken remotely by the satellites.

Another important future prospect is further achievements that can be obtained by constructing linkages between the surface data and those measured from satellites which will lead to new revenues in physical modeling of our atmosphere.

An important conclusion was reached by Gueymard (2003a, 2003b) which seems still valid: "No further improvements in current high performance models will therefore be necessary until more accurate fundamental data becomes available." Another fact is that new validations and comparisons should use larger number of data sets but with higher accuracy and reliability which must be checked with the local organizations and with the existing available data. The future prospects given by Gueymard et al. (1995) seem still important research avenues to be carried out.

Models and the correlations presented in this chapter and of course others that could not be covered here necessitate further validations using accurate and longer surface data sets. Predictions and formalism of these models should be compared and linked by the models and observations obtained from the new generation satellites (Mueller 2004) to attain detailed information about solar irradiation on a spatially denser or even continuously on the surface of the Earth.

Acknowledgements Author would like to thank to Turkish State Meteorological Service for supplying data and to Mr. B. Aksoy for his valuable discussions. Valuable discussions with Dr. K. Yang and also his permission for the inclusion of the computer code of their model in the CD-ROM within the content of this chapter are also kindly acknowledged.

References

Abdalla YAG, Baghdady MK (1985) Global and diffuse solar radiation and in Doha. Solar Wind. Technol. 3: 267.

Akinoglu BG (1992a) On the random measurements of Robitzsch pyranograps. Proc. Of the 2nd World Renewable Energy Congress. Reading, UK, Pergamon Press, Oxford, p. 2726–2730.

Akinoglu BG (1992b) Quadratic variation of global solar radiation with bright sunshine hours. Proc. Of the 2nd World Renewable Energy Congress. Reading, UK, Pergamon Press, Oxford, p. 2774–2778.

Akinoglu BG (1991) A review of sunshine-based models used to estimate monthly average global solar radiation, Renewable Energy 1: 479–497.

Akinoglu BG (1993) A physical formalism for the modified Angström Equation to estimate solar radiation. Doga-Tr. J. of Physics 17: 345–355.

Akinoglu BG (1990b) A further comparison and discussion of models to estimate to estimate global solar radiation. Energy 15: 865–872.

Akinoglu BG (2004) Effect of climate type and latitude on monthly average daily global solar radiation for five different climates in USA. Proc. Of ISES Asia-Pacific, October, Korean Solar Energy Society. GwangJu, South Korea, p689–695.

Akinoglu BG and Ecevit A (1989) Comparison and discussion of quadratic models to estimate global solar radiation. Proc. Of 9th Int. Conf. on Energy and Env., Miami-USA.

Akinoglu BG and Ecevit A (1990a) Construction of a quadratic model using modified Angström coefficients to estimate global solar radiation, Solar Energy 45: 85–92.

Akinoglu BG and Ecevit A (1993) Comparison and Discussion of the Eight Sunshine-based Correlations of Global Radiation. Doga-Turkish Journal of Physics 17: 79–95.

Akinoglu BG, Oguz C, Oktik S (2000), Relations between monthly average hourly solar radiation, hourly bright sunshine and air mass for Mugla-Turkey. Proc. World Renewable Energy Congress VI (WREC 2000), July, Brighton, UK, p2469–2472.

Aksoy B (1997), Estimated monthly average global solar radiation for Turkey and its comparison with observations. Renewable Energy 10: 625–633.

Aksoy B (1999) Analysis of Changes in Sunshine Duration Data for Ankara, Turkey. Theor. Appl. Climatol. 64: 229–237.

Aksoy B (1997) Variations and trends in global solar radiation for Turkey, Theor. Appl. Climatol. 58: 71–77.

Angström A (1924) Solar and terrestrial radiation. Quart. J. Roy. Met. Soc. 50: 121–126.

Angström A (1956) On the computation of global radiation from records of sunshine, Arkiv För Geofysik, Band 2 nr 22, 471–479.

Atwater MA and Ball JT (1978) A numerical solar radiation model based on standard meteorological observations. Solar Energy 21: 163–170.

Badescu V (1999) Correlations to estimate monthly mean daily solar global irradiation: application to Romania. Energy 24: 883–893.

Badescu V (1988) Comment on the statistical indicators used to evaluate the accuracy of solar radiation computing models. Solar Energy 40: 479–480.

Bird RE (1984) A simple solar spectral model for direct-normal diffuse horizontal irradiance. Solar Energy 32: 461–471.

Brunt D (1934) Physical and Dynamical Meteorology. Table 2, p. 100.

Chen LX, Li LW, Zhu WQ, Zhou X J, Zhou ZJ and Liu HL (2006) Seasonal trends of climate change in the Yangtze Delta and its adjacent regions and their formation mechanisms. Meteorol. Atmos. Phys.: 92, 11–23.

Chen R, Kang H, Ji X, Yang J, Zhang Z (2006) Trends of the global solar radiation and sunshine hours in 1961–1998 and their relationships in China. Energy. Conv. Mgmt. 47: 2859–2866.

Coulson KL (1975) Solar and Terrestrial Radiation. Academic Press, New York.

Davies AJ and McKay DC (1982) Estimating solar irradiance and componenets. Solar Energy 29: 55–64.

Davies JA, Abdel-Wahab M and McKay DC. (1984) Estimating Solar Irradiation on Horizontal Surface, Int. J. Solar Energy 2: 405–424.

Duffie JA and Beckman WA (1991) Solar Engineering of thermal processes. John Wiley and Sons, New York.

Fritz S (1949) The albedo of the planet earth and clouds. J. Meteor. 6: 277–282.

Gopinhathan KK (1988) A general formula for computing the coefficients of the correlation connecting global solar radiation to sunshine duration. Solar Energy 41: 499–503.

Gueymard C (1993a) Analysis of monthly average solar radiation and bright sunshine for different thresholds at Cape Canevral, Florida, Solar Energy 51: 139–145.

Gueymard AC (2003a) Direct solar transmittance and irradiance with broadband models. Part I: detailed theoretical performance assessment. Solar Energy 74: 355–379.

Gueymard AC (2003b) Direct solar transmittance and irradiance predictions with broadband models. Part II: validation with high quality measurements. Solar Energy 74: 381–395.

Gueymard C, Jindra P and Estrada-Cajigal V (1995), A critical look at recent interpretations of the Angström approach and its future in global solar radiation prediction, Solar Energy 54:357–363.

Gueymard C (1993b) Mathematically integrable parameterization of clear-sky beam and global irradiances and its use in daily irradiation applications. Solar Energy 50: 385–397.

Gueymard C (1989) A two-band model for the calculation of clear sky solar irradiance, illuminance, and photosynthetically active radiation at the Earth's surface. Solar Energy 43: 253–265.

Handbook of Methods of Estimating Solar Radiation (1984), Swedish Council for Building research. International Energy Agency, Solar R. and D., Task V, Subtask B, Stockholm, Sweden.

Hay JE and Wardle DI (1982) An assessment of the uncertainty im measurements of solar radiation. Solar Energy 29, 271–278.

Houghton HG (1954) The annual heat balance of the northern hemisphere, J. Meteor. 11: 1–9.

Ineichen P, Guisan O and Perez R (1990) Ground-reflected radiation and albedo. Solar Energy 44: 207–214.

Iqbal M (1979) Correlation of average diffuse and beam radiation with hours of bright sunshine. Solar Energy 23: 169–173.

Jain PC (1990) A model for diffuse and global irradiation on horizontal surface. Solar Energy 45: 301–308.

James F and Roos M (1977) MINUIT: A system for function minimization and analysis of the parameter errors and correlation, Cern Computer Center Program Library, Long Write-Up, D-506.

Kasten F and Duffie JA (1993) Editorial, Solar Energy 50, 383.

Kimball HH (1919) Variations in the total and luminous solar radiation with geographical position in the United States. Monthly Weather Review 47: 769–793.

Lacis AA. and Hansen JA. (1974) A parametrization for the absorption of solar radiation in the Earth's atmosphere. J. Atmos. Sci. 31: 118–133.

Leckner B (1978) The spectral distribution of solar radiation at the Erath surface-Elements of a model. Solar Energy 20: 143–150.

Madkour MA., El-Metwally M, Hamed AB (2006) Comparative study on different models for estimation of direct normal incidence (DNI) over Egypt Atmosphere. Renew. Energy 31: 361–382.

Martinez-Lozano JA, Tena F, Onrubia JE and De La Rubia J (1984) The historical evolution of the Angström formula and its modifications: Review and bibliography, Agric. Forest. Meteorol., 33: 109–128.

Monteith JL (1962) Attenuation of solar radiation. Quart. J. Roy. Meteorol. Soc. 88: 508–521.

Mueller RW et al. (2004) Rethinking satellite-based solar irradiance modeling-The SOLIS clear sky model. Remote sensing of Environment 91: 160–174.

Ogelman H, Ecevit A and Tasdemiroglu E (1984) A new method for estimating solar radiation from bright sunshine data. Solar Energy 33: 619–625.

Page J (2005) First conference on measurement and modeling of solar radiation and daylight "Challenges for the 21st Century. Energy 30: 1501–1515.

Painter HE (1981) The performance of a Campbell-Stokes sunshine recorder compared with a simultaneous record of the normal incidence irradiation. Meteor. Mag. 110: 102–109.

Paulescu M and Schlett Z (2003) A simplified but accurate spectral solar irradiance model. Theor. Appl. Climatol. 75: 203–212.

Prescott JA (1940) Evaporation from a water surface in relation to solar radiation. Trans. Roy. Soc. S. A. 64: 114–118.

Rietveld HR (1978) A new method to estimate the regression coefficients in the formula relating radiation to sunshine. Agric. Meteorol. 19: 479–480.

Suehrcke H (2000) On the relationship between duration of sunshine and solar radiation on the Earth's surface: Angström's equation revisited. Solar Energy 68: 417–425.

Tasdemiroglu E and Sever R (1989) Estimation of total solar radiation from bright sunshine hours in Turkey. Energy 14: 827–830.

Trewartha GT (1968) An introduction to climate, McGraw Hill, New York.

Yang K, Huang GW and Tamai N (2001) A hybrid model for estimating global solar radiation, Solar Energy 70: 13–22.

Yang K, Koike T, Ye B (2006) Improving estimation of hourly, daily, and monthly solar radiation by importing global data sets. Agric. Forest. Meteorol. 137: 43–55.

Yang, K. (2007) Private communication, April 2007.

Yorukoglu M, Celik AN (2006), A critical review on the estimation of daily global solar radiation from sunshine duration. Energy Conv. Mgmt. 47: 2441–2450.

Chapter 6
Solar Irradiation Estimation Methods from Sunshine and Cloud Cover Data

Ahmet Duran Şahin and Zekai Şen

1 Introduction

Solar irradiation is dependent on different causes including astronomical and meteorological factors. In practical studies it is not possible to consider all the factors, and therefore, so far simple but effective models for its prediction from a few numbers of factors are presented. The first of such models takes into consideration only the sunshine duration measurements for the solar irradiation estimation, and unfortunately, it is still under use without critical assessment of the underlying restrictive assumptions and simplifications in model parameter estimation methodology. In addition to these criticisms the classical models are in the form of linear mathematical expressions and the parameter estimation procedure is also a linear procedure, which leads to constant estimations based on all the data. The sunshine duration and solar irradiation data have seasonal and random effects especially for durations less than one year. The basic concern in this chapter is weather seasonal effects are also reflective in the parameter estimations? Classical approaches consider only the random errors and by using the least squares methodology tries to provide parameter estimations on the basis of the minimum squared error. Although some researchers proposed addition of a non-linear term into the basic Angström equation but again without consideration of the restrictive assumptions and finally they also obtained constant model parameter estimations. Unfortunately, all over the world the same estimation procedure is under current use, and therefore, in this chapter the attention is drawn on the pitfalls in the model parameter estimations and accordingly some innovative approaches are suggested without restrictive assumptions although the same simple solar estimation model is used. A dynamic model estimation procedure is proposed, which leads to a sequence of parameters and hence it is possible to

Ahmet Duran Şahin
Istanbul Technical University, Turkey, e-mail: sahind@itu.edu.tr

Zekai Şen
Istanbul Technical University, Turkey, e-mail: zsen@itu.edu.tr

look at the frequency distribution function (probability distribution function, PDF) of the model parameters and decide whether the arithmetic average of the parameters or the mode (the most frequently occurring parameter value) should be used in further solar irradiation estimations. It is shown on the basis of some solar irradiation and sunshine duration data measurements on different locations in Turkey that the model parameter estimations abide by the Beta PDF. Besides, it is also possible to find the relationship between the model parameters at a single station, which shows temporal parameter variations. Apart from the dynamic model parameter estimation procedure, an unrestricted solar irradiation parameter estimation procedure is also presented which considers the conservation of the model input and output variables' arithmetic mean and the standard deviations only, without the use of least squares technique. The quantitative comparisons of all the methodologies proposed in this chapter are presented with the classical model results on the basis of different error assessments.

2 Basic Equations for Sunshine Duration and Extraterrestrial Solar Irradiation

Extraterrestrial solar irradiation (H_0) and length of the day (S_0) can be estimated deterministically by taking into consideration basic geographic and astronomic quantities including latitude, (ϕ), declination, (δ), surface azimuth angle, (γ), hour angle, (ω), zenith angle, (θ_z), solar altitude angle, (α_s), solar azimuth angle, (γ_s) and solar constant (I_{SC}). Interrelationships among these variables are presented either in the form of equations or tables in many solar energy books (Iqbal 1983; Duffie and Beckman 2006). Figure 6.1 presents the astronomical configuration of these quantities.

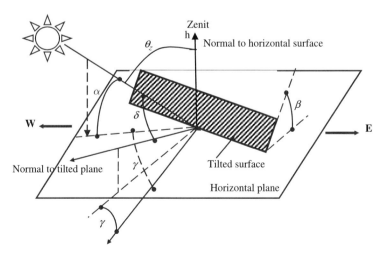

Fig. 6.1 Basic solar angles

In the following, the physical determinations and valid equations for such a configuration are given.

Latitude (ϕ): The angular location north (positive) or south (negative) of the equator, ($-90° \leq \phi \leq 90°$).

Declination (δ): The angular position of the sun at solar noon (i.e., when the sun is on the local meridian) with respect to the equator plane, north direction has positive value and its variation range is $-23.45° \leq \delta \leq 23.45°$. For its calculation first θ_o is expressed as,

$$\theta_o = \frac{2\pi d_n}{365} \tag{6.1}$$

where θ_o is the sun's position angle, which depends on day of the year, d_n that is zero in January first. In that case declination angle could be estimated according to

$$\delta = 0.006918 - 0.399912\cos\theta_o + \sin\theta_o - 0.006759\cos 2\theta_o$$
$$+ 0.000907\sin 2\theta_o - 0.002697\cos\theta_o + 0.00148\sin 3\theta_o \tag{6.2}$$

Surface azimuth angle (γ): The deviation of the projection on a horizontal plane of the normal to the surface from the local meridian, with zero due south, east negative and west positive, ($-180° \leq \gamma \leq 180°$).

Hour angle (ω): The angular displacement of the sun east or west of the local meridian due to rotation of the earth on its axis at $15°$ per hour as morning negative and afternoon positive.

Zenith angle (θ_z): The angle between the vertical and the line to the sun i.e., the angle of incidence of beam radiation on a horizontal surface. At solar noon zenith angle is zero, in the sunrise and sunset this angle is $90°$.

$$Cos(\theta_z) = \cos(\phi)\cos(\delta)\cos(\omega) + \sin(\phi)\sin(\delta) \tag{6.3}$$

Solar altitude angle (α_s): The angle between the horizontal and the line to the sun.

Solar azimuth angle (γ_s): The angular displacement from south of the projection of beam radiation on the horizontal plane. Displacements east of south are negative and west of south are positive. Solar azimuth angle can be estimated as,

$$\sin\gamma_s = \frac{\cos\delta.\sin\omega}{\sin\theta_z} \tag{6.4}$$

Length of the day (S_0): This term is described as time duration between sunrise and sunset, and it is one of the extraterrestrial variables of classical Angström (1924) equation and could be estimated as,

$$\cos S_0 = -\tan\phi\tan\delta \tag{6.5}$$

Length of the day (S_0) could be estimated in hours by dividing degree value of S_0 to 15 that represents timely one as degree.

$$S_0 = \frac{Arc(\cos S_0)}{15} \tag{6.6}$$

Solar constant (I_{SC}): It is equivalent to the energy from the sun, per unit time, received on a unit area of surface perpendicular to the direction of propagation of the radiation, at mean earth-sun distance, outside of the atmosphere. Solar constant has been adopted as $1.94\,\text{cal/cm}^2/\text{sec}$ or $1367\,\text{W/m}^2$ (Duffie and Beckman 2006). Recently, this value is considered with its physical meaning and estimated as $1366.1\,\text{W/m}^2$ that covers solar spectrum between 0 and $1000\,\mu\text{m}$, (Gueymard 2004).

Extraterrestrial solar irradiation (H_0): Extraterrestrial solar irradiation is important not only for solar engineering calculation but also for energy balance of the earth. It is a function of solar constant, sun-earth distance ratio and declination angle. Sun-earth distance ratio changes with time of the year and it is expressed as,

$$\frac{\overline{R}}{R} = \frac{1}{1 - 0.033\cos\dfrac{2\pi.n}{Y}} \tag{6.7}$$

where $\overline{R}$ is the mean value of sun-earth distance which is equal to 1.49×10^{11} m; R is the actual sun-earth distance; n is the day number in the year and Y is the total day of year. All these angles and aforementioned equations take part in the determination of H_0 and S_0. Consequently, H_0 received from the sun on a unit area of surface perpendicular to the radiation direction of propagation, at any earth-sun distance, outside of the atmosphere, can be estimated as,

$$H_0 = \left(\frac{\overline{R}}{R}\right)^2 I_{sc}\cos\theta_z \tag{6.8}$$

as the total extraterrestrial solar irradiation from sunrise to sunset becomes

$$H_0 = \int_{S_{Sunrise}}^{S_{Sunset}} \left(\frac{\overline{R}}{R}\right)^2 I_{sc}\cos\theta_z dt \tag{6.9}$$

It is obvious that for solar irradiation estimation extraterrestrial solar irradiation and sunshine duration have the same importance, i.e., they are directly and functionally related to each other. Equation (6.9) indicates total extraterrestrial solar irradiation intensity change depending on the length of day and sun-earth astronomical position.

3 Measured Global Solar Irradiation and Sunshine Duration Properties

Global solar irradiation and sunshine duration on the horizontal surface are measured with radiometers, pyranometers and sunshine recorders. Depending on atmospheric conditions, differences occur between total extraterrestrial and measured global solar irradiation for the same time and unit area. The meteorological events affect directly on the solar energy calculations in a random manner due to weather conditions which give rise to temporal and spatial variations. For these reasons, randomness occurs in the solar irradiation and sunshine duration evolutions. The meteorological variability reduces the astronomical daily H_0 and S_0 values in two ways.

1. The astronomical H_0 and S_0 are shortened due to meteorological events, which are measured at a solar station as global solar irradiation, H, and sunshine duration, S. In other words, $S < S_0$ and $H < H_0$,
2. The shortening effect is not definite but might be in the form of different random amounts during a day or month depending on the climate and atmospheric conditions. These are solar irradiation scattering by air molecules, water, dust and aerosols as well as absorption by O_3, H_2O, CO_2 and other greenhouse gasses.

Consequently, ratios of measured solar energy variables at surface to their astronomical counterparts H/H_0 and S/S_0 assume values between zero and one in a random manner depending on the cloud cover and atmospheric turbidity of the period concerned.

4 Angström Equation

Solar irradiation and sunshine duration records depend on the combined effects of astronomical and meteorological events. The first relationships occurred in the form of a linear expression as suggested by Angström (1924). His formula has been used in practical applications for many years to estimate the daily, monthly and annual global solar irradiation, H, from the comparatively simple measurements of sunshine duration, S, according to the following expression

$$\frac{H}{H_0} = a + b\frac{S}{S_0} \qquad (6.10)$$

where H and S are the daily global irradiation received on a horizontal surface at ground level and sunshine duration, respectively, and a and b are model parameters. As explained above, although H and S vary temporally in a random manner, H_0 and S_0 have fixed values that are given by deterministic expressions and the question is whether the model parameters a and b also vary temporally and randomly at a given station. In most applications so far in the literature, a and b are considered as constants for the time period used in the application of Eq. (6.10). For instance, if daily values are used then a straight-line is matched through the scatter of solar irradiation versus sunshine duration plots which minimizes the sum of square deviation from this line. On the other hand, estimation of Angström coefficients by the application of regression technique yields constant values as

$$b = \frac{\sum\limits_{i=1}^{n}\left[\left(\frac{H}{H_0}\right)_i - \overline{\left(\frac{S}{S_0}\right)_i}\right]\left[\left(\frac{H}{H_0}\right)_{i-1} - \overline{\left(\frac{S}{S_0}\right)_{i-1}}\right]}{\sqrt{\sum\limits_{i=1}^{n}\left[\left(\frac{H}{H_0}\right)_i - \overline{\left(\frac{S}{S_0}\right)_{i-1}}\right]^2 \left[\left(\frac{H}{H_0}\right)_{i-1} - \overline{\left(\frac{S}{S_0}\right)_{i-1}}\right]^2}} \qquad (6.11)$$

and

$$a = \overline{\left(\frac{H}{H_0}\right)} - b\overline{\left(\frac{S}{S_0}\right)} \qquad (6.12)$$

Angström linear model relates global irradiation to sunshine duration by ignoring other meteorological factors such as the rainfall, relative humidity, maximum temperature, air quality, elevation above mean sea level, etc. The effects of other meteorological variables appear as deviations from the straight-line fit to the scatter diagram. In order to cover these errors to a certain extent, it is necessary to assume that the model coefficients are not constants, but random variables that change with meteorological conditions (Şahin and Şen 1998). On the other hand, many researchers have considered additional meteorological factors in order to increase the accuracy of estimations (Prescott 1940; Swartman and Ogunlade 1967; Rietveld 1978; Soler 1990; Şen et al. 2001). However, a common point to all these studies is that parameter estimates are obtained by the least squares method with minimum but remaining error. Many researchers (Sabbagh et al. 1977; Dogniaux, Lemonie 1983; Gopinathan 1988; Jain 1990; Akinoglu and Ecevit 1990; Lewis 1989; Samuel 1991; Wahab 1993; Hinrichsen 1994) have considered additional parameters increasing the estimation accuracy. For instance, Ögelman et al. (1984) incorporated the sunshine duration standard deviation for better model parameter estimations.

Soler (1986, 1990) has shown that monthly variations of $(a + b)$ are meteorologically sound and similar for different locations. It has been shown by Hinrichsen (1994) that, physically $a > 0$. Furthermore, Gueymard et al. (1995) showed that a corresponds to the relative diffuse radiation on an overcast day, whereas $(a + b)$ corresponds to the relative cloudless-sky global irradiation. Depending on weather and atmospheric conditions H and S vary temporally and spatially in a random manner but H_0 and S_0 have fixed values. The question is whether the model parameters, a and b also vary temporally at a given station. In most applications, a and b are considered as constants for the time periods used in practice. However, it is shown by Şahin and Şen (1998) that a and b also change temporally and spatially.

A detailed historical evolution of Angström equation is explained by Martinez-Lozano et al. (1984) with further criticisms are presented by Gueymard et al. (1995) and accordingly some authors suggested alternative methods (Suehrcke 2000; Şen 2001; Şen and Şahin 2001; Şahin et al. 2001).

4.1 Physical Meaning of Angström Equation

This equation is used generally in the radiative transfer and solar engineering studies. Details of radiative transfer are given by Akinoglu in Chapter 5. Logically, if sunshine duration is high then received solar irradiation on the surface will be high. There are four related variables and two parameters in Eq. (6.10). The variables are extraterrestrial, terrestrial global solar irradiation, length of the day and sunshine duration properties, which are discussed to a certain extent above. Relationship between the ratio variables (H/H_0 and S/S_0) is achieved through parameters (a and b).

In Eq. (6.10) the ratio on the left hand side is H/H_0 called as clearness index and it gives additional information about the astronomic position of the earth, conditions of the atmosphere and the characteristics of surroundings stations. It also depends on seasonal variations of the sun-earth distance. Similar seasonal sun characteristics, the time variation of this ratio has approximate periodicity. The ratio on the right hand side S/S_0 gives information about atmospheric characteristics and conditions of the study area. It is referred to as cloudless index and it is as important as clearness index. This ratio is directly proportional to hydrometeors, especially water vapour content of atmosphere.

The first parameter a represents diffuse component of the global solar irradiation and generally approximates to zero at clear atmospheric conditions. When the sunshine duration ratio, S/S_0, is zero at the overcast conditions, then the solar irradiation ratio is equal to this parameter. The second parameter b represents variation and relation of ratios in the Angström equation, which has four different conditions as shown in Fig. 6.2. In the case of no atmosphere, total extraterrestrial solar irradiation reaches the horizontal surface without any reduction and hence the relationship between the two ratios occurs along the 45° straight line. This is statistically possible but physically impossible situation on the earth.

The second regression line, corresponding to "observed$_1$" situation could represent some observable conditions in the earth. In this case a value represents "not high" total diffuse irradiation and hence there are cloudy and cloudiness separations

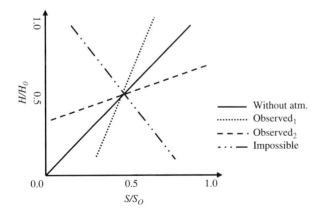

Fig. 6.2 Relationship between ratios

due to conditions in the area. The third curve, "observed$_2$", represents a very cloudy location with high amount of diffuse irradiation. This could be a good condition for electrical solar applications but not for thermal perspectives. In reality, the last curve cannot be represented physically by Angström equation because of inverse proportional linearity.

4.2 Assumptions of the Classical Equation

There is a set of assumptions that are necessary for the validation of Angström equation. These assumptions can be given as follows,

1. The model parameters are assumed invariant with time on the average as if the same sunshine duration appears on the same days or months of the year in a particular location,
2. Whatever the scatter diagram of H versus S, the regression line is automatically fitted leading to constant a and b estimates for the given data. In fact, these coefficients depend on the variations in the sunshine duration during any particular time interval. Since sunshine duration records have inherently random variability so are the model parameters, but in practice they are assumed as constants,
3. Angström approach provides estimations of the global solar irradiation on horizontal surfaces, but unfortunately it does not give clues about global solar irradiation on a tilted surface because diffuse and direct irradiations do not appear in the Angström model separately.
4. In this approach many meteorological factors are ignored such as the relative humidity, maximum temperature, air quality, and elevation above mean sea level. Each one of these factors contributes to the relationship between H and S and their ignorance causes some errors in the prediction and even in the model identification. For instance, classical Angström equation assumes that the global solar irradiation on horizontal surfaces is proportional to the sunshine duration only. The effects of other meteorological variables always appear as deviations from the straight line fit on any scatter diagram all,
5. The physical meanings of the model coefficients are not considered in most of the applications studies but only the statistical linear regression line is fitted and parameters estimations are obtained directly. The regression method does not provide dynamic estimation of the coefficients from available data, and
6. Statistically linear equations have six restrictions such as the normality, linearity, conditional distribution means, homoscedascity (variance constancy), autocorrelation and lack of measurement error.

4.3 Angström Equation and Its Statistical Meaning

In classical approaches, this equation does not lead to analyze statistical properties of a and b parameters. As a result of this restriction temporal variation of these parameters could not be obtained because of least square technique assumptions.

However, statistical properties of these assumptions are managed by considering two successive months as one linear equation, (Şahin and Şen 1998). It is shown that the average coefficient values are not enough to represent the whole variability in the meteorological factors, and therefore, their variance and still better distribution functions should be taken into consideration in future global solar irradiation estimations Şen (2001). A simple successive substitution method is proposed by Şahin and Şen (1998).

4.4 Non-linear Angström Equation Models

Most of the sunshine based non-linear solar irradiation estimation models are defined as the modifications of the Angström expression in Eq. (6.10). Some authors have suggested changes in the model parameters, a and b, seasonally, for arriving at better estimations. On the other hand, Ögelman et al. (1984) added a non-linear term, which appears as a quadratic expression,

$$\frac{\overline{H}}{H_0} = a + b\frac{\overline{S}}{S_0} + c\left(\frac{\overline{S}}{S_0}\right)^2 \tag{6.13}$$

Akinoglu and Ecevit (1990) have observed that this model is superior to others in terms of global applicability. They have applied it to some Turkish data, and finally, obtained a suitable model as,

$$\frac{\overline{H}}{H_0} = 0.195 + 0.676\frac{\overline{S}}{S_0} - 0.142\left(\frac{\overline{S}}{S_0}\right)^2 \tag{6.14}$$

After these preliminary non-linear methods, higher order polynomial type of non-linear models are also proposed into the solar energy literature, and especially, Zabara (1986) correlated the Angström parameters to third power of the sunshine duration ratio.

Akinoglu and Ecevit (1990) found a global relationship between the Angström parameters by using the published a and b values for 100 locations from all over the world and the relationship is expressed in quadratic form as

$$a = 0.783 - 1.509b + 0.89b^2 \tag{6.15}$$

Details of these proposals and models are given in Chapter 5 of this book.

5 Enhancing Statistical Meaning of Angström Equation with Two Methodologies

In this chapter, especially two methodologies that enhance statistical meaning of Angström equation parameters and variables are explained in detail. These are successive substitution methodology (SSM) and unrestricted method (UM).

5.1 Successive Substitution Methodology (SSM)

A simple substitution method was proposed by Şahin and Şen (1998) for dynamic estimation of Angström's coefficients which play significant role in relation of the global solar irradiation to the sun shine duration through a linear model. Their mathematical estimation procedures are presented on the basis of successive global radiation and sun shine duration record substitutions into the model. This procedure yields a series of parameter estimations and their arithmetic averages are closely related to the classical regression method estimates. The series of model parameter estimations provide an ability to assess these parameters statistically. Consequently, such a dynamic parameter estimation procedure evaluates and enables one to make interpretations with their normal and extreme values. Additionally, necessary relative frequency distribution functions of these parameters appear in the form of Beta probability distribution function. Routinely recorded daily and monthly global irradiation and sunshine duration values are used by the regression technique for determining the coefficients in Angström equation. The use of such a deterministic model provides linear unique predictions of global solar irradiation given the sunshine duration. In order to consider effects of unexplained part, it is necessary to estimate coefficients from the successive data pairs "locally" rather than "globally" as in the classical regression approach (Fig. 6.3).

Parameter b represents variation and relation of ratios $d(H/H_0)$ and $d(S/S_0)$ which corresponds to the slope of the linear relationship defined as,

$$b = \frac{d(H/H_0)}{d(S/S_0)} \tag{6.16}$$

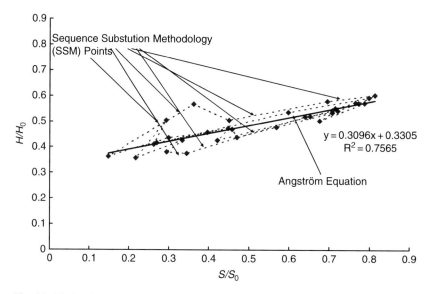

Fig. 6.3 SSM and Angström representation

This first-order differential equation can be written in terms of backward finite difference method as,

$$b'_i = \frac{\left(\frac{H}{H_0}\right)_i - \left(\frac{H}{H_0}\right)_{i-1}}{\left(\frac{S}{S_0}\right)_i - \left(\frac{S}{S_0}\right)_{i-1}} \qquad (i = 2,3,4,\ldots,n) \qquad (6.17)$$

Herein, n is the number of records and b'_i is the rate of global solar irradiation change with the sunshine duration between time instances $i-1$ and i. For daily data, these are successive daily rates or in the case of monthly records, they are monthly rates. Arrangement from Eq (6.10) by considering Eq. (6.17) leads to the successive time estimates of a'_i as,

$$a'_i = \left(\frac{H}{H_0}\right)_i - b'_i\left(\frac{S}{S_0}\right)_i \qquad (i = 2,3,4,\ldots,n) \qquad (6.18)$$

The application of these last equations to actual relevant data yields $(n-1)$ coefficient estimations. Each pair of the coefficient estimate (a'_i, b'_i) explains the whole information for successive pairs of global solar irradiation and corresponding sunshine duration records. Comparison of Eq. (6.17) with Eqs. (6.11), (6.18) and (6.12) indicates that regression technique estimation does not allow any randomness in the coefficient calculations. However, the proposed finite differences method coefficient estimations assume the regression technique estimations and it provides flexibility in the parameter calculations.

Furthermore, it is possible to obtain the relative frequency distribution of a'_i's and b'_i's. In addition to any statistical parameter such as variance or standard deviation. Confidence limits can also be stated at a certain significance level as 5% or 10%. Extreme values of a'_i and b'_i also become observable by finite difference method solution.

Taking the average values of both sides in Eq. (6.18) leads to finite difference averages of coefficients as,

$$\overline{a'} = \overline{\left(\frac{H}{H_0}\right)} - \overline{b'\left(\frac{S}{S_0}\right)} \qquad (6.19)$$

The difference of this expression from Eq. (6.12) results in,

$$\overline{a'_i} - a = b\overline{\left(\frac{S}{S_0}\right)} - \overline{b'_i\left(\frac{S}{S_0}\right)} \qquad (6.20)$$

since $\overline{ab} \leq \overline{a}\overline{b}$, this expression can be written in the form of an inequality as,

$$\overline{a'_i} - a \geq (b - \overline{b'})\overline{\left(\frac{S}{S_0}\right)} \qquad (6.21)$$

As a result, it is shown that by SSM, a dynamic behavior can be given to Angström equation.

5.2 Data and Study Area

Turkey is located between latitudes $36°$ N and $42°$ N and longitudes $26°$ E and $45°$ E (Fig. 6.4) It has significant solar energy potential especially in the southern parts including the Mediterranean region. Twenty eight global solar irradiation and sunshine duration measurement stations scattered all over the country are considered for regionalization studies in this chapter (see Fig. 6.4). At each station, daily records are available concurrently for 12 or 11 years, including 1993.

Monthly mean values of solar irradiation and sunshine duration for 11 year are used in calculation; hence each station has 132 monthly mean data. All data are measured with classical actinographs and sunshine duration recorders by State Meteorological Service, which is a member of ECMWF (DMI 2005). Detailed calculations and presentations are provided only for three sites as Adana, Ankara and Istanbul.

Adana is located in the south Turkey, where Mediterranean Sea effects occur with moderate, severe and extreme drought magnitudes at high level sunshine duration and around high solar irradiation values in this region. These extreme values show that sunshine duration is more effective in the area than solar irradiation. In this area, almost in each day of the year, water is heated with solar collectors and hence solar power plants can be built in the Mediterranean areas.

Ankara is located in the central part of Turkey where there are variations depending on the continental climate effects due to highest summer month topographic condition effects on the rainfall occurrences. Generally, in this semi-arid region, moderate and severe droughts many occur due to occasional effects, which rise to high level degree at this region.

Istanbul has both continental and maritime climatic effects. It is located in northwestern part of Turkey. Due to maritime effect cloud amounts are high and sunshine duration values are lower than southern and central regions of Turkey (Şen and Şahin 1998; Şaylan et al. 2002).

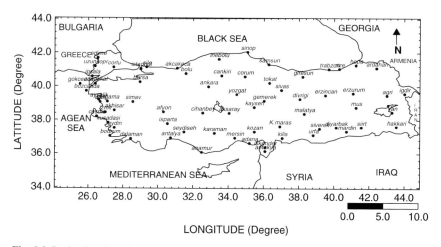

Fig. 6.4 Station locations in Turkey

5.3 Case Study

SSM is applied independently for each station and finally parameter estimation series are obtained for a'_i and b'_i. The lower order statistics that could not be managed with Angström equation are shown in Table 6.1 together with the classical Angström parameters. It is to be noted that in the application of the SSM, mode values are considered rather than arithmetic averages as used in the classical methods. It is easy and practical to do statistical analysis of Angström equation parameters and variables with SSM. Also included in Table 6.1 are the relative error (RE) percentages between the classical method arithmetic average and mode values of SSM. In the CD accompanying this book the reader will find all a' and b' values

Table 6.1 Statistical properties of SSM and Angström equation

Station Name	Angström Equation		Mode, SSM		Stand. Dev., SSM		Relative Error (%)	
	a	b	a'	b'	a'	b'	a, a'	b, b'
Adana	0.33	0.29	0.31	0.31	1.27	1.75	6.05	6.42
Adiyaman	0.30	0.22	0.27	0.26	0.80	1.10	9.93	17.90
Afyon	0.40	0.28	0.34	0.29	0.78	1.95	15.20	5.78
Amasya	0.30	0.38	0.27	0.37	2.44	6.79	6.72	3.92
Anamur	0.36	0.25	0.29	0.35	1.51	2.04	21.40	26.90
Ankara	0.31	0.32	0.30	0.48	3.03	6.09	4.74	32.00
Antalya	0.33	0.38	0.33	0.32	1.98	2.78	0.60	16.20
Aydin	0.32	0.42	0.33	0.42	1.88	3.38	4.50	0.95
Balikesir	0.23	0.37	0.22	0.34	1.39	3.98	0.88	6.28
Bursa	0.27	0.33	0.24	0.35	1.25	3.09	9.62	4.03
Çanakkale	0.31	0.33	0.31	0.45	3.81	4.58	2.50	27.70
Çankiri	0.35	0.32	0.11	0.25	5.92	6.86	57.30	21.60
Diyarbakir	0.23	0.48	0.43	0.42	2.09	3.06	45.00	12.73
Elazig	0.32	0.32	0.24	0.40	1.74	2.46	25.40	16.38
Erzincan	0.44	0.15	0.40	0.25	0.83	2.42	9.15	36.05
Eskişehir	0.39	0.26	0.34	0.43	0.74	1.56	13.81	39.60
İstanbul	0.30	0.35	0.28	0.55	0.90	3.34	5.08	35.50
Isparta	0.36	0.16	0.28	0.16	0.58	1.03	21.32	0.00
İzmir	0.33	0.33	0.32	0.42	1.16	1.71	1.80	22.00
Kars	0.50	0.12	0.74	0.41	1.32	2.22	33.30	70.00
Kastamonu	0.32	0.24	0.19	0.31	0.70	2.04	41.17	21.20
Kayseri	0.36	0.23	0.31	0.30	2.87	4.02	13.73	24.10
Kirsehir	0.43	0.20	0.18	0.25	1.31	2.15	57.30	21.81
Konya	0.38	0.27	0.31	0.39	1.79	3.80	20.00	32.20
Malatya	0.31	0.37	0.24	0.47	2.04	0.47	23.45	21.80
Mersin	0.33	0.40	0.27	0.48	0.87	1.25	16.60	17.30
Samsun	0.34	0.31	0.22	0.40	2.78	6.23	33.70	22.40
Trabzon	0.28	0.38	0.26	0.46	6.82	23.69	6.27	17.86
Van	0.51	0.14	0.40	0.23	34.07	41.92	21.76	36.60

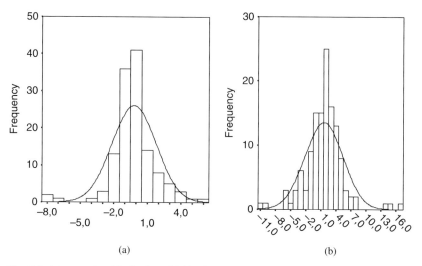

(a)   (b)

Fig. 6.5 Adana station (a) a' values (b) b' values

for Adana, Ankara and İstanbul stations separately during 129 months. In addition, the CD-ROM includes relation between average a' and b' values for mentioned three stations.

It is possible to obtain empirical frequency distribution functions or any other time variation features of the parameters from estimations, where Angström approach by the classical regression technique application does not give such an opportunity at a fixed point. Figures 6.5a and 6.5b present the empirical and theoretical histograms of a'_i and b'_i for Adana, respectively. The theoretical histograms appear as normal distribution functions.

Figure 6.6 presents the monthly average regional relationships between a'_i and b'_i for Adana and Ankara stations. These two figures can be arranged as the

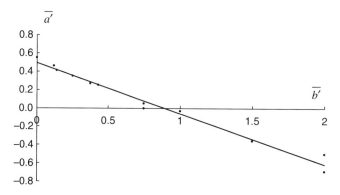

Fig. 6.6a Relationship between $\overline{a'}$ and $\overline{b'}$ at Adana

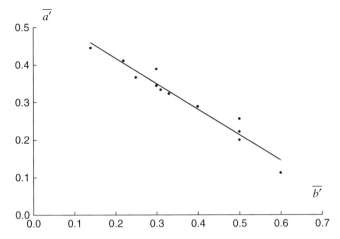

Fig. 6.6b Relationship between $\overline{a'}$ and $\overline{b'}$ at Ankara

temporal variation graph between $\overline{a'}$ and $\overline{b'}$. It is seen that high (low) values of $\overline{a'}$ follow low (high) values of $\overline{b'}$. Additionally, SSM provides sequences of coefficient estimations at any station that constitutes the basis of temporal histogram and this is useful in setting up the confidence limits in future global irradiation estimations.

6 Unrestricted Methodology (UM)

An alternative unrestricted method is proposed by Şen (2001) for preserving the means and variances of the global irradiation and the sunshine duration data. In the restrictive regression approach (Angström equation), the cross-correlation coefficient between H/H_0 and S/S_0 represents linear relationship only. By not considering this coefficient in the UM, some non-linearity features in the solar irradiation–sunshine duration relationship are taken into account. Especially, when the scatter diagram of solar irradiation versus sunshine duration does not show any distinguishable pattern such as a straight-line or a curve, then the use of UM is recommended for parameter estimations.

In practice, the estimation of model parameters is achieved most often by the least squares method and regression technique using procedural assumptions and restrictions in the parameter estimations. Such restrictions, however, are unnecessary because procedural restrictions might lead to unreliable biases in the parameter estimations. One critical assumption for the success of the regression equation is that the variables considered over certain time intervals are distributed normally, i.e. according to Gaussian PDF. As the time interval becomes smaller, the deviations from the Gaussian (normal) distribution become greater. For example, the relative frequency distribution of daily solar irradiation or sunshine duration has more skewness compared to the monthly or annual PDFs. The averages and variances of the solar irradiation and sunshine duration data play predominant role in many calculations and the conservation of these parameters is regarded as more important than the cross-correlation coefficient in any prediction model. In Gordon and Reddy (1988), it is stated that a simple functional form for the stationary relative frequency distribution for daily solar irradiation requires knowledge of the mean and variance only. Unfortunately, in almost any estimate of solar irradiation by means of computer software, the parameter estimations are achieved without caring about the theoretical restrictions in the regression approach. This is a very common practice in the use of the Angström equation all over the world.

The application of the regression technique to Eq. (6.10) for estimating the model parameters from the available data leads to new statistical approach (Şen 2001)

$$b = r_{hs} \sqrt{\frac{Var\overline{(H/H_0)}}{Var\overline{(S/S_0)}}} \tag{6.22}$$

and

$$a = \overline{\left(\frac{H}{H_0}\right)} - r_{hs} \sqrt{\frac{Var\overline{(H/H_0)}}{Var\overline{(S/S_0)}}} \overline{\left(\frac{S}{S_0}\right)} \tag{6.23}$$

where r_{hs} is the cross-correlation coefficient between global solar irradiation and sunshine duration data, $Var(.)$ is the variance of the argument; and the overbar $(-)$ indicates arithmetic averages during a basic time interval. Most often in solar engineering, the time interval is taken as a month or a day and in rare cases as a season or a year. As a result of the classical regression technique, the variance of predictand, given the value of predictor is

$$Var\left[\overline{(H/H_0)}/\overline{(S/S_0)} = S/S_0\right] = \left(1 - r_{rs}^2\right) Var\overline{(H/H_0)} \tag{6.24}$$

This expression provides the mathematical basis for interpreting r_{rs}^2 as the proportion of variability in $\overline{(H/H_0)}$ that can be explained by knowing $\overline{(S/S_0)}$ from Eq. (6.24), one can obtain after arrangements

$$r_{rs}^2 = \frac{Var\overline{(H/H_0)} - Var\left[\overline{(H/H_0)}/\overline{(S/S_0)} = S/S_0\right]}{Var\overline{(H/H_0)}} \tag{6.25}$$

In this expression, if the second term in the numerator is equal to 0, then the regression coefficient will be equal to 1. This is tantamount to saying that by knowing $\overline{(S/S_0)}$ there is no variability in $\overline{(H/H_0)}$. Similarly, if it is assumed that $Var\left[\overline{(H/H_0)}/\overline{(S/S_0)} = S/S_0\right] = Var\overline{(H/H_0)}$ then the regression coefficient will be

0. This means that by knowing $\overline{(S/S_0)}$ the variability in $\overline{(H/H_0)}$ does not change. In this manner, r_{rs}^2 can be interpreted as the proportion of variability in $\overline{(H/H_0)}$ that is explained by knowing $\overline{(S/S_0)}$. In all the restrictive interpretations, one should keep in mind that the cross-correlation coefficient is defined for joint Gaussian (normal) PDF of the global solar irradiation and sunshine duration data. The requirement of normality is not valid, especially if the period for taking averages is less than one year. Since, daily or monthly data are used in most practical applications, it is over-simplification to expect marginal or joint distributions to abide with Gaussian (normal) PDF. As mentioned before, there are six restrictive assumptions in the regression equation parameter estimations such as used in the Angström equation that should be critically taken into consideration prior to any application. The UM parameter estimations require two simultaneous equations since there are two parameters to be determined. The average and the variance of both sides in Eq. (6.10) lead without any procedural restrictive assumptions to the following equations,

$$\overline{\left(\frac{H}{H_0}\right)} = a' + b'\overline{\left(\frac{S}{S_0}\right)} \tag{6.26}$$

and

$$Var\overline{(H/H_0)} = b'^2 Var\overline{(S/S_0)} \tag{6.27}$$

where for distinction, the UM parameters are shown as a' and b', respectively. These two equations are the basis for the conservation of the arithmetic mean and variances of global solar irradiation and sunshine duration data. The basic Angström equation remains unchanged whether the restrictive or unrestrictive model is used. Equation (6.26) implies that in both models the centroid, i.e. averages of the solar irradiation and sunshine duration, data are preserved equally, hence both models yield close estimations around the centroid. The deviations between the two model estimations appear at solar irradiation and sunshine duration data values away from the arithmetic averages. Simultaneous solution of Eqs. (6.26) and (6.27) yields parameter estimates as,

$$b' = \sqrt{\frac{Var\overline{(H/H_0)}}{Var\overline{(S/S_0)}}} \tag{6.28}$$

and

$$a' = \overline{\left(\frac{H}{H_0}\right)} - \sqrt{\frac{Var\overline{(H/H_0)}}{Var\overline{(S/S_0)}}}\overline{\left(\frac{S}{S_0}\right)} \tag{6.29}$$

Physically, variations in the solar irradiation data are always smaller than the sunshine duration data, and consequently, $Var\overline{(S/S_0)} \geq Var\overline{(H/H_0)}$ and for Eq. (6.28) this means that $0 \leq b' \leq 1$. Furthermore, Eq. (6.28) is a special case of Eq. (6.22) when $r_{sh} = 1$ and the same is valid between Eqs. (6.23) and (6.29). In fact, from these explanations, it is clear that all of the bias effects from the restrictive assumptions are represented globally in r_{sh}, which does not appear in the UM parameter estimations. The second term in Eq. (6.29) is always smaller than the first one, and hence a' is always positive. The following relationships are valid between the restrictive and UM parameters

$$b' = \frac{b}{r_{hs}}$$ (6.30)

and

$$a' = \frac{a}{r_{hs}} + \left(1 - \frac{1}{r_{hs}}\right)\overline{\left(\frac{H}{H_0}\right)}$$ (6.31)

These theoretical relationships between the parameters of the two models imply that b and b' are the slopes of the restrictive models. The slope of the restricted (Angström) equation is larger than the UM ($b' > b$) according to Eq. (6.30) since always $0 \le r_{hs} \le 1$ for global solar irradiation and sunshine duration data scatter on a Cartesian coordinate system. As mentioned previously, two methods almost coincide practically around the centroid (averages of global solar irradiation and sunshine duration data). This further indicates that under the light of the previous statement, the UM over-estimates for sunshine duration data greater than the average value and under-estimates the solar irradiation for sunshine duration data smaller than the average. On the other hand, Eq. (6.31) shows that $a' < a$. Furthermore, the summation of model parameters is,

$$a' + b' = \frac{a+b}{r_{hs}} + \left(1 - \frac{1}{r_{hs}}\right)\overline{\left(\frac{H}{H_0}\right)}$$ (6.32)

These last expressions indicate that the two approaches are completely equivalent to each other for $r_{hs} = 1$. The UM is essentially described by Eqs. (6.26), (6.30) and (6.31). Its application supposes that the restricted model is first used to obtain a', b' and r. If r is close to 1, then the classical Angström equation coefficient estimations with restrictions are almost equivalent to a' and b'. Otherwise, the UM results should be considered for application as in Eq. (6.26).

6.1 Case Study

This methodology is applied to twenty eight stations which are already shown in Fig. 6.4 for Turkey. Also for interpretations three stations (Adana, Ankara and Istanbul) are considered in detail. Parameter estimation according to restricted and UMs are given in Table 6.2 for all considered sites.

Through the UM, it has been observed that in the classical regression technique, requirements of normality in the frequency distribution function and of linearity and the use of the cross-correlation coefficient are imbedded unnecessarily in the parameter estimations. Assumptions in the restrictive (Angström) model cause

Table 6.2 Estimated Angström parameters with restricted and UMs

Station Name	Restricted (Angström Equation		Unrestricted Methodology (UM)	
	a	b	a'	b'
Adana	0.33	0.29	0.20	0.50
Adiyaman	0.30	0.22	0.25	0.28
Afyon	0.40	0.28	0.36	0.34
Amasya	0.30	0.38	0.26	0.41
Anamur	0.36	0.25	0.26	0.40
Ankara	0.31	0.32	0.28	0.38
Antalya	0.33	0.38	0.22	0.55
Aydin	0.32	0.42	0.25	0.53
Balikesir	0.23	0.37	0.20	0.41
Bursa	0.27	0.33	0.21	0.46
Çanakkale	0.31	0.33	0.27	0.41
Çankiri	0.35	0.32	0.32	0.40
Diyarbakir	0.23	0.48	0.16	0.61
Erzincan	0.44	0.15	0.34	0.33
Eskişehir	0.39	0.26	0.33	0.39
İstanbul	0.30	0.35	0.27	0.41
Isparta	0.36	0.16	0.30	0.26
İzmir	0.33	0.33	0.25	0.45
Kars	0.50	0.12	0.34	0.45
Kastamonu	0.32	0.24	0.29	0.29
Kayseri	0.36	0.23	0.30	0.36
Kirsehir	0.43	0.20	0.36	0.30
Konya	0.38	0.27	0.30	0.40
Malatya	0.31	0.37	0.26	0.45
Mersin	0.33	0.40	0.36	0.45
Trabzon	0.28	0.38	0.23	0.51
Van	0.51	0.14	0.38	0.34

over-estimations in the solar irradiance amounts as suggested by Angström for small (smaller than the arithmetic average) sunshine duration and under-estimations for large sunshine duration values. Around the average values solar irradiation and sunshine duration values are close to each other for both models, however, the UM approach alleviates these biased-estimation situations. Additionally, the UM includes some features of non-linearity in the solar energy data scatter diagram by ignoring consideration of cross-correlation coefficient. Finally, in Fig. 6.7 straight-lines obtained separately from the classical regression and UM approaches are presented for Ankara, Adana and İstanbul stations. Especially, at Adana, Fig. 6.7b, UM and classical approach deviates significantly. This means that at Adana all points are scattered more randomly than others. In the CD accompanying this book the reader will find details of Fig. 6.7.

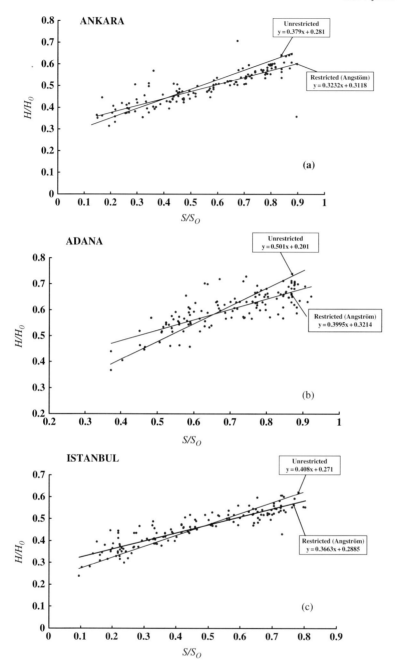

Fig. 6.7 Restricted and unrestricted methodologies (a) Ankara; (b) Adana; (c) İstanbul

7 An Alternative Formulation to Angström Equation

In this approach, relation between extraterrestrial variables (length of day and solar irradiation) ratio and terrestrial variables (measured sunshine duration and global solar irradiation) ratio are taken into consideration. This approach has not any restrictive assumptions and the main idea is to suggest a practical formulation for solar irradiation and sunshine duration estimations.

As mentioned before, solar irradiation and sunshine duration records depend on combined effects of astronomical and meteorological events. Meteorological events effects on the solar energy calculations introduce random behaviors. Meteorological solar irradiation (terrestrial) H and sunshine duration S variables have randomness in their temporal and spatial evolutions due to the shortening effect, in other words, the reduction amount of solar irradiation and sunshine duration is not definite but might be in the form of random amounts during a day or month depending on the atmospheric composition, climate and weather conditions.

7.1 Physical Background of Proposed Methodology

Through the classical approaches, it is difficult to find atmospheric effect to extraterrestrial solar irradiation and length of day. As seen in Fig. 6.8 global irradiance mostly reaches the surface depending on sun position's, sunrise and sunset times. In a partially cloudy day, there would be some discontinuities for measurements during the day, these detailed reduction and variation appear as randomness

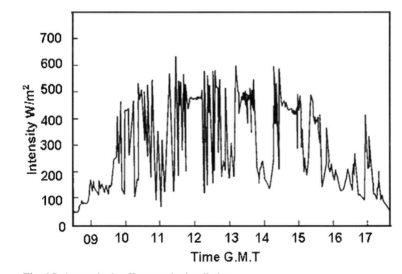

Fig. 6.8 Atmospheric effect to solar irradiation

in Fig. 6.8 which is a good example for atmospheric conditions and variability in order to receive solar irradiation. According to the suggestion in this equation, extraterrestrial variable ratio S_0/H_0 is assumed to have a reduction amount, R_e, due to cloud cover, dust, humidity, etc.

Such reductions in sunshine duration and solar irradiation are measured on the horizontal surface. There is a relation between extraterrestrial and terrestrial ratios due to atmospheric effects. This reduction effect can be expressed as

$$\frac{S_0}{H_0}(1 - R_e) = \frac{S}{H} \tag{6.33}$$

where R_e represents extraterrestrial ratio reduction amount. The reduction factor results from Eq. (6.33) as,

$$R_e = \frac{\left(\frac{S_0}{H_0} - \frac{S}{H}\right)}{\frac{S_0}{H_0}} = 1 - \frac{S/H}{S_0/H_0} \tag{6.34}$$

Given the astronomical calculations of H_0 and S_0 together with measurements of the H and S, R_e can be calculated easily from Eq. (6.34). If R_e is known then terrestrial sunshine duration S and solar irradiation H can be estimated as

$$S = \frac{HS_0(1 - R_e)}{H_0} \quad \text{or} \quad H = \frac{SH_0}{S_0(1 - R_e)} \tag{6.35, 36}$$

This formulation has the following advantages,

1. Atmospheric effect to extraterrestrial solar components can be explained easily. In other words, reduction amount in solar irradiation or length of day can be evaluated by proposed method,
2. Angström equation parameters (a and b) need for each period (month, day or hour) a long time measurements for each station. This method provides reduction in the parameter evaluations for each month, day or hour. In other words, atmospheric effects to extraterrestrial solar variables are monitored for each period, and
3. In the proposed method, there is no need for least square technique parameter estimation procedure and no restrictive assumption. As mentioned earlier, there are six assumptions in the least square methodology like Angström linear equation applications. However, in the proposed approach there is no assumption.

7.2 Case Study

R_e parameter of each month at each station are calculated and it is seen that some R_e's are negative. As can be concluded from Eq. (6.33) when extraterrestrial ratio is higher than terrestrial ratio, a positive R_e value occurs, otherwise a negative R_e value is obtained. Generally, in Adana station terrestrial ratios are higher than extraterrestrial ratios except for few months. At this station, sunshine duration values are high but solar irradiations are not as high as expected. As a result, terrestrial ratios are higher than extraterrestrial ratios.

In Ankara station, most of the characteristic R_e's have negative values. These ratios also show atmospheric effect to extraterrestrial ratio. In some months, extraterrestrial ratio reduced by 60% and 40% is received by horizontal surface. Under all circumstances, R_e values represent atmospheric effect irrespective of their signs. Positive and negative values must be considered for comparison between terrestrial and extraterrestrial ratios. Monthly mean values indicate that in the first and the last two months of the year, terrestrial ratios are higher than extraterrestrial ratios. On the contrary, in other months terrestrial ratios are smaller than extra terrestrial ratios. Hence, negative R_e values occur during eight months in this station.

In contrast to Adana and Ankara stations, majority of R_e values at Istanbul are positive (Fig. 6.9). Four months atmospheric effects to extraterrestrial ratios are higher than 0.6. It is estimated that average monthly terrestrial ratio values are higher than extraterrestrial ratios during seven months (Şen and Şahin 2001). In this figure polynomial connections occur below the straight line of extraterrestrial ratios. In the CD accompanying this book the reader will find details of this figure.

By using R_e values, measured terrestrial variables (sunshine duration and solar irradiation) can be estimated with Eqs. (6.35) and (6.36). Although Angström parameters, a and b, are constants, in the proposed method there are different R_e values for each month with a sequence of R_e values and hence it is possible to make probabilistic estimations, which provide an opportunity for the temporal prediction of R_e values and solar irradiation reductions. If R_e value of each month is used for estimation then there might be very little error. For optimal usage, constant R_e value must be considered with minimum estimation error. Herein, positive and negative R_e values are estimated. Hence, average positive R_{ep} value is calculated from positive

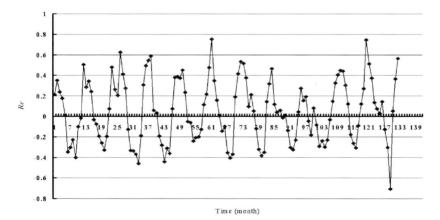

Fig. 6.9 R_e values for İstanbul station

R_{ep} values and average negative R_{en} is calculated from negative R_e values. Three stations have R_{ep} and R_{en} values in addition to Agström parameters are presented in Table 6.3. Other variables are taken into account directly.

If R_e value is positive then R_{ep} is taken into account for terrestrial sunshine duration and solar irradiation estimations from Eqs. (6.35) and (6.36), otherwise R_{en} is used.

Estimations for Istanbul by both methods are compared and it is seen that R^2 value for Angström method estimation is 0.97. This means that correlation coefficient of this representation is 0.98 which is a very good result. However, the proposed method has R^2 value as 0.87 which is also a good representation, but not better than Angström approach. In the case of time series graphics, regression technique gives some misleading information. All maximum values of Angström estimations are higher than measurements. In other words, over-estimations occur in H values in this station (Fig. 6.10a). Sometimes physically impossible values are estimated by Angström equation. Finally, measured sunshine duration, S values are estimated by both methods. It is observed that the proposed method gives better results than the Angström approach (see Table 6.3). In this station, some sunshine duration estimations have negative values (Fig. 6.10b). In addition, details of this figure could be maximized in the accompanying CD.

Adana station H estimations with proposed method are compared by measurements through the regression technique on the basis of the coefficient of determination (R^2). The same procedure is also used for terrestrial variables estimation through Angström equation. Proposed and Angström method estimations of H with high R^2 are given in Table 6.3. It is observed that Angström equation estimations are better than proposed method in Adana station. On the other hand, if time series comparison is considered then generally measured data are represented better except at maximum and minimum values. At maximum values over-estimations occur, but at minimum values under-estimations exist in Adana station. The same procedure is applied for sunshine duration estimation in Adana, and it is seen that proposed method estimation of sunshine duration is better than Angström equation. Especially, sunshine duration estimation is not meaningfully represented by Angström equation. In other words, R^2 attached with the Angström equation is not meaningful.

Table 6.3 Parameter and variables estimation of classical regression and new equations

Station	R_e		Angström Equation		Proposed Met. Measured H	Angst. Eq. Measured H	Proposed Met. Measured S	Angs. Eq. Measured S
	R_{ep}	R_{en}	a	b	R^2	R^2	R^2	R^2
Adana	0.12	−0.26	0.33	0.28	0.91	0.94	0.84	0.57
Ankara	0.24	−0.26	0.30	0.33	0.94	0.94	0.94	0.89
İstanbul	0.24	−0.26	0.28	0.36	0.87	0.97	0.88	0.87

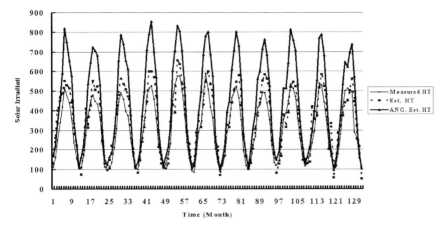

Fig. 6.10a Comparison of measured and estimated solar irradiation H values by both methods in İstanbul

In time series comparison Angström equation appears weak for terrestrial sunshine duration estimation. In one of the months, Angström equation estimates sunshine duration as zero which is physically impossible. In other words, during one month absolutely closed conditions could not be observed at these latitudes especially in this station.

Ankara station estimations by both methods indicate that terrestrial solar irradiation by Angström equation estimation is better than the proposed method. Like Adana, both methods have R^2 values higher than 0.94 and at maximum values in some months over-estimations occur by proposed method. It should be

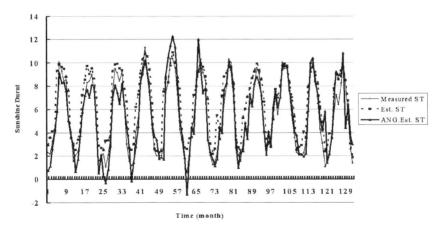

Fig. 6.10b Comparison of measured and estimated sunshine duration S values by both methods in İstanbul

remembered that, R^2 value of the proposed method is 0.94 which means that correlation coefficient between measured and estimated value is 0.96. This result is very representative for estimation purposes. In addition to measured global solar irradiation H and sunshine durations, S values are also estimated by both methods. It is understood that better estimations are possible through the proposed method where R^2 values are higher than Angström equation results (Table 6.3). Time series approach shows that the proposed method is more representative than Angström approach. Similar to Adana station in Ankara one month has physically impossible result by Angström equation estimation.

For the accuracy of the proposed and classical Angström models mean bias error (MBE), root mean square error (RMSE) and relative error (RE) are used and compared (Table 6.4).

It is not easy to see differences between estimated and measured values by using MBE except for İstanbul. In Adana and Ankara MBE values are approximately equal to each other, but in İstanbul MBE of H values estimated from Angström equation, is 42.86% that is very high for engineering approaches. When one looks at RMSE values, it is seen that as a result of summation of square differences, these errors are higher than MBE and generally bigger than 10% except errors of estimated H values for Adana and Ankara by Angström equation. Other RMSE values are higher for H and S than estimations by Angström equation. It is clearly seen that Angström equation is not a good approach for these parameters. Especially, in İstanbul unacceptable errors are estimated by classical approach. One of the other comparison methods is the relative error (RE) approach that is a very useful tool for engineering calculation. It is seen that for all station, RE values of Angström equation are higher than errors of proposed method except H values in Adana and Ankara that are estimated by Angström equation (Table 6.4).

Table 6.4 Different estimation errors for classical regression approach and proposed method

STATION	Error (%)	H		S	
		New Met.	Angström	New Met.	Angström
ADANA	MBE	0.361	0.404	0.171	−0.921
	RMSE	12.627	6.786	12.436	26.019
	RE	9.556	7.513	9.556	20.867
ANKARA	MBE	2.649	0.782	−1.666	−0.815
	RMSE	12.802	5.566	11.877	84.206
	RE	12.016	4.886	12.015	14.182
İSTANBUL	MBE	10.323	42.869	16.065	−0.056
	RMSE	21.869	52.178	26.019	116.745
	RE	16.614	27.198	20.758	21.026

8 Conclusions

In this chapter, different new methods are applied to the Angström equation and a new alternative methodology is proposed to see dynamic behavior of this equation and solar irradiation variables. A dynamic model estimation procedure as the successive substitution method (SSM) is proposed, which leads to a sequence of parameters and hence it is possible to look at the frequency distribution function (probability distribution function, PDF) of the model parameters and decide whether the arithmetic average of the parameters or the mode (the most frequently occurring parameter value) should be used in further solar irradiation estimations. It is shown on the basis of some solar irradiation and sunshine duration data measurements on different locations in Turkey that the model parameter estimations abide by the Beta PDF. Besides, it is also possible to find the relationship between the model parameters at a single station by using SSM, which shows temporal parameter variations. In addition, it is easy and practical to do statistical analysis of Angström equation parameters and variables with SSM.

Apart from the dynamic model parameter estimation procedure, an unrestricted model (UM) for solar irradiation parameter estimation procedure, is also presented which considers the conservation of the model input and output variables' arithmetic mean and the standard deviations only, without the use of least squares technique. Assumptions in the restrictive (Angström) model cause over-estimations in the solar irradiance amounts as suggested by Angström for small (smaller than the arithmetic average) sunshine duration and under-estimations for large sunshine duration values. Around the average values solar irradiation and sunshine duration values are close to each other for both models, however, the UM approach alleviates these biased-estimation situations. Additionally, the UM includes some features of non-linearity in the solar energy data scatter diagram by ignoring consideration of cross-correlation coefficient.

Finally, an alternative formulation to Angström equation is proposed for sunshine duration and solar irradiation variables estimation. According to the suggested formulation, extraterrestrial variable ratio S_0/H_0 is assumed to have a reduction amount, R_e, due to cloud cover, dust, humidity, etc. Such reductions in sunshine duration and solar irradiation are measured on the horizontal surface. There is a relation between extraterrestrial and terrestrial ratios due to atmospheric effects. This reduction amount, R_e represents atmospheric effect to extraterrestrial solar irradiation. Given the astronomical calculations of H_0 and S_0 together with measurements of the H and S, R_e can be calculated easily from proposed formulation. This methodology and Angström equation procedure are compared and it is shown that there are some physical problems with classical Angström approach.

References

Akınoğlu BG, Ecevit A (1990) Construction of a Quadratic Model Using Modified Angström Coefficients To Estimate Global Solar Radiation. Solar Energy, 45, 2: 85–92.

Dogniaux R, Lemonie M (1983) Classification of radiation sites in terms of different indices of atmospheric transparency. In Proc. EC Contactor's Meeting on Solar Radiation Data, Solar Energy R of D in the EC, series F, Vol. 2. Reidel, Dortrecht, pp. 94–105.

Duffie JA, Beckman W.A. 2006. In 3nd Edition. Solar engineering of thermal processes. Wiley, New York.

Gopinathan KK., 1986. A general formula for computing the coefficients of the correlation connecting global solar radiation to sunshine duration. *Solar Energy*, 41(6): 499–502.

Gordon, JM, Reddy TA (1988) Time series analysis of daily horizontal solar radiation. Solar Energy, 41: 215–226.

Gueymard CA, Jindra P, Estrada-Cajigal V (1995). A critical look at recent interpretations of the Angström approach and its future in global solar radiation prediction. Solar Energy 54 (5): 357–363.

Gueymard CA (2004) The sun's total and spectral irradiance for solar energy applications and solar radiation models. Solar Energy 76: 423–453.

Hinrichsen K (1994) The Angström Formula with Coefficients Having a Physical Meaning. Solar Energy, 52, 6: 491–495.

Iqbal M (1983) An Introduction to Solar Radiation, Academic Press, Toronto.

Jain PC (1990) A model for diffuse and global irradiation on horizontal surfaces. Solar Energy 45, 5: 301–306.

Lewis G (1989) The Utility of the Angström-Type Equation for the Estimation of Global Radiation. Solar Energy, 43, 5: 297–299.

Martinez-Lozano JA, Tena F, Onrubia JE, Delarubia J (1984) The historical evolution of the Angström formula and its modifications: review and biography. Agricultural and Forest Meteorology 33 (2–3): 109–128.

Ögelman H, Ecevit A, Tasdemiroglu E (1984). A New Method for estimating solar radiation from bright sunshine data. Solar Energy 33(6): 619–625.

Prescott JA (1940) Evaporation from water surface in relation to solar radiation. Trans. Royal Soceity Australia. 40: 114–116.

Rieltveld MR (1978) A New Method for Estimating the Regression Coefficients in the Formula Relating Solar Radiation to Sunshine. Agricultural Meteorology, 19: 243–252.

Sabbagh JA, Saying AAM, El-Salam EMA (1977) Estimation of the Total Solar Radiation from Meteorological Data. Solar Energy 19: 307–311.

Şahin AD, Şen Z (1998) Statistical analysis of the Angström formula coefficients and application for Turkey. Solar Energy 62: 29–36.

Sahin AD, Kadioglu M, Sen Z (2001) Monthly clearness index values of Turkey by harmonic analysis approach. Energy Conversion and Management, 42: 933–940.

Samuel TDMA (1991) Estimation of Global Radiation for Sri Lanka, Solar Energy 47: 5, 333.

Şaylan L, Sen O, Toros H, Arısoy A (2002) Solar energy potential for heating and cooling systems in big cities of Turkey. Energy Conversion and Management 43: 1829–1837.

Şen Z (2001) Angström equation parameters estimation by unrestricted method. Solar Energy 71(2): 95–107.

Şen Z, Öztopal A, Sahin AD (2001) Application of genetic algorithm for determination off Angström equation coefficients. Energy Conversion and Management 42: 217–231.

Şen Z, Şahin AD (2001) Solar irradiation polygon concept and application in Turkey. Solar Energy 68 (1): 57–66.

Soler A (1990) Monthly specific Rietveld's Correlations. Solar and Wind Technology, 7 (2–3): 305–306.

Suehrcke H (2000) On the relationship between duration of sunshine and solar radiation on the earth's surface: Angström's equation revisited. Solar Energy. 68(5): 417–425.

Swartman RK, Ogunlade O (1967) Solar Radiation Estimates from Common Parameters. Solar Energy 11: 170–172.
Wahab AM (1993) New Approach to Estimate Angström Coefficients. Solar Energy 51(4): 241–245.
Zabara K (1986) Estimation of the global solar radiation in Greece. Solar Energy and Wind Technolog, 3: 267.

Chapter 7
Solar Irradiation via Air Temperature Data

Marius Paulescu

1 Introduction

In estimating the amount of solar energy that can be used in applications, the selection of a suitable algorithm takes into account the availability of meteorological data as input. The air temperature is an all-important parameter recorded by all meteorological stations around the world, but it is not a common parameter for the computation of solar radiation. However, temperature maxima, minima, mean or amplitude have been included in solar energy modeling as a task in crop simulation models, developed recently for agriculture. Because temperature measurements are simple and robust, there is a reason for such models to be adapted for estimating daily solar energy with reasonable accuracy in various applications, such as photovoltaics.

The chapter is organized as follows. In the first section models which use air temperature together with cloudiness as parameters are described while in the second section self-contained air temperature daily irradiation models are presented. The third part is dedicated to estimation of solar radiation inside fuzzy logic. Two verified recipes for drawing up temperature based solar radiation models, one in the frame of classical statistics and the other one inside fuzzy logic, are outlined. A C program included on the CD-ROM, which enable fuzzy calculation for daily global solar irradiation is presented. Finally the accuracy of all the enumerated models is assessed under Romanian climate (Eastern Europe) in comparison with models which use sunshine duration or cloudiness at input.

The arguments that follow are leading to the conclusion that air temperature can be used with success in the estimation of the available solar energy.

Marius Paulescu
West University of Timisoara, Romania, e-mail: marius@physics.uvt.ro

2 Prediction from Air Temperature and Cloud Amount

Extinction of solar radiation due to the clouds is more important than that due to any other atmospheric constituents. The majority of solar irradiation models take into account the extinction of radiation in relation to cloud cover via the Ångström-Prescott equation (Ångström 1924; Prescott 1940). Customary derivation (Jain 1990) leads to relative sunshine duration as a *natural* parameter in this type of correlation. Over time, in order to increase accuracy, the original Ångström's equation has been modified and related to other surface meteorological parameters. The fractional cloud amount N is often used instead of relative sunshine duration (Haurvitz 1945, Kasten and Czeplak 1980).

Many previous modeling efforts have been conducted to include daily extreme of air temperature t_{max}, t_{min} besides daily mean of cloudiness N in empirical solar irradiation estimation. Taking air temperature and cloudiness in computation is motivated by the usual availability of both meteorological parameters. Embedding the air temperature in models is meant to increase prediction quality, having the practical experience that accuracy decays with increasing cloudiness. The cause of increasing error with increasing N mainly derives from the definition of cloudiness, which for a given N does not take into account whether the sun is shinning or is behind the clouds. But, the drawback of including air temperature in the fitting process leads to a closer connection of the model to parental geographical location.

The equation from Supit and Van Kappel (1998):

$$H(N,\Delta t) = H_e \left(a \left(t_{max} - t_{min} \right)^{1/2} + b \left(1 - N \right)^{1/2} \right) + c \tag{7.1}$$

is a typical model which linearly relate daily global solar irradiation at ground level H to its extraterrestrial value H_e. Practically, Eq. (7.1) combines the square root of temperature amplitude dependence (Hargreaves et al. 1985) and a non-linear cloudiness dependence of the daily solar irradiation (Kasten and Czeplak 1980). The coefficients a, b and c are provided by Supit and Van Kappel (1998) for 95 various location in Europe from Finland to Spain exhibiting large dispersions that reveals the model local specificity. For example the temperature coefficient a takes values between 0.028 at Goteborg, Sweden (latitude 57.7°N; longitude 12.0°E; altitude 2m) and 0.115 at Murcia, Spain (38.9°N; 1.23°W; 62m), but it is also irregular with latitude - $a = 0.086$ at Lund, Sweden (55.7°N; 13.2°W; 73m).

The three models from (El-Metwally 2004) derived with data coming from northern Africa use other expressions for the correlation. Apart from Supit and Van Kappel model, it incorporates separately t_{max} and t_{min} either in a linear relation, such as:

$$H(N,t_{max},t_{min}) = aH_e + bt_{max} + ct_{min} + dN + e \tag{7.2}$$

or in an unfamiliar exponential one, as long as H_e is under the exponent. The regression coefficients in Eq. (7.2) given for seven Egyptian locations are also distinctly different.

To explain the air temperature role in this type of equations, we computed the second order polynomial regression coefficients with data (daily t_{max}, t_{min}, N, H recorded in 1998–2000) from Timisoara, Romania (45.76°N; 21.25°E; 85m), using the least square method, for the following two equations:

$$H(N) = -0.01H_e^2 + (-0.42N + 0.9)H_e - 3.948N^2 + 4.577N - 1.685 \quad (7.3)$$

$$H(N, \Delta t) = -0.014H_e^2 + (0.025\Delta t - 0.256N + 0.56)H_e - 2.64N^2 +$$
$$+ (0.06\Delta t + 1.804)N - 0.05\Delta t - 4.723 \cdot 10^{-3}\Delta t^2 + 0.1 \quad (7.4)$$

The difference between these two equations consists of the presence of daily temperature amplitude $\Delta t = t_{max} - t_{min}$ in the Eq. (7.4).

The effect of taking temperature into consideration in the H model is depicted in Fig. 7.1 where the Eqs. (7.3) and (7.4) are plotted in respect to cloudiness. The $H(N)$ curve from Eq. (7.3) is shifted by Eq. (7.4) in a band $H(N, \Delta t)$, which can be interpreted that for a given N, Δt acts as a refinement according to weather conditions. It makes sense if we bear in mind that the air temperature amplitude is lower in the cloudy days than in the sunny days.

All of these models natively have been designed to avoid the task of solar irradiance computation, being focused on direct computation of daily solar irradiation. However solar irradiation basically represents a sum over time of the irradiance. Therefore it is possible to include air temperature in a solar irradiance model.

2.1 Mathematically Integrable Solar Irradiance Model

An approach in two steps for including air temperature on the input parameters list of global solar irradiation models follows. First, an empirical solar irradiance model with air temperature besides cloudiness is employed. Subsequently, daily global solar irradiation is computed by integration of the irradiance model between sunrise and sunset.

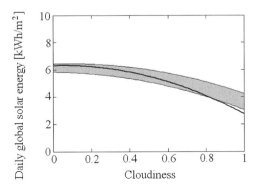

Fig. 7.1 Daily global solar irradiation computed with Eqs. (7.3) (bold line) and (7.4) (a band delimited by two curves corresponding to $\Delta t = 10°C$, lower one, and $\Delta t = 20°C$, upper one) in respect to cloudiness

We start with searching for an appropriate correlation between global solar irradiance at the ground $G(N, t, h)$ and outside atmosphere G_e, with air temperature t and solar elevation angle h as parameters: $G(N,t,h) = G_e f(N,t,\sin h)$. An *acceptable response* yields from the following function:

$$f(N, T_r \cdot \sin h) = c_1(N) \cdot \left[c_2(N) - N^{c_3(N)}\right] \cdot [T_r \sin h]^{c_4(N)} \tag{7.5}$$

where $T_r = 1 + t/273$ with temperature t in degree Celsius.

The coefficients c_i, $i = 1,2,3,4$ are the subject of a fitting process that is running for different classes of cloudiness. The result is a discrete set of coefficients which could be carried on in a secondary fitting process to approximate it with continuous functions. Thus, in addition to the determination coefficient, the monotone behavior of the discrete coefficients with respect to cloudiness is used as a selection criterion in the first fitting process. Turning $f(N, t, h)$ into a continuous function is a requirement for solar irradiation computation by mathematically integration over time. As an example, Fig. 7.2 shows the coefficients c_1, c_2, c_3 fitted with data coming also from Timisoara (Paulescu and Fara 2005) and the corresponding approximation functions $f_i(N)$, $i = 1,2,3$ $(c_4(N) = f_4(N) = 1.16)$:

$$f_1(N) = e^{-0.1341 + 2.44181 - 4.66676N^2 + 3.83066N^3}$$
$$f_2(N) = 0.7988 + 0.27829 N^{0.73642}$$
$$f_3(N) = (1.46112 - 0.91168 N)^{-1} \tag{7.6 a,b,c}$$

It is remarkable that the coefficients are along regular curves. Therefore, for all range of cloudiness the correlation can be expressed as:

$$f(N, T_r \cdot \sin h) = \begin{cases} f_1(N) \left(f_2(N) - N^{f_3(N)}\right)(T_r \sin h)^{1.16}, & 0 \le N < 0.95 \\ 0.19831 \cdot (T_r \sin h)^{1.16}, & N \ge 0.95 \end{cases} \tag{7.7}$$

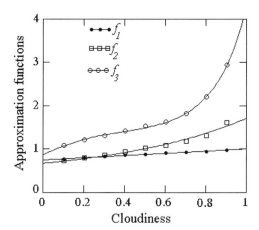

Fig. 7.2 Discrete coefficients from Eq. (7.5) and the continuous approximation functions, Eqs. (7.6 a,b,c) in respect to cloudiness

Introduction of a discontinuity near to $N = 1$ is an ordinary practice in solar energy estimation to improve the model accuracy in most cloudy or overcast situation. Even if these corrections are present, such models cannot estimate global solar irradiance at a high level of accuracy in predominantly clouded conditions. However the results are acceptable when the solar irradiation is computed. It can be done by integrating between sunrise and sunset of irradiance:

$$H_j = C \int_{-\omega_0}^{\omega_0} G(N, \omega, T_a(j, \omega)) \, d\omega \tag{7.8}$$

where cloudiness is replaced with its daily average. For $T_a(j, \omega)$ a suitable model based on daily air temperature extreme is described as:

$$t(j, \omega) = a \cdot t_0(j, \omega) + b$$

$$t_0(j, \omega) = \begin{cases} t_{max}(j) - [t_{max}(j) - t_{min}(j)] \cdot \left(1 - \cos\left(\dfrac{\pi}{2}\dfrac{\omega_m - \omega}{\omega_m + \omega_0(j)}\right)\right) & \omega \leq \omega_m \\[2ex] t_{max}(j) - [t_{max}(j) - t_{min}(j+1)] \cdot \sin\left(\dfrac{\pi}{2}\dfrac{\omega_m - \omega}{\omega_m - \omega_0(j+1)}\right) & \omega > \omega_m \end{cases}$$

$$\tag{7.9 a,b}$$

Equation (7.9a) empirical adjust the Eq.(7.9b) to a local meteo-climate ($a = 0.99$ and $b = -0.41$, at Timisoara). $t(j, \omega)$, in $^\circ$C, is the estimated air temperature in the Julian day j at hour angle ω. $\omega_0(j)$ is the sunset hour angle and ω_m is the hour angle at which the maximum air temperature is reached. In this model we assume ω_m to be the same in every day. C in Eq. (7.8) is accordingly to the unit of H: for $C = 12/\pi$, H is in Wh/m^2.

The quality of the air temperature estimation using the sine-cosine Eqs. (7.9 a,b) can be assessed from Fig. 7.3. It is a scatter plot of estimated versus observed air temperature at the station of Timisoara. Figure 7.3a shows the scattering of instantaneous values at 9.00, 12.00 and 15.00 local standard time in the year 2000 while Fig. 7.3b shows the daily mean air temperature in the years 1998–2000. It can be

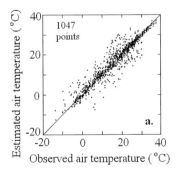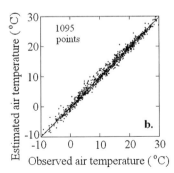

Fig. 7.3 Estimated air temperature with Eq. (7.9 a,b) versus observed: a. Instantaneous values; b. Daily average

seen that, when a mathematical integration is performed, the prediction accuracy increase. It is due to the fact that the model performance is high in the middle of the day when maximum of solar energy is collected.

There are a variety of methods with various degrees of complexity developed to approximate diurnal temperature from its maxima and minima. We emphasize here an empirical model from (Cesaraccio et al. 2000) as being acceptably accurate for estimating the hourly mean of air temperature. It is useful when Eq (7.8) is applied for the computation of hourly solar radiation.

All these models demonstrate that the daily global solar irradiation can be related to the corresponding extraterrestrial value using at input only daily minimum and maximum air temperature besides daily mean of cloudiness. But, from common observations, daily extremes of air temperature encapsulate information about weather condition. Consequently the solar energy estimation can be made straightforward by eliminating the cloudiness from the input.

3 Models for Daily Solar Irradiation from Daily Extremes of Air Temperature

Solar radiation controls the temperature and moisture profile of soil and provides energy for photosynthesis. For assessing the potential productivity in agriculture recently there are proposals for modeling seed germination, crop-weed interaction and crop growth (Sirotenko, 2001; Cheeroo-Nayamuth, 1999) where solar energy is a major variable. This is a segment of solar energy computation where the most popular models have been developed using only minimum and maximum air temperature as input parameters.

Bristow and Campbell (1984) established an empirical equation for daily global irradiation using air temperature amplitude $\Delta t = t_{max} - t_{min}$:

$$H/H_e = a\left(1 - e^{-b(\Delta t)^c}\right) \tag{7.10}$$

The coefficients a, b and c have been found to be distinct for every location. Moreover this model demands calibration which involves a solar energy database – or such models are applied just to search out this quantity. Despite the disadvantage of requiring local calibration, the Bristow and Campbell scheme has been used as a core by many other models. Thornton and Running (1999) refine the Bristow and Campbell model over a wide geographic area with the aim to eliminate the need for locally calibration. A comprehensive evaluation of different 14 variations of the Bristow and Campbell method can be read in Wiss et al. (2001). We note here the updating done by Donatelli and Bellocchi (2001) which accounts for seasonal effects on cloudless transmittance using a sine function:

$$H/H_e = a\left[1 + b\sin\left(c\frac{\pi}{180}j\right)\right]\left(1 - e^{\frac{-d(\Delta t)^2}{\Delta t_w}}\right) \tag{7.11}$$

j is the Julian day, Δt_w is the weekly Δt and a, b, c, d are empirical constants. This model is one of the basic in the RadEst3.00 application (Donatelli et al. 2003), a useful tool to estimate global solar radiation in a given location.

In Eqs. (7.10) and (7.11) the clear sky model is implicitly embedded. But the usual modeling approaches for solar irradiation run in two steps: first the solar irradiation under clear sky condition H_0 is computed and second, cloud cover is accounted via the Ångström-Prescott equation. For H_0 the integration of a clear sky model between sunrise and sunset and summing up these results to daily, monthly or yearly irradiation is the ordinary approach. There is a large number of solar irradiance models elsewhere reported, having either empirical or physical basis that can be used in the computation of the daily solar irradiation with a reasonable level of accuracy. Apart from the Eqs. (7.10) and (7.11), in the following model daily global solar irradiation is related to its maximum possible value using daily air temperature extremes. The input is the daily temperature amplitude range Δt and the 5-day average of daily mean air temperature $\bar{t}_5$ computed as $\bar{t} = (t_{max} + t_{min})/2$. A range of several days for the calculation of average air temperature, centered in the day for which H is estimated, is introduced as for a good estimation of daily mean air temperature in a certain period of the year. In addition to Δt, the deviation of $\bar{t}$ from $\bar{t}_5$ is an appropriate measure of weather condition on the day: higher Δt and $t \approx \bar{t}_5$ indicate a sunny day while a lower Δt and $t < \bar{t}_5$ an overcast day.

Practically the model considers a linear dependence of H with respect to H_0, having slope and interception as functions of Δt and $\bar{t}_5$, respectively.

$$H = H_0 f_1(\Delta t) + f_2(\bar{t}_5)$$
$$f_1(\Delta t) = a_1 + b_1(\Delta t)^{c_1}$$
$$f_2(\bar{t}_5) = a_2 + b_2 \sin\left(\frac{2\pi \bar{t}_5}{d_2} + c_2\right) \qquad \text{(7.12 a,b,c)}$$

Figure 7.4 displays a 3D graph of the solar irradiation estimated with Eq. (7.12) showing the way in which the functions $f_1(\Delta t)$ and $f_2(t)$ act on H_0. Two surfaces are plotted corresponding to $H_0 = 2\,\text{kWh/m}^2$ (in the winter days) and $H_0 = 8\,\text{kWh/m}^2$ (in the summer days) between which H are enfolded when $(\Delta t, t)$ vary in the usual

Fig. 7.4 Plots of the Eq. (7.12 a,b,c) in the usual range of temperature for $H_0 = 2\,\text{kWh/m}^2$ (lower surface) and $H_0 = 8\,\text{kWh/m}^2$ (upper surface)

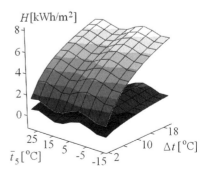

range. The graphic points out the role of sine function as a seasonal adaptor for H_0 depending only on t.

The coefficients used are $a_1 = -0.324$, $b_1 = 0.366$, $c_1 = 0.424$, $a_2 = 0.00576$, $b_2 = 0.372$, $c_2 = 1.832$ and $d_2 = 26.35$ which particularize the model for Western Romania (Paulescu et al. 2006). The sensitivity to origin location is due to the fact that the daily amplitude of air temperature and daily mean air temperature are parameters influenced in a complex manner by local meteo-climate. An increase of the model generality concerning the application area is possible by introduction a factor, denoted ξ, which adapt the coefficients in Eqs. (7.12 a,b,c) taking into account the behaviour $\Delta t = \Delta t(t)$ over a period extending to several years. The approach has been introduced in (Paulescu et al. 2006) and points out that the corrective factor is characteristic to each location. This coefficient in its simplest form can be considered a constant. A practical implementation of ξ will be described in the next sub-section for fuzzy models.

The greatest benefit of the model (7.12 a,b,c) results from the synergism among the possibility of using simplified but accurate clear sky empirical irradiance models which require as input only geographical and temporal coordinates, and an Ångström type equation which require at input air temperature, a worldwide measured parameter.

4 Fuzzy Models

Fuzzy sets theory was introduced in 1965 by Lotfy A. Zadeh (Zadeh, 1965) and basically replaces the Aristotelian YES/NO logic with a multi-valued logic. In other words, the interval between Boolean elements 0 and 1 is filled with real numbers. In this way, fuzzy logic can provide an algorithm with the strength to capture uncertainties with a flexibility that resembles human reasoning and facilitates a heuristic approach in modeling phenomena otherwise complicated natural systems. In many applications fuzzy logic has been adopted as a standard method (see examples from Passino and Yurkovich, 1998) but it is still an emerging field in solar energy estimation. A literature survey shows that there are only few models concerning fuzzy modeling of solar radiation (Sen 1998; Santamouris et al. 1999; Gomez and Casanovas, 2003).

Since fuzzy logic is quit different from Boolean logic a short introduction is in place, followed by an application to solar irradiation computation based on air temperature. More about fuzzy logic can be read in Zimmermann (1996) or Passino and Yurkovich (1998).

4.1 Fuzzy Logic Introduction

A fuzzy logic model is a functional relation between two multidimensional spaces containing the *fuzzy sets*, which are the central concept of Zadeh theory and are defined as:

$$A = \{(x, m_A(x)) : x \in X\} \tag{7.13}$$

where $m_A(x)$ is the *membership function* expressing the degree of elements x in the fuzzy set A.

Different sets are distinguished by different membership functions. Let's see an intuitive example. Assuming the Julian days set $\{1, 2 \ldots 172\}$ corresponds to X from the definition Eq. (7.13). We are familiar with the division of X at $j_0 = 80$ (March, 21) in two sets (seasons): WINTER for $j \in W = \{1, 79\}$ and SPRING for $j \in S = \{1, 171\}$. We are used to express that j belongs to W by an application $f : X \rightarrow \{0, 1\}$, showed in Fig. 7.5. It is what we call a crisp set. Fuzzy logic relaxes the crossing from W to S by replacing the step-like separation between WINTER and SPRING with a slow passing in a finite interval around j_0. Thus, the binary domain $\{0, 1\}$ is filled with real numbers being turned into a continuous domain and the function $f(t)$ is replaced with the membership functions $m_{WINTER}(j)$ and $m_{SPRING}(j)$. From Fig. 7.5 a day up to February 20 ($j < 51$) certainly is a WINTER day while a day after April 20 ($j = 111$) certainly is a SPRING day; 1 March, $j = 60$, is assigned of $m_{WINTER}(60) = 0.83$ degree to be WINTER and $m_{SPRING}(60) = 0.17$ degree to be SPRING. Therefore, the membership function reads out the level of confidence for a day to be in the one of the sets WINTER or SPRING.

In fuzzy sets theory a physical variable, as Julian day in previously example is named *linguistic variable*. The values of a linguistic variable are not numbers, as in the case of deterministic variables, but linguistic values, called *attribut*, expressed by words or sentences (e.g. WINTER and SPRING). The number of attributes of a linguistic variable and the shape of membership functions depends on the application, being specified in a heuristic way. Theoretically, the membership function can have any form; in practice triangular and trapezoidal forms are widely used.

Different fuzzy sets are combined through membership functions:

$$\text{Fuzzy intersection (AND)}: m_{A \cap B} = \min(m_A(x), m_B(x)), \forall x \in X \tag{7.14a}$$

$$\text{Fuzzy reunion (OR)}: m_{A \cup B} = \max(m_A(x), m_B(x)), \forall x \in X \tag{7.14b}$$

Equations (7.14) define the Zadeh fuzzy operators (Zadeh 1965). There are also others definitions of fuzzy logic operators (Zimmermann 1996) but we will use only the definitions Eqs. (7.14) in operations with fuzzy sets.

The map between the input and the output fuzzy spaces is a collection of associative rules, each reading:

$$\textbf{IF} \, (premises) \, \textbf{THEN} \, (conclusions) \tag{7.15}$$

Fig. 7.5 The characteristic function $f(j)$ of a crisp set and membership functions $m_{WINTER}(j)$ and $m_{SPRING}(j)$ of the Julian day attributes WINTER and SPRING, respectively

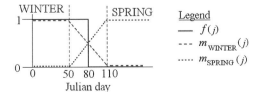

Every premise or conclusion consists of an expression as:

$$(variable) \text{ IS } (attribute) \tag{7.16}$$

connected through fuzzy operator **AND**.

The information is carried out from input to output of a fuzzy system in three steps:

1. *Fuzzification* is a coding process in which each numerical input of a linguistic variable is converted in membership function values of attributes.
2. *Inference* is a process itself in two steps:

   – The computation of a rule fulfilled by the intersection of individual premises, applying the fuzzy operator **AND**.
   – Often, more rules drive to the same conclusion. To yield the conclusion (i.e., the membership function value of a certain attribute of output linguistic variable) the individual confidence levels are joined by applying the fuzzy operator **OR**.

3. *Deffuzification* is a decoding operation of the information contained in the output fuzzy sets resulted from the inference process, with the purpose of providing an output crisp value. There are more methods for deffuzification (Zimmermann, 1996); we apply the *Center Of Gravity* (COG) method, one of the most popular:

$$y_{crisp} = \frac{\sum_i c_i \int m_{y_i}(x)dx}{\sum_i \int m_{y_i}(x)dx} \tag{7.17}$$

In the Eq. (7.18), c_i is the center of the membership function (generally, where it reaches its peak), the integral $\int m_{y_i}(x)dx$ represents the area under the membership function $m_{y_i}(x)$ corresponding to the attribute i of the output linguistic variable y.

4.2 A Fuzzy Model for Daily Solar Irradiation

The algorithm outlined below has been designed for the computation of daily global solar energy and is a slightly modified variant of the one reported in (Tulcan-Paulescu and Paulescu 2007). The model is conducted with two input linguistic variable: daily amplitude of air temperature $\Delta t_j = t_{max,j} - t_{min,j}$ and Julian day j. The output variable is the clearness index $k_T = H_j/H_{e,j}$ (Liu and Jordan 1960) where H_j represents the daily solar irradiation in the j day while $H_{e,j}$ is its extraterrestrial value.

In fuzzy modeling practice the number of attributes associated to a linguistic variable increase with increasing of input/output data spreading. Because of relative higher than medium scattering of Δt to k_T, these variables are characterized by eight attributes. It is emphasized that the attributes T_i and K_i often are described in terms as LOW, MEDIUM or HIGH. For a simplified notation the numeral subscript i is more suitable, assign with ascending i attributes ranging from VERY LOW to VERY HIGH. For Julian day linguistic variable only two attributes have been considered, WINTER (W) and SUMMER (S)

All the input membership functions are plotted in Fig. 7.6, where the notation for every attribute is specified.

The membership functions for Δt_j ($i = 1 \ldots 8$) attributes are triangular:

$$m_{\Delta t, i}(\Delta t, \xi) = \begin{cases} \max\left(0, \frac{\Delta t - a_i \xi}{c_i \xi - a_i \xi}\right) & \text{if} \quad \Delta t < c_i \xi \\ \max\left(0, 1 - \frac{\Delta t - c_i \xi}{b_i \xi - c_i \xi}\right) & \text{otherwise} \end{cases} \quad (7.18)$$

The coefficients a_i, b_i and c_i have the signification depicted in Fig. 7.6. The membership function of attributes T_1 and T_8 are saturated towards zero ($m_{\Delta t,1} = 1$ if $\Delta t < c_1$) and infinite ($m_{\Delta t,8} = 1$ if $\Delta t > c_8$), respectively. The factor ξ fits the algorithm to the territory. As was introduced in Eq. (7.18) it compresses or expands the membership functions associated to Δt attributes to overlay the specific Δt range in a given location. A recipe for the computation of ξ as function of yearly mean of air temperature $\bar{t}$ and yearly mean of daily air temperature amplitude $\overline{\Delta t}$ is reported in (Tulcan-Paulescu and Paulescu 2007):

$$\xi(\overline{\Delta t}, \bar{t}) = 0.00413\bar{t}^3 - 0.964\bar{t}^2 + 1.078\bar{t} - 0.00565\bar{t}^2 \overline{\Delta t} - \\ -0.023\bar{t}\overline{\Delta t} + 0.009476\bar{t}\overline{\Delta t}^2 + 0.495\overline{\Delta t} - 0.0468\overline{\Delta t}^2 + \\ -0.002223\overline{\Delta t}^3 - 3.581 \quad (7.19)$$

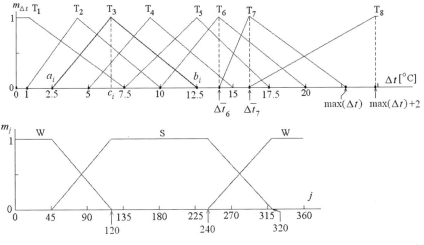

Fig. 7.6 The membership functions of the input linguistic variable: daily temperature range (up) and Julian day (down). The notation for triangular shape of membership functions is indicated to the attribute T_3 ($i = 3$). For $i = 1$ to 7 the c_i is equal to the mean value of elements in set T_i, denoted $\overline{\Delta t_i}$

Equation (7.9 a,b) is not applicable in every location; the condition to use it in a given location characterized by the pair $(\overline{\Delta t}, \overline{t})$ is $0.6 < \xi(\overline{\Delta t}, \overline{t}) < 1.4$.

The results from our study can be regarded as a starting point for future developments of increasing the generality level of temperature based models. The model universality and versatility is determined by the way in which the factor ξ can be related to the local meteo-climate.

The role of the Julian day linguistic variable is to enhance model prediction in cold season, when the irradiation models accuracy decays. Thus it is allowed to enable specific rules for days characterized with WINTER attribute. On the other hand, everyone knows from routine observations that some spring or autumn days are sometimes closer to the summer one and other times to the winter ones; this behavior is well accounted for by the trapezoidal membership functions of Julian day attributes:

$$m_{j,s} = \begin{cases} \max\left(0, 1 - \frac{j-c_1}{b_1-c_1}, \frac{j-c_3}{b_3-c_3}\right) & \text{if} \quad c_1 < j < b_3 \\ 1 & \text{otherwise} \end{cases}$$

$$m_{j,w} = \begin{cases} \max\left(0, \frac{j-a_2}{c_2-a_2}\right) & \text{if} \quad j < c_2 \\ \max\left(0, 1 - \frac{j-c_3}{b_3-c_3}\right) & \text{if} \quad j > c_3 \\ 1 & \text{otherwise} \end{cases} \quad \text{(7.20 a,b)}$$

The membership functions of K_i attributes are fixed as triangular, symmetric and equidistant:

$$m_{k_T,i}(k_T) = \begin{cases} \max\left(0, \frac{k_T-a_i}{c_i-a_i}\right) & \text{if} \quad k_T < c_i \\ \max\left(0, 1 - \frac{k_T-c_i}{b_i-c_i}\right) & \text{otherwise} \end{cases} \quad \text{(7.21)}$$

We underline that the potential users can not tune the output membership functions, as they have no measurements of the daily solar irradiation. They apply the numerical algorithm exactly for obtaining these data. The coefficients a_i, b_i and c_i with the signification from Fig. 7.6, are depicted in Fig. 7.7 where $k_{T,j}$ membership function are displayed.

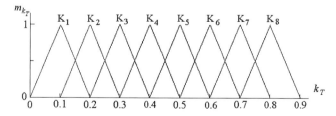

Fig. 7.7 The membership functions of the output linguistic variable $k_{T,i}$ attributes

Table 7.1 Input/output associative rules of the fuzzy algorithm

Rule#	1	2	3	4	5	6	7	8	9	10	11	12	13	14	15	16
Δt	T_1	T_2	T_3	T_4	T_5	T_6	T_7	T_8	T_1	T_2	T_3	T_4	T_5	T_6	T_7	T_8
j	S	S	S	S	S	S	S	S	W	W	W	W	W	W	W	W
k_T	K_1	K_2	K_3	K_4	K_5	K_6	K_7	K_8	K_1	K_3	K_4	K_5	K_5	K_5	K_6	K_7

The input/output mapping of the fuzzy system is presented in Table 7.1. Every rule is encompassed in a column meaning a fuzzy implication in Eq. (7.15). By example the rule #7 is reading:

$$\text{IF } \Delta t \text{ IS } T_7 \text{ AND } j \text{ IS S THEN } k_T \text{ IS } K_7 \tag{7.22}$$

Thus the rules are expressed closer to the human thinking if we bear in mind that the attributes notation with numeral subscript replaces *words*. As a matter of fact, the rule Eq. (7.22) has to be understood as: *If daily temperature amplitude is high in a summer day then also the clearness index is high*, with the assumption that HIGH is associated to T_7 and K_7 attributes.

With the input/output mapping listed as a matrix in Table 7.1, the fuzzy algorithm is ready for use. A handling example of fuzzy model application and its implementation in a computer program are presented in the following.

4.3 Computation Examples

Let's see the functioning of the model by hand-working k_T over the model assumption that the triangle peak coordinate of k_T membership functions is computed as an arithmetical mean of the other two coordinates $c_i = (a_i + b_i)/2$ and $b_7 = 22.5°C$.

The fuzzy model is running for the inputs: $\Delta t = 14°C$ and $j = 90$. The process is illustrated graphically in Fig. 7.8.

1. Fuzzyfication. Crisp inputs are transformed into confidence levels of input linguistic variable attributes, being computed with the equations (7.18) and (7.20 a,b). For $\Delta t = 14°C$, the linguistic variable air temperature amplitude is characterized by three attributes with the corresponding confidence levels:
   T_4: $m_{\Delta t,4} = 0.2$
   T_5: $m_{\Delta t,5} = 0.7$
   T_6: $m_{\Delta t,6} = 0.8$
   Julian day $j = 90$ have both attributes SUMMER and WINTER:
   S: $m_S = 0.4$
   W: $m_W = 0.6$
2. Inference. According to the rule-base from Table 7.1, six rules are set-up. At this step the fuzzy inputs are combined logically using the operator AND (Eqn. 7.14a) to produce the output values:

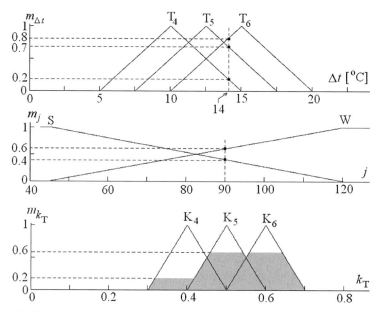

Fig. 7.8 Membership functions associated to the attributes of input linguistic variables $m_{\Delta t}$ and m_j and output linguistic variable m_{k_T}. Only the attributes with nonzero confidence level are plotted. The area corresponding to the integral of output membership functions, truncated at the corresponding degree appears in gray shading

Rule#4 $\quad m_{k_T,4} = \min\left(m_{\Delta t,4}, m_{j,S}\right) = \min\left(0.2, 0.4\right) = 0.2$

Rule#5 $\quad m_{k_T,5} = \min\left(m_{\Delta t,5}, m_{j,S}\right) = \min\left(0.7, 0.4\right) = 0.4$

Rule#6 $\quad m_{k_T,6} = \min\left(m_{\Delta t,6}, m_{j,S}\right) = \min\left(0.8, 0.4\right) = 0.4$

Rule#12 $\quad m_{k_T,5} = \min\left(m_{\Delta t,4}, m_{j,W}\right) = \min\left(0.2, 0.6\right) = 0.2$

Rule#13 $\quad m_{k_T,5} = \min\left(m_{\Delta t,5}, m_{j,W}\right) = \min\left(0.7, 0.6\right) = 0.6$

Rule#14 $\quad m_{k_T,5} = \min\left(m_{\Delta t,6}, m_{j,W}\right) = \min\left(0.8, 0.6\right) = 0.6$

Each rule leads to an attribute of output linguistic variable clearness index. But the rules Rule#5, Rule#12, Rule#13, Rule#14 sum up to the same conclusion, attribute K_5. The different degree of fulfillment K_5 needs to be summarized in just one conclusion, which is achieved by unifiying the individual results with the fuzzy operator OR. Thus the confidence level of output linguistic variable attribute K_5 is obtained as:

$$m_{k_T,5} = \max\left(0.4, 0.2, 0.6, 0.6\right) = 0.6$$

3. Defuzzyfication. The result of the inference process is translated from fuzzy logic into a crisp value using the COG method (Eq. 7.17). After simple manipulation it writes:

$$k_T = \frac{c_4 m_{k_T,4}\left(1 - \frac{m_{k_T,4}}{2}\right) + c_5 m_{k_T,5}\left(1 - \frac{m_{k_T,5}}{2}\right) + c_6 m_{k_T,6}\left(1 - \frac{m_{k_T,6}}{2}\right)}{m_{k_T,4}\left(1 - \frac{m_{k_T,4}}{2}\right) + m_{k_T,5}\left(1 - \frac{m_{k_T,5}}{2}\right) + m_{k_T,6}\left(1 - \frac{m_{k_T,6}}{2}\right)} \quad (7.23)$$

and, using the numerical values from the inference task, the optimal k_T predicted by the fuzzy algorithm is equal to 0.532.

4.4 Program Description

A C program (ProgFuzzy.c) that computes the daily global solar irradiation using these fuzzy procedures, is included on the CD. The membership functions of the linguistic variables *air temperature amplitude* and *Julian day* are defined as in Fig. 7.6. To compute the membership function of Δt attributes Eqs. (7.18) and (7.19) are used. For this a data file "stationtemperatures.prn" is read from disk. It should contain 365 rows with the daily air temperatures organized in 4 tab-delimited columns as follows:

Julian day   Mean   Maxima   Minima

The data is used to calculate the coefficients $c_i = \overline{\Delta t_i}$ and $c_7 = \max(\Delta t)$ in Eq. (7.18) for Δt attributes and yearly average $\overline{\Delta t}$ and $\bar{t}$ in Eq. (7.19) for ξ factor. This file (stationtemperatures.prn) should be prepared by the user. For this, a large on-line database, Global Surface Summary of Day Data, from National Climatic Data Center –NCDC, Asheville, USA, which contains surface meteorological parameters collected over 8000 stations around the world, including air temperature mean maxima and minima is available at http://www.ncdc.noaa.gov.

The program has been designed to compute the global solar irradiation in a given day and for a given air temperature amplitude. The user is asked to input the local latitude (in degrees), Julian day, air temperature maxima and minima (in Celsius). The program will return the global solar energy (in $\mathrm{KWh/m^2/day}$).

The C source file ProgFuzzy.c can be easy modified to meet user requirements. For example, the stationtemperatures.prn file in given example could be extended for a better account of local meteo-climate particularities by adding data of several years. One can build a loop to compute the solar irradiance in a period, by reading the data from a new input file instead of asking for the input from keyboard.

5 Accuracy of Solar Irradiation Models Based on Air Temperature Data

The results synthesized below are based on results reported in a lot of paper concerning the testing of temperature based models under Romanian climate.

The database here considered contains daily global solar irradiation, maximum and minimum air temperature, sunshine duration and daily mean of cloudiness, all recorded in the year 2000. The stations belong to the grid of Romanian Meteorological Agency: Bucuresti (44.5°N; 22.2°E; 131m), Constanta (44.2°N, 28.6°; at the Black Sea seacoast), Craiova (44.3°N; 23.8°E; 110m), Iasi (47.2°N; 27.6°E; 130m) and Timisoara (45.7°N; 21.2°E; 85.5m), Galati (45.48°N; 28.01°E; 72m).

The accuracy of different models is compared using two statistical indicators: Relative Root mean Square Errors (*rrmse*) and Relative Mean Bias Errors (*rmbe*) which are reading as:

$$rrmse = \sqrt{n \sum_{i=1}^{n} (F_i - y_i)^2 \bigg/ \sum_{i=1}^{n} y_i}, \quad rmbe = \sum_{i=1}^{n} (F_i - y_i) \bigg/ \sum_{i=1}^{n} y_i \quad (7.24)$$

where y_i and F_i are the i-th measured and computed values of radiation quantities, while n is the number of measurement taken into account.

In Table 7.2, air temperature based models are compared with models that do not include air temperature as input. The models have been run with the following parameters: $a = 0.075$, $b = 0.428$ and $c = -0.283$ in equation (7.1) being appropriate for 45°N latitude; $a = 0.71$, $b = 0.112$, $c = -6.72 \cdot 10^{-3}$ and $d = -0.283$ in Eq. (7.11) as mean values at 45°N latitude, provided by the author (http://www.isci.it). The empirical irradiance A model (Adnot et al. 1979) has been used for the clear sky global solar irradiation which was carried along in the Kasten and Czeplak (1979) equation for daily solar irradiation. The parametric Hybrid model proposed by Yang et al. (2001) has been run with an Ångström – Prescot type equation provided by them. The input used local recorded parameters with two exceptions: the depth of ozone layer equal to $0.35 \, \text{cm} \cdot \text{atm}$ and the Ångström turbidity coefficient $\beta = 0.089$ computed as a mean value after Yang et al. (2001). We place these models in Table 7.2 because it was proved that they are appropriate for Romania (Badescu, 1997; Paulescu and Schlett, 2004).

The results from Table 7.2 demonstrate that the estimation of monthly mean of daily global solar irradiation can be performed with an acceptable accuracy using temperature based models. This is comparable with the accuracy of estimation using classical models. The adding of air temperature to cloudiness in equations like Eq. (7.1) is not leading to significant improvements. But the solar irradiation can be computed via cloudiness and air temperature if sunshine duration data is missing. The models which use only air temperature as parameter (including the fuzzy model) shows the same accuracy but have the merit to use for input the highest

Table 7.2 Range of statistical indicators of accuracy of monthly mean daily global solar irradiation. The models have been applied at the mentioned stations for the year 2000. Statistical indicator range includes results from Paulescu and Schlett (2004); Paulescu et al. (2006)

Model	Input parameters	*rrmse*	*rmbe*
H/H_e, Eq. (7.1)	N, Δt	[0.109, 0.211]	[−0.132, 0.123]
H/H_e, Eq. (7.7)	N, t_{max}, t_{min}	[0.081, 0.134]	[−0.078, 0.064]
H/H_e, Eq. (7.11)[a]	Δt, Δt_w	[0.095, 0.144]	[−0.129, 0.107]
H/H_0, Eq. (7.12 a,b,c)	Δt, $\bar{t}$	[0.079, 0.175]	[−0.118, 0.094]
Fuzzy	t_{max}, t_{min}	[0.082, 0.062]	[−0.020, 0.006]
A	N	[0.062, 0.118]	[−0.029, 0.082]
Hybrid	σ	[0.065, 0.121]	[−0.045, 0.069]

[a]*rrmse* range not includes data from the station of Constanta where it is >0.3

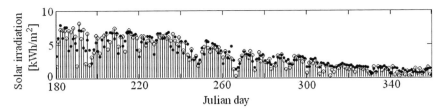

Fig. 7.9 Estimated with the fuzzy model and observed daily global solar irradiation at the station of Timisoara in the last six months of the year 2000

spatial recorded meteorological parameter. Since these models are close connected to the origin location need careful calibration when are applied in location with air temperature special regimes (seacoast or higher altitudes). Regarding the details of daily global solar irradiation, Fig 7.9 shows that the estimation of daily solar irradiation with a temperature model tracks actual measurements with good accuracy.

6 Conclusions

Simple formulae that can be used to calculate daily global solar irradiation based on air temperature data have been exposed. These models either using air temperature as additional parameter to cloudiness or using only air temperature are both viable alternatives to the classical equations based on sunshine duration. These equations may be useful in many locations where sunshine duration measurements are missing but air temperature measurements are available in many-year database. Thus, the number of sites where the estimation can be performed is much higher. The methods based on temperature database comparison are able in many cases to exceed the sensitivity of temperature models to origin location. A distinct case is the model built inside fuzzy logic, which may exhibit the flexibility needed in solar energy forecast. The readers can test the presented fuzzy model included on the CD and potential users are encouraged to modify the fuzzy procedures in order to customize particular applications.

References

Adnot J, Bourges B, Campana D, Gicquel R (1979) Utilisation des courbes de frequence cumulees pour le calcul des installation solaires. In Analise Statistique des Processus Meteorologiques Appliquee a l'Energie Solaire, Lestienne R. Paris

Ångström A (1924) Solar and terrestrial radiation. Quart J Roy Meteor Soc 50:121–126

Badescu V (1997) Verification of some very simple clear and cloudy sky models to evaluate global solar irradiance. Sol Energy 61: 251–264

Bristow KL, Campbell LC (1984) On the relationship between incoming solar radiation and daily maximum and minimum temperature. Agric For Meteor 31:159–166

Cesaraccio C, Spano D, Duce P, Snyder RL (2001) An improved model for determing degree-day values from daily temperature data. Int J Biometeorol 45:161–169

Cheeroo-Nayamuth BF (1999) Crop modeling/simulation: An overview. In Proc of 4th AMAS, Reduit, Mauritius, pp. 11–26

Donatelli M, Bellocchi G (2001) Estimate of daily global solar radiation: new developments in RadEst3. In Proc of 2nd International Symposium Modeling Cropping Systems, Florence, pp. 213–214

Donatelli M, Bellocchi G, Fontana F (2003) RadEst3.00: Software to estimate daily radiation data from commonly available meteorological variables. Eur J Agron 18:363–367; http://www.isci.it/

Gomez V, Casanovas A (2003) Fuzzy modeling of solar irradiance on inclined surfaces. Sol Energy 75:307–315

Hargreaves GL, Hargreaves GH, Riley P (1985) Irrigation water requirement for the Senegal River Basin. J Irrig Drain E-ASCE 111:265–275

Haurvitz B (1945) Insolation in relation with cloudiness and cloud density. J Meteorol 2:154–166

Jain PC (1990) A model for diffuse and global irradiation on horizontal surfaces. Sol Energy 45:301–308

Kasten F, Czeplak G (1980) Solar and terrestrial radiation dependent on the amount and type of clouds. Sol Energy 24:177–189

Liu BY, Jordan RC (1960) The inter-relationship and characteristic distribution of direct, diffuse and total solar radiation. Sol Energy 4:1–19

El-Metwally M (2004) Simple new methods to estimate global solar radiation based on meteorological data in Egypt. Atmospheric Research 69:217–239

Passino KM, Yurkovich S (1998) Fuzzy Control. Addison Wesley Longman Menlo Park

Paulescu M, Schlett Z (2004) Performance assessment of of global solar irradiation models under Romanian climate. Renew Energy 29:767–777

Paulescu M, Fara L (2005) On the relationship between global solar radiation and daily maximum and minimum air temperature. UPB Sci Bull A 67:41–50

Paulescu M, Fara L, Tulcan-Paulescu E (2006) Models for obtaining daily global solar irradiation from air temperature data. Atmos Res 79:227–240

Prescott JA (1940) Evaporation from water surface in relation to solar radiation. Trans Roy Soc Austr 64:114–118

Santamouris M, Mihalakakou G, Psiloglou B, Eftaxias G, Asimakopoulos DN (1999) Modeling the Global Solar Radiation on the Earth's Surface Using Atmospheric Deterministic and Intelligent Data Driver Techniques. J Climate 12(10):3105–3116

Sirotenko OD (2001) Crop Modeling: Advances and Problems. Agron J 93:650

Sen Z (1998) Fuzzy algorithm for estimation of solar irradiation from sunshine duration. Sol Energy 63:39–49

Supit I, Van Kappel RR (1988) A simple method to estimate global radiation. Sol Energy 3:147–160

Thorton PE, Running SW (1999) An improved algorithm for estimating incident daily solar radiation from measurements of temperature, humidity and precipitation. Agric For Meteor 93;211–228

Tulcan-Paulescu E, Paulescu M (2007) A fuzzy model for solar irradiation via air temperature data. Theor Appl Climatol, DOI: 10.1007/s00704-007-0304-6

Wiss A, Hays CJ, Hu Q, Easterling WE (2001) Incorporating Bias Error in Calculating Solar Irradiance: Implications for Crop Simulations. Agronom J 93:1321–1326

Yang K, Huang G W, Tamai N (2001) A hybrid model for estimating global solar irradiance. Sol Energy 70:13–22

Zadeh LA (1965) Fuzzy sets. Inf Control 8:338–353

Zimmermann HJ (1996) Fuzzy set theory and its application 3rd-ed Kluwer Academic Publishers, Boston, MA

Chapter 8
Models of Diffuse Solar Fraction

John Boland and Barbara Ridley

1 Introduction

This chapter continues the work of Boland and Scott (1999) and Boland, Scott and Luther (2001) who developed models for some Australian locations using the clearness index and time of day as predictors. More recently, Boland and Ridley (2007) have presented the theoretical basis for a generic model for diffuse radiation, and additionally, a methodology for identifying possibly spurious values of measured diffuse. There is strong motivation for undertaking this study, wherein a number of Australian locations have been included. Spencer (1982) adapted Orgill and Hollands (1977) model and tested it on a number of Australian data sets for the reason that most of the work in the field has been performed using higher latitude North American and European data sets.

The evaluation of the performance of a solar collector such as a solar hot water heater or photovoltaic cell requires knowledge of the amount of solar radiation incident upon it. Solar radiation measurements are typically only for global radiation on a horizontal surface. They may be on various time scales, by minute, hour or day. Additionally, one can infer global radiation from satellite images. We have used inferred daily totals of global radiation. Presently, there is some satellite inferred data available at the three hour time scale, and it is expected that this will become more widespread in the future. At present we will only assume daily data available for a wide range of locations.

These global values comprise two components, the direct and the diffuse. "I_{DN}, the direct normal irradiance, is the energy of the direct solar beam falling on a unit area perpendicular to the beam at the Earth's surface. To obtain the global irradiance

John Boland
University of South Australia, Mawson Lakes, e-mail: john.boland@unisa.edu.au

Barbara Ridley
University of South Australia, Mawson Lakes, e-mail: barbara.ridley@unisa.edu.au

the additional irradiance reflected from the clouds and the clear sky must be included" (Lunde 1979, p. 69). This additional irradiance is the diffuse component.

Typically solar collectors are not mounted on a horizontal surface but tilted at some angle to it. Thus it is necessary to calculate values of total solar radiation on a tilted surface given values for a horizontal surface. It is not possible to merely employ trigonometric relationships to calculate the solar radiation on a tilted collector. This is because the diffuse radiation is anistropic over the sky dome and the "radiative configuration factor from the sky to the tilted solar collector is not only a function of the collector orientation, but is also sensitive to the assumed distribution of the diffuse solar radiation across the sky" (Brunger 1989). There are two different approaches to calculating the diffuse radiation on a tilted surface; using analytic models (Brunger 1989) or empirical models such as that of Perez et al. (1990). Each relies on knowledge of the diffuse radiation on a horizontal surface. The diffuse component is not generally measured. Consequently, it is very useful to have a method to estimate the diffuse radiation on a horizontal surface based on the measured global solar radiation on that surface.

Numerous researchers have studied this problem and have been successful to varying degrees. Liu and Jordan (1960) developed a relationship between daily diffuse and global radiation which has also been used to predict hourly diffuse values. The predictor typically used in studies is not precisely the global radiation but the "hourly clearness index k_t, the ratio of hourly global horizontal radiation to hourly extraterrestrial radiation" (Reindl et al. 1990). Orgill and Hollands (1977) and Erbs (1982) correlate the hourly diffuse radiation with k_t, but Iqbal (1980) extended the work of Bugler (1977) to develop a model with two predictors, k_t and the solar altitude. Skartveit and Olseth (1987) also use these two predictors in their correlations. Reindl et al. (1990) use stepwise regression to "reduce a set of 28 potential predictor variables down to four significant predictors: the clearness index, solar altitude, ambient temperature and relative humidity." They further reduced the model to two predictor variables, k_t and the solar altitude, because the other two variables are not always readily available. Another possible reason was that some combinations of predictors may produce unreasonable values of the diffuse fraction, eg. greater than 1.0 (Reindl et al. 1990). Skartveit et al. (1998) developed a model which in addition to using clearness index and solar altitude as predictors, have added a variability index. This is meant to add the influence of scattered clouds on the sky dome.

As well, Gonzalez and Calbo (1999) stress the importance of including the altitude and the variability of the clearness index in any predictions of the diffuse fraction. Aguiar (1998) fitted an exponential model to Mediterranean daily data using only the clearness index and found a consistency of fit amongst locations of similar climate.

Boland et al. (2001) developed a validated model for Australian conditions, using a logistic function instead of piecewise linear or simple nonlinear functions. Recently, Jacovides et al. (2006) have verified that this model performs well for locations in Cyprus. Their analysis includes using moving average techniques to demonstrate the form of the relationship, which corresponds well to a logistic relationship. Suehrcke and McCormick (1988) and McCormick and Suehrcke (1991) present some significant work on modelling diffuse radiation, including pointing out that "instantaneous diffuse fraction correlation differs markedly from the correlations obtained for integrated diffuse fractions". However, in most instances, it is integrated values that are normally available for modelling purposes, and indeed it is integrated values that are used in performance estimation software. Thus, we are responding to this specific need in providing understanding of the modelling issues on an hourly time scale.

We have made significant advances in both the physically inspired and formal justification of the use of the logistic function. In the mathematical development of the model utilising advanced non-parametric statistical methods, we have also constructed a method of identifying values that are likely to be erroneous. The method, using quadratic programming, will be described. Using this method, we can eliminate outliers in diffuse radiation values, the data most prone to errors in measurement. Additionally, this is a first step in identifying the means for developing a generic model for estimating diffuse from global and other predictors. Examples for both Australian and locations in other parts of the world will be presented.

2 Defining the Problem

We shall begin the discussion by limiting ourselves to using one predictor variable only, the clearness index. It is the most significant predictor, and as such is the basis for the earlier models. In Section 8, we discuss the use of additional explanatory variables.

There are two data sets for consideration in the development of the model and its subsequent use. The first, Adelaide, is a good quality data set recorded by the Australian Bureau of Meteorology (BOM). The second, Geelong, is a set of data collected at a private weather station at Deakin University, near Melbourne. The reading apparatus is known to fail from time to time and will give infeasible values for diffuse radiation. We deal specifically with an hourly time scale in this investigation. Many of the simulations used to model the performance of systems under the influence of solar inputs for which this estimation technique is required, such as house energy ratings scheme software, are hourly based. The first step in analysing the data is to construct two new variables. The variables are

$$k_t = \frac{I_g}{H_0}, \quad d = \frac{I_d}{I_g} \tag{8.1}$$

where I_g, I_d, H_0 are the global, diffuse and extraterrestrial radiation integrated over the hour in question. Figures 8.1 and 8.2 display hourly values of diffuse fraction against the clearness index.

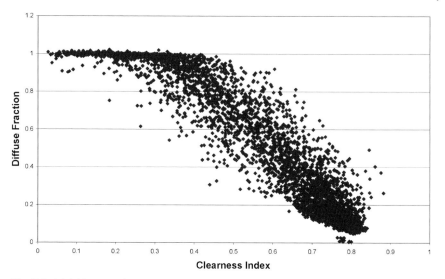

Fig. 8.1 Adelaide – raw data

In Fig. 8.2, the points in the top right hand corner may be considered suspect due to the high clearness index combined with a high diffuse fraction. These values may be spurious and are best removed from the data set before continuing with model fitting. So we thus define the basis for our initial investigation. We examine a model relating diffuse fraction, and thus diffuse radiation, to clearness index. Subsequently, we develop a methodology for rejecting data values with low probability of being feasible.

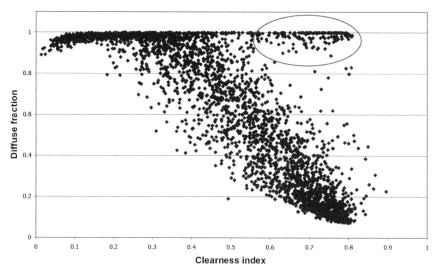

Fig. 8.2 Geelong – raw data

3 Constructing a Model of the Diffuse Fraction

We begin with examining qualitatively the rationale for using a logistic function to model the diffuse fraction as a function of clearness index. We then derive the mathematical framework supporting the suitability of this form of relationship.

3.1 Justification for a Logistic Function – Experimental Data Analysis

As mentioned previously, we desire to investigate the form of a relationship for diffuse fraction as predicted by clearness index for several reasons. We propose a better one-predictor variable (k_t) model for Australian conditions because the models developed elsewhere have not proven adequate for Australian conditions (Spencer, 1982). This then leads to a proposal of another one-predictor functional form for site-specific models. Also, we propose a parsimonious diffuse model, with fixed coefficients, for general application, which may be used to estimate diffuse solar radiation in the absence of measured data. Such a model may well be able to be parameterised to suit diverse climates.

To understand the relationship better, we construct a moving average of the diffuse fraction, as a function of the clearness index, rather than the standard version, a function of time. Specifically,

$$d_{ave} = \frac{1}{N} \sum_{i=1}^{N} d_i(k_t) \tag{8.2}$$

The functionality with respect to clearness index comes as a result of ordering the diffuse fraction for increasing clearness index. Figure 8.3 displays the moving average for a moving window of length $N = 101$ for the Adelaide data. It seems that a logistic type function will suit the modelling of the relationship as it can be spatially flexible for modelling purposes and requires a minimum number of parameters. Polynomials of orders 3 and 4 were checked for suitability, but gave a weaker fit than the logistic function.

A logistic growth model, growth being the most common phenomenon modelled in this fashion, resembles exponential growth in the early stages but in the later stages there is a reversal of concavity to reflect either a limited amount of resources or a maximum life span or size for example. However, logistic decay is also a possibility in the biological realm. For instance, in predator-prey models logistic decay can prevail in modelling predator numbers when multiple sources of food are present.

The concept of what drives logistic growth or decay was examined by Nash (1975). He describes a system where individuals occupy one of two states, one with growth level 0 and one with growth level h. The transition from level h to level 0 is not possible and the transition probability for $0 \rightarrow h$ is proportional to

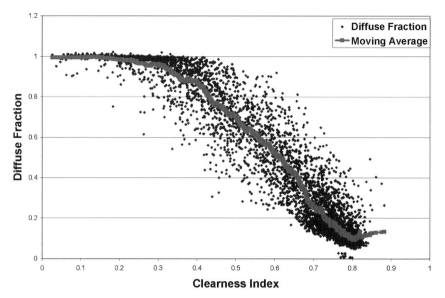

Fig. 8.3 Moving average of diffuse fraction as a function of clearness index

the percentage of individuals in level h. We tend to get a so-called "band wagon" effect where the probability increases as the number of individuals increases, but of course, there is not an infinity of individuals available, so the transition probability must begin at some stage to decrease. It is essentially a non-stationary Markov process.

Perhaps we can envisage the dependence of diffuse fraction on clearness index in a similar "band wagon" manner, but with decay. As the atmosphere becomes clearer there is an increasing tendency to clear but of course there is a saturation effect here as well – the atmosphere can only tend towards perfectly clear. However, one would expect that this changing probability would progress in a smooth fashion, rather than in a piecewise linear version, as taken by the earliest diffuse fraction models. The subsequent section will describe the theoretical justification.

3.2 The Theoretical Development of the Relationship

The first step is to develop the procedure for a systematic fit of a function that best describes the data for Adelaide – seemingly data of reasonable quality. We want to fit a linear function of the form

$$y_i = \beta_0 + \beta_1 x_i + \varepsilon_i \tag{8.3}$$

to the data. This requires that the errors, ε_i, be independent and identically distributed. This involves transforming the data into a form wherein there is homogeneity of the data as the clearness index varies. As can be seen in Figure 8.1, there is a much greater spread for the diffuse fraction for middle values of the clearness index.

3.2.1 The Homogenising Transformation

In this subsection we give the derivation of the transformation that will take the original data set into a homogeneous band wherein we can use regression techniques to estimate parameters β_0, β_1 in Eq. (8.3).

Define a set of ordered pairs (x_i, y_i), $i = 1, \ldots, n$, where x_i represents the clearness index k_t and y_i represents the diffuse fraction d_t. Discretise the y values into k sub-intervals of equal size. The value of k ranged between 50 and 100 for various locations, depending on the number of points and the ability of the software we used, Matlab, to handle the transformation. Write

$$y = Ua \tag{8.4}$$

with constraints

$$y^T \cdot 1 = 0 \tag{8.5}$$

$$y^T \cdot y = 1 \tag{8.6}$$

where a, which is $k \times 1$, records the bin scores and U, which is $n \times k$, is the incidence matrix. An incidence matrix is a matrix that shows the relationship between x and y, the matrix has one row for each element i of x and one column for each element j of y. The entry in row i and column j is 1 if x and y are related (called incident in this context) and 0 if they are not.

The constraints Eqs. (8.5) and (8.6) are present to allow the data to be standardised. This means the data is centred around zero (Eq. (8.5)) with a constant variance of unity (Eq. (8.6)).

The regression model in matrix form is $\mathbf{y} = X^T \beta + \varepsilon$, and using least squares, the residual sum of squares is

$$\varepsilon^T \varepsilon = (y - X\beta)^T (y - X\beta) \tag{8.7}$$

The ordinary least squares problem is to find the best estimate $\hat{\beta}$ to minimise $\varepsilon^T \varepsilon$. Therefore the problem is to find

$$
\begin{aligned}
\min \varepsilon^T \varepsilon &= (y - X\hat{\beta})^T (y - X\hat{\beta}) \\
&= (y^T - \hat{\beta}^T X^T)(y - X\hat{\beta}) \\
&= y^T y - \hat{\beta}^T X^T y - y^T X\hat{\beta} + \hat{\beta}^T X^T X\hat{\beta} \tag{8.8}
\end{aligned}
$$

But, from the theory of regression, the best estimate for the parameters is $\hat{\beta} = (X^T X)^{-1} X^T y$, from which we also get $\hat{\beta}^T = y^T X[(X^T X)^{-1}]^T$. Substituting these into Eq. (8.8), we get

$$
\begin{aligned}
\min \varepsilon^T \varepsilon &= y^T y - y^T X (X^T X)^{-1} X^T y - y^T X [(X^T X)^{-1}]^T X^T y \\
&\quad + y^T X [(X^T X)^{-1}]^T X^T X (X^T X)^{-1} X^T y \\
&= y^T y - y^T X (X^T X)^{-1} X^T y - y^T X [(X^T X)^{-1}]^T X^T y \\
&\quad + y^T X [(X^T X)^{-1}]^T X^T y \\
&= y^T y - y^T X (X^T X)^{-1} X^T y \\
&= y^T [I - X (X^T X)^{-1} X^T] y
\end{aligned}
\tag{8.9}
$$

If we write $P_X = X(X^T X)^{-1} X^T$, then we obtain

$$
\begin{aligned}
\min \varepsilon^T \varepsilon &= y^T (I - P_X) y \\
&= y^T y - y^T P_X y \\
&= 1 - y^T P_X y
\end{aligned}
\tag{8.10}
$$

P_X is called the projection matrix. The problem is to minimise the residual sum of squares $\varepsilon^T \varepsilon$, but $\varepsilon^T \varepsilon \geq 0 \Rightarrow 1 - y^T P_X y \geq 0$. Therefore, $y^T P_X y \leq 1$ and the problem becomes to

$$
\max y^T P_X y
\tag{8.11}
$$

The method of Lagrange multipliers takes a problem with objective function plus constraints and converts it such that the constraints enter the objective function that is to be minimised or maximised. Now, by introducing the Lagrange multipliers λ_1 and λ_2 for each constraint Eqs. (8.5), (8.6) a linear combination is formed involving the multipliers as coefficients. The objective for the optimisation is expressed as

$$
\max \ y^T P_X y + \lambda_1 y^T 1 + \lambda_2 (y^T y - 1)
\tag{8.12}
$$

In terms of **a**, the constraints are

$$
\begin{aligned}
\mathbf{a}^T U^T 1 &= 0 \\
\mathbf{a}^T U^T U \mathbf{a} &= 1
\end{aligned}
\tag{8.13}
$$

Now, in order to denote the constraints as a function of **a**, let $v_i = \sqrt{n_i}$ and $b_i = \sqrt{n_i} a_i$, thus

$$
\begin{aligned}
\mathbf{b} &= \Lambda^{1/2} \mathbf{a} \\
\mathbf{a} &= \Lambda^{-1/2} \mathbf{b}
\end{aligned}
\tag{8.14}
$$

where

$$
\Lambda^{1/2} =
\begin{bmatrix}
\sqrt{m_1} & 0 & \cdots & 0 \\
0 & \sqrt{m_2} & \cdots & 0 \\
0 & \vdots & \ddots & \vdots \\
0 & 0 & \cdots & \sqrt{m_n}
\end{bmatrix}
\tag{8.15}
$$

and m_j is the number of elements in bin j. Thus

$$
\begin{aligned}
y^T P_X y &= \mathbf{a}^T U^T X (X^T X)^{-1} X^T U \mathbf{a} \\
&= \mathbf{b}^T \Lambda^{-1/2} U^T X (X^T X)^{-1} X^T U \Lambda^{-1/2} \mathbf{b}
\end{aligned}
\tag{8.16}
$$

Let

$$G = \Lambda^{-1/2} U^T X (X^T X)^{-1} X^T U \Lambda^{-1/2} \tag{8.17}$$

Now, by substituting these equations into the objective function it can now be written

$$\text{maximize } z = b^T G b + \lambda_1 v^T b + \lambda_2 b^T b \tag{8.18}$$

Further, we can derive $v^T G = 0$. Hence v is an eigenvector of G for eigenvalue zero, and it can be shown that b is a leading eigenvector of G. Thus b can be found, hence a, the bin scores, is calculable.

3.2.2 Use of Transformed Data

The transformation is now applied and the data, as shown in Fig. 8.4, is seen as forming an approximately homogeneous band to which a linear function can be fitted.

The linearly transformed data is now analysed in the statistical analysis package Minitab using Sen's method (Sen, 1968) to determine the slope for the linear function. This is a non-parametric method of estimating the slope by taking all possible pairs of points and calculating the slope between each of them. Ordinary least squares should be sufficient after the transformation, but we ensure we obtain robust estimators in the case where the errors are identically distributed but not necessarily normal. The optimal slope is the median of these values. However, Minitab has an upper limit on the number of slopes that it can calculate, so, where the size of the data set is greater than 4000, a random sample of size 4000 is taken from the two

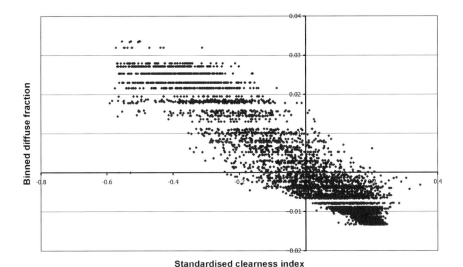

Fig. 8.4 Adelaide data transformed to give a homogeneous band

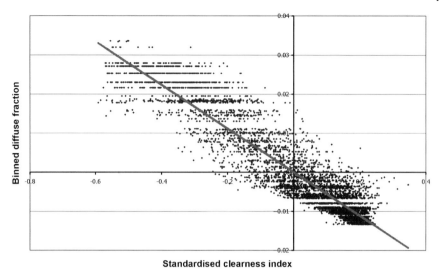

Fig. 8.5 Adelaide transformed data with line of best fit

columns *x* and *y*, which are the output from linearization. Figure 8.5 illustrates the
transformed data with the fitted line.

Figure 8.6 depicts the data and the fitted line back-transformed. This figure provides the pictorial justification that indeed the logistic curve model as described in
Boland et al. (2001) is the representation of the data that is most suitable. The model
depicted here is not smooth since the transformation is not performed on continuous
data but on the data collected into bins or sub-intervals.

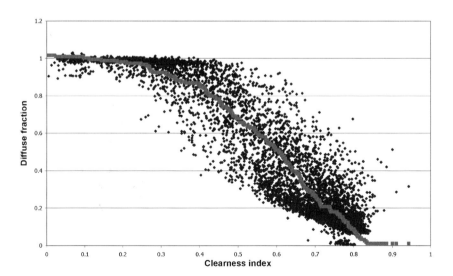

Fig. 8.6 Adelaide data and fitted line transformed back to the original range

The final step in determining the model equation is to fit a logistic function to the back-transformed line of best fit. The form of this equation is

$$d = \frac{1}{1 + e^{\beta_0 + \beta_1 k_t}} \qquad (8.19)$$

There are various methods for performing the fit. One method is to transform Eq. (8.19) into a linear equation in β_0, β_1 and apply linear regression techniques. To do so, however, we need to make a slight alteration.

$$\ln\left(\frac{1 - d_i}{d_i}\right) = \beta_0 + \beta_1 k_{ti} \qquad (8.20)$$

In order to perform this last step, $d_i < 1 \forall i$, since $\ln(0)$ is undefined. Therefore, all diffuse fraction values equal to unity have to be slightly adjusted to something like $1 - 1 \times 10^{-5}$. An alternative method that doesn't involve this alteration uses the Solver utility in Excel (see Section 10 for details on how to implement it for this problem). Using this tool, we define a function involving the unknown parameters, construct the sum of squared deviations between the model and the data values, and then sum these. We minimise the sum of squared deviations by picking the best estimates of β_0, β_1. This is performed by the method of steepest descent or similar in Solver.

Figure 8.7 gives the comparison of the broken curve and the logistic function, both plotted against the data. Thus, we now have the functional form of the model for the diffuse fraction as a function of clearness index.

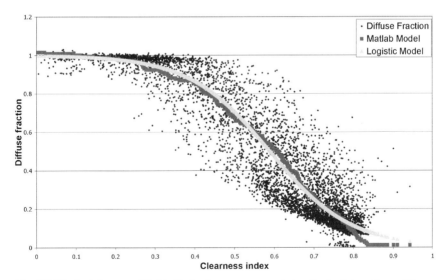

Fig. 8.7 The logistic curve fitted to the broken curve generated as the back-transform of the non-parametric line of best fit

This derivation has been specific for data from one location, Adelaide. There are many questions arising from this description that must be addressed. For example, there are less than 20 BOM locations in Australia where diffuse radiation is measured. Thus, this sort of derivation can be performed only for those sites. For these locations, one could use a model like this to predict what the diffuse fraction, and thus the diffuse radiation would be if the global radiation values are available. This could be useful for filling in missing values of the diffuse component when there is equipment failure. Or, if diffuse radiation had been measured for some time, and then discontinued, the diffuse could be predicted for either measured global solar or global inferred from satellite data. Another very important instance is related to "weather generators" and the production of Reference Meteorological Years by statistical and /or stochastic methods.

However, an important use for such a model is in the prediction of diffuse for locations where there are no diffuse measurements available, only global. How good is this model for use with sites other than the one where the model building has proceeded? In other words, can we somehow construct a generic model, since the main problem we are trying to deal with is the lack of transportability of models constructed for other climates? Another question is whether there are ways to incorporate other predictors to enhance the fit of the model. Finally, how can we make use of this model to help identify data values that have a high probability of being the result of some problems with the recording equipment, and thus eliminate them from the data set? This refers back to the problem with the Geelong data. In the next section, we will go towards answering the first and last questions. We will give some information about our progress with adding other predictor variables in Section 8. We will also discuss how to deal with situations where only daily totals of solar radiation are available.

4 Results for Various Locations

We have applied the algorithm to various locations for which we have data, in various parts of the world. Table 8.1 gives values we have estimated for the parameters, β_0 and β_1, and also a measure of goodness of fit of the model. We have used

Table 8.1 Parameter estimates and NRMSD for various locations

	AUSTRALIA		AFRICA	EUROPE			ASIA	
	Adelaide	Darwin	Maputo	Bracknell	Lisbon	Uccle	Macau	Average
$\hat{\beta}_0$	−8.83	−4.53	−8.18	−4.38	−4.80	−4.94	−4.87	−4.94
$\hat{\beta}_1$	9.87	8.05	8.80	6.62	7.98	8.66	8.12	8.30
NRMSD(%)	26.7	24.7	28.7	13.1	22.2	18.9	17.1	

the normalised root mean square difference (NRMSD) – see Chapter 11 for the definition.

The estimates given in the table are not so dissimilar as to make us believe that we have to have a separate model for each separate location. Inspection of figures constructed using average values of the estimates of the parameters leads us to believe that there is scope for use of a so-called generic model. This model could be used to predict the diffuse radiation for any location necessary. Note that Geelong has not been included in determination of the average values. We can now use the generic model as a model for Geelong. We would argue that this is a better approach than building a separate model for Geelong anyway, since we believe that the Geelong data contains many infeasible values.

To construct the generic model, we aggregate the data from the various locations, apart from Geelong, and apply the algorithm for estimating the parameters. In this way, we obtain

$$d = \frac{1}{1 + e^{-5.0 + 8.6k_t}} \tag{8.21}$$

5 Validation of the Model

We present validation of this model in three separate ways. In the first instance, Figure 8.8 gives a visual comparison of the moving average given in Fig. 8.3 and the generic model.

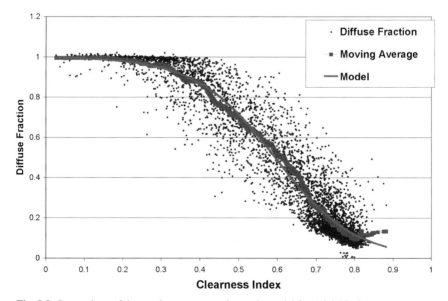

Fig. 8.8 Comparison of the moving average and generic model for Adelaide data

These curves are not dissimilar, thus lending credence to the idea that the generic model is suitable. The second method of validation is given by the work of Jacovides et al. (2006) who compared 10 models in the literature for predicting diffuse fraction. One of these was a precursor to this generic model (Boland et al. 2001), wherein we used some data from the Geelong weather station to estimate parameters. The estimated β_0, β_1 were similar to the present ones. Jacovides and his co-workers found that our model performed well for data from Cyprus in a reasonably exhaustive study.

In fact, two of the other models that were tested in that study, that of Reindl et al. (1990) and Karatasou et al. (2003) have been used here for a further validation of the generic model. The reason that the Reindl model has been chosen for this validation is that the study in question appears to have been quite well performed, with a list of 28 possible predictor variables being examined for their worth. We will refer to this study in Section 8 on adding more explanatory variables. The Karatasou model was one of the best performing models in the study – as it constructs the model with data from that region. Figure 8.9 gives a comparison between their one predictor models and the present model. The Karatasou model is the lowest curve, the Reindl one is the piecewise linear and our model is the third. Additionally, we calculated the NRMSD and normalised mean bias difference (NMBD) for all three models and obtained results given in Table 8.2. It can be seen from these results that the new model performs very well compared to these models. Further comparisons of this type will be reserved for the model developed with more predictors.

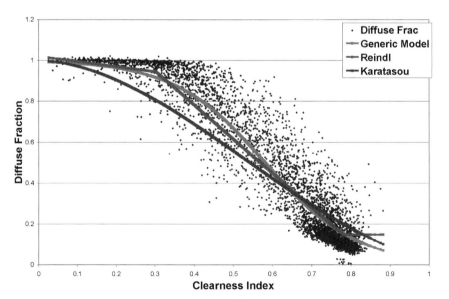

Fig. 8.9 The generic model and the Reindl and Karatasou models applied to the Adelaide data

Table 8.2 Comparison of statistical measures – the present model and the Reindl and Karatasou models

	Boland and Ridley	Reindl et al.	Karatasou et al.
NRMSD (%)	22.9	24.1	30.8
NMBD (%)	1.4	2.1	8.1

6 Identifying Outliers

We use quadratic programming to identify the data values with low probability of occurrence under the assumption that we have found a valid expression for the fitted curve. A linearly constrained optimisation problem with a quadratic objective function is called a quadratic programming problem. It may have no solution, a unique solution, or more than one feasible solution. If there are n points then we would expect that feasible values would have the probability of occurrence $p_i \approx 1/n$, and those infeasible will have a probability close to zero. Mathematically we rely on the fact that the least squares empirical likelihood for simple linear regression takes the observations $x_i =$ clearness index and $y_i =$ diffuse fraction, and will choose $\{p_i\}$ to minimise $\sum (p_i - 1/n)^2$ subject to

$$\sum p_i(y_i - \hat{y}_i) = 0$$
$$\sum p_i x_i(y_i - \hat{y}) = 0$$
$$\sum p_i = 0$$
$$p_i \geq 0 \qquad (8.22)$$

Here, $\hat{y}_i = (1 + e^{\hat{\beta}_0 + \hat{\beta}_1 k_t})^{-1}$ and $p_i = \Pr(y_i = \hat{y}_i)$.

The constraints once again are of the standardising type. The first requires that the sum of the departures from the best estimate must be zero, while the second one forces the variance to be unity. Figure 8.10 gives the histogram of the probabilities associated with the Geelong data values. We hope to obtain a number of $p_i \equiv 0$, but this is not always the case. Here $n = 3166 \Rightarrow 1/n = 0.000316$. If we delete the lowest 5% of the probabilities, we obtain the "cleaned" data in Figure 8.11, with the generic model overlaid. It is a comprehensive display of the effectiveness of the approach we are using. The data now resembles the sort of scatter that one would expect from this type of graph.

As a second exercise of this type, we chose a climate dissimilar to what one might find in Australia, Bracknell in England. We used the generic model once again as the best estimate, and then applied the quadratic programming algorithm to quality assure the data. In this instance, we obtained 157 probabilities exactly zero out of 3462 values, or 4.53%. Figures 8.12 and 8.13 give the quality assured data and also the deleted data respectively. One can see from Fig. 8.13 that the quadratic programming algorithm has rejected only data values in the upper right

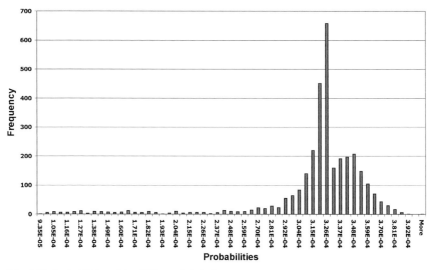

Fig. 8.10 Probabilities of feasibility and their frequency

corner, the principal area of concern. We believe this is a rigorous statistical process, and if by chance – this is a statistical determination – some valid data points are eliminated, it is not a significant problem. We are "cleaning" the data in order to better construct a model, and the loss of a few valid points will not affect that process.

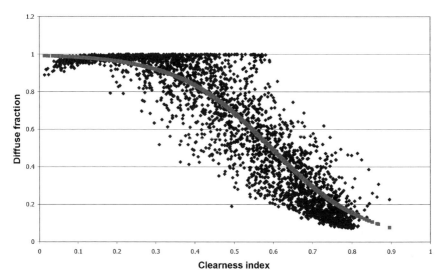

Fig. 8.11 Geelong data with outliers removed, and the generic model superimposed

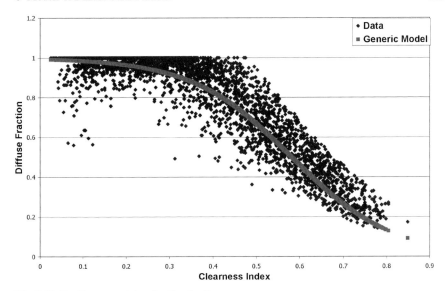

Fig. 8.12 Quality assured data for Bracknell

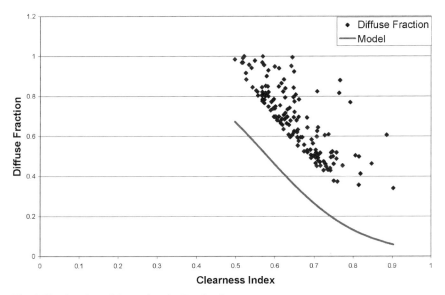

Fig. 8.13 The rejected data points for Bracknell

7 Conclusion for the Single Predictor Variable Model

We have demonstrated a statistically rigorous method of constructing a closed form function model for the diffuse fraction. Additionally, we have shown how it, along with an innovative quadratic programming formulation, can be used to identify values that have a high probability of being infeasible. The model has been deemed suitable for modelling in general since we checked it for a number of locations in different climates. We have checked the data cleaning capability for other locations and it performed well, but we have only used Geelong as an example. We continue to work on improvements to this modelling in the following ways:

- it is not certain if we should be using the 5% limit for probabilities to identify the outliers for all locations,
- we need to confirm the use of the generic model for more locations, and also refine its construction.
- we have in Boland et al. (2001) identified other predictors to enhance the fit, including solar altitude and daily clearness index. The next section deals with a preliminary discussion about the identification procedure that we are presently undertaking.

8 Identification of Further Explanatory Variables

Reindl et al. (1990) presented a comprehensive study of the prediction of the diffuse fraction of solar radiation from other ground variables, including clearness index, relative humidity, solar altitude angle and so on (a total of 28). The clearness index is the proportion of extraterrestrial irradiation reaching a location and thus is a measure of 'cloudiness'. However, in this paper we will consider only one of their models. They found that most of the possible predictors gave insignificant benefit to the prediction. The four that they used in the final model included relative humidity and ambient temperature. These obviously are measured variables. In Australia, measurements of ambient temperature are not taken with the same frequency as global solar radiation. Historically solar radiation was taken $1/2$ hourly in Australia and ambient temperature 3-hourly, currently both are taken on the $1/2$ hour – when recorded. Also, there are many locations for which the humidity would not be recorded. Our objective was to be able to predict the diffuse fraction with as few measured predictors as possible. Thus we will base our comparisons between our work and that of Reindl et al. (1990) on their model which uses only clearness index and solar altitude (solar altitude being a calculated rather than measured variable). In fact, if one examines the results from that work, an analysis of the efficacy of adding the extra two variables may well argue that they do not add sufficiently to the predictability to consider them. This may be in contradiction to what some will consider is essential as some atmospheric models are quite sensitive to solar elevation, but we are reporting and comparing the results of Reindl. The (Schwarz) Bayesian Information Criterion (BIC) (Tsay 2005), Eq. (8.23), includes a penalty function to ensure parsimony, ie that the positive effect of adding more predictor variables is balanced by the need to estimate more parameters. The added explanation of variability in the Reindl model gained by the addition of two extra variables is not great, and may well have been rendered unnecessary if the BIC criterion had been used.

$$BIC = -\frac{2}{T}\ln(\text{likelihood}) + \frac{l\ln(T)}{T} \tag{8.23}$$

Here l is the number of parameters estimated and T the number of data points. The form of Reindl et al. model we will be using for comparison is

$$\frac{I_d}{I_g} = \eta_1 + \gamma_1 k_t + \delta_1 \sin(\alpha) \quad 0 \le k_t \le 0.3, \quad \frac{I_d}{I_g} \le 1.0,$$

$$\frac{I_d}{I_g} = \eta_2 + \gamma_2 k_t + \delta_2 \sin(\alpha) \quad 0.3 \le k_t \le 0.78, \quad 0.1 \le \frac{I_d}{I_g} \le 0.97,$$

$$\frac{I_d}{I_g} = \gamma_3 k_t + \delta_3 \sin(\alpha) \quad k_t > 0.78, \quad \frac{I_d}{I_g} \ge 0.1. \tag{8.24}$$

Here I_g is the global solar radiation on the horizontal plane, I_d is the diffuse radiation on the same plane, k_t is the clearness index, α is the solar altitude, and the $\eta's$, $\gamma's$ and $\delta's$ are parameters to be determined.

The Skartveit et al. (1998) model is too complicated to reproduce in full. It is sufficient to note that it uses three explanatory variables, including clearness index and solar altitude. The third predictor is called the hourly variability index σ_3 and is defined as the "root mean squared deviation between the 'clear sky' index of the hour in question (ρ) and, respectively, the preceding hour (ρ_{-1}) and the deceding (ρ_{+1})":

$$\sigma_3 = \{[(\rho - \rho_{-1})^2 + (\rho - \rho_{+1})^2]/2\}^{0.5} \text{ or}$$
$$\sigma_3 = |\rho - \rho_{\pm}| \tag{8.25,26}$$

In this relationship the latter expression is used whenever the preceding or following hour is missing (start or end of the day). Also, $\rho = k_t/k_1$ where $k_1 = 0.83 - 0.56e^{-0.06\alpha}$, a measure of the cloudless clearness index.

As mentioned previously, Reindl et al. (1990) identified 28 possible predictor variables and through statistical analysis determined that four of these (clearness index, ambient temperature, the sine of the solar angle and relative humidity) gave the best results. We will consider the solar angle out of this grouping, as well as a number of other possible predictors. We also consider apparent solar time (AST) as well as solar angle since it, unlike the altitude, is asymmetric about solar noon, and this may aid in explaining differences in the atmosphere between morning and afternoon. Satyamurty and Lahiri (1992) point out this asymmetry in their work on similar diffuse fraction models, Zelenka (1988) presents work on monthly direct beam radiation which refers to the asymmetry about solar noon and Bivona et al. (1991) also allude to this phenomenon. We thus took into account time of day, as well as solar angle as a possible predictor, inherently capturing the asymmetry, which is caused by the fact that the cloud size generally grows towards the afternoon (and secondarily, also aerosol depth) as soil heating by the Sun progresses and atmospheric convection increases.

We consider a type of variability predictor as Skartveit et al. (1998), but in a different form. Instead of using a measure of how much the present hour's clearness differs from surrounding hours, we take a point from Erbs et al. (1982). In it, after determining the dependence of the hourly diffuse fraction on clearness index, they take the error, or residual values, and model them as a first order autoregressive model. This serial dependence concept has intuitive appeal, since it could be argued that there is some inertia in the atmosphere that can be picked up in this manner. It could be that this inertia can be encapsulated in using values of the lagged clearness index as a predictor. However, since we are not attempting to forecast the diffuse fraction, we take as an extra predictor both a lag and a lead of the clearness index and average them. As well, we consider that there may well be a case for the daily clearness index to be used as a predictor – the whole day may have a common characteristic. Note that what we are trying to do is to find as many possible predictors as we can of a type that requires as little as possible recorded data. The number of sites in Australia that are recording even global solar radiation at sub-diurnal time scales is dropping, with a greater dependence on satellite inferred daily totals. A daily profile can be inferred from that data, but then diffuse values will have to be estimated from a model relying possibly on very few measured values.

Spearman's correlation coefficients were calculated for the diffuse fraction paired with all the possible predictors, and the results are given in Table 8.3. This is for use when one cannot assume that each variable is normally distributed. The raw scores are converted to ranks, and then the sample correlation coefficient is given by $r = (6 \sum_{i=1}^{n} d_i^2)/n(n^2 - 1)$, where n is the sample size, and d_i is the difference in rank for the i the subject.

We must also consider the possibility of multicollinearity between predictors influencing the selection. Multicollinearity occurs when two or more explanatory variables are correlated. The inclusion of both in the model may result in redundancy. To check for this partial F tests are often run. We are led to entertain the use of solar altitude in some form (on its own seems as good as in terms of the sine of the angle – adopted to moderate the effects at high angles), daily clearness index, variability and possibly AST. All but AST seem to fit with including them in the exponential form, while the form of the possible addition of AST will have to be determined.

We also inspected the correlations between possible predictor variables. There are a number of relatively high correlations, including for instance, between k_t, K_t and the two variability series. From this, since the coefficient for the correlation between the diffuse fraction and our variability variable is much higher than the Skartveit one, we discarded the Skartveit variable as a possible predictor. AST has only a small, albeit significant coefficient, but it is not correlated with any other predictor except very slightly with k_t, and therefore it turns out to be a contributor to the predictability. The solar altitude is correlated with the other predictors, but not

Table 8.3 The correlation of the diffuse fraction with the various predictors

	AST	α	k_t	K_t	Variability	Skartveit Variability
Correlation	−.051	−0.331	−0.931	−0.838	−0.73	0.427
p-value	<0.01	<0.01	<0.01	<0.01	<0.01	<0.01

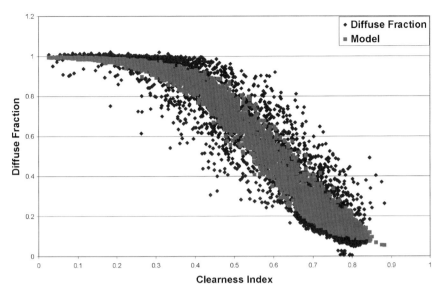

Fig. 8.14 The model fit with added predictors

to a great degree, and it thus is a significant contributor. Even though there is a high degree of correlation between k_t, K_t and variability, they all provide a contribution. Figure 8.14 gives a depiction of the total predictability from using the group of predictors. It should be noted that this is somewhat of an exploratory analysis. We will continue the work to determine if we can construct a sensible generic model using the multiplicity of predictors. Figures 8.15 and 8.16 give an idea of exactly where

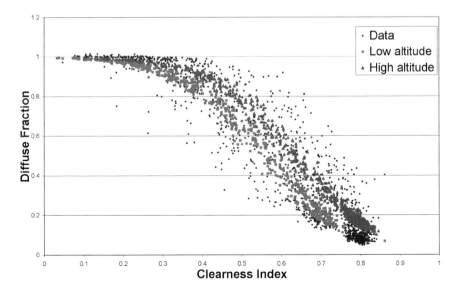

Fig. 8.15 The effect of adding the solar altitude as a predictor

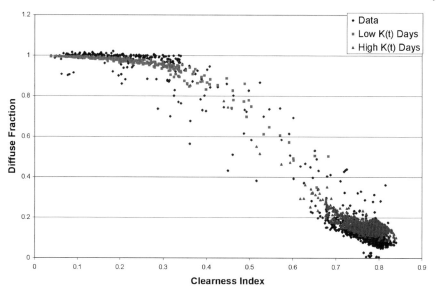

Fig. 8.16 The effect of adding the daily clearness index as a predictor

two of the added predictors make their contribution. Low values of the solar altitude add to the predictability in the bottom part of the scatter and the opposite for high values. As for the daily clearness index variable, it performs as one might expect. Low values of this variable, corresponding to a generally cloudy day, will add to the predictability for generally cloudy hours. The opposite effect occurs as well.

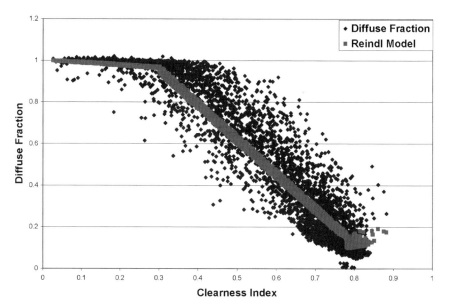

Fig. 8.17 The Reindl multiple predictor model fit

Table 8.4 Comparing the present model with that of Reindl et al.

	Boland-Ridley	Reindl et al.
NRMSD (%)	20.6	23.1
NMBD (%)	−0.2	2.35

We have compared our model with the added variables to the Reindl et al. (1990) model, which we view as the result of a thorough investigation. Skartveit et al. (1998) added the variability predictor, but since we have added a similar variable, the remainder of their model is similar in nature to that of Reindl. Since we have not yet developed a generic model with a multiplicity of predictors, thus estimating parameters specifically for the single location, we thought it would not be fair to compare these results with Reindl's generic two parameter model. Actually, it should be noted that their model is actually an eight parameter model, since there is the estimation of parameters for separate intervals of k_t - see Eq. (8.24). Figure 8.17 gives a depiction of their model result. Table 8.4 also gives a comparison of the statistical measures comparing the two models. The NRMSD is similar in both cases, with the present model being somewhat better. The major difference is in the bias difference, with the present model displaying a much lower degree of bias.

We have begun examining the use of many predictors for several locations. It is instructive to show that the predictability improves for other locations apart from Adelaide. As an example, we show in Fig. 8.18 the result of adding the other predictor variables for a location that has a distinctly different climate from Adelaide,

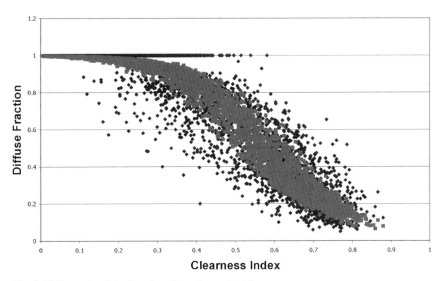

Fig. 8.18 Example of the fit using all predictors for Macau

Macau. This gives impetus to the idea that we should in future work concentrate on developing a generic predictive model using all the predictor variables.

9 The Daily Solar Profile

There are several situations where we will only have data sets that consist of daily total solar radiation. This can be the case if that is all that is gathered or more significantly, if we are interested in, for instance, simulating the performance of a system dependent on solar radiation for a location where there is no data collection at all. For such a location, we can infer the daily total on a horizontal plane from satellite data. We have constructed an algorithm (see Section 10 for details on how to implement it) to obtain a smoothed average daily profile of hourly values corresponding to the daily total, from knowledge only of that total and of course being able to calculate sunrise and sunset. We rely also on knowledge gained from the time series modelling described in Chapter 11 of this book, defining exactly what intra-day cycles exist. We make the quite reasonable assumption that the solar radiation peaks at solar noon. All of this leads to a simple constrained optimisation problem to determine the necessary Fourier coefficients for the embedded cycles. Once we have this daily profile for the global, we can use Eq. (8.21) to estimate the diffuse fraction, and thus the diffuse at any time of the day. Figure 8.19 gives an illustration of this calculation, along with the direct normal radiation (calculated from the other two components).

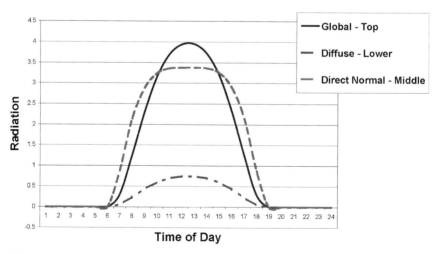

Fig. 8.19 Estimating the diffuse radiation when we have had to infer hourly values from daily totals

10 Algorithms

Please note that there are Excel files containing the algorithms for modelling the diffuse fraction and for constructing the daily profile on the CD accompanying the volume. The algorithms are briefly described below.

10.1 Diffuse Fraction Model Parameter Estimation

We have constructed software in Excel to allow a user to input a set of solar radiation data, global, direct and diffuse and construct their own diffuse fraction model. This will allow the user to determine their own parameter estimates if they have a set, albeit small, of radiation data. They then can apply their model to infill when there is missing data for their location, or to estimate diffuse radiation for a location nearby. The appropriate file is *Diffuse_model.xls*. Following is a set of instructions for using the software:

1. Open the Excel file – you will be accessing a sheet called **Data**. It is here that you will run the macro to organise the data. **Note that you will have to be able to activate the macros embedded in the file.** However, beforehand you will have to copy your data into the sheet **RawData** under the format given in the file **SurfaceSolarFormat.pdf**.

2. By clicking on the button **Data Collect**, you will run the software. You will be asked to enter the number of days of data you have, and then the latitude and longitude, both being the absolute values of these quantities. After this has run, you will have a number of rows of data, all sorted so that there are no rows for before sunrise or after sunset.

3. Highlight the data (not the titles), and copy it to the sheet (starting in row 10) you wish to use for estimation, either **Single Predictor** or **Multiple Predictors.**

4. In cell **U6** on **Single Predictor** or **U7** on **Multiple Predictors**, you will have an objective function. You will be invoking the optimisation tool in Excel, called **Solver**, to perform a least squares optimisation to find the best estimates of the parameters to minimise the sum of squared deviations between your model and the data. Before you invoke **Solver**, you must ensure it is available. If you are unsure, go to the **Tools** menu, and **Add-Ins**. Make sure the **Solver** option is ticked. If it isn't, do so, and it will be added, unless it was not loaded as part of the installation of Excel. You may have to re-install it.

5. You will find that we have set up **Solver** in each instance to perform the parameter estimation. **However**, before you begin, you will have to make a slight alteration to the sheet. Since we don't know how many data points you have, we can't set everything up. So;

6. Highlight **T10:U10** and fill these formulae down to the end of the data. Then, alter the sum in **U6** to sum the cells from **U10** to the end of the data.

7. Go to the **Tools** menu and **Solver**, and hit the button **Solve**. The parameter estimation will be performed.

10.2 Daily Profile

The file *DailyProfiling.xls* will allow the user to construct a profile over the day for a whole year's daily total solar radiation values. **Note that you will have to be able to activate the macros embedded in the file.** If only the profile for a few days is wanted, one can then leave the other days blank. The file will open up on the sheet **Data**. If the user inspects the sheet **RawData**, they will see an example data set for Darwin, with some days missing. It can be seen that the data is in column **G**. By hitting the button **Construct Profile**, you will generate a whole year's profile (of course just missing the days when there is no total).

11 Conclusions

We have been able to demonstrate two important results. We have constructed a statistically rigorous generic diffuse fraction model. As well, we can confidently estimate which values of a set of diffuse radiation measurements are in all probability spurious. If, in this determination, we reject some values that are actually feasible, at least we have a model to estimate a replacement for that time period. As yet, our generic model is only validated for use with a single predictor, the clearness index. We continue to work on developing a generic model with added explanatory variables to add to the predictability.

References

Aguiar, R. (1998) CLIMED Final Report. JOULE III. Project No. JOR3-CT96-0042. INETI-ITE, Dep. Renewable Energies, Lisbon. pp.53–54.

Bivona, S., Burton, R. and Leone, C. (1991) Instantaneous distribution of global and diffuse radiation on horizontal surfaces, Solar Energy 46(4), 249–254.

Boland J. and Scott L. (1999) Predicting the Diffuse Fraction of Global Solar Radiation using Regression and Fuzzy Logic, Proceedings of the ANZSES Conference, Geelong, Nov.

Boland J., McArthur (formerly Scott) L., and Luther M. (2001) Modelling the Diffuse Fraction of Global Solar Radiation on a Horizontal Surface, Environmetrics, 12: 103–116.

Boland J. and Ridley B.H. (2007) Models of diffuse solar radiation, Renewable Energy, (in press), available online June 13, 2007.

Brunger A.P. (1989), Application of an Anistropic Sky Model to the Calculation of the Solar Radiation Absorbed by a Flat Plate Collector, Proceedings Solar World Congress, Kobe. Biennial Meeting International Solar Energy Society, Kobe, Japan, September.

Bugler, J. M. (1977) The determination of hourly insolation on an inclined plane using a diffuse irradiance model based on hourly measured global horizontal insolation, Solar Energy 19: 477–491.

Erbs, D.G., Klein S.A. and Duffie J.A. (1982) Estimation of the diffuse radiation fraction for hourly, daily and monthly average global radiation, Solar Energy, 28: 293–302.

Iqbal, M. (1980) Prediction of hourly diffuse solar radiation from measured hourly global solar radiation on a horizontal surface, Solar Energy 24: 491–503.

Jacovides C.P., Tymvios F.S., Assimakopoulos V.D. and Kaltsounides N.A. (2006) Comparative study of various correlations in estimating hourly diffuse fraction of global solar radiation, Renewable Energy, 31: 2492–2504.

Liu B.Y.H. and Jordan R.C. (1960), The Interrelationship and Characteristic Distribution of Direct, Diffuse and Total Solar Radiation, Solar Energy, 4: 1–19.

Karatasou S., Santamouris M. and Geros V. (2003) Analysis of experimental data on diffuse solar radiation in Athens, Greece for building applications, Int J. Sustainable Energy, **23** (1–2): 1–11.

Lunde P.J. (1979), Solar Thermal Engineering, John Wiley and Sons, New York.

McCormick, P. G. and H. Suehrcke (1991). Diffuse fraction correlations. Solar Energy, 47: 311–312.

Nash, J C (1975) A discrete alternative to the logistic growth function, Appl. Statist., 26: 9–14.

Orgill J.F. and Hollands K.G.T. (1977), Correlation Equation for Hourly Diffuse Radiation on a horizontal Surface, Solar Energy, 19: 357.

Perez R., Ineichen P., Seals R., Michalsky J. and Stewart R. (1990), Modeling Daylight Availibility and Irradiance Components From Direct and Global Irradiance, Solar Energy, 44: 271–289.

Riendl D.T., Beckman W.A. and Duffie J.A. (1990), Diffuse Fraction Correlations, Solar Energy, 45: 1–7.

Satyamurty, V. V. and Lahiri, P. K. (1992) Estimation of symmetric and asymmetric hourly global and diffuse radiation from daily values. Solar Energy 48(1): 7–14.

Sen, P.K. (1968), Estimates of the regression coefficient based on Kendall's tau. Journal of the American Statistical Association. 63:1379–1389.

Skartveit A. and Olseth J.A. (1987), A Model for the Diffuse Fraction of Hourly Global Radiation, Solar Energy, 38: 271–274.

Skartveit A., Olseth J.A. and Tuft, M. E. (1998), An Hourly Diffuse Fraction Model with Correction for Variability and Surface Albedo, Solar Energy, 63: 173–183.

Spencer J.W. (1982), A Comparison of Methods for Estimating Hourly Diffuse Solar Radiation From Global Solar Radiation, Solar Energy, 29: 19–32.

Suehrcke, H. and P. G. McCormick (1988). The diffuse fraction of instantaneous solar radiation. Solar Energy, 40: 423–430.

Ruey Tsay (2005) Analysis of Financial Time Series (2nd Edition), Wiley Series in Probability and Statistics.

Zelenka, A. (1988) Asymmetrical analytically weighted R_b factors. Solar Energy 41(5): 405–418.

Chapter 9
Estimation of Surface Solar Radiation with Artificial Neural Networks

Filippos S. Tymvios, Silas Chr. Michaelides and Chara S. Skouteli

1 Neural Networks: An Overview

An Artificial Neural Network (ANN) is an interconnected structure of simple processing units, whose functionality can graphically be shown to resemble that of the biological processing elements, the neurons, organized in such a way that the network structure adapts itself to the problem being considered. The processing capabilities of this artificial network assembly are determined by the strength of the connections between the processing units, the specific architecture pattern followed during the construction of the network and some special set of parameters adopted during the training of the network. Haykin (1994) states that:

> A neural network is a massively parallel distributed processor that has a natural propensity for storing experiental knowledge and making it available for use. It resembles the brain in two respects: 1. Knowledge is acquired by the network through a learning process; 2. Interconnection strengths between neurons, known as synaptic weights or weights, are used to store knowledge.

During the last two decades, ANN have proven to be excellent tools for research, as they are able to handle non-linear interrelations (non-linear function approximation), separate data (data classification), locate hidden relations in data groups (clustering) or model natural systems (simulation). Naturally, ANN found a fertile ground in solar radiation research. A detailed survey about the applicability of ANN to various Solar Radiation topics is given in section 8 of this Chapter.

Filippos S. Tymvios
Meteorological Service, Nicosia, Cyprus, e-mail: ftymvios@spidernet.net

Silas Chr. Michaelides
Meteorological Service, Nicosia, Cyprus, e-mail: silas@ucy.ac.cy

Chara S. Skouteli
University of Cyprus, Nicosia, e-mail: chara@ucy.ac.cy

2 Biological Neurons

Nature's own basic information processing unit is the neuron. A highly simplified schematic of a neuron is shown in Fig. 9.1. A neuron is an individual cell characterized by architectural features that represent rapid changes in voltage across its membrane as well as voltage changes in neighboring neurons (Churchland and Sejnowski 1992).

Biological neurons are organized and structured in a very complex three-dimensional morphology (Fig. 9.2). The result is a construction, capable of processing information, to analyze and solve problems, remember, compose, dream and feel (in manners and mechanisms which are largely unknown). It consists of a massive number of cells with a high degree of interconnectivity that process information in parallel. There are over one hundred billion neural cells, and each neuron, on the average, receives information from thousands of neighboring neurons (five to ten thousands). Overall, there are typically over 10^{15} connections (synapses, as explained below) in the brain. The anatomic morphology of these neurons and their connections are what make the brain so complex and also so systematic in conducting the various cognitive tasks.

From the neurobiological point of view, a neuron consists of four key elements: the Soma, the Dentrites, the Axon and the Synapses. These four elements contribute in providing the whole network structure with the neuronal attributes, such as the past experience memory signals and the mapping operation of neuronal information. In a simplified way, the operation of a neural network is described below:

The soma receives short electrical pulses from neighboring neurons, processes the information received and produces output signals that are pushed through the axon for further processing by other neurons. The connection is accomplished at a button-like terminal, called the synapse (Fig. 9.2). The strength (weight) of the

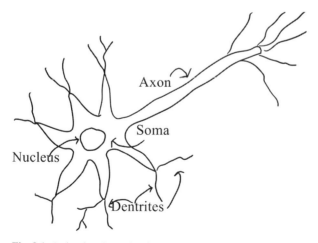

Fig. 9.1 A simple schematic of a single biological neuron

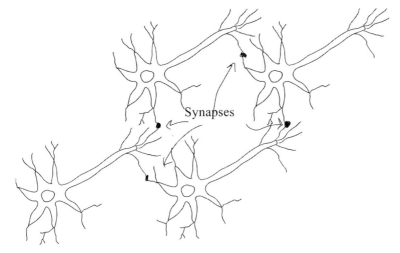

Fig. 9.2 A simplified schematic of a neural network

synapse is a representation of the storage of knowledge and thus the memory for previous knowledge. The synaptic operation assigns a relative weight (significance) to each incoming signal according to the past experience (knowledge) stored in the synapse. Soma adds all the received signals multiplied by the weight given to each specific synapse and if the weighted aggregation of the inputs exceeds a certain threshold it is forwarded to the axon from where it is guided to other neurons.

3 Artificial Neurons

The building brick of any neural computing system is an artificial representation of the fundamental cell of the brain: the neuron. A schematic model of an artificial neuron is illustrated in Fig. 9.3. Artificial neurons (or Processing Elements - PE's) are designed to respond to the applied inputs and to behave consistently. The original artificial neuron is considered to be the TLU (Threshold Logic Unit), proposed by W. McCulloc and W. Pitts (1943). The inputs and outputs are both binary while the activation function (explained later in this section) is the threshold function, taking the values of 0 and 1. A modern artificial neuron, the perceptron, is illustrated in Fig. 9.3.

The artificial implementation of a biological neuron is, in reality, an algorithm or an electronic circuit whose operation can be summarized in a few simple steps:

– All the input values are multiplied by a predetermined weight and summed
– A bias is aggregated to the result
– The sum is introduced to the activation function and is altered accordingly
– The signal flows to the next neuron

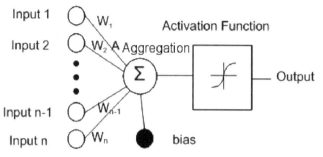

Fig. 9.3 A simplified schematic of a multiple input artificial neuron (Perceptron)

Using mathematical notation, the output of a neuron can be written as

$$Y = f(b + \sum_i w_i x_i) \qquad (9.1)$$

Here, b is the bias for the neuron, w_i is the weight for the specific input, x_i is the actual value that is passed to the neuron for the summation and f is the activation function.

The bias input to the neuron algorithm is an offset value that helps the signal to exceed the activation function's threshold. There are many choices for the neuron transfer function. In general, these functions would better be bounded, continuous and differentiable between the upper and lower limits for the following two reasons: They have to be bounded in order to protect the network from extreme values passing through the nodes, regardless of the magnitude of the inputs and they have to be differentiable in order to be able to implement the popular "Delta Rule" (McClelland and Rumelhart 1986), a procedure essential during the weight adjusting phase (the training of the network). Many modern learning algorithms are based on the "Delta Rule". A significant number of activation functions used in Multi-Layer Perceptrons (MLP) architectures originate from the family of the sigmoid functions (Table 9.1). The bounded term makes these functions improper for use in the output layer since the result of the network has to be scaled to the range of the desired output. For this reason, for the output layer, a linear function is usually preferred. Some common activation functions are given in Table 9.1.

4 Artificial Neural Networks

An artificial neural network is a collective set of such neural units, in which the individual neurons are connected through complex synaptic joints characterized by weight coefficients and every single neuron makes its contribution towards the computational properties of the whole system. As models of specific biological computational structures, ANN consist of distributed information processing elements,

Table 9.1 Common activation functions used in ANN

Activation function	Graphical Illustration	Formula
Threshold		$f(x) = 1,\ x > 0$ $f(x) = 0,\ x \leq 0$
Linear		$f(x) = x$
Hyperbolic Tangent Sigmoid		$f(x) = \dfrac{e^x - e^{-x}}{e^x + e^{-x}}$
Sine or Cosine		$f(x) = \sin(x),$ $f(x) = \cos(x)$
Logistic Sigmoid		$f(x) = \dfrac{1}{1 + e^{-x}}$

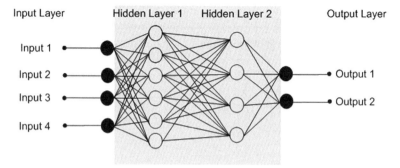

Fig. 9.4 An example of a four layer neural network

possessing an inherent potential for parallel computation. In fact, parallel processing is operating in the brain but not as yet in ANN: in personal computers the processes are performed serially.

A usual setup in solar energy applications (section 8) consists of a three layer MLP (as presented in section 6): input, hidden and output layer. Occasionally, more than one hidden layer is used, increasing the total number of layers from three to four. More hidden layers will result to an unnecessary increase of the network's complexity without improving results (except for function approximation setups). An example of such a network is given in Fig. 9.4.

5 The Perceptron

The simplest ANN is the Perceptron (Fig. 9.3). It was invented in 1957 at Cornell Aeronautical Laboratory by Frank Rosenblat (1958). The perceptron is a pattern-recognition machine primarily intended for character recognition. As Rosenblatt showed, the Perceptron can be trained to recognize linearly separable patterns with the help of special training algorithms, often called learning rules. Learning rules have the ability to tweak the network's weights in such a way that the algorithm will converge to a solution in a finite number of steps, assuming that such weights exist (Hagan et al. 2006, pp. 4–15). The Perceptron has also the ability to generalize (to respond to unknown input data).

A Perceptron has multiple inputs. It is able to classify between linearly separable data groups. An example of a linear classification problem is given in Fig. 9.5 where a perceptron is used to classify a set of data into four groups. The perceptron is capable of classifying the input data by dividing the input space into four regions. The principal limitation of the Perceptron is that the Perceptron, as originally proposed by Rosenblat, is not a general purpose processing device. It cannot separate nonlinearly separable input patterns such as the eXclusive-OR function (XOR) (Minsky

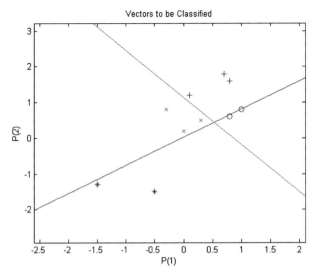

Fig. 9.5 Classification of input data in four groups

and Papert 1968; 1988). The Perceptron's limitations are overcome by the MLP, presented in section 6.

5.1 Perceptron Learning Algorithm (Delta Rule)

The perceptron learning algorithm (Delta Rule, McClelland and Rumelhart 1986) can be summarized as follows:

− All the synaptic weights and biases are initialized to small random values
− The input data are fed to the network and the output is then compared to the desired one
− The synaptic weights (Wi_{OLD}, Wi_{NEW}) are adapted according to the following:

$$Wi_{NEW} = Wi_{OLD} + \lambda \left(y_{computed} - y_{desired}\right) \tag{9.2}$$

where $y_{computed}$ is the computed output, $y_{desired}$ is the desired output and λ is a constant, usually called the "learning rate" of the algorithm. If it is too small, the whole procedure becomes slow; if it is too high the algorithm becomes "insensitive" and may fall into an infinite loop going back and forth into opposite directions trying to locate the optimum weights for the given problem.
− The procedure is repeated until the difference of the computed and desired result falls below a certain, predefined threshold (goal) or when a certain amount of epochs is reached (an epoch is one sweep through all the records in the training set).

6 Multi-Layer Perceptron (MLP)

The Multi-Layer Perceptron (MLP) extends the Perceptron model with hidden layers between input and output layers. It is a feedforward network, typically trained with backpropagation (to be discussed in more detail in the next section). The MLP is the most popular nonlinear ANN architecture, employed into a wide variety of problems in applied sciences. Even the simplest kind of a MLP network with a sufficient number of processing elements is called a universal approximator due to its ability to approximate any nonlinear relationship between inputs and outputs to any degree of accuracy (Hornik et al. 1989; Leshno et al. 1993).

Conceptually, the MLP architecture consists of an input layer, one or more hidden layers of neurons with non-linear activation functions and the output layer. Hidden layers are not exposed to input vectors, as input and output layers are. A MLP network is illustrated in Fig. 9.4.

The MLP's main advantages compared to other neural model structures is that it is easy to implement and that it can approximate any input/output map. The key disadvantages are that they are trained slowly and require large amounts of training data. One rule of thumb is that the training set size should be 10 times the network weights to accurately classify data with 90% accuracy (Principe et al. 1999).

6.1 Backpropagation

Backpropagation has been by far, the most popular and widely used learning technique for training ANN. It is widely accepted that it was first proposed by Paul Werbos (Werbos 1974), and further developed by David E. Rumelhart, Geoffrey E. Hinton and Ronald J. Williams (Rumelhart et al. 1986) who proposed the backward propagation of the errors as a model of the learning process.

In principle, a supervised (controlled output) training algorithm, as the backpropagation is, repeatedly applies a set of input vectors to a network, while continually updates the network's synaptic weights, until a stop criterion is met (e.g., a maximum number of epochs or an error goal).

The training of a backpropagation network typically starts with random weights on its synapses. It is then exposed to a training set of input data. The output of the network is compared to the example (supervised training) and a learning procedure alters the network interconnections (weights). The connections are adjusted so that the inputs are associated more strongly towards the expected answer. As the training proceeds, the network's response to the input data becomes better and better. The training is repeated for a set of examples until the network learns to produce accurate results. Once the network is trained using the preselected inputs and outputs, all the synaptic weights are frozen and the network is ready to be tested on new input information (simulation).

A summary of the technique is given below:

- Feed the MLP with a training sample
- Compare the output to the desired output from that sample and calculate the error in each neuron (this is the local error)
- The local error is assumed to be caused by the neurons of the previous level, proportionally to the weight value of each neuron connection arriving to the level under investigation
- Adjust the weights arriving to each neuron to minimize the local error
- Repeat the steps above to the previous level (backwards) using as error the neurons' contribution to the local error of the previous step.

6.2 Error Surface

Although a MLP is able to find solutions for difficult problems, as discussed above, the results cannot be guaranteed to be perfect or even correct. Results obtained from MLP are just approximations of a desired solution and a certain error is always present. Slightly different network setups may produce disproportional deviation of the results to the one or the other direction.

A backpropagation network consists of m-synapses. This means that at any time, the state of the network can be represented by an m-dimensional state vector. The squared error of the network at a certain state is a scalar quantity. For every specific vector pattern used (e.g. p), the squared error (E_p) is defined by:

$$E_p = \frac{1}{2} \sum_{k=1}^{n} \left(y_{computed_k} - y_{desired_k} \right)^2 \tag{9.3}$$

where the summation is taken over all output units (n), $y_{compuded}$ is the calculated value and $y_{desired}$ is the desired value.

This error can be generalized by summing it over all input vectors in the training space. This generates a global error function (also called total sum squared error), defined as :

$$GError = \sum_{p} E_p \tag{9.4}$$

The global error is a one-dimensional representation of the overall error computed by the network at an m-state. The errors obtained from all the possible synaptic weights and all possible input vectors, form an "error surface" which is generally very complex and uneven, with peaks and craters, valleys and slopes (and this is just for a three dimensional representation of the error surface). It is constant with time for specific network architecture and the input vectors. The goal of the backpropagation algorithm is to locate, via gradient descent, a global minimum in the error surface. This global minimum refers to a specific set of synaptic weights which define the network for the given dataset/architecture. In other words, a learning algorithm behaves like an explorer searching the error surface for the network that will produce the minimum global error to the input dataset.

Let's consider a neuron with a sigmoid activation function (Table 9.1) with two one element input vectors (P) and two associated one dimensional target values (T).

```
P = [2 4] , T = [0.5 0.8]
```

If we calculate the errors produced by all different networks composed with the same neuron within a combination of all possible weight and bias values (we used ranges of −4 to 4 for both weight and bias), then we have the error surface that results from this specific problem. This error surface with a contour plot underneath is illustrated in Fig. 9.6. The best weight and bias values are considered to be those that result the lowest point on the error surface. The network that will use these specific weight and bias will produce the minimum error.

6.3 Data Preparation (Preprocessing)

Although a properly designed neural network is able to deal with most problems, the training of the network is, usually, not a trouble free procedure. Pre-processing the network's inputs and targets improves the efficiency of neural network training. Problems that may emerge could involve a range of qualities or a composition of several different types of data, whereby one piece of information is represented

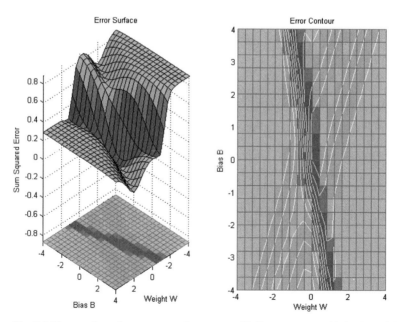

Fig. 9.6 Error surface of a perceptron (input vector [2 4], output vector [0.5 0.8], weight and bias limits −4 to 4)

by a combination of several different qualities. Such a representation reduces the number of input or output neurons needed. By performing Principal Component Analysis (PCA) methods (Demuth and Beale 2004, p. 5–62 and an excellent example on p. 5–65) on the input data it is expected that no unnecessary variables are inserted into the network. The obvious benefit of the analysis of the input data is the reduction of input variables and consequently the network's complexity (less neurons). Applying a normalization procedure before presenting the input data to the network is generally a good practice, since mixing variables with large magnitudes in combination with those with small magnitudes will result in confusing the learning algorithm about the importance of each variable and forcing it to finally reject (minimize the associated weights) the variable with the smaller magnitude (Samarasinghe 2007 p. 259; Demuth and Beale 2004 p. 5–59).

6.4 Overfitting - Underfitting the Network

In most cases, overfitting a neural network is an undesirable situation where the error of the network during the training phase becomes minimal, but when new data, not previously presented to the network are used, the error is large compared to the error acquired from the learning phase. If overfitting occurred, the training of the network has failed since the network has memorized the training data, produces excellent results within the known input dataset but it is incapable of generalizing the output to unknown datasets. The result is a neural network inadequate for general use. Overfitting usually occurs when the size of the network is too large for the specific application or when the training dataset is small. Function approximation is considered a deviation from the above rule; for function approximation we seek overfitting to capture a function. By minimizing the size of the network we are in danger to underfit the data. Underfitting is the case where the ANN implemented is not sufficiently complex to correctly detect the pattern in a noisy data set. One way to overcome this problem (if the size of the input dataset can afford it), is to split the input set to three subsets: the training set, the validation set and the testing set. In the training phase, we check both the training set error and the validation set error and we update the network weights as usual, using the error from the training set. When the network tends to overfit the input data, the training set error continues to decrease, while the validation data set error will increase. At this point, the training stops and the network freezes at the validation minimum stage. This procedure is called "early stopping" of the network.

7 Building Neural Networks

The reader is presented in the following with two proposals (amongst several available) that may be used to build ANN models for solar radiation estimation. Two such software proposals are the ANN dedicated Neurosolutions software package and MATLAB which is a general purpose high-level language and interactive environment suitably equipped for building ANN. Two examples with real data are given together with some guidelines and instructions. The data and the respective files referred to in this Chapter can be found on the accompanying CD-ROM.

7.1 Dataset

A small dataset is provided for the reader to experiment with. The observations were performed at Athalassa (35°8'27" North, 33°23'47" East, height 161m above mean sea level), which is the site of the main radiometric station of the Meteorological Service of Cyprus. The station location is at a semirural site on the outskirts of Nicosia, the largest city in the centre of the island. The data covers the period from 1/5/2003 to 31/5/2003 and it consists of 4 files:

– cyprus_data_daily.txt
– cyprus_data_hourly.txt
– cyprusdaily.mat
– cyprushourly.mat

The first two are text files and the latter two are MATLAB variable files. As implied by the filenames, there are two sets of data; one with daily values and one with hourly values (notice the underscore characters in the text filenames).

The variables contained in the files are explained in Table 9.2.

Table 9.2 A list of the input parameters provided in the dataset

Parameter	Description	Instrument	Wavelength	units
Year	2003	–	–	–
Month	05 (May)	–	–	–
Day	1 to 31	–	–	–
Sunshine	Minutes of bright sunshine (hour,day)	Campbell datalogger	–	minutes
Global	Global radiation (hour,day)	Kipp & Zonen CM21	(305–2800)nm	$(J/m^2/h, day)$
Diffuse	Diffuse radiation (hour,day)	Kipp & Zonen CM21	(305–2800)nm	$(J/m^2/h, day)$
PAR_tot	Par radiation (hour,day)	LI-COR, Inc. LI-190	(400–700)nm	$(J/m^2/h, day)$
PAR_dif	Diffuse Par (hour,day)	LI-COR, Inc. LI-190	(400–700)nm	$(J/m^2/h, day)$
CUVA	UV A radiation (10nm window)	Kipp & Zonen CUVA1	(363–373) nm	$(J/m^2/h, day)$
CUVB	UV B radiation (2nm window)	Kipp & Zonen CUVB1	(305–307) nm	$(J/m^2/h, day)$

7.2 An Outline of Cyprus Climate

The island of Cyprus lies between latitude circles 34.6° and 35.6° North and between meridians 32° and 54.5° East, surrounded by the eastern Mediterranean sea. Its Mediterranean climate is characterized by the succession of a single rainy season (November to mid-March) and a single longer dry season (mid-March to October). This generalization is modified by the influence of maritime factors, yielding cooler summers and warmer winters in most of the coastline and low-lying areas. Visibility is generally very good. However, during spring and early summer, the atmosphere is quite hazy, with dust transferred by the prevailing southeasterly to southwesterly winds from the Saharan and Arabian deserts, usually associated with the development of desert depressions (Michaelides et al. 1999).

7.3 Applying ANN with Neurosolutions

Neurosolutions is a very popular software package in the neural networks application world. It is equipped with an icon-based graphical user interface that provides a powerful and flexible development environment through simple wizards. As an example of the methodology needed, we will implement Neurosolutions to model solar radiation at the earth's surface from the experimental data described in section 7.1.

Neurosolutions provide the user with three different ways to design neural networks:

– method 1: Manually
– method 2: The Neural Expert Wizard
– method 3: The Neural Builder Wizard

Each method corresponds to the user's experience: novice users should follow method 3 while more experienced users should prefer method 2. Method 1 is not for the beginner / medium level user since Neurosolutions' approach to the Neural Network design is proprietary, using an icon-based user interface and special terminology rules about network components. Based on our own experience with Neurosolutions, it seems that method 2 with manual modifications after the creation of a network is the fastest way to build custom setups of Neural Networks.

• Choosing the appropriate network

Initially, the user needs to choose a network setup (see Fig. 9.7).

For our example we choose the "Multi-Layer Perceptron" option. We browse to the Chapter's data folder and we load `cyprus data daily.txt` file. Neurosolutions uses ASCII files in columns with a heading on top (the variable). If the header is not present, Neurosolutions inserts one using a predefined string structure. Then, the user has to define which columns represent the input variables and which

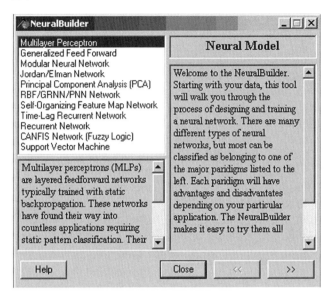

Fig. 9.7 Neurosolutions: Choosing the architecture and structure of the network

columns represent the target variables. It is also possible to use different files for inputs and targets. The same applies for the validation and the testing set, as the user is able to give new test files or let Neurosolutions to choose the necessary data from the input data, as a percentage of the initial input file. The validation / test members are selected randomly. The normalization / denormalization is done by Neurosolutions automatically and the whole procedure is completely non-transparent to the user.

• Structuring the network

The user now needs to define how many hidden layers the network will have. By fast-preprocessing the dataset, Neurosolutions is also able to make an initial guessing of the network's architecture and will suggest the number of the neurons per layer. Everything on this panel can be altered: the activation function of each layer, the number of PE's in each hidden layer and the learning rule and its attributes (see Fig. 9.8).

• Learning phase

The last panel deals with the learning control and the way of the output presentation. Once defined, the software builds the model and everything is ready for the training. Neurosolutions is also able to modify the network characteristics according to genetic algorithms (GA) implemented into the software (GA tick box wherever it is possible to apply). It can reject input variables as not important, find the optimum number of neurons in a layer and also find the best values for training parameters of the chosen learning algorithm (see Fig. 9.9).

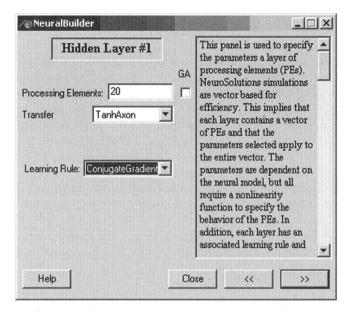

Fig. 9.8 Neurosolutions: Specifying the network's structural characteristics and learning rule

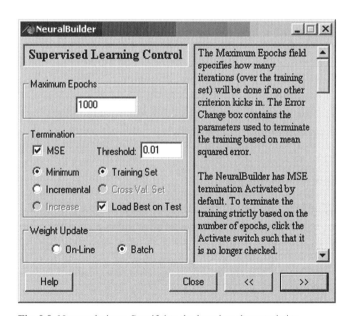

Fig. 9.9 Neurosolutions: Specifying the learning characteristics

7.4 Applying ANN with MATLAB

MATLAB is a numerical computing environment and also a programming language. It allows easy matrix manipulation, plotting of functions and data, implementation of algorithms, creation of user interfaces and interfacing with programs in other languages. The Neural Network Toolbox extends MATLAB with tools for designing, implementing, visualizing and simulating neural networks. It also provides comprehensive support for many proven network paradigms, as well as graphical user interfaces (GUIs) that enable the user to design and manage neural networks in a very simple way (http://www.mathworks.com/ products/neuralnet/).

Using the data provided on the accompanying disk, we will create, step by step, a neural model able to calculate the daily Photosynthetically Active Radiation, commonly called as PAR (Jacovides et al. 2004) from sunshine duration data.

(The m-file is available on the accompanying Compact Disk under the name ...\matlab_examples\ch12_bookexample1.m)

– Import data into MATLAB's workspace.

```
load ('cyprusdaily.mat') ;
```

– Let's Plot all the available data for the month (see Fig. 9.10)

```
Hold;
plot(Diffuse, '-o');
plot(Global, ':x');
plot(PAR_dif,'--+');
plot(PAR_tot, '-.d');
title ('Input data');
xlabel('Days (May 2003)');
ylabel('Jm(-2)/day');
legend ('Difuse','Global','PAR_dif',
'PAR_tot');
```

– Preprocessing of the data (normalization)

```
[xn,xmin,xmax,yn,ymin,ymax]=premnmx(Global,PAR_tot);
```
`premnmx` is preprocessing (normalizing) the data so that input and target values fall into the interval $[-1,1]$

– Create the network

```
net=newff([-1 1],[10 1],{'tansig','purelin'});
```
Using function `newff` we have created a new network named "net"
Function `newff` takes as input :

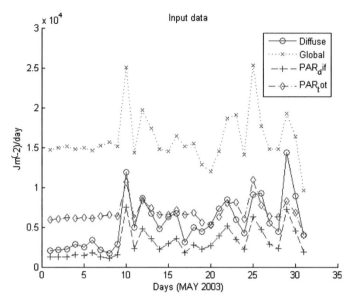

Fig. 9.10 Input Daily data

[xmin, xmax] : the input range. Since the input data are
normalized, we use [-1 1]

[10,1] : the network structure: two layers, first layer
(hidden) consists of 10 neurons and the second
(output) layer of 1 neuron)

{'tansig', : the activation function in each layer. Tansig in
'purelin'} hidden layer, purelin in the output layer

The network created using MATLAB's notation method is illustrated in Fig. 9.11.

– Initialize network

```
net=init(net);
xn=xn';
yn=yn';
```

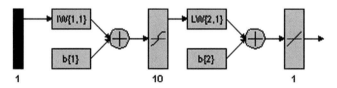

Fig. 9.11 The neural network implementation using MATLAB's notation

Now, all weights and biases are randomized and the input / output variables are converted to rows (MATLAB requires that all input data must be presented as row vectors)

– Simulate the network without training to show the initial response to the input data

```
Simulation=sim(net,xn);
plot(xn,yn,xn,Simulation,'r+');
title ('Untrained Network Response');
legend('Measured PAR,'Simulated PAR')
```

The result of the simulation of the untrained network is presented in Fig. 9.12.

– As an example of altering the training parameters, we set different number of epochs and an early stopping goal:

```
net.trainParam.epochs=200;
net.trainParam.goal=.0015;
```

In this way we set the maximum number of epochs and as an early stopping goal, the performance function to be equal to 0.0015.

– Training the network

```
net=train(net,xn,yn);
anorm=sim(net,xn);
```

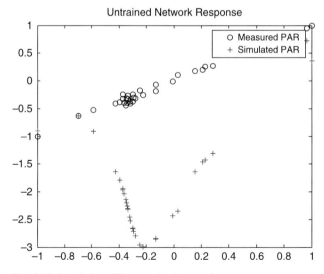

Fig. 9.12 Simulation of the untrained network

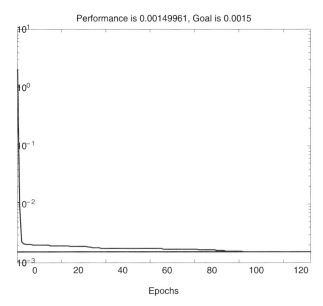

Fig. 9.13 Network performance

The network is trained. "anorm" is the normalized output of the training set (see Fig. 9.13).

The goal is reached early: just 132 epochs.

– Postprocessing of the data

```
a=postmnmx(anorm,ymin,ymax);
```

The output of the simulation is de-normalized and stored in row vector "a"

– Plot the output (see Fig. 9.14)

```
plot (x,y,x,a);
title ('Network Response');
legend('input data','network output');
```

From Fig. 9.13 we can falsely assume that by increasing the training epochs, the overall error will be reduced. This is not the case though. As discussed in section 6.4, the network that we created is overfitted. Its performance will probably increase with increasing epochs and will eventually approach zero but the network will be incompetent to unknown data; it cannot generalize. To train the network more efficiently and to improve performance against unknown data, we could split the input dataset into three subsets: The training set, the testing set and the validation set with 14, 7 and 7 members, respectively. The population of the dataset seems to be small to train the neural network properly but since the relationship we are investigating is a simple linear function (Jacovides et al. 2007), the network will find no trouble to converge to a solution.

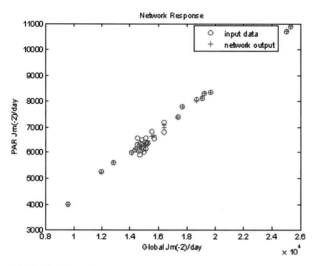

Fig. 9.14 Network response

The validation procedure is capable of reducing the total number of epochs needed, from 132 to just 9 epochs. In a complex network this improvement is significant.

The output results for the new network trained with the smaller dataset are given in Fig. 9.15.

In Fig. 9.16 we see the performance of the network for the three datasets for each epoch. The relationship between the input/output data is simple enough to have the network converge to a near optimum solution possible, while retaining the ability

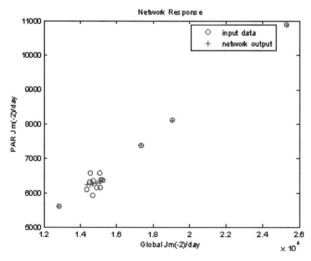

Fig. 9.15 Network response with smaller dataset

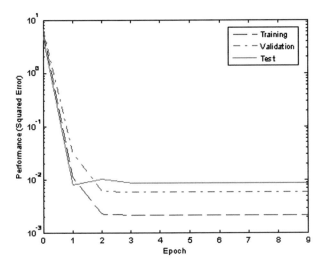

Fig. 9.16 Network performance using training, testing and validation dataset as input during training phase

to generalize results in just 9 epochs, even though a limited number of input data is used for the training.

Another example of an ANN approach to solar radiation modeling is also provided for the reader, in the accompanying disk under the name `ch12_bookexample2.m`. This example models the daily total solar radiation as a function of the sunshine duration (Tymvios et al. 2002; 2005), using MATLAB code to split the datasets into training, testing and validation datasets following the methodology presented above.

8 A Survey of Neural Network Modeling Approaches

The efforts for the introduction of ANN techniques in estimating solar irradiance by using as inputs recorded weather parameters and other environmental variables started in the late 1990's. A large spectrum of approaches has been covered in this endeavor. The problem was approached from different perspectives: the modeling approaches employed are characterized by a large variety of neural network models with varying architectures and a diversity of input and output parameters.

This section surveys the neural networks approaches that were employed by various investigators. For a more comprehensive reading, a grouping of the available literature is made. The grouping of this published research was made considering the primary objective of the models and the output parameter on which the major focus of each paper was made. This output parameter determines the intended use of the respective model. The first group consists of works on estimating the hourly solar radiation, the second group those for the daily solar radiation and the third group those for monthly mean values. A model for predicting the maximum solar irradiance is outlined in the fourth group. Models for time series prediction are discussed in a fifth group. The final sixth group comprises modeling studies that were developed to provide estimated solar irradiance values at places where no direct measurements are available. In each group, the respective studies have been arranged in a chronological order.

The examples presented in the following give just a general idea of the application of ANN in the study of solar irradiance. A more complete account can be found in the original publications; here, only a brief description of the methodology is given together with the data base that was used in the process and a short summary of the results. The above are provided here to allow the reader to locate the respective literature for an elaborate study.

What is very interesting to note is that in less than 20 years of applications of ANN to solar radiation studies, a great variety of neural training approaches have been used (different learning algorithms, architectures etc) and a multitude of input variables have been explored (meteorological, geographical etc). What is common in almost all of these studies is the validation of the respective proposed methodology with independent data.

8.1 Hourly Solar Irradiance Models

Hontoria et al. (1999; 2002) made use of the concept of atmospheric transmittance (or clarity index, defined as the ratio of the global irradiation and the extra-atmospheric irradiation) in an effort to generate hourly solar radiation series by usingl ANN. In particular, they propose a feedforward – feedback architecture; first, they decompose the atmospheric transmittance into a trend component (comprising the mean) and a random component (comprising the random fluctuations about the mean). Subsequently, they used a MLP model for the trend component; for building models of the random component's autocorrelation coefficient and standard deviation they have also used a MLP (see Lippmann 1987; Haykin 1994).

The data used by Hontoria et al. (1999; 2002) consist of records for eight years from seven stations in Spain, representative of various climatic areas of the country. Seven years of these data were used for the training of the ANN; the remaining was used for validation testing. In their results, they have shown that with the models they constructed, both the trend and the random components were reproduced satisfactorily.

A neural network based on backpropagation techniques, a deterministic atmospheric model and a fuzzy logic method were used by Santamouris et al. (1999) in order to estimate hourly values of global solar radiation by using data collected at a hill at the center of Athens in Greece. Training of the neural model was performed with hourly values of the input climatic parameters for the estimation of integrated hourly global solar radiation values for 11 years (1984–94). The input parameters of the neural network model were the air temperature, relative humidity, sunshine duration and calculated extraterrestrial radiation and the output was the global solar radiation.

In this work, Santamouris et al. (1999) implemented a backpropagation learning procedure and a neural network architecture consisting of one hidden layer of 15 log-sigmoid neurons followed by an output layer of one linear neuron. The accuracy of the neural network was tested by comparing the real data in the testing set of the year 1995, with the estimated values reached with the neural network approach. The mean-square error was found to be quite acceptable (e.g. $0.22 \, \text{MJ}/\text{m}^2$ for July and $0.20 \, \text{MJ}/\text{m}^2$ for January). They concluded that the presented results are quite encouraging for developing a feedforward backpropagation neural network approach to simulate and predict future values of global solar radiation time series by extracting knowledge from their past values. Also, they investigated larger architectures constructed by adding more hidden layers or nodes and they concluded that they had longer converging times but did not significantly improve the network's prediction accuracy.

Sfetsos and Coonick (2000) focused on estimating hourly solar radiation by using two artificial intelligence based techniques. These include linear, feed-forward, recurrent Elman and Radial Basis neural networks, together with the adaptive neuro-fuzzy inference scheme. In a univariate approach, they make use of and predict raw solar radiation values. They also studied the development of multivariate models which can make use of other meteorological variables as potential inputs (e.g. temperature, pressure and wind speed and direction). The data used in the study are mean hourly solar radiation values on a horizontal level (in W/m^2) recorded on the island of Corsica for a period of 63 days during late spring and early summer of 1996. The data were separated into training, evaluation and prediction subsets. The training set is used for the training of the models, whereas, the evaluation set, which was unknown during the training phase, was used to check the progress of the network. The model that is used for the prediction set is the one whose parameters minimize the Root Mean Squared Error of the evaluation set. The performance of each forecasting method is subsequently evaluated with the prediction set.

Sfetsos and Coonick (2000) conclude that a comparison between the various models has indicated as the optimum prediction model a feed-forward model using the Levenberg-Marquardt algorithm (see Hagan and Menhaj 1994).

Dorvlo et al. (2002) have made use of a clearness index (defined as the ratio of the observed solar radiation to the maximum solar radiation). They approached the problem by using Radial Basis Functions (RFB) and MLP models. More specifically, they implemented three MLP neural networks (with 1, 2 and 3 hidden layers) and one RBF network. The MLP and RBF networks were trained and validated with data from eight meteorological stations in Oman of at least ten years (1986–1998). Six out of these eight stations were used to train the networks; the data from the remaining two stations were used for validation. To randomize the training and validation data sets, they based the selection of the training and the validation stations on all twenty-eight possible combinations of training-validation stations.

The input parameters used in this study, were a parameter representing the sunshine ratio, (i.e. the ratio of sunshine hours to maximum sunshine hours), a time parameter (month of the year) and geographic parameters (latitude, longitude and altitude) and the output of the network was the clearness index, as defined above.

Multiplying the clearness index (as this is produced by the network) by the maximum solar radiation yields an estimate of the solar radiation. Statistical testing of the models has shown that all the four networks (i.e. three MLP and one RBF) performed very well. However, although the performance of the four networks was similar, Dorvlo et al. (2002) suggest that the use of the RBF network requires less computational effort compared to the MLP neural networks.

Very often, the modelers are faced with a very important issue that is common to all stochastic approaches, namely the choice of the proper predictors (i.e. the set of features that can contribute to good predictions). In this respect, the work by López et al. (2005) is considered as a contribution to resolve this issue. More specifically, they experimented with a Bayesian method (see MacKay 1994, Neal 1996) in order to determine the more relevant input parameters in modeling hourly direct solar irradiance by using ANN. The hourly averaged data that they have used refer to a radiometric station in the USA, namely Desert Rock, between 1998 and 1999 and the experiments were implemented for a set of MLP models. The variables that they considered are: measured values of global and direct solar irradiance, temperature, relative humidity, surface pressure and wind speed; derived parameters like the dew point temperature and precipitable water; the cosine of the solar zenith angle representing the solar position; the relative optical air mass. The methodology yielded that the most relevant variables for estimating the direct irradiance are the clearness index (ratio of horizontal global irradiance and horizontal extraterrestrial irradiance) and the relative optical air mass.

Elminir et al. (2005) address the problem of determining solar radiation data in different spectrum bands from meteorological data using ANN modeling. The meteorological data employed consisted of daily values for wind direction, wind velocity, ambient temperature, relative humidity and cloudiness as they were generated from the hourly average values for Helwan city, an urban area in Egypt. The solar radiation components used in this study are the infra-red, ultra-violet and global insolation. A backpropagation based ANN was adopted.

In their study, Elminir et al. (2005) used three sets of data. The first is a training set comprising data recorded at Helwan site over the period from January to December 2001. These data were used for the network adjustment, in order to reach the best fitting of the nonlinear function representing the phenomenon. The second is the test set comprising data for the same site over the period from January to December 2002; these data were given to the network still in the learning phase for error evaluation and to update the best thresholds and weights. The third is the validation set referring to data recorded at Aswan monitoring station over the period from January to November 2002. This latter set was used to further evaluate the model's prediction capability: the model that was developed was used to estimate the same components of insolation for Aswan site, namely infra-red, ultra-violet and global insolation; the ANN model predicted these components with an accuracy of 95%, 91% and 92%, respectively.

8.2 Daily Solar Irradiance Models

Elizondo et al. (1994) reported one of the earliest efforts to estimate daily solar radiation by using ANN. The aim of their research was the development of a neural network model which could predict daily solar radiation, by using as input variables the *observed* weather variables (e.g. maximum and minimum air temperatures and daily precipitation amount) and calculated *environmental* variables (e.g. daylength and daily total clear sky radiation). The backpropagation neural network algorithm was implemented in this study.

Training and evaluation was performed with twenty-three years of data from four sites in the southeastern USA. These data were split into a training data-set consisting of eleven years and a testing dataset, consisting of the remaining twelve years.

The evaluation of the neural network model was made by comparing the predicted daily solar radiation (output) resulting from the application of the predictants' values (input) in the testing set with the measured radiation observations. The neural network model had a tendency to over predict solar radiation at low values and under predict it at high values. Nevertheless, average errors and coefficients of variation were low and the coefficients of determination were high. The root mean square error was found to be as low as $2.92 \, \text{MJ}/\text{m}^2$.

Kemmoku et al. (1999) reported on their approach to develop a set of ANN to forecast the total insolation of the next day. The neural networks were trained with daily weather data of 6 years (1988–1993) for Omaezaki in Japan. The model built was then tested to forecast the daily insolation for four months in 1994, namely April, August, October and December. The set of meteorological parameters used in this neural network modeling approach consist of the difference between the first and the second local maxima of atmospheric pressure, the difference between the first and the second local minima of atmospheric pressure, the average atmospheric pressure, the clearness index, the temperature and atmospheric pressure at 18:00 h, and the insolation at a time before sunset.

They followed a multi-stage neural network approach that can briefly be described in the following. The first stage, is based on the consideration that there is a correlation between the insolation of the next day and the difference between average atmospheric pressures of the next day and previous day. This is because solar irradiance depends on weather conditions, which is closely related to the atmospheric pressure (i.e. change of weather conditions depends on the changes of the atmospheric pressure). Therefore, the average atmospheric pressure can be forecast by the first stage neural network from meteorological data. In the second stage neural network, the insolation level of the next day is forecast from this average atmospheric pressure and meteorological data. A third stage refines the insolation prediction depending on three classes of the insolation level forecast above. Therefore, this final stage considers high, middle and low levels of insolation and proceeds to the final prediction with three different neural networks, accordingly.

For the testing data, the percentage mean error statistic was used and the performance of the multistage neural network model was estimated to be around 20% for all the four months comprising the evaluation dataset.

8.3 Models for Monthly Mean Daily Solar Radiation

Mohandes et al. (1998) performed an investigation for modeling monthly mean daily values of global solar radiation on horizontal surfaces; they adopted a backpropagation algorithm for training several multi-layer feedforward neural networks. Data from 41 meteorological stations in Saudi Arabia were employed in this research: 31 stations were used for training the neural network models; the remaining 10 stations were used for testing the models. The input nodes of the neural networks are: latitude (in degrees), longitude (in degrees), altitude (in meters) and sunshine duration (as the ratio of the actual values divided by the maximum possible values).

The output of the network is the ratio of monthly mean daily value of the global solar radiation divided by extraterrestrial radiation received at the top of the atmosphere. The results from the 10 test stations indicated a relatively good agreement between the observed and predicted values.

Along the same line is the research by Mohandes et al. (2000), in another research for simulating monthly mean daily values of global solar radiation (the output of the model is the ratio of monthly mean daily value of the global solar radiation divided by extraterrestrial radiation outside the atmosphere). They retained the same input parameters as above (latitude, longitude, altitude, sunshine duration) but they added a new one, namely, the month number. They made use of the same data sets, which were also separated into the same training and testing sub-sets. In this research, they use Radial Basis Functions (RBF) neural networks technique (Wassereman 1993; Bishop 1996) and compare its performance with that of the MLP as used in their previous study (see Mohandes et al. 1998).

The performance of both the RBF and MLP networks was tested against the independent set of data from 10 stations by using the mean absolute percentage error as the testing statistic. The test has indicated mixed results for individual stations but, overall, RBF performs better than MLP.

In a more recent endeavor, Mellit et al. (2006) studied a wavelet network architecture and its suitability in the prediction of daily total solar radiation. Wavelet networks are feedforward networks using wavelets as activation functions and have been used successfully in classification and identification problems. This architecture provides a double local structure which results in an improved speed of learning. The objective of this research was to predict the value of daily total solar radiation from preceding values; in this respect, five "structures" were studied involving as input various combinations of total daily solar radiation values.

The meteorological data that have been used in this work are the recorded solar radiation values during the period extending from 1981 to 2001 from a meteorological station in Algeria. Two datasets have been used for the training of the network. The first set includes the data for 19 years and the second dataset comprises data for one year (365 values) which is selected from the database. In both cases, the data for the year 2001 are used for testing the network.

The validation of the model was performed with data which the model had not seen before and predictions with a mean relative error of 5% were obtained. This is considered as an acceptable level for use by design engineers.

8.4 *Maximum Solar Irradiance Models*

Kalogirou et al. (2002) describe the development of a neural network approach for the prediction of maximum solar radiation. This model uses as input a set of simple data: the month of the year, the day of month, the Julian day, the season and the mean of ambient temperature and relative humidity. As such, these data affect the availability and intensity of solar radiation. The data used in this study were for Nicosia in Cyprus and refer to a one year period. Data for eleven months were used for training and testing and the rest were used as the independent set for validation.

As part of this study, Kalogirou et al. (2002) experimented with various network architectures in an attempt to determine which one could perform better. In this respect, recurrent type networks and feedforward ones have been investigated. Also networks with different sizes and learning parameters have been tried. The architecture that was selected for best performance was the "Jordan Elman recurrent network" and is composed of four layers, one of which is hidden and one is used for dampened feedback. The extra layer is connected to the hidden layer.

The network was used to predict the maximum solar radiation. The overall correlation coefficient obtained for the validation dataset is 0.9867. This model was expected to perform better on clear days. Indeed, the predictions were found to be more accurate during the summer period.

8.5 *Time Series Prediction Models*

Mihalakakou et al. (2000) presented a simulation of time series for the total solar radiation in Athens, Greece by using neural networks. Hourly values of total solar radiation for twelve years and for various months of the year are used. Nine years (1984–1992) were used for training the neural network and three years (1993–1995) for testing the network. The training data set was used to provide a fitting approximation function, and the testing set to validate the ability of prediction of the previously trained network. The zero-valued nighttime records of total solar radiation were omitted from both the training and testing sets and therefore they are not used in the processes.

In their research to predict future values of total solar radiation time series, they have considered two versions of models:

(a) a one-lag scheme was used for predicting the next value of the time series, given a number of past values, thus generating a one-step-ahead prediction; this part of their study can be considered as falling under the hourly solar irradiance models, described above.

(b) a multi-lag scheme was used for the prediction of several time steps in the future; in this version, the predicted output is fed back to the input for the next prediction, thus allowing the generation of a new prediction two time steps ahead.

In their study for predicting the total solar radiation time series, Mihalakakou et al. (2000) adopted a multi-layer feedforward network based on backpropagation learning procedure. The neural network architecture that was selected as giving the better convergence consists of one hidden layer of 16 log-sigmoid neurons followed by an output layer of one linear neuron.

For the one-lag predictions it was found that the neural network approach is able to predict the total solar radiation with remarkable success For the multi-lag predictions, it was found that, for the warm period of the year, it is possible to predict with sufficient accuracy the total solar radiation ten to twenty days in advance; for the cold period of the year, however, the predictions were not so promising. The authors note that for a data-driven method such as the neural network, the results depend strongly on the training data set; therefore, during the cold season, the larger variability of the weather makes the prediction very difficult.

8.6 Models for Solar Potential

The term "model for solar potential" is adopted here to embrace a series of models that were constructed in an attempt to estimate solar radiation at locations for which no radiation data are available. Generally, direct solar radiation measurements are rather sparse; therefore, for the estimation of solar radiation at places with no direct measurements, deterministic models have long been developed having as inputs meteorological variables such as sunshine duration, cloud cover, temperature and humidity. The first example of this kind is the widely known Ångström's approach (Ångström 1924) that was traditionally considered as a standard procedure. The introduction of ANN techniques has opened new horizons in the effort of estimating solar irradiance where direct measurements are not available, by using as inputs other recorded meteorological or environmental parameters or simply "extrapolating" the available data of nearby locations.

An early study was carried out by Williams and Zazueta (1996) in order to determine whether a neural network could adequately estimate the solar radiation for a location that is not equipped with radiation instrumentation. In their study, they used actual weather data collected at Gainesville, FL in the USA to train a backpropagation neural network which was subsequently used to estimate solar radiation. Their results indicated that the correlation coefficient for the neural network was 0.86 compared to 0.77 for the results from an analytical approach.

Al-Alawi and Al-Hinai (1998) develop an ANN model in an effort to analyze the dependence of global radiation on climatological variables with the aim to estimate the latter for locations not used during the training of the model. They employed data from six weather stations in the Sultanate of Oman, covering the period 1987 to 1992.

The backpropagation paradigm was adopted for the training of a multilayer feedforward network. The neurons in the input layer receive eight input signals representing the location, month, mean pressure, mean temperature, mean vapor pressure, mean relative humidity, mean wind speed and mean duration of sunshine.

The results from the study of Al-Alawi and Al-Hinai (1998) indicate that, for the data used in the training process, the model can estimate the global radiation with high accuracy. Furthermore, in order to demonstrate the generalization capability of this approach was also tested to predict global radiation for a location the archived data of which were not included in the training dataset. The monthly predicted values of the ANN model compared to the actual global radiation values for this independent dataset produced an accuracy of 93% and a mean absolute percentage error of 5.43.

Another application of ANN for the estimation of the solar potential is presented by Tambouratzis and Gazela (2002). Their tasks in their research included the estimation of solar horizontal radiation at a target location (a sub-urban location in Athens, Greece) by using the corresponding values from a reference location (in the centre of the town, 15km away). In the same study, they also performed a similar task for the temperature. The dataset employed consist of hourly values of ambient temperature and solar horizontal radiation, which have been accumulated during 1997 and 1998 at the two locations. It is very interesting to note that in this study a bi-modal approach has been adopted in building up what they call an "ANN estimator". First, an unsupervised learning ANN, namely the Self-Organising Maps that was proposed by Kohonen (1995), was used to cluster the dataset into four partitions, representing four combinations of seasonality and cloudiness. Second, one single-layer backpropagation ANN was assigned to each partition for estimating the solar radiation at the target location. The ANN estimator has been found capable of accurately estimating solar radiation at the target location from the corresponding measurements collected at the reference location.

In a more recent study, Sözen et al. (2005), used ANN with meteorological (e.g. mean sunshine duration, mean temperature), geographical (e.g. latitude, longitude, altitude) and other (e.g. month) data as inputs, to establish the solar potential in Turkey. Data from twelve stations considered to be representative of the different climatic regions of the country were used. To train and test the neural network, they made use of data recorded in the period 2000–2003. Data from nine of the above stations were used for training; data for the remaining three stations were used for testing the proposed methodology. The backpropagation learning algorithm has been used in a feedforward single hidden layer neural network; a number of variants of the algorithm were used in the study.

The proposed ANN methodology was tested with statistical measures. The results indicated that the ANN methodology is better than the classical regression models in establishing the solar potential at new sites which are not outfitted with solar radiation equipment.

Also, recently, Mellit et al. (2005) developed a hybrid model to predict the daily global solar radiation. They combined ANN and a library of Markov transition matrices. Their experiments were made with data from 60 meteorological stations in Algeria during 1991–2000. The ANN model constructed was a feedforward neural network with three inputs (latitude, longitude and altitude), a hidden layer and an output layer consisting of 12 neurons representing monthly radiation.

The combined ANN and Markov transition matrices proceed as follows: firstly, a neural network block was trained based on 56 known monthly solar radiation data from the database; secondly, these data were divided by the corresponding extraterrestrial value, thus yielding monthly values of the clearness index; subsequently, by using a library of Markov transition matrices and these latter values, sequences of daily clearness indices were generated.

The above model was trained with the data from 56 stations, leaving 4 stations aside for validation. The unknown validation data set produced very accurate predictions, with a root mean squared error not exceeding 8%; also, a correlation coefficient between predicted and observed data ranging from 90% and 92% was obtained. These findings suggest that the proposed model can be used for the estimation of the daily solar radiation by using simple geographical parameters as input: altitude, longitude and latitude.

9 Conclusions

Artificial Neural Networks comprise a nonlinear statistical method that in the past two decades have become very popular in tackling a variety of problems related to atmospheric science. The increasing interest in the application of ANN to atmospheric sciences has been reviewed by many authors. For example, Gardner and Dorling (1998) discuss the applications of the MLP in the atmospheric sciences; Hsieh and Tang (1998) discuss issues related to ANN applications in meteorology and oceanography; more recently, Michaelides et al. (2007) reviewed the adoption of ANN in connection to meteorological aspects of energy and renewable energy applications. The survey presented above provides evidence of a wide use of ANN in solar radiation studies. Nevertheless, the usefulness of ANN employment in solar radiation science can be appreciated by providing answers to questions on their *practical* and *theoretical* advantages as well as on their plausible superior *performance*, when compared to other methodologies.

The first question that can be asked in connection to the intensive use of ANN in atmospheric science and in solar radiation studies, in particular, is: "has the application of ANN yielded any *theoretical* or *practical* advantages over other traditional methods?" Advantages in using this modern computational approach are discussed by almost every author who publishes results of an ANN paradigm. Some authors even do this in a rather extensive manner and in ways pertained to the methodology adopted. The following are just some of the most important advantages of ANN, as regards their application to solar radiation studies.

Many of the atmospheric processes are not fully comprehensible and also cannot be expressed in deterministic terms: the highly complex nature underlying many of such processes goes beyond human comprehension. For this reason, in many occasions, such processes have been treated statistically. In this respect, ANN appear to be appropriate in by-passing the need for a total understanding of underlying physical processes. Moreover, the non-linear character of many of the atmospheric processes renders ANN a suitably qualified tool for their study.

ANN models are considered to learn the key information patterns within a domain, which is usually multidimensional, avoiding complex rules. This makes ANN attractive in studies on solar energy which require data for a specific location but are seldom available and need to be estimated either from data for other parameters (climatic, geographic, astronomic etc) or from data for other locations at some distance away. The ANN approach is particularly suitable in tackling problems in the presence of noisy atmospheric data (Rumelhart et al. 1986). This is quite important also in cases where data are missing or are multi-modal in nature (e.g. atmospheric pressure).

As in many scientific fields, ANN have been considered as an alternative way that can be employed to tackle ill-defined problems. Also, ANN form a non-parametric methodology which is based on training without any statistical assumptions or relations regarding the input data: they are simply trained to yield output results from examples.

The second question that can be put forward refers to whether "ANN exhibit a demonstrated superior performance than other methods?" The surge of interest in using the neural approach in solar radiation investigations, as presented in this section, marks a definite trend in the related research in this area towards the adoption of ANN. However, it is interesting to note that most of the published works are comparative in nature: almost all authors place a considerable effort in their investigations to prove that ANN perform better than other methods, thus justifying the adoption of this modern methodology. The three comparative approaches identified in the literature, are given in the following, together with typical examples:

(a) Several researchers apply the use of ANN in solar radiation studies but it is very interesting to note that in this new trend they do not disregard the traditional statistical methodologies; on the contrary, in many of the published research papers which are discussed in this Chapter, there has been a rather systematic parallel investigation of both the traditional and the modern ANN methodologies.

An example of this comparison is given by Mohandes et al. (2000). They compare two ANN based networks (namely, RBF and MLP) to the empirical Ångström's regression model, as it was later modified by Rietveld (1978). From the comparative results, no clear advantage of either method is obvious when the results are looked upon station by station. However, when the mean absolute percentage error is averaged for all the stations in the testing dataset, the RBF appears to outperform the other two; the regression model follows, thus performing better than the other ANN model (i.e. MLP).

Along the same lines is the work of Tymvios et al. (2002; 2005) who compared ANN models trained with daily values of the measured sunshine duration, the theoretical sunshine duration, the month of the year and the daily maximum temperature in order to estimate total global radiation at Athalassa radiometric station in Cyprus. The comparison of the performance of the ANN models with that of various Ångström-type models revealed that the latter are outperformed by most of the former.

(b) Few researchers perform comparisons of ANN with other non-traditional statistical methodologies and in particular with other modern computational methodologies. Santamouris et al. (1999) performed a comparison of a deterministic atmospheric model and the two intelligent techniques, namely an ANN and a fuzzy

Table 9.3 Seven combinations of input variables and hidden layers as used by Tymvios et al. (2002, 2005) and their respective performance scores: Mean Bias Error (MBE) and Root Mean Square Error (RMSE).

Model	Input variables	Number of hidden layers	*Neurons*	MBE %	RMSE %
1	Measured daily sunshine duration, theoretical daily sunshine duration	1	77	−0.68	6.28
2	Month, measured daily sunshine duration	1	61	−0.53	6.64
3	Month, Daily maximum temperature	1	61	0.78	10.15
4	Measured daily sunshine duration, Month, Daily maximum temperature	1	46	−0.35	6.11
5	Measured daily sunshine duration, Month, Daily maximum temperature	2	46 − 23	−0.27	6.57
6	Measured daily sunshine duration, Theoretical daily sunshine duration,Daily maximum temperature	1	46	−0.30	5.97
7	Measured daily sunshine duration, Theoretical daily sunshine duration, Daily maximum temperature	2	46 − 23	0.12	5.67

logic method. The comparison of the ANN and fuzzy logic models revealed that the former perform better than the latter.

(c) It is quite common in solar radiation studies, to experiment with various ANN architectures and algorithms. The ANN models constructed by Tymvios et al. (2002; 2005), mentioned in (a) above, consist of multilayer feedforward networks with one and two hidden layers which were trained with the backpropagation learning algorithm. The performance of the seven ANN models which were developed with various combinations of input parameters and hidden layers was tested by using the Mean Bias Error (MBE) and Root Mean Square Error (RMSE) (see Table 9.3).

The first three are models with only two input variables and only one hidden layer. The fourth and sixth models were trained by using three input variables and two hidden layers. Finally, the fifth and seventh are models built with three input variables and two hidden layers; more hidden layers were examined but with little change in their performance.

To complete this review on the retrieval of solar irradiance by using ANN some cautionary statements are put forward, regarding the application of this methodology which is increasingly adopted in a large spectrum of scientific disciplines.

1. ANN is a data-driven method and as such requires a sufficient amount of data. Normally, as regards radiation and other data that can be used as input should span over a period of a few years. Experimentation with ANN on small databases may lead to misleading results. There is no definite recipe as to the size of the data needed but confidence to the models is enhanced as more data become available in the learning phase and a subsequent validation. To this end, it must be stressed that most of the published works presenting applications of ANN methodologies to estimate solar irradiance include some kind of validation against an independent set of data (i.e. data not used for the training of the respective neural models). This is very important for establishing the validity of a model (constituting a widely accepted approach for this reasoning) but inevitably reduces the size of the database that can be used in the learning phase.

2. The development of an ANN model that performs satisfactorily in one location, does not necessarily guarantee the transferability of the model to another location, irrespective of any apparent similarities. This site specificity of the ANN approach must be borne in mind when attempting to transfer a model that has been built for one location to another. Although the same input variables may be available, the model must be retrained to ensure adaptation to the unseen data for the new location.

3. The easiness in employing an unrestricted number of input variables has led many investigators to give erroneous information about the variables that rationally affect the dependent one or about variables which seem to have no justifiable place in the whole process. The inclusion of any superfluous input variable to the model should be avoided if possible. This matter of selection of predictors is common to predictive statistical methodologies. The selection of the proper input variables can be made in an objective or subjective (intuitive) manner. A combination of the two is given by Bremnes and Michaelides (2007) who suggest a greedy forward search algorithm suitable for use with ANN procedures in choosing from a large number of potential input variables.

4. Many investigators are tempted to employ an unnecessary architectural complexity in building up their ANN models. However, increasing the complexity does not necessarily lead to better performing models. Indeed, in several of the studies presented in this Chapter, (but also in studies dealing with issues other than solar radiation), investigators tested whether the prediction capability of the networks built is enhanced by increasing their complexity. It is concluded from several published papers that adding more hidden layers or nodes does not always improve the network's predictive power but it can rather only slow the convergence (see Mihalakakou et al. 2000). Also, an excessive number of hidden layers is often unproductive: models with more than two hidden layers tend to memorize the training sets and although they can yield excellent results for known inputs, they perform poorly on unknown data (Kalogirou et al. 1997; 2002). Briefly it can be claimed that one or two hidden layers should be adequate for most applications. As a matter of fact, Hornik et al. (1989) showed that "multilayer feedforward networks with as few as one hidden layer are capable of universal approximation in a very precise and satisfactory sense".

References

Al-Alawi SM, Al-Hinai HA (1998) An ANN-based approach for predicting global radiation in Locations with no direct measurement instrumentation. Renew Energ 14:199–204

Ångström A (1924) Solar and terrestrial radiation. Q J Roy Meteor Soc 4:121–126

Bishop CM (1996) Neural networks for pattern recognition. Oxford University Press

Bremnes JB, Michaelides SC (2007) Probabilistic visibility forecasting using neural networks. Pure Appl Geophys 164: 1365–1381

Churchland PS, Sejnowski TJ (1988) Perspectives on Cognitive Neuroscience. Science 242: 741–745

Demuth H, Beale M (2004) Neural Network Toolbox User' Guide-version 4. The Mathworks Inc

Dorvlo ASS, Jervase JA, Al-Lawati A (2002) Solar radiation estimation using artificial neural networks. Appl Energ 71:307–319

Elizondo D, Hoogenboom G, McClendon R (1994) Development of a neural network to predict daily solar radiation. Agr Forest Meteorol 71: 115–132

Elminir HK, Areed FF, Elsayed TS (2005). Estimation of solar radiation components incident on Helwan site using neural networks. Sol Energy 79: 270–279

Gardner MW, Dorling SR (1998) Artificial neural networks (the multilayer perceptron) - A review of applications in the atmospheric sciences. Atmos Environ 32:2627–2636

Hagan M, Menhaj M (1994) Training feed-forward networks with the Marquardt algorithm. IEEE T Neural Networ 5:989–993

Hagan M, Demuth H, Beale M (2006) Neural Network Design. Colorado University Bookstore, ISBN 0-9717321-0-8

Haykin S (1994) Neural Networks: A comprehensive foundation. MacMillan College Publishing, New York

Hontoria L, Aguilera J, Zufiria P (2002) Generation of hourly irradiation synthetic series using the neural network multilayer perceptron. Sol Energy 72:441–446

Hontoria L, Riesco J, Zufiria P, Aguilera J (1999) Improved generation of hourly solar irradiation artificial series using neural networks. In: Proceedings of Engineering Applications of Neural Networks, EANN99 Conference. Warsaw, Poland, 13–15 September 1999, pp 87–92

Hornik K, Stinchcombe M, White H (1989) Multilayer feedforward networks are universal approximators. Neural Networks 2-5:359–366

Hsieh WW, Tang B (1998) Applying neural network models to prediction and data analysis in meteorology and oceanography. B Am Meteorol Soc 79:1855–1870

Kalogirou S, Neocleous C, Michaelides S, Schizas C (1997) A time series reconstruction of precipitation records using artificial neural networks. In: Proceedings of EUFIT'97. Aachen, Germany, pp 2409–2413

Jacovides CP, Tymvios FS, Assimakopoulos VD, Kaltsounides NA (2007) The dependence of global and diffuse PAR radiation components on sky conditions at Athens, Greece. Agr Forest Meteorol 143:277–287

Jacovides CP, Tymvios FS, Papaioannou G, Assimakopoulos DN, Theofilou CM (2004) Ratio of PAR to broadband solar radiation measured in Cyprus, Agr Forest Meteorol 121:135–140

Kalogirou S, Michaelides S, Tymvios F (2002) Prediction of maximum solar radiation using artificial neural networks. In: Proceedings of the World Renewable Energy Congress VII (WREC 2002) on CD-ROM. Cologne, Germany

Kemmoku Y, Orita S, Nakagawa S, Sakakibara T (1999) Daily insolation forecasting using a multi-stage neural network. Sol Energy 66:193–199

Kohonen T (1995) Self-organising maps. Springer Verlag

Leshno M, Ya Lin V, Pinkus A, Schocken S (1993) Multi-layer feed-forward networks with a non-polynomial activation function can approximate any function. Neural Networks 6:861–867

López G, Batlles FJ, Tovar-Pescador J (2005) Selection of input parameters to model direct solar irradiance by using artificial neural networks. Energy 30:1675–1684

Lippmann RP (1987) An introduction to computing with neural nets. IEEE ASSP Magazine, April, pp 4–22

McClelland J, Rumelhart D, PDP Research Group (1986) Parallel Distributed Processing: Explorations in the Microstructure of Cognition. Foundations 1 Cambridge, MA: MIT Press

McCulloch WS, Pitts WH (1943) A Logical Calculus of the Ideas Imminent in Nervous Activity. B Math Biophys 5:115–133

MacKay DJC (1994) Bayesian non-linear modelling for the energy prediction competition. ASHRAE Transactions 100:1053–1062

Mellit A, Benghanem M, Hadj Arab A, Guessoum A (2005) A simplified model for generating sequences of global solar radiation data for isolated sites: Using artificial neural network and a library of Markov transition matrices approach. Sol Energy 79:469–482

Mellit A, Benghanem M, Kalogirou SA (2006) An adaptive wavelet-network model for forecasting daily total solar radiation. Appl Energ 83:705–722

Michaelides SC, Evripidou P, Kallos G (1999) Monitoring and predicting Saharan desert dust transport in the eastern Mediterranean. Weather 54:359–365

Michaelides SC, Tymvios FS, Kalogirou S (2006) Artificial Neural Networks for Meteorological Variables Pertained to Energy and Renewable Energy Applications.[Chapter 2] : Artificial Intelligence in Energy and Renewable Energy Systems. Kalogirou, S. (Editor). Nova Science Publishers, Inc

Mihalakakou G, Santamouris M, Asimakopoulos DN (2000) The total solar radiation time series simulation in Athens, using neural networks. Theor Appl Climatol 66:185–197

Minsky M, Papert S (1969, reprint 1988) Perceptrons - Expanded Edition: An Introduction to Computational Geometry (Paperback). The MIT Press

Mohandes AM, Rehman S, Halawani TO (1998) Estimation of global solar radiation using artificial neural networks. Renew Energ 14:179–184

Mohandes M, Balghonaim A, Kassas M, Rehman S, Halawani TO (2000) Use of radial basis functions for estimating monthly mean daily solar radiation. Sol Energy 68:161–168

Neal RM (1996) Bayesian learning for neural networks. Lecture notes in Statistic. Springer-Verlag, New York

Principe JC, Euliano NR, Lefebvre CW (1999) Neural and Adaptive Systems. John Wiley & Sons

Rietveld MR (1978) A new method for estimating the regression coefficients in the formula relating solar radiation to sunshine. Agr Meteorol 19:243–352

Rosenblatt F (1958) The Perceptron: A probabilistic model for information storage and organization in the brain. Psychol Rev 65:386–408

Rumelhart DE, Hinton GE, Williams RJ (1986) Learning internal representation by error propagation. In: Rumelhart DE, McClelland JL (Eds). Parallel distributed processing: Explorations in the microstructure of cognition 1, MIT Press. Cambridge, MA, pp. 318–362

Samarasinghe S (2007) Neural Networks for Applied Sciences and Engineering. Auerbach Publications

Smith M (1996) Neural Networks for statistical modeling. International Thompson Computer Press, London UK

Santamouris M, Mihalakakou G, Psiloglou B, Eftaxias G, Asimakopoulos DN (1999) Modeling the global solar radiation on the earth surface using atmospheric deterministic and intelligent data driven techniques. J Climate 12:3105–3116

Sfetsos A, Coonick AH (2000) Univariate and multivariate forecasting of hourly solar radiation with artificial intelligence techniques. Sol Energy 68:169–178

Sözen A, Arcaklıoğlu E, Özalp M, Çağlar N (2005) Forecasting based on neural network approach of solar potential in Turkey. Renew Energ 30:1075–1090

Tambouratzis T, Gazela M (2002) The accurate estimation of meteorological profiles employing ANNs. Int J Neural Syst 12:1–19

Tymvios FS, Jacovides CP, Michaelides SC (2002) The total solar energy on a horizontal level with the use of artificial neural networks. In: Proceedings of the 6th Hellenic Conference of Meteorology, Climatology and Atmospheric Physics. Ioannina, 26-28 Sep., Greece, pp. 468–475

Tymvios FS, Jacovides CP, Michaelides SC, Scouteli C (2005) Comparative study of Angstrom's and artificial neural networks' methodologies in estimating global solar radiation. Sol Energy 70:752–762

Wassereman PD (1993) Advanced methods in neural networks. Van Nostrand Reinhold

Werbos PJ (1974) Beyond Regression: New Tools for Prediction and Analysis in the Behavioral Sciences. Ph.D. thesis, Harvard University

Williams D, Zazueta F (1996) Solar radiation estimation via neural network. In: Proceedings of the 6th International Conference on Computers in Agriculture. Cancun, Mexico, pp. 1143–1149

Chapter 10
Dynamic Behavior of Solar Radiation

Teolan Tomson, Viivi Russak and Ain Kallis

1 Introduction

This chapter is addressed mainly to engineers working on utilization of solar energy converted from the global radiation. Authors expect that the reader is acquainted with fundamentals and terminology of solar engineering (explained for instance in Duffie and Beckman 1991). Global (or: total) solar radiation is the sole energy carrier for the whole nature. Fossil fuels are in fact chemically stored primeval solar radiation. Yet more - thermal stresses and fatigue due to changing insolation involve the destruction of the lithosphere and they also participate in the development of (desert) landscape. Variability of the insolation has to be considered in the solar engineering too and it is analyzed in this chapter with different approaches. Utilization of the solar energy is mostly supported and limited with its storing, which has to be based on the consideration of the dynamical behavior of solar radiation. Fatigue effects mentioned above assess the life-time of materials used and should be considered in solar engineering (Koehl 2001; Carlson et al. 2004).

Solar radiation on the infinitely (in practice – sufficiently) long time axis is a stationary ergodic process that includes both periodical and stochastic components. It remains always in the interval between zero and some upper value not exceeding the solar constant. Still, solar radiation could be a non-stationary process during some shorter time interval, intended for practical problem-solving.

Periodical component has the astronomical and stochastic component has the meteorological origin. Figure 10.1 shows the yearly diagram of the (relative) normal

Teolan Tomson
Tallinn University of Technology, Estonia, e-mail: teolan@staff.ttu.ee

Viivi Russak
Tartu Observatory, Tõravere, Estonia, e-mail: russak@aai.ee

Ain Kallis
Tallinn University of Technology, Estonia, e-mail: kallis@aai.ee

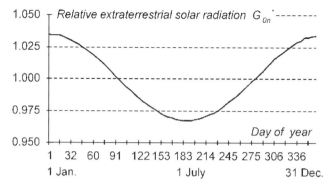

Fig. 10.1 Normal extraterrestrial solar irradiance G_{0n}^{*} is a periodical variable

extraterrestrial irradiance (Kondratyev 1969), which is a result of the elliptic trajectory of the Earth around the Sun. Other small variations of the solar constant have second order meaning (e.g. Duffie and Beckman 1991).

Declination caused by the slope of the Earth's axis with regard to the elliptic path about the Sun and rotation of the Earth involves additional periodical changes of solar radiation. These processes assess the yearly periodical component. Diurnal periodical component is assessed by rotation of the Earth.

The state of the atmosphere involves both stochastic and periodical changes. The turbidity of the atmosphere and cloud cover has mainly stochastic origin, but not only. Periodical monsoon seasons in tropical areas are well known. Less attention has been paid to the trajectories of Atlantic (Prilipko 1982) and Arctic (Brümer et al. 2000) cyclones (Fig. 10.2) over Northern Europe (Scotland, Scandinavia, the Baltic states and North-West Russia), which have also seasonally periodical behavior.

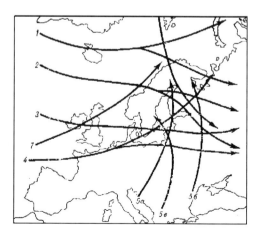

Fig. 10.2 Typical routes of cyclones over North-Europe

Numerous varying cyclones involve fast and frequent changes in solar radiation. Therefore, this area requires attention from the point-of-view of transient effects of solar radiation. Frequent and crucial changes in the radiation level will lead to problems concerning the power network, if numerous grid-connected photovoltaic (PV) plants will realized (Jenkins 2004). Cyclones emphasize the share of diffuse radiation. The annual ratio of diffuse G_d fraction to the global G radiation (irradiance) is well correlated with the cyclonic activity. In Estonia $G_d/G = 0.5$ (Russak and Kallis 2003), but in Israel it is significantly lower $G_d/G \leq 0.4$ (Lyubansky et al. 1999). The character of clouds has indirect impact on the variability of solar radiation. High clouds Ci, Cc, Cs reduce the value of direct radiation and it's increments too. High variability of solar radiation and high values of its fluctuations are most probable in the occurrence of convective clouds Cu and Cb (Mullamaa 1972). Commonly, the instant values of diffuse radiation and the mentioned ratio are determined by the simultaneous existence of several cloud layers.

To solve practical problems of solar engineering, the infinite long-time axis has to be divided into finite intervals. In some cases, during the mentioned interval, solar radiation could be considered a constant. Figure 10.3 shows the behavior of direct beam irradiance G_b in the clear-sky conditions (Riihimaki and Vignola 2005), which can be well approximated with a constant value in the significant share of a day.

On the other hand, the same variable has infinitely high changes around sunrise and sunset moments, although these time intervals are out of scope for engineering purposes.

As energy supply is based on global radiation G, below we will consider it as the basic variable. While some technological solutions perform differently from direct and diffuse components, these have to be also mentioned. Depending on the averaging interval, the solar radiation data sets are mostly compound processes that contain both periodical and random components. A trend of an unknown data set may be a fragment of a periodical component with the period longer than the used set.

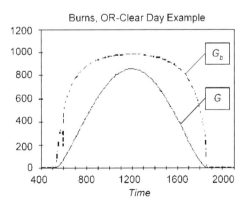

Fig. 10.3 Beam G_b and global G irradiance in a clear-sky day in Burns, Oregon, (OR) USA

2 Averaged Data Sets

Solar data recorders use different sampling intervals. The sampled data are mostly mean (average) values of the measured variable. If the recorded data set is capacious (long and hard to observe), it is additionally concentrated calculating average values over an excerpt of recorded data. This averaging process involves a hidden estimation that the averaged data set is a stationary process or can be considered to be such without a major error. Conditions of stationarity are the following: a moving average along the data set is a constant (the process has no trend) and the constant (in general) is also the value of the moving variance (standard deviation) (Bendat and Piersol 1974). Averaged data sets can be presented in the form of tables or diagrams. The latter are preferred in the present chapter as they are easily observed. Data set diagrams below are mostly paired with the diagrams of their autocorrelation functions (ACF) to comment on their main quality. The same information should equally be presented via probability density spectrograms which are invertible by the Fourier transformation from ACF. The examples are mainly based on the data recorded at the Tartu-Tõravere Meteorological Station (TOR) of the Estonian Meteorological and Hydrological Institute (http://www.emhi.ee/index.php?ide = 8,74). This station is a high-quality radiation monitoring station in the Baseline Surface Radiation Network (BSRN) and is located in the south-eastern part of Estonia ($58.25°$ N and $26.5°$ E, 70 m a.s.l.). Global irradiance on the horizontal plane was measured as the average value during one-minute intervals. Up to 2001, a Russian-made M-115 type Yanishevsky's pyranometer was used, which was replaced by the Kipp & Zonen pyranometer CM21 in 2002. The time constant of the M-115 pyranometer is 8 s, and for the CM21 it is less than 12 s. All the other data sets with longer sampling intervals are averaged data over the primary data set.

2.1 Annual Sums of Global Radiation

Annual sums of global radiation are suitable for long-term process analysis and mainstream trend development (Rönnelid 2000). Figure 10.4 shows the set of annual

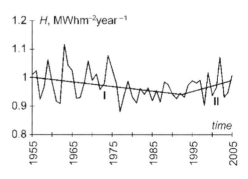

Fig. 10.4 Annual sums of the global irradiance in TOR

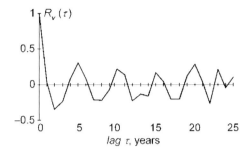

Fig. 10.5 ACF of annual sums of the global irradiance in TOR

sums of global radiation at TOR during the last 50 years. Linear trend line "I" shows decreasing solar radiation in 1955–1990. Since 1990 the trend line "II" has been increasing. Such a behavior of global radiation is possibly the result of the changing circulation of the atmosphere particularly in Northern Europe. The effects of air-quality regulations and the decline of the Eastern European economy mentioned in (Wild et al. 2005) should have second-order meaning. Both of the trends shown in Fig. 10.4 were eliminated in the ACF calculation (Fig. 10.5). ACF shows that a certain year is practically independent of the preceding year: the correlation time is less than one year. The correlation time is defined below.

2.2 Monthly Sums of Global Radiation

Monthly sums of global radiation (Vaniček 1992) are widely used for several purposes. It is a good tool to calculate seasonal storages (Oliveti et al. 2000) or analyze (seasonal, annual) efficiency of solar installations (Tepe et al. 2003). Figure 10.6 shows the time diagram of monthly sums of global radiation at TOR for 2003–2005, which has no trend for the selected interval. The ACF of the mentioned data set in Fig. 10.7 has mainly periodical character with a 12-month period. The stochastic component in this example is negligible. Due to the low share of the stochastic component and practically no higher harmonic components, the 12-month period is

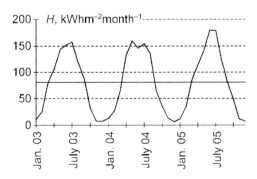

Fig. 10.6 Monthly sums of global radiation at TOR during years 2003–2005

Fig. 10.7 ACF of monthly
sums of global radiation at
TOR during years 2003–2005

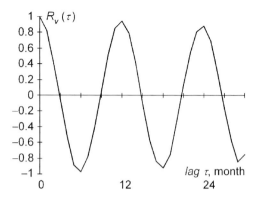

presented without distortions. Otherwise, the period of the first harmonic component
at the beginning of the ACF will be plotted approximately.

2.3 Daily Sums of Global Radiation

Daily sums of radiation are the most frequently used average values (Hassan 2001;
Fesla et al. 1992; Callegari et al. 1992). Variance of this variable has to be considered
in short-time storage design (Markvart 2006) used in PV systems. Figure 10.8 shows
the diagram of the daily sums of global irradiance in TOR for 2005.

The trend line is demonstrated for the first half of the year only, but ACF Fig. 10.9
is calculated separately for both the first and the second half of the year, considering
the trend for both.

Some specific intervals may be highlighted in the diagram in Fig. 10.8: most
stable days (with minimum of variability) occur in the interval 12–28 April and
18–27 May. Most variable days occur between 8–7 May and 28–14 June. After 8
November, for the rest of the year, it is also very stable without any beam radiation.
That is characteristic of this season and this area in general. The decrease of ACF
is less sharp in the second half of the year. This indicates that the variability of the

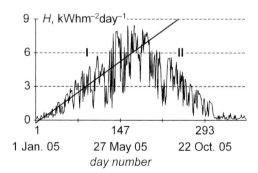

Fig. 10.8 Daily sums of
global radiation at TOR during 2005

Fig. 10.9 ACF of the daily
sums of global radiation
at TOR in the first (I) and
second (II) half of the 2005

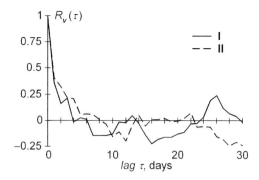

observed daily sums has reduced. To see diurnal averaging, one needs shorter time
interval data.

2.4 Hourly Sums of Global Radiation

Diurnal periodicity can be shown perfectly well at the one-hour averaging interval.
Figure 10.10 presents the time diagram of global radiation between 15 and 25 May
2005 in TOR and Fig. 10.11 – its ACF. In the chosen example, diurnal periodicity prevails to a large extent over the stochastic component that is shown by the
decreasing (upper) envelope line of Fig. 10.11. In the example, successive days correlate well and this example can be classified as solar radiation "stable in general".
The trend line shown in Fig. 10.10 supports such conditions in Fig. 10.8. Hourly averages are widely used for analysis (Gueymard 2000; Craggs et al. 1999; Gonzalez
and Calbo 1999; de Miguel and Bilbao 2005) and solar equipment design (Amato
et al. 1988). Due to the influence of higher (mainly third) harmonics in the diurnal diagram, the period of the first harmonic component in ACF may be a slightly
different - not exactly 24 hours.

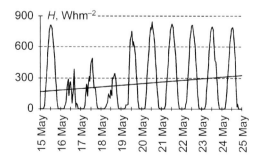

Fig. 10.10 Hourly sums
of global radiation at TOR
between 15 and 25 May 2005

Fig. 10.11 ACF of hourly
sums of global radiation at
TOR between 15 and 25 May
2005

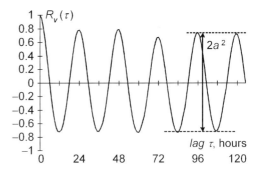

2.5 Non-Standard Averaging Intervals in Scientific Literature

Utilities in the United States use 15 minute data as standard time intervals. The GEWEX dataset consist monthly averages and 15 minute average data. The ten-minute sampling interval is widely used in recorded data sets, e.g. in road and airport service and wind speed monitoring. In some cases, these data sets are available in web-sites and can be used for analysis (e.g. in the study of spatial correlation) of solar radiation. The BSRN data are recorded in three minute intervals or shorter. Five- and ten-minute sampling intervals also have been used in the scientific literature also (Craggs et al. 1999; Gonzalez and Calbo 1999), however, these are mainly exceptions. One minute long intervals are used below (§ 5) to analyze transient processes of solar radiation.

3 Processing Data Sets Expected to be Stationary

The datasets above characterize stationary processes. Let's study how they could be analyzed.

The task is to decompose the recorded unknown data set into both the periodic and stochastic component and assess the main quality of both of them (Boland 1995). We expect a discrete data set and will use processing in the digital form.

Below we will show how the widely known standard program EXCEL can be easily used for data set analysis that is expected to be stationary. Example is shown in the CD-ROM attached to the book. The procedure is explained based on the flow chart in Fig. 10.12 and is divided into two rather independent parts:

1. Template preparation to calculate the autocovariation ACVF and autocorrelation functions ACF ($R_v(\tau)$).
2. Recorded dataset preparation appropriate to calculate the ACVF form.

An EXCEL worksheet is a matrix of cells itself from which we will use up to "N" columns and up to "M+1" rows to build the template. Original recorded variables

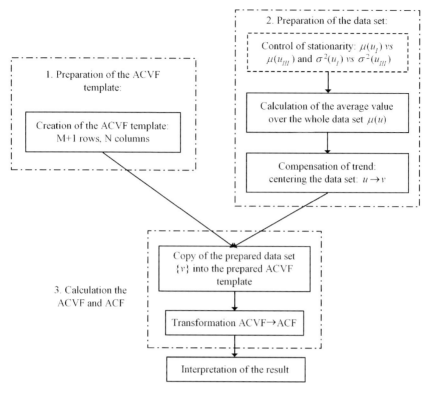

Fig. 10.12 Data processing procedure of a data set expected to be stationary

written in cells are denoted by small characters u_1, u_2, u_3, ..., u_m and v_1, v_2, v_3, ..., v_m, if they are preliminarily processed. Here small characters m, n denotes the sequence number of data, capital letters M, N denote row and column numbers in the template, correspondingly.

In reference to an address in the worksheet, we will use square brackets. For instance, [A;15] means a cell in column "A", row "15". In most cases, the examples above have a periodical component, hidden in the data set of the recorded values u_1, u_2, u_3, ..., u_m. According to the Nyquist criterion, the periodical component can be detected if the sampling interval is minimally two times shorter than the period of the (hidden) periodical component. It is essential to ensure that the maximal lag time (measured with the number of processed data n) is significantly longer than some sampling intervals and the expected period of the (hidden) periodical component $n \gg (5-10)$. The number of the recorded data m should be larger than the number of processed data n, i.e. $m > n$. The larger m and n are, the more exact the result of the analysis is.

If doubts arise about the stationarity of the data set, stationarity control should be followed by comparing the average value and variance during the first (index "I" in subscript Fig. 10.12) and the last third (index "III" in subscript Fig. 10.12) of the set (Brendat and Piersol 1974) Fig. 10.12 –dashed block. They must be close to each other; differences depend on the experience and discretion of the operator. Equality of means can be controlled via the installed "AVERAGE" function. Variance can be easily controlled with the help of the installed functions "STDEV" for the first and last third of the set. In most cases, data sets have a linear trend over the whole set or its share (Figs. 10.4 and 10.8). The linear trend has to be eliminated before the calculation of the ACVF. Obviously, EXCEL allows the linear trend line be created by clicking on the created diagram of the data set $\{u\}$ and $\{v\}$ using the command "Add trend line".

3.1 Preparation of the Recorded Data Set

Preparation of the recorded data set is illustrated in Figs. 10.13 and 10.14. The first column A consists of the sequence numbers of the recorded data 1, 2, ..., j, ..., m with the values u_1, u_2, u_3, ..., u_m in column B. In the cell [B;(M+1)] the mean value $\mu(u)$ of the recorded data set is calculated via the "AVERAGE" function. Cell [\$C;\$1] is reserved for the expected slope of the trend line α. Variables v_1, v_2, ..., v_m are centered data calculated via the formula:

$$v_j = u_j - \mu(u) - \alpha(j - m/2), \text{ where } j \in \{1, 2, \ldots, m\} \tag{10.1}$$

It is sufficient to create the formula in the cell [D;1] with the addresses [B;1] and [\$C;\$1]. [NB: Character "\$" in the address fixes the cell, *i.e.* the said address will be stored independent of the "mark and drag" operation.] All the other cells in column D will be calculated automatically with the help of the "mark and drag" operation.

For data sets $\{u\}$ and $\{v\}$, their diagrams (example: Fig. 10.14) as well the trend lines ("Add trend line") are created. In the next step, concerning the trend line $\{v\}$ α in the cell [\$C;\$1] will be varied. After a short iteration process, we will find a fitted value of α, which warrants a suitable trend line of the variable $\{v\}$ as a horizontal line (without jumps). The mean of the dataset $\{v\}$ will obtain a small value $\mu(v) \rightarrow 0$.

	A	B	C	D
1	1	u_1	α	$v_1 = u_1 - \mu(u) - \alpha(1 - m/2)$
2	2	u_2		$v_2 = u_2 - \mu(u) - \alpha(2 - m/2)$
3	3	u_3		$v_3 = u_3 - \mu(u) - \alpha(3 - m/2)$
...	...	...		...
M–1	$m-1$	u_{m-1}		$v_{m-1} = u_{m-1} - \mu(u) - \alpha(m-1) - m/2$
M	m	u_m		$v_m = u_m - \mu(u) - \alpha(m - m/2)$
M+1		$\mu(u) = \text{AVERAGE}(u_1 \ldots u_m)$		$\mu(v) = \text{AVERAGE}(v_1 \ldots v_m) \rightarrow 0$

Fig. 10.13 Preparation of the data set appropriate for ACVF calculation

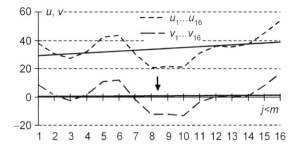

Fig. 10.14 Recorded and centered data sets prepared for calculation of ACF

Figure 10.14 shows the example of the diagrams. (Bold in Fig. 10.14) trend line of the variable $\{v\}$ is shown **before** the last step of iteration to demonstrate its step-wise structure.

3.2 Template Preparation to Calculate Autocovariation and Autocorrelation Functions

Template preparation to calculate ACVF and ACF is shown in Fig. 10.15. The first row "1" shows the lag step numbers $\{1, 2, \ldots, i, \ldots, n\}$. Column "A" gives the numbers of sampling intervals $\{1, 2, \ldots, j, \ldots, m\}$. Column "B" provides the centered data set $v_1, v_2 \ldots v_m$, which employs cells $\{[B;2], [B;3] \ldots [B;(M-1)]\}$. Cells [B;M] and [B;(M+1)] stay empty. Beginning with the row "2", columns C, D,..., i, ..., N consist of the products of sampled values $v_i \cdot v_j \in \{[B;2]*[B;3] \ldots [B;2]*[B;4] \ldots [B;2]*[B2;n]\}$. [NB: in EXCEL asterisk marks the multiplication operation]. Row M shows the instantaneous values of the ACVF depending on the lag step number "i" and it is created as the sum of the mentioned products, divided by the sum of the used rows (M$-i$). The first of them in the cell [C;M] is the square of the compound process

$$\Psi^2 = A^2 + \sigma^2 \tag{10.2}$$

and consists sum of the variance of the stochastic component and amplitude squared of the periodical component A, the assessment of which is explained below.

Row M+1 consists of successive instantaneous values of the ACF depending on the lag step number "i". ACF is the preferred form of (stationary) stochastic process presentation. It is generated by dividing each cell in row M to the first of them: cell [\$C;\$M]. The algorithm for the template of products $v_i \cdot v_j$ is given with the help of the corresponding expressions. We will write them into row 2 (columns C, D, ..., N) only, all the other rows will be generated automatically with the help of the "mark and drag" operation (basic cells for the "mark and drag" operation in row 2 (Fig. 10.15) are highlighted by enlarged font size). While the template is prepared, we have to transport (fixed with the operation "Ctrl") columns A and D (Fig. 10.13) into columns A and B (Fig. 10.15). ACVF and correspondingly ACF

	A	B	C	D	E	...	N
1	0		0	1	2	...	n
2	1	v_1	$v_1 \cdot v_1$	$v_1 \cdot v_2$	$v_1 \cdot v_3$	...	$v_1 \cdot v_n$
3	2	v_2	$v_2 \cdot v_2$	$v_2 \cdot v_3$	$v_2 \cdot v_4$	...	$v_2 \cdot v_n$
4	3	v_3	$v_3 \cdot v_3$	$v_3 \cdot v_4$	$v_3 \cdot v_5$	...	$v_3 \cdot v_n$
...	...	...	...	...	...	...	...
M-2	m−1	v_{m-1}	$v_{m-1} \cdot v_{m-1}$	$v_{m-1} \cdot v_m$	0	...	0
M-1	m	v_m	$v_m \cdot v_m$	0	0	...	0
M	$R_v(\tau)$		$(\Sigma(v_1 \cdot v_1 \ldots v_m \cdot v_m))/(m-0)$	$(\Sigma(v_1 \cdot v_2 \ldots v_{m-1} \cdot v_m))/(m-1)$	$(SUM([E:2]:[E:(M-1)]))/(m-2)$	...	$(SUM([N:2]:[N:(M-1)]))/(m-n)$
M+1	$R_v(\tau)$		[C:M]/[$C:$M]	[D:M]/[$C:$M]	[E:M]/[$C:$M]	...	[N:M]/[$C:$M]

Fig. 10.15 Template for the calculation of autocovariation and autocorrelation function

will be calculated automatically. This template is a universal tool and can be used repeatedly, each time correcting the number of sampling intervals m and lag steps n.

3.3 Interpretation of the Results

From the autocorrelation function we can look for the main quality of the expected stationary process:

1. Period of the (fundamental) harmonic component.
   The period of the fundamental harmonic component is equal to the lag interval between two characteristic points of the ACF, considering the cyclical behavior of the autocorrelation "waves" out of the beginning of ACF. These characteristic points may be local maximums or points where ACF crosses the zero value (in one direction). The first few "waves" may be distorted due to the influence of the stochastic component. Attention is to be paid to the scale ("weight") of the lag axis.
2. Amplitude of the (fundamental) harmonic component.
   The relative value of the square of the mentioned amplitude a^2 can be found as the mean value of local maximums $\mu(a^2)$ out of the beginning of ACF. This applies because of the exponential character of the ACF of the (ergodic) stochastic component, which eventually tapes to zero far from the beginning. Amplitude of the (fundamental) harmonic component A is

$$A = \sqrt{(\Psi^2 \cdot \mu(a^2))} \tag{10.3}$$

   where Ψ^2 can be found in the cell [C;M] (Fig. 10.15).

3. Variance (and standard deviation) of the stochastic component.
   The variance (and standard deviation) of the stochastic component can be found as

$$\sigma^2 = \Psi^2 \cdot \left(1 - \mu\left(a^2\right)\right) \tag{10.4}$$

   If the periodical component is filtered out (compensated in the ACF), we can see the "clean" stochastic component (rest of the ACF $R_{vs}(\tau_\rho)$) and assess the correlation interval.

4. Correlation interval.
   We consider the correlation interval τ_ρ equal to the lag τ, during which the value of the cleaned rest of the ACF (stochastic component of ACF R_{vs}) has decreased $\lg(e)$ times; also τ_ρ if $R_{vs}(\tau_\rho) = 0.4343$ That is the most frequently used definition of the correlation interval and it shows the degree of correlation of the following value of the variance with the instant value of thus.
      Unfortunately, from the ACF it is impossible to find the phase of the periodical component. We need other methods here.

4 Typical Regimes of Solar Radiation

Different radiation conditions are possible. Let's try to classify them. In Fig. 10.8 some specific regimes of solar radiation were mentioned. These regimes can be characterized by the correlation between global irradiance and its diffuse fraction (Tomson and Mellikov 2004). In clear-sky conditions, both of them, G and G_d are periodical processes determined for each moment of time. Even diffuse irradiance is determined and can be easily calculated, considering the current time and the geographical location (Duffie and Beckman 1991). Temporary clouds or upper clouds (*Cirrostratus*) have no significant influence on radiation, which is stable in general.

Correlation between them is close to the functional relation in Fig. 10.16 and this symptom characterizes radiation "stable in general".

The same symptom is also valid for the overcast conditions. In this case, global irradiance G and its diffuse fraction G_d are identical (Fig. 10.17). Such a regime occurs on overcast days with clouds *St, Sc,* and *Ns*. Some random cracks have no significant influence on the radiation level, which is of low global value. This value

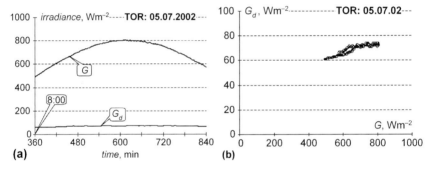

Fig. 10.16 Example of a day with clear-sky conditions: (**a**) diagram of irradiance; (**b**) correlation between global irradiance and its diffuse fraction

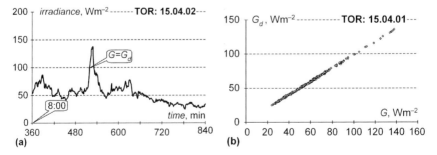

Fig. 10.17 Example of an overcast day: (**a**) diagram of irradiance; (**b**) correlation between global irradiance and its diffuse fraction

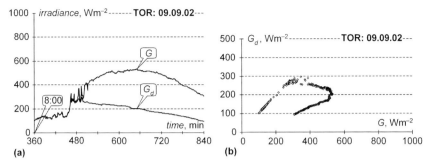

Fig. 10.18 Example of a day with transient conditions: (**a**) diagram of irradiance; (**b**) correlation between global radiation and its diffuse fractions

depends on the number and mutual position of the layers of clouds and has a wide range $50 < G < 300\,\mathrm{Wm^{-2}}$.

Naturally, transient events are possible, as shown in Fig. 10.18 (a), while the first stable regime is transformed to another. The figure shows that stable regimes exist in the time intervals *time* < 450 and *time* > 600 and in both intervals G and G_d are in good correlation (Fig. 10.18 (b)). Here numbers on the abscise axis show the sequence of minutes of the day in the example. However, they do not correlate during the transient time ($450 < time < 600$). Below we will highlight two stable regimes "clear sky" and "overcast". Lack of correlation between G and G_d is a clear symptom of instability of radiation.

(Highly) variable radiation due to clouds is shown in Fig. 10.19 (a). In this case, increments of global irradiance ΔG are close to the absolute values of global irradiance $\Delta G \approx G$, such events being frequent. The time interval during which irradiance is changing is less than 10 min. Correlation between the global irradiance and its diffuse fractions exists only for the envelope lines in Fig. 10.19 (b). Shadows of clouds perform like a switch, which commutates radiation between the two levels G and G_d. Switching operates like a stochastic telegraph signal (Morf 1998). Such a

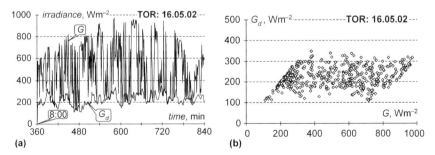

Fig. 10.19 Example of a day with highly variable conditions: (**a**) diagram of irradiance; (**b**) correlation between global irradiance and its diffuse fraction

regime corresponds to clouds *Cumulus* and *Stratocumulus translucidus* (Mullamaa 1972).

The presented example in TOR is not any exception, similar diagrams can be found in other regions too (Vijayakumar et al. 2005; Soubdhan and Feuillard 2005).

5 Minute-Long Averages of Global Irradiance

Examples presented in Figs. 10.17–10.19, which show the dynamic behavior of radiation during a day, were built up on the basis of minute-long averages of irradiance. This time basis is necessary if we describe fast varying radiation referred to Soubdhan and Feuillard (2005), Gansler et al. (1995), Tovar et al. (1998), Skartveit and Olseth (1992), Suehrcke and McCormick (1988), Vijayakumar et al. (2005), Walkenhorst et al. (2002), Tomson and Tamm (2006). To investigate solar devices that have higher frequency dependence, fast changes in solar radiation must be taken into consideration. Such investigations are essential in the analysis of the dynamic losses of solar collectors or the thermal fatigue of materials due to solar radiation. Prospective development of PV electricity generation will evoke a problem in the dynamical cooperation of (dispersed) PV-farms with the grid (Jenkins 2004). The transient behavior of solar radiation is particularly important in regions prone to high cyclonic activity (variable cloudiness). The analysis below is based on the data sampled from 1999–2002 in TOR, considered for the study during April–September only, as solar radiation at this latitude cannot be utilized effectively during the long winter season. Measurements were taken from 7:00–17:00 (solar time) during which the incidence angle, Θ_T, was less than $75°$. This is considered the daily performance period. If the incidence angle increases greater than $75°$, then the optical efficiency, $\tau\alpha$, of any solar collector will rapidly approach zero and the device cannot convert solar radiation effectively. Therefore, sunrise and sunset times with fast changing direct radiation in clear-sky conditions are not important from the point-of-view of engineering. The minute-long range studied below is much shorter than the year-long or even diurnal period of solar radiation. Therefore, any periodical instability within the minute-long range is not considered here. In the examples below, the magnitude of radiation and its increments are presented mainly in real power units as the temperature of collectors and output power changes may be calculated immediately. We will demonstrate that it does not differ from the traditional presentation form in relative units (clearness index).

5.1 Studied Regimes of Variable Solar Radiation

Solar radiation during the summer season which is technologically important at high latitudes, can vary to different extents, which was illustrated by some sample of stable and unstable radiation Figs. 10.17 – 10.19. Naturally, differentiation of "stable" or "instable" radiation can be subjective, without a clear breaking point.

To classify solar radiation as "stable" or "unstable", data should be filtered pre-liminarily (Tomson and Mellikov 2004). Accordingly, in the used approach, any 10-minute time interval is considered unstable if the sum of the absolute values of global irradiance increments $\Delta G = G_{i+1} - G_i$ exceeds $1000\,\mathrm{Wm}^{-2}$ during the 10-minute interval. Here, i and $i+1$ are two subsequent minutes and N is the total number of unstable 10-minute intervals during the day.

$$\sum_{i=j}^{j+10} |G_{i+1} - G_i| > 1000\,\mathrm{Wm}^{-2}\mathrm{min}^{-1}, \; j \in \{1\ldots N\} \qquad (10.5)$$

10 minute interval was chosen because there is a considerable amount of meteo-rological data available in 10 minute intervals. The critical sum of solar irradiance increments is taken roughly $1000\,\mathrm{Wm}^{-2}$. Daily radiation is considered highly un-stable if the sum of time intervals with unstable radiation exceeds 50% of the total performance period of 10 hours. Daily radiation is considered stable if the sum of time intervals with unstable radiation does not exceed 10% of the total performance period. Measurements during 1999–2002 showed that the share of clear summer days in TOR with stable radiation was 10.7%, including 16.7% of solar energy. The share of overcast days was 8.8%, including 2.7% of solar energy. The remaining 80.5% of days come under other classifications, implying that solar radiation is not merely strictly unstable. These days include 81.1% of solar energy, thus meriting greater attention.

5.2 Methodical Approach

Irregular changes of global irradiance due to stochastic cloud cover, quantified as fluctuations ΔG, are studied to characterize the instability of solar radiation. For clear days and days with various degrees of cloudiness, the behavior of these fluc-tuations will be different and can be characterized with the average value of solar irradiance increments. Increment is defined as difference of average values of the global radiation between two subsequent one minute intervals. The relative num-ber of increments or the frequency of the fluctuations over an extended period can be plotted as a frequency distribution function $F(\ldots)$ depending on the magnitude of the increments. The fluctuation includes positive increments of solar irradiance while radiation is increasing and negative increments while it is decreasing.

To prove that the distribution function of irradiance increments does not depend on the method of data presentation, an analysis of four variable days, 3–6 August 2002, was made for absolute ΔG (Fig. 10.20 (a)) and relative ΔG^* (Fig. 10.20 (b)) values of irradiance. It is evident that the distribution functions of positive incre-ments are similar for both methods.

The figures show two trend-lines with highly different slopes. The steeper slope corresponds to the range of low values of increments $\Delta G < 50\mathrm{Wm}^{-2}\mathrm{min}^{-1}$ ($\Delta G^* < 5\%$). The expressions for the trend lines are as follows:

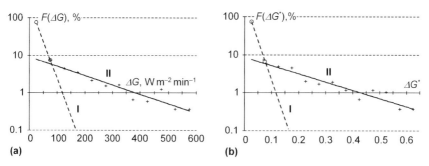

Fig. 10.20 Distribution function of beam irradiance increments in terms of absolute **(a)**, and relative **(b)** (clearness index), values

Fig. 10.20 (a) **I** $F(\Delta G) = 770 \cdot \exp(-2.3\Delta G)$ Fig. 10.20 (b) **I** $F(\Delta G^*) = 819 \cdot \exp(-2.4\Delta G^*)$
Fig. 10.20 (a) **II** $F(\Delta G) = 10.5 \cdot \exp(-0.29\Delta G)$ Fig. 10.20 (b) **II** $F(\Delta G^*) = 9.6 \cdot \exp(-0.25\Delta G^*)$

Between two subsequent periods of fluctuations, a time interval with stable irradiance may occur and if so, its distribution function is studied also. The behavior of terrestrial solar radiation varies such that irradiance intervals can be grouped based on the parameters addressed in this chapter. The classifications do not have distinct borders and those suggested here can be debated.

5.3 Definition of the Fluctuation of Solar Irradiance

Fluctuations are assessed by solar irradiance increments, defined as the difference between average global irradiances measured over consecutive one-minute intervals. These natural increments may be positive $\Delta G > 0$ or negative $\Delta G < 0$. The distribution of these natural increments is shown in Fig. 10.21, created by data from five summer months in 2002 ($\sim 10^5$ samplings). For this figure, increment values less than $50\,\mathrm{W m^{-2}\,min^{-1}}$ are ignored. They will be discussed later. The frequency of intervals with no fluctuations is small. Figure 10.21 verifies that the frequency of

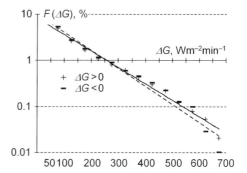

Fig. 10.21 Distribution of positive and negative increments of solar irradiance in summer 2002, excluding low values

Fig. 10.22 Distribution of
the magnitude of increments
of solar irradiance during
summer seasons 1999–2002,
including low values

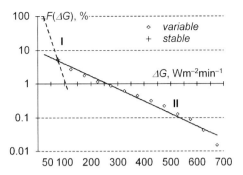

positive and negative natural increments is equal and therefore we define the universal *increment* of solar irradiance as an absolute value.

Figure 10.22 shows the distribution of increments of solar irradiance based on data from 1999–2002. The figure shows also two ranges with distinct trends. The range "stable" ($\Delta G < 50\,\mathrm{Wm^{-2}\,min^{-1}}$) coincides largely with clear-sky conditions with stable irradiance. The range "variable" ($\Delta G > 50\,\mathrm{Wm^{-2}\,min^{-1}}$) covers all the other conditions. The frequency of increments $F(\Delta G)$ is the number of increments of the indicated magnitude. The expressions for the trend lines are as follows:

Fig. 10.21, $\Delta G > 0$ $F(\Delta G) = 10 \cdot \exp(-4.1\Delta G)$ Fig. 10.22, **I** $F(\Delta G) = 1440 \cdot \exp(-2.8(\Delta G)$

Fig. 10.21, $\Delta G < 0$ $F(\Delta G) = 13.7 \cdot \exp(-4.5\Delta G)$ Fig. 10.22, **II** $F(\Delta G) = 11.5 \cdot \exp(-0.43(\Delta G)$

5.4 Fluctuation of Solar Irradiance During Stable in General Time Intervals

The range of small increments $\Delta G < 50\,\mathrm{Wm^{-2}\,min^{-1}}$ is witnessed in both clear and overcast conditions. In clear conditions, measurement errors and high clouds (*Cirrostratus*) may result in low level variability. Stability in overcast conditions occurs in the presence of opaque clouds, e.g. at *Nimbostratus*. Simultaneous occurrence of clouds on several layers smoothes out the contrast between sunshine and shade. Generally, the magnitude of fluctuations of solar irradiance in cloudy conditions is larger and the frequency of each occurrence decreases with an increase in their magnitude, as shown in Fig. 10.23.

Average values of measurements during clear and cloudy days are shown with lines in Fig. 10.23. Clear days are represented with the function $F(\Delta G) = 1411 \cdot \exp(-5.0\Delta G)$, dashed line "I". During clear days, most of the increments have low values of $\Delta G < 50\,\mathrm{Wm^{-2}\,min^{-1}}$, and 16.2% have $\Delta G = 0$. During stable overcast days, presented with a solid line "II", most of the increments are in the range $\Delta G < 150\,\mathrm{Wm^{-2}}$, and 4.8% have $\Delta G = 0$. The slope of the solid line varies throughout and no critical point exists to denote where the regime of

Fig. 10.23 Distribution of
increments of the solar irradiance for stable radiation

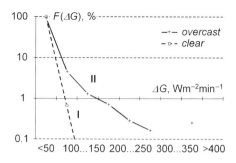

unstable radiation begins. Therefore, the measure for stability is taken arbitrarily as
$\Delta G < 50\,\mathrm{Wm}^{-2}\mathrm{min}^{-1}$ for clear days and $\Delta G < 150\,\mathrm{Wm}^{-2}\mathrm{min}^{-1}$ for cloudy days.

5.5 Fluctuation of Solar Irradiance During Unstable in General Time Intervals

An example of extremely unstable radiation is presented in Fig. 10.19 and this
regime appears with *Cumulus, Stratocumulus translucidus* and *Altocumulus translucidus* clouds. The measured solar radiation data with the first two cloud covers have
been studied (Mullamaa 1972), but excluding short intervals. This regime is presented with the line "variable" in Fig. 10.22. The frequency of large increments
with $\Delta G \gg 500\,\mathrm{Wm}^{-2}\mathrm{min}^{-1}$ is low and the few measurements in this range are
spread out. The analysis shows that in terms of absolute values, the average mag-
nitude of fluctuations has the same daily time dependence as the global irradiance
itself.

5.6 Duration of Stable Irradiance During Generally Unstable Time Intervals

It is of interest to examine intervals of stable radiation during generally unstable
periods of radiation. The condition of stability $\Delta G < 50\,\mathrm{Wm}^{-2}\mathrm{min}^{-1}$ was used
to select stable intervals during 10 of the most unstable summer days of 2002 in
Table 10.1.

The method of analysis is illustrated in Fig. 10.24. The data are filtered, consider-
ing that $\Delta G = 0$ is used for all values in the range $\Delta G < 50\,\mathrm{Wm}^{-2}\mathrm{min}^{-1}$. For each
stable interval exceeding one minute with $\Delta G = 0$, its duration is shown. Such in-
tervals are between minutes 482 and 487 with $\Delta t = 487 - 482 = 5\,\mathrm{min}$, and minute
489 with $\Delta t = 1\,\mathrm{min}$.

Table 10.1 Examples of irradiance in summer 2002 with high variability

Date	May 16	May 17	June 9	June 10	June 15	June 22	July 6	July 7	July 16	Aug. 3
G, Wm^{-2}	451.7	494.5	585.8	487.8	487	525.1	472.1	595.8	558.7	516.6
Average ΔG, Wm^{-2}min^{-1}	129	115	88.1	97.5	91.3	91.2	101.2	76.9	71.9	91.1
N	39	29	22	26	25	24	30	22	23	24
Δt_{max}, min	17	39	39	72	60	69	48	69	184	23
S, %	12.8	24.2	26.7	25.3	33.3	20.5	24.5	45.2	53	28.3

G is average global irradiance per day; *Average* ΔG is average increment per day; N is number of 10-minute long intervals per day for which the preliminary condition of unstable radiation $\Sigma_{10min}(\Delta G > 1000\,\text{Wm}^{-2}\,\text{min}^{-1})$ is satisfied; Δt_{max} is duration of the longest daily time interval with stable radiation; S is percent of time per day with stable radiation.

All of the data in Table 10.1 with stable intervals less than 10 minutes have been taken into account. The results are presented in Fig. 10.25, and the behavior of the distribution function $F(\Delta t_{st})$ for the average values of stable intervals is exponential as well. The first of them, "I", is mainly caused by a methodical fault of the analysis, as slowly varying solar irradiance is considered as a stepwise sequence of different one-minute increments. It results in an artificially increased number of time intervals with $\Delta t_{st} = 1$ minute. The equation for the dashed line is $F(\Delta t_{st}) = 300.9 \cdot \exp(-1.48\Delta t_{st})$.

A more reliable range of the distribution of time intervals with stable irradiation lies at $2 < \Delta t < 10$ minutes and is presented with the line "II". The equation for the solid line is $F(\Delta t_{st}) = 22.4 \cdot \exp(-0.42\Delta t_{st})$.

Time intervals of duration $\Delta t_{st} > 10$ minutes are ignored by the analysis as they do not correspond to the preliminary criterion of unstable radiation.

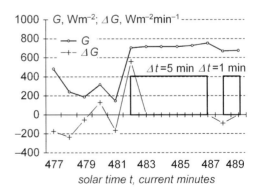

Fig. 10.24 Method to consider time intervals with stable radiation during generally unstable radiation

Fig. 10.25 Distribution of
time intervals with stable
radiation during generally
unstable radiation

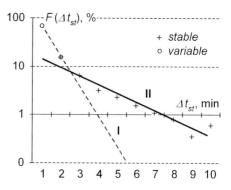

5.7 Periodicity of Fluctuations of Unstable Radiation

As we are interested here in unstable irradiance, we analyze the days in Table 10.1
within the limitation $\Delta G > 150\,\text{Wm}^{-2}\,\text{min}^{-1}$ and exclude stable intervals which
do not satisfy (10.5). The time interval between two sequential large increments
$\Delta G > 150\,\text{Wm}^{-2}\,\text{min}^{-1}$ of the same sign can be considered as the *period* of increments $T_{\Delta G}$. Naturally, this *period* is a stochastic variable too and has a distribution
given in Fig. 10.26.

Figure 10.26 (a) shows the whole range, which is constructed by time steps of
10 minutes. Figure 10.26 (b) shows the selected range of very variable irradiance,
which is constructed by time steps of 1 minute. Both of them have an exponential character of the distribution of stochastic periods $F(T_{\Delta G})$, but with different
powers of the exponent. In Fig. 10.26 (a), $F(T_{\Delta G}) = 232.8 \cdot \exp(-1.35 T_{\Delta G})$, while
in Fig. 10.26 (b), $F(T_{\Delta G}) = 18.9 \cdot \exp(-0.19 T_{\Delta G})$. Nearly 80% of these stochastic

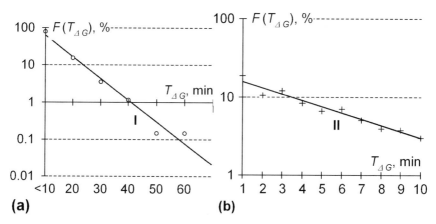

Fig. 10.26 Distribution diagrams of stochastic periods of solar irradiance fluctuations at unstable
conditions: **(a)** – whole periods, **(b)** – short periods less than 10 minutes

periods are shorter than 10 minutes and there is a significant number of very short periods $T_{\Delta G} \leq 1$min (19% of them).

5.8 Application of the Statistical Model of Short-Term Stability

To exemplify the model described above, operation of a solar hot water system that uses pumped circulation will be discussed. A controller switches the pump on, once the irradiance increases above a certain threshold. During the transient time, the solar collector produces hot water with a cooler outlet temperature than normally. Then the collector outlet temperature is stabilized on the level determined by the efficiency of the current collector at stable solar irradiance. If irradiance becomes variable again and decreases below another (lower) threshold value, circulation stops and the collector cools to the ambient temperature, which means that dynamic losses will become significant. These transient losses are the function of the dynamic behavior of the fast and frequently changing irradiance, which can be predicted using the statistical models for the magnitude and frequency of irradiance increments as presented in this chapter.

Inverters for solar electric systems have to handle rapid changes described in the chapter and may be more affected by rapid variations in solar radiation.

6 Conclusions

Dynamical behavior of solar radiation has been characterized as a compound stationary process including stochastic and periodical components and/or a transient phenomenon. Methods of investigation of dynamical effects have been described. We found, that:

1. Due to very wide range of frequencies in changes of solar radiation different approaches for investigations have to be used.
2. There is no certain provision when any of the the recommended approaches has to be used. Up to now that depends on the experience of the investigator.
3. The very fast changes in solar radiation (in the second-long range) are up to now out of attention and have to be investigated as they complicate the performance of PV systems.

Acknowledgement

These studies are partially supported by Estonian Science Foundation, grant No. 6563.

References

Amato U, Cuomo V, Fontana F, Serio C, Silvestrini P (1988) Behavior of hourly solar irradiance in the Italian climate. Solar Energy 40 : 65–79

Bendat JS, Piersol A (1974) Random Data: Analysis and Measurement Procedures. Mir, Moscow, (in Russian, translated)

Boland J (1995) Time-series analysis of climatic variables. Solar Energy 55: 377–388

Brümmer B, Thiemann S, Kirchgäßner A (2000) A cyclone statistics for the Arctic based on European Centrere-analysis data. Meteorology and Atmospheric Physics 75, Numbers 3–4

Callegari M, Festa R, Ratto CF (1992) Stochastic modeling of daily beam irradiation. Renewable Energy 2: 611–624

Carlson B, Brunold S, Gombert A, Heck M (2004) Assessment of durability and service lifetime of some static solar energy materials. Proc. of EuroSun2004, PSEGmbH, Freiburg . 2, pp 2–774–2–783

Craggs C, Conway E, Pearsall NM (1999) Stochastic modeling of solar irradiance on horizontal and vertical planes at a northerly location. Renewable Energy 18: 445–463

Duffie JA, Beckman WA (1991) Solar engineering of thermal processes. 2-nd ed. J.Wiley & Sons inc

Fesla R, Jain S, Ratto CF (1992) Stochastic modeling of daily global irradiation. Renewable Energy 2: 23–34

Gansler A, Beckman WA, Klein SA (1995) Investigation of minute solar radiation data. Solar Energy 55: 21–27

Gonzalez J, Calbo J (1999) Influence of the global radiation variability on the hourly diffuse fraction correlations. Solar Energy 65: 119–131

Gueymard C (2000) Prediction and Performance Assessment of Mean Hourly Global Radiation. Solar Energy 68: 285–303

Hassan AH (2001) The variability of the daily solar radiation components over Hewlan. Renewable Energy 23: 641–649

Jenkins N (2004) Integrating PV with the power system. Keynote lecture on the conference EuroSun2004, 23.06.2004, Freiburg, Germany

Kohel M (2001) Durability of solar energy materials. Renewable Energy 24: 597–607.

Kondratyev KY (1969) Radiative Characteristcs of the Atmosphere and Earth Surface. Gidrometeoizdat, Leningrad (in Russian)

Lyubansky V, Ianetz A, Seter I, Kudish A, Evseev EG (1999) Characterization and intercomparison of the global and beam radiation at three sites in the southern region of Israel by statistical analysis. Proc. of ISES SWC1999 on CD-ROM, 1, pp 419–426

Markvart T, Fragaki A, Ross JN (2006) PV system using observed time series of solar radiation. Solar Energy 80: 46–50

Miguel A, Bilbao J (2005) Test reference year generation from meteorological and simulated solar radiation data. Solar Energy 78: 695–703.

Morf H (1998) The Stochastic Two-Stage Solar Irradiance Model (STIM). Solar Energy 62: 101–112

Mullamaa Ü-AR (Ed) (1972) Stochastic structure of cloud and radiation fields, Tartu (in Russian, res. English)

Oliveti G, Arcuri N, Ruffolo S (2000) Effect of climatic variability on the performance of solar plants with interseasonal storage. Renewable Energy 19: 597–607

Prilipko G (1982) Climate of Tallinn. Gidrometeoizdat, Leningrad, (in Russian)

Riihimaki L, Vignola F (2005) Trends in direct normal solar irradiance in Oregon from 1979–2003. Proc. of ISES SWC2005 on CD-ROM, paper 1487.pdf

Rönnelid M (2000) The origin of an asymmetric annual irradiation distribution at high latitudes. Renewable Energy 19: 345–358

Russak V, Kallis A (2003) Handbook of Estonian Solar Radiation Climate. Ed by Tooming H, EMHI, Tallinn (mostly in Estonian)

Skartveit A, Olseth JA (1992) The probability density and autocorrelation of short-term global and beam irradiance. Solar Energy 49: 477–487

Soubdhan T, Feuillard T (2005) Preliminary Study of One Minute Solar Radiation Measurements Under Tropical Climate. Proc. of ISES SWC2005 on CD-ROM, paper 1511.pdf

Suehrcke H, McCormick PG (1988) The frequency distribution of instantaneous insolation values. Solar Energy 40: 423–430

Tepe R, Rönnelid M, Perers B (2003) Swedish solar systems in combinations with heat pumps. Proc. ISES SWC2003 on CD-ROM, paper P5-25, 8 p

Tomson T, Mellikov E (2004) Structure of Solar Radiation at High Latitudes. Proc. of Euro-Sun2004, PSEGmbH, Freiburg, 3, pp 3–899–3–904

Tomson T, Tamm G (2006) Short-term variability of solar radiation. Solar Energy 80: 600–606

Tovar J, Alados-Arboledas L, Olmo FJ (1998) One-minute global irradiance probability density distributions conditioned to the optical air mass. Solar Energy 62: 387–393

Vanicek K (1992) The estimation of solar radiation data in Czechoslovakia. Renewable Energy 2: 457–460

Vijayakumar G, Kummert M, Klein SA, Beckman WA (2005) Analysis of short–term solar radiation data. Solar Energy 79: 495–504, Erratum: Solar Energy 80: 139–140

Walkenhorst O, Luther J, Reinhart C, Timmer J (2002) Dynamic annual daylight simulations based on one-hour and one-minute means of irradiance data. Solar Energy 72: 385–395

Wild M, Gilgen H, Roesch A, Ohmura A, Long CN, Dutton EG, Forgan B, Kallis A, Russak V, Tsvetkov A (2005) From Dimming to Brightening: Decadal Changes in Solar Radiation at Earth's Surface. Science 308, (5723): 847–850

Chapter 11
Time Series Modelling of Solar Radiation

John Boland

1 Introduction

In order to derive a mathematical model of a physical system, one should know as much as possible about the inputs to the system to be able to decide the form of responses possible by the system to these inputs. This is true for estimating the performance of systems like photovoltaic cells, solar hot water heaters and passive solar houses. As pointed out by Kirkpatrick and Winn (1984), 'Due to the sensitivity of a passive solar building to the environment, appropriate modelling of the environment is equally as important as modelling of the building system.' Thus one can tailor the model to meet the needs of the inputs, trimming the model of any extraneous features. This demand side approach to the problem should by definition be more efficient.

Classical time series modelling structures are used to first describe the behaviour of global solar radiation on both daily and hourly time scales. Subsequently, procedures for generating synthetic sequences are presented, as well as procedures for generating sequences on a sub-diurnal time scale when only daily values (or inferred daily values) are available. In Chapter 8 of this book, methods for estimating the diffuse radiation component in the latter situation will be described.

Solar irradiation can be represented as a combination of two components, a deterministic one and a stochastic one. The deterministic component comprises cycles at frequencies of 0,1,2 cycles per year, 1,2,3 cycles per day and sidebands of the daily harmonics. Residual time series are formed by subtracting the aggregation of the contributions at these significant frequencies from the original measured time series and then dividing by a similar model of the standard deviation. It is shown that the daily total solar irradiation standardised residuals can be represented by a first-order autoregressive progress (AR(1)). Similarly, the standardised hourly residuals follow

John Boland
University of South Australia, Mawson Lakes, e-mail: john.boland@unisa.edu.au

an AR(1) process. There is the question of how one caters for this double time scale. That will also be addressed.

There are numerous benefits of approaching the problem of modelling the solar radiation in this manner. We will see that the identification of the significant cycles not only gives a sense of the physical processes involved, but also allows us to relatively easily alter the solar radiation time series to cater for global warming effects. As well, by approaching the problem in this way, one should be able to provide better data sets for those who solve the problems for systems dependent on solar radiation via simulation techniques. The construction of data sets in this manner provides synthetic input for testing models of system performance of other systems either utilising or affected by climatic inputs such as solar hot water heaters, photovoltaic cells and solar thermal systems. The modellers do not use real data sets for their work, but instead use a *typical* year's data to test their model.

It is necessary to define what a *typical* year means. A Typical Meteorological Year (TMY) or a Test Reference Year (TRY) or a Design Reference Year (DRY) has to be constructed to 'correspond to an "average" year, regarding both the occurrence and the persistence of warm/cold, sunny/overcast and/or dry/wet periods in all months or seasons.' (Festa and Ratto 1993). There is no guarantee that the TMY thus generated will exhibit the long-term statistical characteristics of the weather for the chosen locality. Nor would it necessarily include sequences of extreme conditions embedded within it. If, on the other hand, it was engineered to contain extreme conditions, the method of construction would conceivably make it less likely to match the long-term statistical characteristics required.

Synthetically generated data sets are more useful for testing models of system performance than either short-term measured data or TMY data. With an algorithm for generating synthetic data, one can produce any number of yearly data sets, the vast majority of which will exhibit characteristics representative of the long-term measured time series. Thus one will have a multitude of possible input sets and this will enhance the testing in that the performance can be judged on a variety of inputs, all typical of the location. On the other hand, a measured year can be significantly different from the long-term time series. A TMY, even if representative of the location, will only provide one realisation of the time series for testing the model.

The treatment of the time series analysis presented here relies heavily on preliminary work by the present author (Boland 1995). It also contains the results of more recent developments in conjunction with colleagues (Magnano and Boland 2007, Boland and Ridley 2004), as well contemporary investigations.

2 Characteristics of Climatic Variables

Measured solar irradiation is an integrated quantity, being the total energy received on a square metre horizontal surface in a certain time period, typically an hour or a half hour. Conversions can be made to obtain an approximate instantaneous value.

There are two components to these variables - a deterministic one composed of all the cycles that these variables follow, yearly, daily, and so on – defining the climate; and fluctuations about this deterministic one which give the day-to-day weather variations. These two components are studied separately and the characteristics of each are determined.

2.1 The Cyclical or Steady Periodic Component

The climatic variables under examination exhibit a cyclical nature. Some cycles are obvious such as yearly and daily but there are also a number of cycles present which are not so obvious. Phillips (1983) has demonstrated the presence of certain important cycles and the present author has confirmed the major thrust of his conclusions (Boland 1995). To determine the relative importance of different cycles, it is appropriate to calculate the Fourier Transform of the continuous function representing solar irradiation. Then, the location of peaks in the amplitude of the transformed function gives the frequencies which are significant. These significant peaks can be identified by examining the power spectrum, and it will be shown how this is defined, as well as the determination of the amount of variance explained by the significant peaks. The Fourier Transform is given by Boland (1995)

$$F(v) = \int_{-\infty}^{\infty} f(t)e^{-2\pi i v t} dt \tag{11.1}$$

where t is time and v is frequency. The function of time $f(t)$ can be also expressed in terms of its Fourier Transform.

$$f(t) = \int_{-\infty}^{\infty} F(v)e^{2\pi i v t} dv \tag{11.2}$$

In such cases where $f(t)$ is a periodic, or almost periodic, band-limited function of time, sampled 2N times over some integer multiple of the fundamental period with a sampling frequency more than two times the largest frequency component of $f(t)$, then the discrete Fourier Transform is a set of 2N Fourier coefficients (Boland 1995)

$$F_n = \frac{1}{2N} \sum_{k=0}^{2N-1} f(kt_s)e^{-i\pi nk/N}; \; n = 0, 1, \ldots, 2N - 1 \tag{11.3}$$

where t_s is the sample interval. The function $f(t)$ can be approximated in terms of these Fourier coefficients as

$$f(t) = \sum_{n=0}^{2N-1} F_n e^{i\pi nt/Nt_s} \tag{11.4}$$

If $f(t)$ is a real-valued function, then it may be written as

$$f(t) = \sum_{n=0}^{N-1} (A_n \cos(\pi nt/Nt_s) + B_n \sin(\pi nt/Nt_s)) \tag{11.5}$$

where A_n is twice the real part of F_n, B_n is the negative of twice the imaginary part of F_n for $n > 0$ and A_0 is the real part of F_0.

When we are dealing with a real series, the determination of the Fourier coefficients is more straightforward.

$$A_0 = \frac{1}{2N} \sum_{k=1}^{2N} f(kt_s) \tag{11.6}$$

$$A_n = \frac{1}{N} \sum_{k=1}^{2N} f(kt_s) \cos(\pi nk/Nt_s)$$

$$B_n = \frac{1}{N} \sum_{k=1}^{2N} f(kt_s) \sin(\pi nk/Nt_s)$$

We use the magnitudes of the coefficients to determine the *power spectrum*, a plot of the power of the signal that lies at each frequency. The power spectrum is the Fourier transform of the autocovariance function and shows how the variance of the stochastic process is distributed with frequency. We can thus examine this plot and determine which of the frequencies contribute most to the series. More precisely, we can determine the contribution to the variance of the series at each frequency. The variance is given by

$$V = \frac{1}{2N} \sum_{k=1}^{2N} (f(kt_s) - \bar{f})^2 \tag{11.7}$$

From use of the orthogonality conditions of the cosine and sine terms (we are using these as a basis set for the function space), we can rewrite the variance as

$$V = \frac{1}{2} \sum_{k=1}^{N-1} (A_k^2 + B_k^2) + A_n^2 = \frac{1}{2} \sum_{k=1}^{N-1} \rho_k^2 + A_n^2 \tag{11.8}$$

Thus, $\rho_k^2/2$ is the contribution of the jth harmonic to the sample variance. Figure 11.1 gives the power attributable to the various harmonics for an example site – Mt. Gambier in South Australia. Note that the major spikes are at the obvious frequencies, once a year and once a day. However, there is a significant contribution at a frequency of twice a day. This means that the morning is different from the afternoon. Additionally, there are significant contributions at frequencies close to once a day and twice a day. If we examine the spectrum about once a day more closely we can get a clearer picture. The frequency of one cycle per day is, of course, 365 cycles per year. We can also see from Fig. 11.1 that there are significant contributions to the variance at 364 and 366 cycles per year.

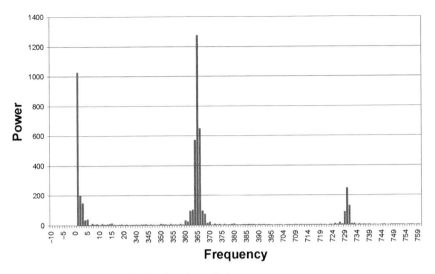

Fig. 11.1 Power spectrum for hourly solar radiation

2.2 Sidebands

The contributions at the frequencies to either side of dominant cycles are called sidebands. In acoustics, "radio transmission involves putting audio frequency information on a much higher frequency electromagnetic wave called a carrier wave" (Hyperphysics 2007). The process of superimposing the sound information on the carrier wave is called modulation, either frequency or amplitude. Either type produces frequencies at the sum and difference of the two frequencies, called sidebands. We see a similar phenomenon, wherein they modulate the amplitude of the oscillation over the day to suit the time of year, thus representing the interference of the once a year and once a day cycles. We get a similar set of sidebands at the twice a day frequency. From Figs. 11.2 and 11.3 it is obvious that the sidebands are a necessary component of the Fourier series representation of the data. Otherwise, the model would include values at night that are significantly different from zero.

2.3 How Many Cycles?

The next question is to identify how many cycles we will include, apart from once a year, once and twice a day and the sidebands to either side of the daily harmonics. We examined the various options, and use one illustrative example to show why using only up to two cycles per day is not sufficient. In winter, if we stop at two cycles a day, there is substantial overshoot into the negative range just before sunrise

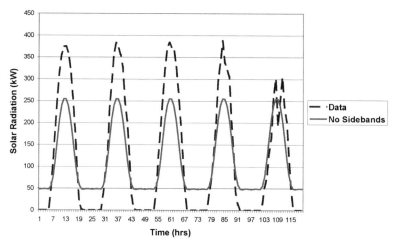

Fig. 11.2 Effect of ignoring the sidebands in summer

and just after sunset – see Fig. 11.4. Of course, one could alleviate this problem by using up to two cycles per day during the daylight hours and simply zeroing the series between sunset and sunrise. However, for simplicity's sake, we will use up to three cycles per day – note that in both seasons, there is not much difference between using three and four. Using the criterion that $\rho_k^2/2$ is the contribution of the jth harmonic to the sample variance, we estimate that the cycles we have identified contribute 84.1% of the variance of the hourly solar radiation.

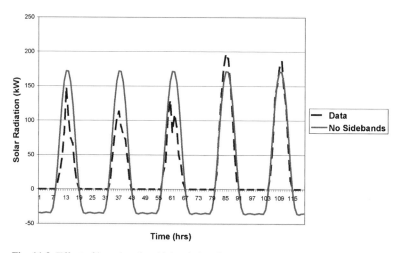

Fig. 11.3 Effect of ignoring the sidebands in winter

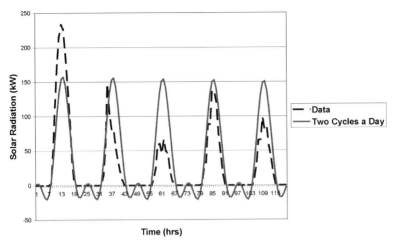

Fig. 11.4 Using up to two cycles per day in winter

2.4 Modelling the Volatility

The standard time series procedure is to simply subtract the contributions of the deterministic component, the embedded cycles, and then to attempt to model the residuals from that process using Box-Jenkins methods (Chatfield 2003). After we have identified and removed trends – including seasonality, we assume that the residuals are a stationary time series. A time series $\{X_t\}$ is *strongly stationary* if the joint distribution of $\left(X_{t_1}, X_{t_2}, \ldots, X_{t_k}\right)$ is identical to that of $\left(X_{t_1} + t, X_{t_2} + t, \ldots, X_{t_k} + t\right)$ for all t. A time series is *weakly stationary* if $E(X_t) = \mu$, a constant, and $Cov(X_t, X_{t-l}) = \gamma_l$, which only depends on the lag l (Tsay 2005). In other words, weak stationarity implies that the series fluctuates with constant variance around a fixed level.

We then utilise particular methods to ascertain whether this residual series is the realisation of a purely random process, or if it contains serial correlation of some nature. Not only are the night values systematically zero – which can be dealt with since it is systematic – but there is a differing variance over the day. This differing variance can be seen more easily if we examine the standard deviation for each hour. In order to do this, we estimate the hourly standard deviation by using eight years of data and calculating the standard deviation for the January 1 values, the January 2 values and so on.

We can deal with this bias since it is systematic. One method is to estimate the variance, or standard deviation, for each hour of the year by examining a number of years of data, and then modelling the standard deviation in the same way as we modelled the original series, with Fourier series. Note that, as stated in Section 1, we are developing the methodology for a location where there is substantial data available. We will specify techniques for dealing with locations where there might only be satellite inferred data in Section 8. After estimating the standard deviation model, we then divide the deseasoned series by this function. This results, in effect,

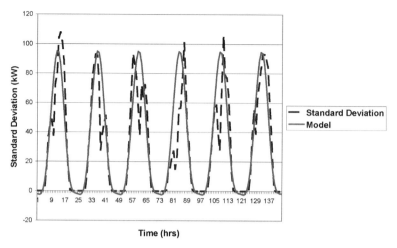

Fig. 11.5 The standard deviation model

in a series of standard scores for each hour of the year – standardised residuals in other words. An example of the standard deviation model fit is given in Fig. 11.5. We can now analyse the standardised residuals using Box-Jenkins methods.

3 ARMA Modelling

We are going to examine the Autoregressive Moving Average (ARMA) process for identifying the serial correlation attributes of a stationary time series (see Bowerman and O'Connell, 1987, Brockwell and Davis 1996, Tsay 2005, Anderson 1976, Chatfield 2003). Another name for the processes that we will undertake is the Box-Jenkins (BJ) Methodology, which describes an iterative process for identifying a model and then using that model for forecasting. The Box-Jenkins methodology comprises four steps:

- Identification of process;
- Estimation of parameters;
- Verification of model, and;
- Forecasting.

3.1 Identification of Process

Assume we have a (at least weakly) stationary time series, ie. no trend, seasonality, and it is homoscedastic (constant variance). Note that since we are essentially discussing the solar radiation during the daytime hours, we are ignoring the systematic zero values at night. Much of the theory to be described in this and the following sections is for a continuous series and will thus have to be modified for our purposes. Later, in Section 4 on the daily model, this theory will apply directly.

The general form of an ARMA model is

$$X_t - \phi_1 X_{t-1} - \cdots - \phi_p X_{t-p} = Z_t + \theta_1 Z_{t-1} + \cdots + \theta_q Z_{t-q} \qquad (11.9)$$

where $\{X_t\}$ are identically distributed random variables $\sim (0, \sigma_X^2)$ and $\{Z_t\}$ are white noise, ie. independent and identically distributed (iid) $\sim (0, \sigma_Z^2)$. ϕ_p and θ_q are the coefficients of polynomials satisfying

$$\phi(y) = 1 - \phi_1 y - \cdots - \phi_p y^p \qquad (11.10)$$
$$\theta(y) = 1 + \theta_1 y + \cdots + \theta_q y^q$$

where $\phi(y), \theta(y)$ are the autoregressive and moving average polynomials respectively. Define the backward shift operator $B^j X_t = X_{t-j}, j = 0, 1, 2, \ldots$ and we may then write Eq. (11.9) in the form

$$\phi(B) X_t = \theta(B) Z_t \qquad (11.11)$$

defining an $ARMA(p,q)$ model. If $\phi(B) = 1$, we then have a moving average model of order q, designated $MA(q)$. Alternatively, if we have $\theta(B) = 1$, we have an autoregressive model of order p, designated $AR(p)$. The question is, how do we identify whether we have an $MA(q)$, $AR(p)$ or $ARMA(p,q)$? To do so, we can examine the behaviour of the autocorrelation and partial autocorrelation functions.

3.2 Autocorrelation and Partial Autocorrelation Functions

We need some definitions to begin with. Suppose two variables X and Y have means μ_X, μ_Y respectively. Then the covariance of X and Y is defined to be

$$Cov(X, Y) = E\left\{ (X - \mu_X)(Y - \mu_Y) \right\} \qquad (11.12)$$

If X and Y are independent, then

$$Cov(X, Y) = E\left\{ (X - \mu_X)(Y - \mu_Y) \right\} = E(X - \mu_X) E(Y - \mu_Y) = 0 \qquad (11.13)$$

If X and Y are not independent, then the covariance may be positive or negative, depending on whether high values of X tend to happen coincidentally with high or low values of Y. It is usual to standardise the covariance by dividing by the product of their respective standard deviations, creating the correlation coefficient. If X and Y are random variables for the same stochastic process at different times, then the covariance coefficient is called the autocovariance coefficient, and the correlation coefficient is called the autocorrelation coefficient. If the process is stationary, then the standard deviations of X and Y will be the same, and their product will be the variance of either.

Let $\{X_t\}$ be a stationary time series. The **autocovariance function** (ACVF) of $\{X_t\}$ is $\gamma_X(h) = Cov(X_{t+h}, X_t)$ and the **autocorrelation function** (ACF) of $\{X_t\}$ is

$$\rho_X(h) = \frac{\gamma_X(h)}{\gamma_X(0)} = Corr(X_{t+h}, X_t). \tag{11.14}$$

The autocovariance and autocorrelation functions can be estimated from observations of $X_1, X_2, \ldots, X_n$ to give the sample autocovariance function (SAF) and the sample autocorrelation function (SACF), the latter denoted by

$$r_k = \frac{\sum\limits_{t=1}^{n-k}(x_t - \bar{x})(x_{t+k} - \bar{x})}{\sum\limits_{t=1}^{n}(x_t - \bar{x})^2} \tag{11.15}$$

Thus the SACF is a measure of the linear relationship between time series separated by some time period, denoted by the lag k. Similar to the correlation coefficient of linear regression, r_k will take a value between $+1$ and -1, and the closer to ± 1, the stronger the relationship. What relationship are we talking about? Consider a lag 1 value close to $+1$ as an example. This means that there is a strong relationship between x_t and x_{t-1}, x_{t-1} and x_{t-2}, and so on. The interesting thing is that what can happen in practice is that because of this serial correlation, it can appear that x_t has a strong relationship with x_{t-d}, d time units away from x_t, when in fact it is only because of this interaction. To sort out this potential problem, one estimates the partial autocorrelation function (PACF). It describes the correlation between observations at some time period d with the influence of the serial correlation removed. It strips away the interconnection and gives only the "pure" correlation. The sample PACF is given by the **Yule-Walker** equations,

$$\begin{bmatrix} 1 & r_1 & r_2 & \cdots & r_{k-2} & r_{k-1} \\ r_1 & 1 & r_1 & \cdots & r_{k-3} & r_{k-2} \\ \cdot & \cdot & \cdot & \cdots & \cdot & \cdot \\ \cdot & \cdot & \cdot & \cdots & \cdot & \cdot \\ r_{k-1} & r_{k-2} & r_{k-3} & \cdots & r_1 & 1 \end{bmatrix} \begin{bmatrix} \hat{\phi}_{k1} \\ \hat{\phi}_{k2} \\ \cdot \\ \cdot \\ \hat{\phi}_{kk} \end{bmatrix} = \begin{bmatrix} r_1 \\ r_2 \\ \cdot \\ \cdot \\ r_k \end{bmatrix} \tag{11.16}$$

The value of $\hat{\phi}_{kk}$ gives the estimate of the PACF at lag k. These equations can be solved using Cramer's Rule from:

$$\begin{aligned} \hat{\phi}_{11} &= r_1 \\ \begin{bmatrix} 1 & r_1 \\ r_1 & 1 \end{bmatrix} \begin{bmatrix} \hat{\phi}_{21} \\ \hat{\phi}_{22} \end{bmatrix} &= \begin{bmatrix} r_1 \\ r_2 \end{bmatrix} \\ \begin{bmatrix} 1 & r_1 & r_2 \\ r_1 & 1 & r_1 \\ r_2 & r_1 & 1 \end{bmatrix} \begin{bmatrix} \hat{\phi}_{31} \\ \hat{\phi}_{32} \\ \hat{\phi}_{33} \end{bmatrix} &= \begin{bmatrix} r_1 \\ r_2 \\ r_3 \end{bmatrix} \end{aligned} \tag{11.17}$$

or, in a more concise recursive representation using

$$\hat{\phi}_{mm} = \frac{r_m - \sum\limits_{j=1}^{m-1} \hat{\phi}_{m-1,j} r_{m-1}}{1 - \sum\limits_{j=1}^{m-1} \hat{\phi}_{m-1,j} r_j} \tag{11.18}$$

Once we have calculated these estimates for a stationary time series, we can use them to give an indication whether we should fit an $AR(p), MA(q)$, or $ARMA(p,q)$ model. The criteria (*) are in general:

- When the SACF dies down gradually and the SPACF has significant spikes at lags $1, 2, \ldots, p$, we should fit an $AR(p)$.
- When the SACF has significant spikes at lags $1, 2, \ldots, q$ and the SPACF dies down gradually, we should fit an $MA(q)$.
- If both die down gradually, we fit an $ARMA(p,q)$. In this case, we will have to progressively increase p, q until we get a suitable model.

The last point brings up an interesting question; how do we decide between competing models? In fact, the situation is often not as simple as these criteria make it seem. Sometimes it is difficult to decide between for instance, an $AR(3)$ and an $ARMA(1,1)$ process. An aid in identifying the appropriate model comes from the principle of parsimony, using criteria from Information Theory. The Akaike Information Criterion (AIC) is one such measure (Tsay 2005). The goal is to pick the model that minimises

$$AIC = -\frac{2}{T} \left\{ \ln(\text{likelihood}) + l \right\} \tag{11.19}$$

Here, l is the number of parameters fitted and T the number of data values. There is a competing criterion, that penalises the number of parameters fitted even more, called the (Schwarz) Bayesian Information Criterion (BIC) (Tsay 2005),

$$BIC = -\frac{2}{T} \ln(\text{likelihood}) + \frac{l \ln(T)}{T} \tag{11.20}$$

Since the hourly data are not in the necessary stationary form due to the zeroing at night, we will not be able to use statistical software to perform the estimates. Section 4, dealing with daily data can be analysed in this manner giving a good example of this procedure. We will continue with the hourly model subsequent to that analysis.

4 The Daily Model

In modelling the solar radiation on a daily basis, the data is first aggregated to daily totals and then the significant cycles are identified using Fourier Transforms. We used two frequencies and we can see from Fig. 11.6 how well the curve fits the

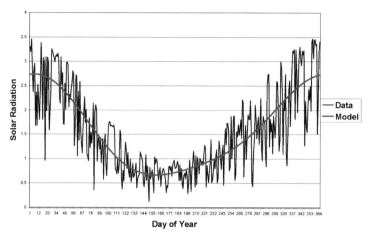

Fig. 11.6 Daily solar radiation and model

general trend of the data. The contribution to the variance is 69.1% from these two cycles.

Subsequent to subtracting the trend from the data, we see that the resulting residuals are not stationary, with a higher variance in the summer than in the winter (see Fig. 11.7). So, similar to the treatment for the hourly data, the values are standardised by dividing by the standard deviation model for each day of the year. To construct this model, we took 30 years of data from the Australian Climatic Database (Energy Partners et al. 2005), calculated the standard deviation for each day of the year for this data set, and then modelled this set with Fourier series (see Fig. 11.8). When the residual data is divided by the model function – see Eq. (11.21), we obtain

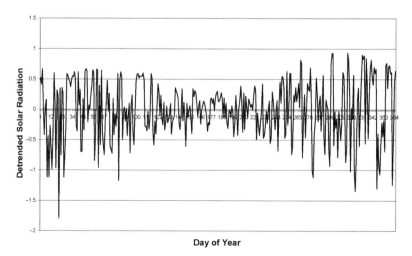

Fig. 11.7 The daily solar radiation with mean function removed

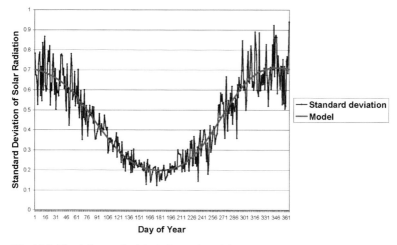

Fig. 11.8 The daily standard deviation and model

Fig. 11.9 showing that we now have a set of what we can assume is stationary data. This means we can analyse this set for serial correlation to determine if it has an ARMA structure. Following the criteria given in (*), we examine the SACF and SPACF using Minitab statistical software.

Figures 11.10 and 11.11 would indicate that the data can be modelled with an AR(1) process. There are a number of ways to ensure that this hypothesis can be substantiated. The first way is to try and overfit the model. We try and fit an AR(2) model to the data, and allow for a constant in the form as well. The statistical output from Minitab for this exercise is given in Table 11.1.

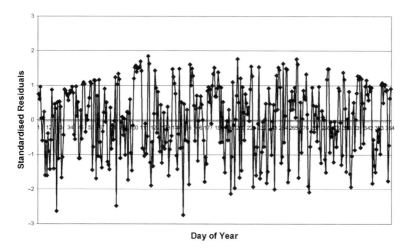

Fig. 11.9 The daily standardised residuals

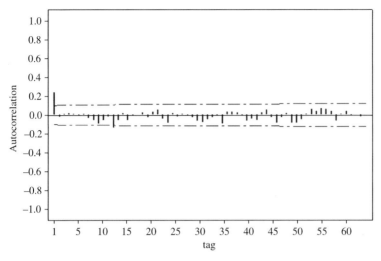

Fig. 11.10 The SACF for daily standardised residuals

The p values indicate that the only significant coefficient is that for an AR(1) model, the standard Markov chain (see Eq. (11.22)). To obtain a reliable estimate of the coefficient, we refit the model ignoring the other terms. The algorithm for best estimation of the coefficients of ARMA models utilised in standard statistical software like Minitab work under the assumption of normality of the noise term Z_t. In fact, the subscript might seem somewhat superfluous since $Z_t \overset{iid}{\sim} N(0, \sigma_Z^2)$. However, it is to indicate that there is a separate distribution for each t, but in this case, they are identically distributed, as well as independent (iid).

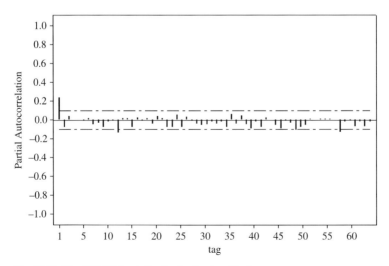

Fig. 11.11 The SPACF for daily standardised residuals

Table 11.1 Overfitting exercise for an AR(1)

Final Estimates of Parameters

Type	Coefficient	SE Coefficient	t	p
AR 1	0.2626	0.052	5.01	<0.0005
AR 2	−0.0798	0.053	−1.52	0.129
Constant	0.0004	0.051	0.01	0.995

$$R_t = \frac{X_t - \overline{X}_t}{S_t} \tag{11.21}$$

$$R_t = \phi R_{t-1} + Z_t \tag{11.22}$$

Grunwald et al. (2000) mention that a simple linear regression estimation, regressing R_t on R_{t-1}, gives a robust estimator for the coefficient. The term robust means that the estimation is performed without any distributional assumptions. In this case, $\phi = 0.2518$, and we can use some diagnostic checking to check the adequacy of the model. If indeed this is a suitable model, we should have that the noise term Z_t is iid. We check the lack of serial correlation by examining the SACF of this series of noise terms – see Fig. 11.12. There are no significant lags, indicating that we have been able to find the appropriate model. However, if we want to be sure, we can use the Ljung-Box measure of lack of fit (Ljung and Box 1978). If the model is appropriate, the quantity

$$Q(r) = n(n+2) \sum_{k=1}^{m} (n-k)^{-1} r_k^2 \tag{11.23}$$

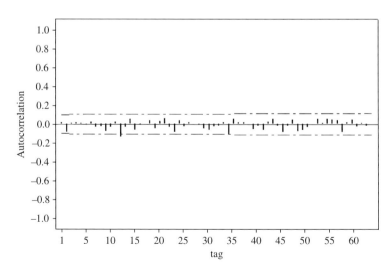

Fig. 11.12 The autocorrelation of the final noise series

Table 11.2 Modified Box-Pierce (Ljung-Box) Chi-Square statistic

Lag	12	24	36	48
Chi-Square	4.9	21.4	32.8	40.6
df	11	23	35	47
p-value	0.846	0.438	0.479	0.658

where r_k are the autocorrelations of the noise, and p, q are the number of parameters estimated in an $ARMA(p, q)$, is distributed as χ^2_{m-p-q}. This allows us to estimate the p-value for the test of an autocorrelation being significant at lag m (under the null hypothesis, if the p-value > 0.05, the autocorrelation at that lag is not significant). If we perform the parameter estimation in Minitab, we do not obtain the robust estimate for ϕ, but the value is not significantly different, and the Ljung-Box test (or portmanteau test as it is also known) results are given automatically. Table 11.2 gives the result, which shows that there is no autocorrelation remaining, as the p-values for all lags examined are greater than 0.05.

A simple calculation will show how much this AR(1) component contributes to the variance of the standardised residuals. Note that from Eq. (11.22), $\sigma_Z^2 = (1 - \phi^2)\sigma_R^2$. In other words, the residual variance is a factor of $1 - \phi^2$ of the variance of the standardised residuals. Therefore, the contribution of the AR(1) modelling to explaining the variance is only 6.34% but remember that the seasonal terms already contributed a great deal to the overall variance.

To validate our procedure, we aggregate all our components for a one step forecast for the original series given by

$$E(X_t) = E(\overline{X}_t + S_t[\phi R_{t-1} + Z_t]) = \overline{X}_t + S_t \phi R_{t-1} \quad (11.24)$$

Note that any contribution of the noise term is zero since $E(Z_t) = 0$. This results in Fig. 11.13, showing the original series with the one step ahead forecast

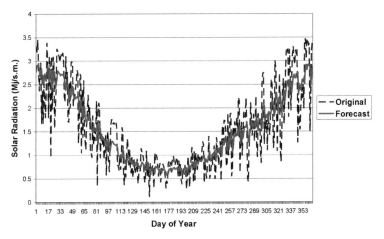

Fig. 11.13 The daily solar radiation and one step ahead forecast

superimposed. The model not only follows the trend, but also displays a higher variance in summer than winter. The contribution to the variance of the forecast series is 71.6%, meaning an unexplained variance of 28.4%. The contribution that was not recorded previously was that of the standard deviation model.

5 Synthetic Generation

For many purposes, we want to be able to generate synthetic sequences of solar radiation values. We will outline the procedure in this section for daily totals and in Section 6 for hourly values. This order is sensible for several reasons. One is that the model is simpler for the daily totals, and thus the procedure is easier to understand, than for the more complex hourly model. Additionally, there is a diminishing list of stations (in Australia at least) where there are ground stations collecting solar data. Thus, we are relying more on satellite inferred solar data, and for the time being at least, this is on a daily time step. In Section 8, we describe a methodology for estimating intra-day values from daily totals.

The synthetic generation procedure follows the fitted model, but including the white noise term as given in Eq. (11.25). To enact the generation, we must construct a model for Z_t. As already stated, $E(Z_t) = 0$, but we must determine its distributional properties. Figure 11.14 shows the histogram of Z_t with a normal curve with the same mean and standard deviation, demonstrating clearly that, unlike the situation desirable for AR processes, Z_t is not normally distributed.

$$X_t = \overline{X}_t + S_t(\phi R_{t-1} + Z_t) \tag{11.25}$$

So, what sort of distribution should we try to fit to this data? Since the original series is physically bounded below by zero and above by the extraterrestrial radiation,

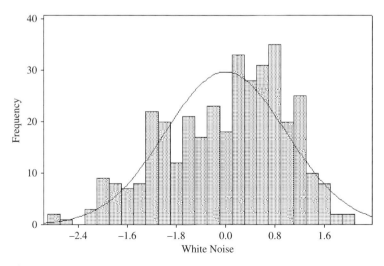

Fig. 11.14 White noise with a normal curve superimposed

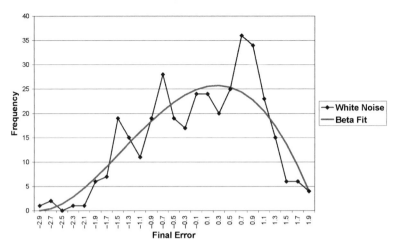

Fig. 11.15 Fitting a Beta distribution to the white noise for the daily model

we are justified in suggesting that a Beta distribution, general form of the probability density function (pdf) given in Eq. (11.26), will be the most appropriate model. The result of fitting this distribution to the white noise, or what can also be called the final error, is given in Fig. 11.15. As can be seen, this distribution gives a very good representation of the noise.

The estimates of the parameters are $\hat{\alpha} = 2.64, \hat{\beta} = 1.89$. It should be noted that these are local parameters. If we were to perform synthetic generation for other locations, a similar analysis would have to be performed. Dense data sets are not available for very many locations, but daily data series inferred from satellite images are available at relatively high resolution, and these can be utilised to perform this analysis.

$$f(x,a,b) = \frac{(x-a)^{\alpha-1}(b-x)^{\beta-1}}{B(\alpha,\beta)(b-a)^{\alpha+\beta-1}}, \; a \le x \le b; \alpha, \beta > 0. \tag{11.26}$$

One now can use the relationship in Eq. (11.25) to generate synthetic sequences of daily solar radiation. The algorithm for this generation (**) follows:

1. Generate 365 values from a Uniform distribution on [0,1] – these will be cumulative probabilities;
2. Use the inverse cumulative distribution function (in Excel in this case) to calculate the corresponding Beta values – these will form the sequence of Z_t values;
3. For the first day this will constitute also the contribution given by $R_t = \phi R_{t-1} + Z_t$, and then we can recursively calculate R_t;
4. Then multiply each day's R_t by the corresponding value of the standard deviation model and add the contribution from the mean model. This will result in the synthetic series.

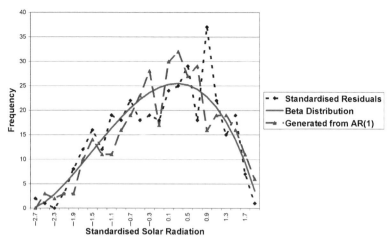

Fig. 11.16 Validating the AR(1) model generation

The accepted wisdom in time series analysis is that one cannot guarantee that the distribution of $\hat{R}_t$ as generated by the use of Step 4 above will match the distribution of the real series unless the real series is normally distributed. Results that we obtained in our modelling suggest that this conjecture should be investigated more deeply to ascertain exactly which types of ARMA models follow this maxim. Figure 11.16 gives the results of a generation of one year's series, along with the original R_t and also a Beta distribution fit to it.

It seems obvious from Fig. 11.16 that the process has been successfully performed to obtain a sensible distribution of synthetic values. As an example, we continue the process of constructing a "synthetic year" and the results are given in Fig. 11.17. We see from this depiction that we obtain a series that has the same general characteristics as a real year's series.

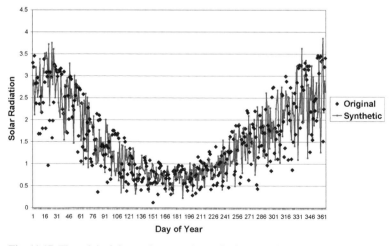

Fig. 11.17 The original data and one year's synthetic generation

6 The Hourly Model

As mentioned previously, the Box-Jenkins modelling of the hourly standardised residuals is not straightforward as the zeroes at night provide essentially a discontinuity which we have to accommodate. Box-Jenkins or ARMA modelling is suitable simply for stationary data and this series does not satisfy this criterion.

However, we can estimate the autocorrelations for particular lags using Eq. (11.15), and then use Eq. (11.17) or (11.18) to estimate the partial autocorrelations. Returning to the comment by Grunwald et al. (2000) we get an even simpler method of determining the significant lags in persistence modelling in this instance. We extend their remark to conjecture that using multiple regression with the standardised residuals at various lags as possible predictors will give us the suitable persistence model. As a preliminary step towards that we will use the calculation of the SACF from Minitab as a guide to which lags to consider. We know that this is technically incorrect but from a simply indicative point of view, we can intuit what may be the situation. We also rely on work with temperature which is continuous (Boland 1995, Magnano and Boland 2007) to conjecture that some combination of up to three lags plus perhaps some connection with the previous day's residuals may be appropriate. Figures 11.18 and 11.19 give the SACF and SPACF for the solar residual series.

From these graphs, we can surmise that y_t, the standardised residual at time t, could depend on y_{t-1}, y_{t-2} and some combination of $y_{t-23}, y_{t-24}, y_{t-25}$ etc. This last set of terms could possibly arise because there is the connection on a daily basis as evidenced in Section 4. Magnano and Boland (2007) found that for modelling half hourly ambient temperature, it was not simply a dependence on the single lagged value y_{t-24}, but rather a moving average of the values around that one. Here, we in-

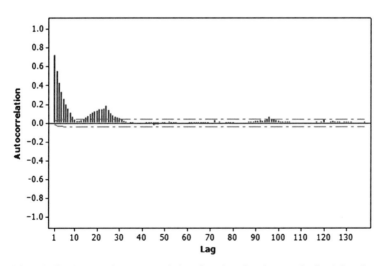

Fig. 11.18 The sample autocorrelation function for the standardised hourly solar radiation residuals

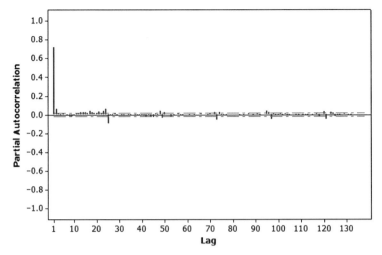

Fig. 11.19 The sample partial autocorrelation function for the standardised hourly solar radiation residuals

vestigate the dependence on lags 1 and 2 as well as the average of the previous day's residuals. Table 11.3 shows that there is a significant relationship, but that the only predictor that should be involved is the lag 1 variable – a similar situation to the daily model. Note that the hypothesis test for multiple regression gives first the p-value for the overall test of significance (in the ANOVA table) with a p-value < 0.05 indicating a relationship, and then reports the extra tests to ascertain the significance of the various candidate predictors, under the $H_0 : \beta_i = 0$. A p-value < 0.05 once again indicates a significant coefficient. However, we decided to investigate this further since we are concerned that there still be some mixing of the two time scales since we do find a significant day-to-day relationship.

To perform this experiment, we regress y_t on y_{t-1} alone and independently on $\bar{y}_{t-24}$ where the latter refers to the average of the previous day. After regressing on the previous day's average, we then regress the residuals from that process on y_{t-1}. In this way, we are trying to determine if we get extra benefit from incorporating the effect of the previous day. We examine the variance of the noise left from both these processes to determine the benefit or not. We perform this exercise for one time of day – midday as an example – to test the mixing of the two time scales.

The results are intriguing – the noise variance is actually higher when performing the two stage approach – incorporating the dependence on the day before. The choice of variable to reflect this daily connection may not be exactly the best one, an average of values around y_{t-24} may be more appropriate. Still, one would not expect that the noise variance would fall dramatically enough to indicate that it should be included. The conclusion we are led to is that the closer, in time, dependence is the dominant influence. We thus restrict our persistence model to one lag. Since we do not have a continuous series, we cannot use standard software to estimate the parameter of this AR(1) process. Thus, the procedure we adopted was to estimate

Table 11.3 Multiple regression output for the hourly standardised residuals

Regression Statistics

Multiple R	0.80
R Square	0.64
Adjusted R Square	0.63
Standard Error	0.62
Observations	364

ANOVA

	df	SS	MS	F	Significance F
Regression	4	245.93	61.48	158.00	8.07E-78
Residual	359	139.70	0.39		
Total	363	385.64			

	Coefficients	Standard Error	t Stat	P-value
Intercept	0.03	0.03	0.95	0.34
Lag 1	0.72	0.05	13.27	0.00
Lag 2	−0.04	0.06	−0.62	0.53
Lag 3	0.11	0.05	2.04	0.04
Average of Previous Day	0.01	0.00	1.57	0.12

the dependence of y_t on y_{t-1} using regression techniques for $t = 9, 10, \ldots, 17$ where in this case t stands for time of day.

Between these time periods and sunrise or sunset, we assume that the deterministic cycles provide a sufficient estimate of the solar radiation. Stochastic departures from this at those times of day will not yield a significant difference in energy on the horizontal surface. So we obtain estimates for ϕ in the model

$$y_t = \phi y_{t-1} + a_t \qquad (11.27)$$

where a_t is white noise. We obtained similar estimates for ϕ for the various times of the day and thus in our model we use the estimate of $\phi = 0.76$ for all time periods.

7 Model Validation

We perform the validation in terms of how well the model performs in one step ahead forecasts for the original data. This entails estimating the value of the standardised residual at time t, using $\hat{y}_t = \phi y_{t-1}$, and then multiplying by the standard

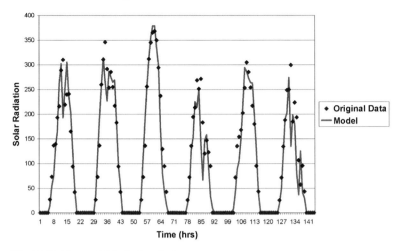

Fig. 11.20 The model fit for hourly data

deviation model, Section 2.4 and then adding the Fourier series model for the mean, Section 2.3. Figure 11.20 shows a typical segment of the series. The normalised root mean square difference (NRMSD) for this fit is 23.2% and importantly the normalised mean bias difference (NMBD) is 0.06% - indicating a very good fit to the data. The error measures are defined as follows. Let the variable being modelled be denoted by η. Then, the normalised root mean square difference (NRMSD) and the normalised mean bias difference (NMBD) are given by, respectively

$$NRMSD = \frac{\left[\frac{1}{m}\sum_{i=1}^{m}(\eta_i - \hat{\eta}_i)^2\right]^{1/2}}{\bar{\eta}}, \quad NMBD = \frac{\frac{1}{m}\sum_{i=1}^{m}(\eta_i - \hat{\eta}_i)}{\bar{\eta}}.$$

As further validation we test how well the model performs the one step ahead prediction for a year that was not part of the model building. In the year we chose at random, the yearly average solar radiation was very close to the year for which the model was built. If it had not been, we would assume that this difference would be easily available, and we could adjust the first term (the average) in the Fourier series component accordingly. Figure 11.21 gives an illustration of the performance of the model. We deliberately selected a sequence of days that are not all clear as that would have been too easy to fit. There are some interesting features of this graph. Except for the second day, the area under the two curves for each day would be similar, even for the fourth day – perhaps the most interesting. The peak in the model when there is a trough in the data reflects the general characteristic of an AR(1), that by its very nature, the prediction will have the same trend as the original data, but with a one step lag. This is obvious since the procedure of $E(X_t) = \phi X_{t-1}$ means you are simply predicting X_t by multiplying X_{t-1} by a constant factor, hence resulting in a lag.

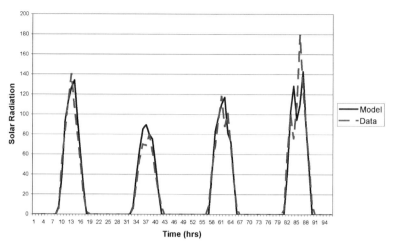

Fig. 11.21 Out of sample model fit

7.1 Synthetic Generation

In a similar manner to that of the daily configuration, we have to model the final
noise term in order to be able to generate synthetic sequences of hourly values. If
we examine the histogram of the a_t (Fig. 11.22) we see that it is not quite symmetric
but for ease of modelling, we shall try and use a Normal approximation to this series.
In the case of the daily model, there is a larger random component, so we had to be

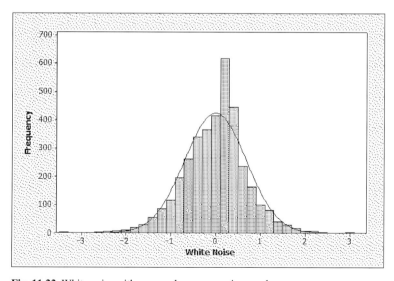

Fig. 11.22 White noise with a normal curve superimposed

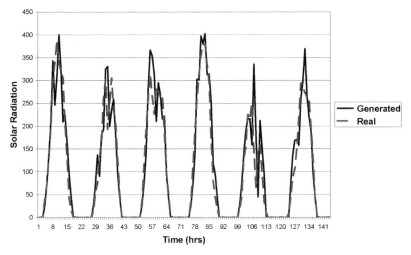

Fig. 11.23 A comparison of a generated series and the original

more precise in modelling the white noise. For the hourly data, the deterministic component explains more of the variance. So, we model $a_t \sim N(0, 0.652^2)$. If we use this for our white noise term, we can compare a set of generated data to the set of original. Obviously, they are not supposed to match, but we use Fig. 11.23 in an illustrative manner, showing that the generated data is of a similar nature as the original data set.

8 Daily Profiling

At locations where only daily total global radiation exists (as estimated through use of satellite derived values or other models using other measured variables), an algorithm is required to form an hourly profile for global radiation. The method uses the following equations to set up a matrix for which the coefficients $(b_0 \ldots b_6)$ are found using regression methods. The equations are based on the basic Fourier series for the mean function using three cycles:

$$y_t = b_0 + b_1 \cos(\omega_1 t) + b_2 \sin(\omega_1 t) + b_3 \cos(\omega_2 t)$$
$$+ b_4 \sin(\omega_2 t) + b_5 \cos(\omega_3 t) + b_6 \sin(\omega_3 t) \qquad (11.28)$$

where $\omega_i = i\pi/12$.

The equations are as follows. At sunrise, sunset and one hour either side, the constraints force the radiation to zero:

$$y_t(\text{sunrise}) = y_t(\text{sunset}) = 0 \qquad (11.29)$$
$$y_t(\text{sunrise} - 1\text{hour}) = y_t(\text{sunset} + 1\text{hour}) = 0$$

Also, the slope of the mean function at midday is equal to zero

$$
\begin{aligned}
y'_{12} &= -b_1\omega_1\sin(\omega_1 t) + b_2\omega_1\cos(\omega_1 t) - b_3\omega_2\sin(\omega_2 t) \\
&\quad + b_4\omega_2\cos(\omega_2 t) - b_5\omega_3\sin(\omega_3 t) + b_6\omega_3\cos(\omega_3 t) \\
&= 0
\end{aligned}
\tag{11.30}
$$

The shape of the mean function is concave, ie. the second derivative is negative

$$
\begin{aligned}
y''_{12} &= -b_1\omega_1^2\cos(\omega_1 t) - b_2\omega_1^2\sin(\omega_1 t) - b_3\omega_2^2\cos(\omega_2 t) \\
&\quad - b_4\omega_2^2\sin(\omega_2 t) - b_5\omega_3^2\cos(\omega_3 t) - b_6\omega_3^2\sin(\omega_3 t) \\
&= -0.05
\end{aligned}
\tag{11.31}
$$

Additionally, the integral of the Fourier series representation of the profile must equal the total global radiation for the day:

$$
\begin{aligned}
\int_{sunrise}^{sunset} y_t\,dt = b_0 t &+ \frac{b_1}{\omega_1}\left[\sin(\omega_1 sunset) - \sin(\omega_1 sunrise)\right] \\
&- \frac{b_2}{\omega_1}\left[\cos(\omega_1 sunset) - \cos(\omega_1 sunrise)\right] \\
&+ \frac{b_2}{\omega_2}\left[\sin(\omega_2 sunset) - \sin(\omega_2 sunrise)\right] \\
&- \frac{b_3}{\omega_2}\left[\cos(\omega_2 sunset) - \cos(\omega_2 sunrise)\right] \\
&+ \frac{b_4}{\omega_3}\left[\sin(\omega_3 sunset) - \sin(\omega_3 sunrise)\right] \\
&- \frac{b_5}{\omega_3}\left[\cos(\omega_3 sunset) - \cos(\omega_3 sunrise)\right] = global_t
\end{aligned}
\tag{11.32}
$$

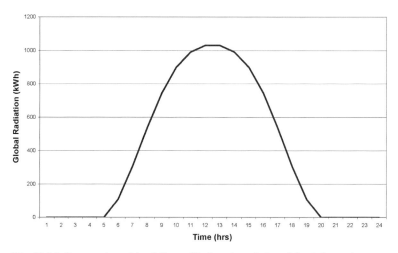

Fig. 11.24 Construction of the daily profile from knowledge of the daily total

The coefficients of b_0 to b_6 are then translated into matrix form and regression is applied to determine the optimal values for b_0 to b_6. Equation (11.28) generated by this method can then be used to create a daily profile for all hours between sunrise and sunset for each day of the data set where a daily global radiation total exists. For values outside these hours, radiation is set to zero. Figure 11.24 gives an example of the profile.

9 Algorithms

In this section, several Excel files are described, wherein appear some of the algorithms to perform the analysis given in this chapter, such as the spectral analysis, Fourier series fitting and the construction of the daily profile given the total daily global radiation. These Excel files are included on the CD accompanying the book.

9.1 Daily Profile

The file *DailyProfiling.xls* will allow the user to construct a profile over the day for a whole year's daily total solar radiation values. **Note that you will have to be able to activate the macros embedded in the file**. If only the profile for a few days is wanted, one can then leave the other days blank. The file will open up on the sheet **Data**. If the user inspects the sheet **RawData**, they will see an example data set for Darwin, with some days missing. It can be seen that the data is in column **G**. By hitting the button **Construct Profile**, you will generate a whole year's profile (of course just missing the days when there is no total).

9.2 Power Spectrum

The file *Power_Spectrum.xls* allows the user to create the power spectrum for a set of data. When the file is opened, the sheet **DFT** is visible. This also involves running a macro, by hitting the **Power Spectrum** button. You will have entered your data in a single column from **A11**. When you run the macro, you will be asked how many data values you have and how many frequencies you want calculated. The graph on the sheet **Power_Spectrum** will provide a visual display of the Discrete Fourier Transform.

9.3 Daily Fourier Series Parameter Estimation

The file *Daily_FS_Estimation.xls* allows the user to estimate the coefficients for the Fourier Series representation for daily data. The file contains some sample data, starting in **A11**. Try this example, and then you can enter your own data. There is a particular facility one must activate in order to perform the estimation. The embedded macro accesses the optimisation tool **Solver** and this must be made available to the Visual Basic macro. Before running the software, go to **Tools, Macro, Visual Basic Editor** and then **Tools, References**. If there is a tickable box for **Solver**, then do so. If not, browse your computer for **solver.xla** and click **Open** when you find it. Hit **Estimate** and you will obtain in the yellow highlighted cells the coefficients (cosine terms on top of the sine terms), and the amount of the variance explained by each frequency in the green highlighted cells (with the total given as well). The orange cell gives the minimised sum of squared deviations between the Fourier Series representation and the data. If you graph columns **A** and **H** together, you will see how well the series models the data.

9.4 Hourly Fourier Series Parameter Estimation

This is very similar to the above on the daily data. All the provisos given in that description hold here as well – particularly to do with the activation of **Solver** within Visual Basic. Here we utilise all the frequencies described in the text, up to three cycles a day plus sidebands for the daily harmonics.

10 Conclusions

A significant segment of this chapter relies on the work in Boland (1995), but there have also been some important additions. The generation of synthetic sequences, while only outlined here, will be very useful in any research where a distribution of performance measures for modelling a solar system is required. Additionally, the demonstration of the generation of an autoregressive process with non Gaussian noise will stimulate further research. It will be of interest to discover where this is possible. The standard wisdom is that it is not necessarily the case that one can ensure the series has the right distributional qualities. However, this procedure may break down only for certain models – for instance, for an $ARMA(p,q)$, rather than an $AR(p)$ model.

We acknowledge what are regarded are the paradigms of development of generation of daily (Aguiar and Collares-Pereira 1988, Graham et al. 1988) and hourly (Aguiar and Collares-Pereira 1992) sequences of solar radiation. The intention in this work is not to attempt to present a better method for such generation. The goal has been to decouple the various processes constituting the time series, in order to better understand the physical underpinnings. This has enabled the research described in Boland and Ridley (2004), wherein we designed algorithms to perform quality assurance for constructing coherent data sets for the Australian Climatic Database. Some of the tasks included infilling of radiation sequences wherein a few values are missing, and also constructing a profile over the day as in Section 8 when a daily total is known or inferred from satellite images, but hourly values are required for a house energy simulation for instance. If one knows the number of significant cycles embedded in a real sequence, then the constructed profile will be more realistic. This knowledge of the Fourier series representation of the daily profile may prove extremely useful when constructing synthetic series to include the influences of climate change. If, for instance, the global solar radiation decreases but the effect is asymmetric about solar noon, then one can re-estimate the Fourier coefficients to cater for this. It may be even more useful in the case of modelling differences in temperature for reconfiguring typical meteorological year data. It is anticipated that a rise in average temperature would not simply mean a translation of the present profile upwards. There would be a greater rise in the daily minimum than in the daily maximum. Once again, the daily profile as expressed in a Fourier series can be adjusted easily to cater for this effect. In summary, this approach to understanding the physical underpinnings by the modelling procedure allows for greater flexibility in adjusting to changes.

References

Anderson O.D. (1974) Time Series Analysis and Forecasting, Butterworths, London.

Aguiar, R.J., Collares-Pereira, M. and Conde, J.P. (1988) Simple procedure for generating sequences of daily radiation values using a library of Markov transition matrices, Solar Energy, 40: 269–279.

Aguiar, R.J. and Collares-Pereira, M. (1992) TAG, a time dependent, autoregressive, Gaussian model for generating synthetic hourly radiation, Solar Energy, 49(3): 167–174.

John Boland (1995) Time-Series Analysis of Climatic Variables, Solar Energy 55(5): 377–388.

John Boland and Barbara Ridley (2004) Quality control of climatic data sets, Proceedings of the 42nd Annual Conference of theAustralia and New Zealand Solar Energy Society, Perth, Dec. 2004.

Bowerman B.L. and O'Connell R.T. (1979) Time Series Forecasting, Duxbury Press, Boston.

Brockwell P.J. and Davis R.A. (1996) Introduction to Time Series and Forecasting, Springer-Verlag, New York.

Chris Chatfield (2003) The Analysis of Time Series, an Introduction, Chapman and Hall.

Energy Partners, Adelaide Applied Algebra and School of Mathematics and Statistics UniSA (2006) Development of Climate Data for Building Related Energy Rating Software, Australian Greenhouse Office, Canberra.

Festa R. and Ratto C. F. (1993) Proposal of a numerical procedure to select reference years, Solar Energy, 50(1): 9–17.

Graham, V.A., Hollands, K.G.T. and Unny, T.E. (1988) A time series model for K_t with application to global synthetic weather generation, Solar Energy, 40: 83–92.

Grunwald GK, Hyndman RJ, Tedesco LM, Tweedie RL. (2000) Non-Gaussian conditional linear AR(1) models. Australian and New Zealand Journal of Statistics, 42: 479–495.

Hyperphysics, http://hyperphysics.phy-astr.gsu.edu/hbase/audio/sumdif.html#c2, accessed June 28, 2007.

Kirkpatrick A. T. and Winn C. B. (1984) Spectral Analysis of the Effective Temperature in Passive Solar Buildings. Transactions of ASME Journal of Solar Energy Engineering; 106:106–119.

G. M. Ljung and G. E. P. Box (1978) On a Measure of a Lack of Fit in Time Series Models, Biometrika, 65: 297–303.

Magnano L. and Boland J.W. (2007) Generation of synthetic sequences of half hourly temperature, Environmetrics, (accepted subject to revisions).

Phillips W. F. (1983) Harmonic analysis of climatic data, Solar Energy, 32(3): 319–328.

Ruey Tsay, 2005, Analysis of Financial Time Series (2nd Edition), Wiley Series in Probability and Statistics.

Chapter 12
A new Procedure to Generate Solar Radiation Time Series from Machine Learning Theory

Llanos Mora-López

1 Introduction

The fundamental idea in this chapter is the use of probabilistic finite automata (PFA) as a means of representing the relationships observed in climatic data series. PFAs are mathematical models used in the machine learning field. Different approaches have been followed to characterize the hourly series of global solar irradiation. Taking into account the nature of these series, it is proposed the use of a new model to characterize and simulate them. This new model is easy to use once it has been built and it allows us to represent the relationships observed in the hourly series of global irradiation. Moreover, it can be embedded in engineering software by including the estimated probabilistic finite automata and the algorithm explained in section 2 in this software. Before giving details about this model, it is reviewed briefly the existing models with special attention to their simplicity, requirements and limitations.

Several studies have been carried out to obtain models which allow us to simulate the hourly series of solar global irradiation. Traditionally, the analysis of time series has been carried out using stochastic process theory. One of the most detailed analyses of statistical methods for time series research was done by (Box and Jenkins 1976). The goal of data analysis by time series is to find models which are able to reproduce the statistical characteristics of the series. These models also allow us to predict the next values of the series from their predecessors. The approach is as follows: first, the model must be identified; to do this, the recorded series are statistically analyzed in order to select the best model for the series. Then the parameters of the model must be estimated. After this, a new series of values can be generated using the estimated model. For example, this approach has been followed in Brinkworth (1977), Bendt et al. (1981), Aguiar et al. (1988), Aguiar and Collares-Pereira (1992) and Mora-Lopez and Sidrach-de-Cardona (1997).

Llanos Mora-López
Universidad de Málaga, Spain, e-mail: llanos@lcc.uma.es

One of the problems with most of these methods is that the probability distribution functions of the generated series are normal when stochastic models are used. This problem can be solved for daily series using first-order Markov models (see Aguiar et al. 1988). For hourly series, to circumvent the problem, a differenced series and ARMA models can be used, e.g., (Mora-Lopez and Sidrach-de-Cardona 1997); however, in this case the simulation of a new series uses a complex iterative process: the use of the differences operator makes it difficult the generation of new series of global irradiation because it is necessary to eliminate the negative values which appear in the series.

Recently some authors have used different types of neural network and finite automata to model values of global solar irradiation on horizontal surfaces; for instance, Mohandes et al. (1998), Kemmoku et al. (1999), Mohandes et al. (2000), Sfetsos and Coonick (2000) and Mora-Lopez et al. (2000). When neural network models have been used, only mean values of daily or hourly global irradiation have been analysed. In the paper by Sfetsos and Coonick (2000) the developed models can be used to predict the hourly solar irradiation time series, but these models are obtained using only data from summer months (63 days). In all cases, the obtained models are "black boxes", and no significant information can be obtained.

2 Probabilistic Finite Automata

A mathematical model called probabilistic finite automata (PFA) will be used to represent a univariate time series. One of the first applications of this model has been proposed in (Rissanen 1983) for universal data compression. Other different practical tasks have been approached with this mathematical model, such as analysis of biological sequences, for DNA and proteins, in (Krog et al. 1993), and the analysis of natural language, for handwriting and speech, in (Nadas 1984), (Rabiner 1994) and (Ron et al. 1994). Probabilistic suffix automata, based on variable-order Markov models, have been used to construct a model of the English language, see (Ron et al. 1994). All these automata allow us to take into account the temporal relationships in a series. We propose the use of this mathematical model –probabilistic finite automata to represent a univariate time series.

Formally, a PFA is a 5-tuple $(\Omega, Q, \tau, \gamma, q_0)$ where:

- Ω is a finite alphabet; that is, a set of discrete symbols corresponding to the different continuous values of the analyzed parameter. The different symbols of Ω will be represented by x_i. For a series, the values observed can be $x_5 x_3, \ldots x_3$ To represent the different observable series for a period t_1 to t_m we will use the symbols $y_1 y_2 \ldots y_m$. So, in the series $x_5 x_3 \ldots x_3$, the symbol y_1 corresponds to the value x_5, the symbol y_2 to x_3 and so on.
- Q is a finite collection of states. Each state corresponds to a subsequence of the discretized time series. The maximum size of a state -number of symbols- is bounded by a value N fixed in advance. This value, also known as order of the

PFA- is related to the number of previous values which will be considered to determine the next value in the series and depends on "memory" of the series.

- $\tau: Q \times \Omega \rightarrow Q$ is the transition function
- $\gamma: Q \times \Omega \rightarrow [0,1]$ is the next symbol probability function
- $q_0 \in Q$, is the initial state

The function γ satisfies the following requirement: For every $q \in Q$ and for every $x_i \in \Omega$, $\Omega_{xi \in \Omega} \gamma(q, x_i) = 1$. In addition the following conditions are required:

- The transition function τ can be undefined only on states $q \in Q$ and symbols $x \in \Omega$, for which $\gamma(q, x) = 0$;
- The function τ can be extended to be defined on $Q \times \Omega^*$ in the following recursive manner:

$$\tau(q, y_1, y_2, \ldots, y_t) = \tau(\tau(q, y_1, y_2, \ldots, y_{t-1}), y_t) \tag{12.1}$$

where $yi \in \Sigma$. Graphically, each state is represented by a node and the edges going out of each state are labeled by symbols drawn from the alphabet. Each state has an associated probability vector which is composed of the probability of the next symbol for each of the symbols of the alphabet. For instance, in Fig. 12.1 a simple PFA is shown.

In this PFA, the alphabet, Ω, is composed of the symbols 0 and 1. The states of the system, Q, are described in each node of the automata: initial (i), 0, 1, 00, 01, 10 and 11. For instance, the state labeled 01 corresponds to the following sequence of values in the series: 1 as the last value and 0 as the previous. The associated vectors at each state (node) are the probabilities which each symbol of the alphabet has to appear in the next moment, after the sequence of symbol that label the node has appeared. For instance, the node labeled with 10, has the associated vector $(0.25, 0.75)$; this means that if the current state is 10, then the next symbol can be 0, with a probability of 0.25 and 1 with a probability of 0.75. The continuous and discontinuous arrows represent the transition function between states (discontinuous for 0, and continuous for 1). For instance, if the current state is 10, and the next symbol is 0, then the following state will be labeled with 00; but if the next symbol is 1, then the following state will be labeled with 01.

In the PFA shown in Fig. 12.1, the states 01 and 11 have the same probability vector as state 1. That is, when the symbol 1 appears, it is not necessary to know

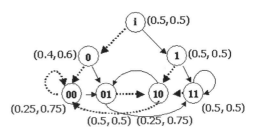

Fig. 12.1 Example of probabilistic finite automata

Fig. 12.2 Simplified probabilistic finite automata

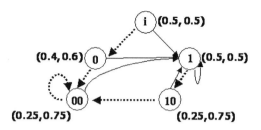

the preceding value to determine the probabilities of the next symbol, since in both
cases, (0 or 1), the probabilities vector of the next symbol is (0.5,0.5). Therefore,
the PFA of Fig. 12.1 can be converted into the PFA shown in Fig. 12.2.

This class of PFA is used to represent variable order Markov models. These simplified automata are the automata proposed in this chapter. They capture the same
information with fewer states than the original automata. Moreover, they allow us to
take into account, for each state, a different number of previous values in the series.

Let us define some concepts that we will use to build the PFAs for hourly global
irradiation time series. Let $\Omega = \{x_1, x_2, \ldots, x_n\}$ be the set of discrete values of the
analized variable and Ω^* denotes the set of all possible sequences which can be
obtained with these values. For any integer N, Ω^N denotes the set of all possible
sequences of length N and $\Omega^{\leq N}$ is the set of all possible sequences with length less
than or equal to N. For any subsequence, Y, represented by $y_1 \ldots y_m$, where $y_i \in \Omega$,
the following notations will be used:

- The longest final subsequence of Y, different from Y, will be $final(Y) = y_2 \ldots y_m$
- The set of all final subsequences of Y will be, $last(Y) = \{y_i \ldots y_m | 1 \leq i \leq m\}$

In the next section it is explained how to build a PFA for a time series.

2.1 Algorithm to Build Probabilistic Finite Automata

The following algorithm can be used to construct the PFA:

1. Compute the series of discrete values.
2. Initialize the PFA with a node, with label null sequence.
3. The set *PSS* -Possible Subsequence Set- is initialized with all sequences of order
   1. Each element in this set corresponds to a sequence of discrete values. Take
   $o = 1$ as the initial value of the order –that is, size of subsequences to consider.
4. If there are elements of order o in *PSS*, pick any of these elements, Y. Using all
   discrete sequences in the series, compute the frequency of Y. If 4.a and 4.b are
   true, then go to 5, else go to 6.

   4.a. The frequency of this sequence is greater than the threshold frequency. This
        frequency depends on the number of sequences used and on the importance
        that an individual sequence has. Usually, if there are many sequences to be

used, a sequence that only happens once or twice is not representative of the series.

4.b. For some $x_p \in \Sigma$, the probability of occurrence of the subsequence Yx_p is *not equal* to the probability of the subsequence $final(Y)x_p$, s, that is:

$$P(x_p|Y) \neq P(x_p|final(Y)) \qquad (12.2)$$

(*not equal*: when the ratio between the probabilities is significantly greater than one)

5. Do

5.a. Add to the PFA a node, labeled with Y, and compute its corresponding probabilities vector.

5.b. For each amplified sequence, Yx_p: if the probability of this augmented sequence is greater than the threshold probability, then include it in *PSS* [Only the sequences with a sufficient probability will be used to build the PFA. This threshold of probability must be defined when the PFA is built and it depends on the amount of data].

6. Remove the analyzed subsequence, Y, from *PSS*.

7. If there are no elements of order o in *PSS*, add 1 to the value of o. If $o \leq N$ and there are elements of length o in *PSS*, then go to 4, else Stop.

2.2 Predicting New Values

A PFA can be used as a mechanism for generating finite sequences of values in the following manner.

- Start from an initial value selected from the alphabet, called the initial state.
- If q_t is the current state, labeled by the sequence $Y = y1 \ldots yt$, then the next symbol is chosen (probabilistically) according to $\gamma(q_t, \cdot)$. A possible way to select this symbol is explained in a next section.
- If $x \in \Sigma$ is the chosen symbol, then the next state, q_{t+1}, is $\tau(q_t, x)$. The label of this new state, Y', will be the longest final subsequence of Yx in the PFA, that is:

$$Y' = Max\{last(Yx)\} \in PFA \qquad (12.3)$$

- The process continues until the length of the required sequence is reached.

In addition, if $P^t(Y)$ denotes the probability that a PFA generates a sequence $Y = y_1 \ldots y_{t-1} y_t$, then:

$$P^t(Y) = \prod_{i=0}^{t-1} \gamma(q_i, y_{i+1}) \qquad (12.4)$$

This definition implies that $P^t(\cdot)$ is in fact a probability distribution over the symbols of sequence, i.e.:

$$\sum_{Y \in \Omega^*} P^t(Y) = 1 \qquad (12.5)$$

2.3 How the Model can be Validated

For a recorded time series, the following steps must be followed to use the proposed model.

- First, if the time series has continuous values, then these values must be discretized. After this, the PFA is built using the discrete series.
- With the PFA and the generation method described above, new values for the time series can be generated.

In order to compare the simulated series to the real ones, several statistical tests can be used. The hypothesis that both series have the same mean and variance will be checked.

The frequency histograms of the recorded and simulated series are also analyzed. To make this comparison, we propose the use of an adaptable goodness-of-fit test, which is based on the two-sample Kolmogorov-Smirnov test, described in (Rohatgi 1976). The objective of this adaptable test is to determine if two distribution functions $F_Y(.)$ and $F_Z(.)$ are the same, except for possible changes in location and scale. Specifically, we have checked the null hypothesis that there exist two unknown values μ and σ such that Z_i and $\mu + \sigma Y_j$ have the same distribution. Using distribution functions it is possible to express our null hypotheses as follows:

$$H_0 : \exists \mu \in \mathfrak{R} \text{ and } \sigma \in (0, +\infty)/\forall u \in \mathfrak{R} \qquad (12.6)$$

$$F_X(u) = F_Y(\frac{u - \mu}{\sigma}) \qquad (12.7)$$

Replacing unknown parameters μ and σ by estimates introduces additional random terms in the statistic. Therefore, to obtain the critical values that must be used in the test, we propose using a bootstrap procedure.

3 Generating Solar Irradiation Time Series Using Probabilistic Finite Automata

The probabilistic finite automata aforementioned can be used to characterize and predict a climatic variable; the hourly global irradiation received on a surface on the ground. For this variable, time series are recorded by meteorological stations at regular time intervals.

It is necessary a stationary time series. From the original series the series of the hourly clearness index must be computed because this series are stationary. The following question - which we have solved- is the discretization of these series. The recorded values are continuous whereas the proposed mathematical model uses discrete values. The discretization method used is explained in section 3.2. The PFAs have been built using the discrete series obtained and new values of the series generated. Finally, these new values have been checked using several tests.

It is important to point out that there are two different processes. The first one is the estimation of the PFAs. For this task it is necessary to use as much as possible sequences of the parameter –in this paper, the hourly clearness index obtained from hourly global irradiation series-; the more series are used the more universal the PFAs will be. Once the PFAs are caculated, the second process is to generate new series of the parameter using as input data the PFAs and the mean value of the parameter. The first process should be done only once –using all the available series- but the second process can be repeat as many times as it is necessary.

3.1 Data Set

The data of the hourly exposure series of global irradiation, $\{G_h(t)\}$, which are used to calculated the PFAs in this chapter were recorded over several years in nine Spanish meteorological stations. In total, 745 months were accounted for. The pertinent latitudes range from 36°N to 44°N. The annual average values of daily global solar irradiation for these locations range from $11\,\mathrm{MJm}^{-2}$ to $18\,\mathrm{MJm}^{-2}$. In order to be able to use more general (universal) PFA it is possible and desirable to obtain PFAs using data from other locations and latitudes and the procedure detailed in this chapter.

The weather characteristics of the used locations are very different. There is a location with an Atlantic moderate climate (Oviedo). There are locations of the interior which have a continental climate, such as Madrid, etc. Finally, the locations of the coast have a Mediterranean climate, with softer temperatures both in winter and in summer (Málaga, Mallorca, etc).

3.2 Discretization of Time Series of Hourly Solar Global Irradiation

The goal is to use an effective and efficient method to transform continuous values into discrete ones using the overall information included in the series and, when possible, feedback with the learning system. To do this, the discrete value which corresponds to a continuous value has been calculated using qualitative reasoning, taking into account the evolution of the series. Qualitative models have been used in different areas in order to obtain a representation of the domain based on properties (qualities) of the systems; see, for instance, (Forbus 1984), (Kleer and Brown 1984) and (Kuipers 1984).

We prosose the use of the qualitative dynamic discrete conversion method described in (Mora-Lopez et al. 2000). It is dynamic because the discrete value associated to a particular continuous value can change over time: that is, the same continuous value can be discretized into different values, depending on the previous values observed in the series. It is qualitative because only those changes which are qualitatively significant appear in the discretized series.

The parameter used to build the PFA is the hourly clearness index, defined as:

$$K_h = G_h/G_{h,0} \tag{12.8}$$

where G_h is the hourly global irradiation and $G_{h,0}$ is the extraterrestrial hourly global irradiation.

The hourly clearness index series have been constructed in an "artificial" way because data from different days have been linked together: the last observation of each day is followed by the first observation of the following day. This assumption has been already done in previous papers, see for instance (Mora-Lopez and Sidrach-de-Cardona 1997), and the obtained results confirm to us the validity of this hypothesis. On the other hand, the number of hours considered for each series (month) is constant and equal for all locations considered. The number of hours considered for each month is: 10 for January, February, November and December; 12 for March, April, September and October; 14 for May, June, July and August.

The alphabet of the PFA is: $\Omega = \{0, 1, \ldots, 7\}$

The relationship between the values of the clearness index and the symbols of the alphabet is the following. For the first symbol of a series, the discrete value of the series will be calculated using the following expression:

$$Y_h = \begin{cases} 0 & 0 \leq K_h < 0.35 \\ \left\lfloor \dfrac{K_h - 0.35}{0.05} \right\rfloor + 1 & 0.35 \leq K_h < 0.65 \\ 7 & K_{h \geq 0.65} \end{cases} \tag{12.9}$$

where $\lfloor A \rfloor$ means the integer value of A. For the following values of the continuous series the discrete values will be obtained using the algorithm described in (Mora-Lopez et al. 2000) and the intervals aforementioned.

3.3 Estimating and Using PFA for Hourly Solar Irradiation Series

Using the expressions presented in section 3.2 and the hourly clearness index series, the discrete series $\{Y_h\}$ are obtained. For instance, if the maximum order of PFA is 4, the set of all possible states will be:

$$Q = \Omega^* = \{0, 1, 2, \ldots, 00, 01, 02, \ldots 77, 000, 001, \ldots 777, \ldots 7777\}.$$

Table 12.1 Example of states and transition probabilities for the PFA corresponding to interval [0.55,0.6[. (column $p(0)$ shows for first row the probability that being in state 46 the following discrete value was 0 and so on for the rest of columns)

state	p(0)	p(1)	p(2)	p(3)	p(4)	p(5)	p(6)	p(7)	p(8)	p(9)
46	0.00	0.00	0.29	0.00	0.00	0.00	0.36	0.35	0.00	0.00
334	0.00	0.00	0.00	0.00	0.30	0.41	0.29	0.00	0.00	0.00
9875	0.00	0.00	0.00	0.00	0.14	0.86	0.00	0.00	0.00	0.00

In this set, the state *6543* can correspond to the following sequence of values for the clearness index: *0.63, 0.56, 0.52, 0.47.*

From all possible subsequences observed in the series, only those with a sufficient probability will be used to build the PFA. This threshold of probability must be defined when the PFA is built.

The monthly series of the hourly clearness index have been grouped using the monthly mean value of the hourly clearness index. The ranges for each group are the same as those defined for the discretization of this parameter. For every interval, one PFA has been built.

To select the best values of the parameters to build the PFAs, we have checked the results obtained with different values of these parameters. The values we have used are:

- Order of the PFA: from *2* to *12.*
- Threshold -minimum number of appearances of a sequence- from *1* to *5.*

For most of the intervals, if the order used for the PFA is *2*, the results are similar to those when the order is *4*; however, for intervals *5* and *6*, using order *4*, the PFA captures the relationship observed in the series better than using order 2. Thus, the selected order (maximum) for the PFA is *4*. The selected minimum number of appearances -required to use a sequence to build a PFA- is *2.*

For instance, three different states of the automata obtained for interval [0.55,0.6[are shown in Table 12.1. The firs column correspond to the state, the following columns show the probability of transition. The state describes the last values of the series that are significant. For example, if the last value is *6* and the previous was *4* the probability that the next value in the series was *2* is *0.29.* In rows three and four the same information is shown for one state or order *3* (state labeled as *334*) and one state of order *4* (state labeled as *9875*).

4 Predicting New Series from PFA

To generate new series we need an initial state. The initial state is the discrete value corresponding to the mean value of the clearness index for each series. Let q_t be the current state. The next symbol, y_{t+1}, is generated as follows: first, a random number

$r \in [0,1]$ is generated. Then, we chose the only component of probabilities vector -for the current state, q_t- which satisfies:

$$y_{t+1} = x_j | \sum_{i=1}^{j} \gamma(q_t, x_i) \geq r \; AND \; x_{j-1} | \sum_{i=1}^{j-1} \gamma(q_t, x_i) < r \qquad (12.10)$$

Once the next symbol is estimated, it is necessary to use the information that is in the PFAs to decide the following state, according to algorithm in 2.1. This new state depends on the sequence of the generated series until now. The number of last values to be considerer for the next state depends of the states that there are in the PFAs. For example, using the abovementioned algorithm, for a PFA or order 5, if the last symbols in the sequence generated are *78654* and there is any state labeled ad *78645*, then it is no necessary to use the value *7* –first value of the sequence; the algorithm continues searching in the PFA for the state *8645*, if this state is not in the PFA either, then the value *8* is not necessary; the process continues until the sequence corresponds to a state in the PFA.

With the PFAs built following the algorithm presented in section 2.1, new sequences of the hourly clearness index have been generated. The original and generated series have been compared using the statistical test described in section 2.3.

The results obtained for each interval of the clearness index are shown in Table 12.2.

In Fig. 12.3, the cumulative probability distribution function of both series of clearness index -recorded and simulated- are shown for a series (data from Madrid).

Moreover, we have calculated the hourly series of global irradiation from the hourly clearness index series. Using statistical tests, the hypothesis that both series have the same mean and variance is not rejected (significance level=*0.05*). For instance, in Figs. 12.4 and 12.5, the recorded and simulated series of two Spanish locations are shown (Malaga, January 1977).

The frequency histograms of the recorded and simulated series of hourly global irradiation have been also analyzed. The frequency histograms have been obtained for each month of the year, using all the recorded and simulated series for that

Table 12.2 Results for each interval of the clearness index defined in column 1. Each interval corresponds to one discrete value estimated from Eq. (12.9). The column Months shows the number of months used for each interval, the column Similar shows the number of months in which the original and generated series are similar (using the test described in section 2.3). The column Perc shows the same information in percentage (Similar months divided by Total months)

Interval	Months	Similar	Perc.
[0.–0.35)	17	17	100
[0.35–0.4)	55	54	98.2
[0.4–0.45)	79	78	98.7
[0.45–0.5)	107	106	99.1
[0.5–0.55)	137	136	99.3
[0.55–0.6)	198	192	97.0
[0.6–0.65)	120	116	96.7
[0.65–1.0)	32	30	93.8

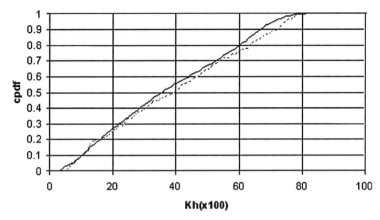

Fig. 12.3 Cumulative probability distribution function of the recorded (*continuous line*) and simulated series (*discontinuous line*) (Madrid)

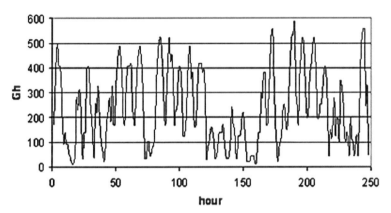

Fig. 12.4 Recorded series of hourly global irradiation, in Wh/m^2 (Málaga, January, 1977)

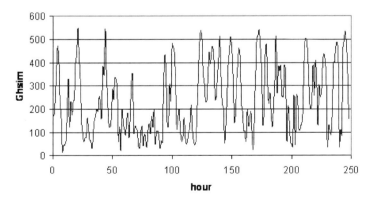

Fig. 12.5 Generated (simulated) series of hourly global irradiation, in Wh/m^2 (Málaga, January 1977)

month over every year. The null hypothesis –the frequency histograms for recorded and simulated series are similar- that the underlying model for both series is the same has never been rejected (significance level=0.05).

5 Software

In the CD-ROM accompanying the book, the reader will find the software HAM supplied with this chapter that allows us to estimate the described PFA. This software was developed by Juan Manuel Manzanares Badía, in the department of "Lenguajes y Ciencias de la computación" of the University of Malaga (Manzanares, 2006). To install the software run the program "Install HAM.exe". After the software is installed, the following directories are created in the PC:

- The directory "C:\Program files\Proyecto Fin Carrera\HAM\", where is the executable program HAM.exe
- The directory "C:\HAM\" that contains some example of files.

An access to program is also created in "Programs" and also a shortcut access in created. The program HAM.exe calculates one PFA for one or more series of clearness index (daily or hourly).

The input data (the hourly or daily clearness index series) must have the following format: each value of the index must be in one line and the file must be ended by the number -2222 (example TEMPORAL.TS in directory "C:\HAM\PRUEBAS", once the program has been installed). It several series are used to estimate one PFA, all the series must be in the same file but each series have to be ended by -1111 and the file have to be ended by -2222 (example TEMPORALES.TS in directory "C:\HAM\PRUEBAS"). The extension for input data files is ".ts".

To calculate one PFA, run HAM.exe, select "Archivo"->"Abrir". In the dialog box, select the file that have the clearness index series. Once the file is loaded, click on "Análisis temporal" -> "Autómatas finites probabilistas". In the dialog box, it is necessary to specify the threshold (in the field "Umbral") and the desired order (in the field "Orden"). Then, the name and directory of file output can be modified (field "Nombre de fichero de salida"). Finally, click on OK and the file is created in the directory that has been specified. This file contains not only the probability finite automata but also additional information about each state such as the number of sequences observed for each state and the absolute values of each transition (see for example the files "TEMPORAL.MK" and "TEMPORALES.MK" in "C:\HAM\PRUEBAS".

6 Conclusions

In this chapter it is proposed a model to generate synthetic series of hourly exposure of global radiation. This model has been constructed using a machine learning approach. It is based on the use of a subclass of probabilistic finite automata which can be used for variable order Markov processes. This model allows us to represent the different relationships and the representative information observed in the hourly series of global radiation; the variable order Markov process can be used as a natural way to represent different types of days, and to take into account the "variable memory" of cloudiness.

A method to generate new series of hourly global radiation, which incorporates the randomness observed in recorded series, has been also proposed. This method only uses, as input data, the mean monthly value of the daily solar global radiation and the probabilistic finite automata.

A software program to estimate PFAs from daily or hourly clearness index series has been also presented.

References

Aguiar RJ, Collares-Pereira M, Conde JP (1988) Simple procedure for generating sequences of daily radiation values using a library of Markov Transition Matrix. Solar Energy 40(3): 269–279.

Aguiar RJ, Collares-Pereira M (1992) Tag: A time-dependent, autoregressive, gaussian model for generating synthetic hourly radiation. Solar Energy 49(3): 167–174.

Bendt P, Collares-Pereira M, Rabl A (1981) The frequency distribution of daily insolation values. Solar Energy 27: 1–5.

Box, GEP, Jenkins GM (1976) Time series analysis forecasting and control. USA: Prentice-Hall.

Brinkworth FJ (1977) Autocorrelation and stochastic modelling of insolation series. Solar Energy 19: 343–347.

Forbus KD (1984) Qualitative process theory. Artificial Intelligence 24: 85–168.

Kemmoku Y, Orita S, Nakagawa S and Sakakibara T (1999). Daily insolation forecasting using a multi-stage neural network. Solar Energy 66(3): 193–199.

Kleer J, Brown JS (1984) A qualitative physics based on confluences. Artificial Intelligence 24: 7–83.

Krog A, Mian SI and Haussler D (1993) A hidden Markov model that finds genes in E.coli DNA. Technical report UCSC-CRL-93-16, University of California at Santa-Cruz.

Kuipers B (1984) Commonsense reasoning about causality deriving behavior from structure. Artificial Intelligence 24: 169–203.

Manzanares Badía, Juan Manuel (2006) Herramienta para el análisis multivariante utilizando técnicas de aprendizaje automático y modelos estadísticos. Proyecto fin de carrera. E.T.S.I. Informática. Universidad de Málaga.

Mohandes M, Balghonaim M, Kassas M, Rehman S and Halawani TO (2000) Use of radial basis functions for estimating monthly mean daily solar radiation. Solar Energy 68(2): 161.

Mohandes M, Rehman S and Halawani TO (1998) Estimation of global solar radiation using artificial neural networks. Renewable Energy 14(1–4): 179–184.

Mora-López L, Fortes I, Morales-Bueno R and Triguero F (2000) Dynamic discretization of continuous values from time series. Lecture Notes in Artificial Intelligence 1810: 280–291.

Mora-López L, Morales-Bueno R, Sidrach-de-Cardona M, Triguero F (2002) Probabilistic Finite Automata and Randomness in Nature: a New Approach in the Modelling and Prediction of Climatic Parameters. Proceeding of the International Environmental Modelling and Software Society Congress. Lugano, Suiza, June 2002.

Mora-López L, Sidrach-de-Cardona M (1997) Characterization and simulation of hourly exposure series of global radiation. Solar Energy 60(5): 257–270.

Nadas A (1984). Estimation of probabilities in the language model of the IBM speech recognition system. IEEE Trans. on ASSP 32(4): 859–861.

Rabiner LR (1994) A tutorial on hidden Markov models and selected applications in speech recognition. Proceedings of the Seventh Annual Workshop on Computational Learning Theory, 1994.

Rissanen J (1983) A universal data compression system. IEEE Trans. Inform. Theory 29(5): 656–664.

Rohatgi VK (1976) An Introduction to Probability Theory and Mathematical Statistics. John Wiley & Sons, USA.

Ron D, Singer Y and Tishby N (1998) On the learnability and usage of acyclic probabilistic finite Automata. Journal of Computer and System Sciences 56: 133–152.

Ron D, Singer Y and Tishby N (1994) Learning probabilistic automata with variable memory length. Proceedings of the Seventh Annual Workshop on Computational Learning Theory.

Sfetsos A, Coonick AH (2000) Univariate and multivariate forecasting of hourly solar radiation with artificial intelligence techniques. Solar Energy 68(2): 169–178.

Chapter 13
Use of Sunshine Number for Solar Irradiance Time Series Generation

Viorel Badescu

1 Introduction

It is a common observation that the amount of solar energy incident on the ground strongly depends on the state of the sky. Larger amounts of radiation are received when the sky is free of clouds. Moreover, when clouds are present, the incident radiation depends on the cloud types.

Two quantities are commonly used to describe the state of the sky. First, there is the total cloud cover amount (sometimes called the cloudiness degree or point cloudiness), which represents the fractional total cloud amount observed by eye. It is expressed in tens (or, sometimes, in oktas) of the celestial vault. The total cloud cover amount is essentially an instantaneous quantity. A daily averaged total cloud cover amount may be computed. The days are sometimes classified according to this average value. This is justified by the observed persistence of cloud cover amount.

For a given time interval S within the daytime, the bright sunshine duration s may be evaluated and the relative sunshine σ (sometimes call the bright sunshine fraction, or the sunshine fraction) may be defined by $\sigma \equiv s/S$. In many cases S is the interval between the sunrise and the sunset (i.e. the day-light duration) in a given day and s is the measured number of daily bright sunshine hours. Shorter S intervals are also used. Of course, a low σ value is an indication for a high cloud cover amount and the relative sunshine is the second common (indirect) indicator of the state of the sky.

Any solar radiation computing model should take account of the state of the sky. This may be done through a variety of means, ranging from very complicated computer codes to empirical relations (for reviews and model classifications see e.g. May et al. (1984), Bener (1984), Davies et al. (1988), Festa and Ratto (1993) and the chapters of this book).

Viorel Badescu
Polytechnic University of Bucharest, Romania, e-mail: badescu@theta.termo.pub.ro

Choosing among the existing models usually takes into account two features: (1) the availability of meteorological and other kind of data required as input by the model and (2) the model accuracy. For many practical purposes and users the first criterion renders the sophisticated programs based on the solution of the radiative transfer equation unusable. As a consequence, the other models (which we call here *simple models*) were widely tested. According to their complexity, these models may be classified as:

- *very simple* models for computing global solar irradiance (and its components, in some particular cases). Here a *very simple model* is defined as follows: (i) *very simple clear sky* models do not require any meteorological parameter as input and (ii) *very simple cloudy sky* models require as input *a single* meteorological parameter associated to the cloudiness degree (e.g. the total cloud cover amount *or* the relative sunshine). The very simple models are important because the majority of the people involved in practical solar energy applications have access (for various reasons) to this kind of models only.
- *simple* models to compute direct, diffuse and global solar irradiance. These models require more than one meteorological parameter as input.

The accuracy of simple and very simple models for computing global solar irradiance was tested and reported in a large number of papers (see, for instance, Badescu (1997)).

A new kind of simple solar radiation model is presented in this chapter. So far, most models use only one parameter describing the state of the sky, i.e. relative sunshine *or* total cloud amount. The original feature of the new model is the usage of two such parameters. First, it is the common total cloud cover amount. Second, a two-value parameter (the sunshine number) stating whether the sun is or is not covered by clouds is also defined and used.

For given radiation component, the kind of data required depends on application and user. For example, average monthly or daily data are required to conduct feasibility studies for solar energy systems. Data for hourly (or shorter) periods are needed to simulate the performance of solar devices or during collector testing and other activities. The new category of models is developed for users in need of "instantaneous" global irradiance).

Actinometric and meteorological data from Romania are used in this work. However, the proposed models are of general interest as they can be easily fitted to data from other countries. The main message of this chapter is the proof that models based on two parameters related to the state of the sky strongly increase the computation accuracy.

2 Meteorological and Actinometric Databases

All the results reported in this chapter were produced by using data measured by Romanian meteorological stations. Thus, a brief presentation of the general climatological aspects associated to these data is necessary. Romania is a small country located in southeast Europe, between $43° 37' 07'' N$ and $48° 15' 16'' N$ and $20° 15' 44'' E$ and $29° 42' 24'' E$. Its area is 237500 km^2 of which 30% is mountains (heights over 800 m), 37% is hills and plateaus (heights between 200 m and 800 m) and 33% is fields. The territory of the country is halved by the Carpathians chain, which stands as a natural border between the three historical provinces: Moldavia, Valahia and Ardeal (Transilvania).

From the point of view of atmospheric circulation, Romania is located in a region where the mean baric field is determined by five action centers (Bazac 1983). Thus, throughout the year the Azores subtropical anticyclone transports damp oceanic air from west to northwest to the center of the continent. Unlike it, the Siberian anticyclone may extend its area from the Far East to the Carpathians. During the winter it brings about continental cold air invasions from east and northeast. Other centers of action are the Iceland and Mediterranean baric depressions, which are more prominent in the cold season. Finally, the warm season increases the influence of the Iranian baric depression. The statistical analysis of the baric centers influence makes evident seven main synoptic situations characteristic of various periods of the year (Bazac 1983).

As a result of the atmospheric circulation and the modifications the Carpathians chain imposes on it, the Romanian territory mostly belongs to the temperate-continental climate. A more detailed classification may differentiate four climatic subtypes and tens of topoclimates (Atlas 1972–1979).

Three databases are used in this chapter. They refer to measured values of several meteorological and actinometric parameters reported by twenty-nine stations owned by the Romanian Meteorological Authority. These stations were selected to give a broader coverage of the country in both latitude and longitude (see Figure 1 and Table 1 of Badescu (1991)). The climate type of these localities was determined in Badescu and Popa (1986) by using the index of continentality (Ivanov), $I(\%)$, given by:

$$I = \frac{E + E_g + 0.25(100 - u)}{0.36L + 14} \cdot 100, \quad (13.1)$$

where $E\,(°C)$ is the difference between the average air temperature from the warmest and coldest months of the year, $E_g\,(°C)$ is the difference between the maximum and minimum air temperature values during the yearly average, u is the yearly average value of the air relative humidity and $L\,(\text{deg})$ is the latitude of the location. The results reported in Badescu (1991) show that twenty-seven locations have a temperate continental climate (I is greater than 120%). The other two locations, situated on the seaside, have a weak continental and weak maritime climate, respectively (I between 100 and 120% and below 100%, respectively). In all cases the climatic type we determined by using meterological data collected during a five years interval is similar to that obtained by using meteorological data from a longer time interval (Atlas 1972–1979).

A first database (ROMETEO) contains average monthly values for various meteorological parameters (bright sunshine hours, total cloud cover amount, wind speed, precipitation, atmospheric temperature and pressure and air relative humidity) in all twenty-nine Romanian localities (IMHR 1964–1972, Badescu et al. 1984). The total cloud amount was estimated by eye by trained weather observers at 01.00, 07.00, 13.00 and 19.00 local time (in tenths of the celestial vault). The daily average value of total cloud amount was obtained as an arithmetic mean of the four estimated values. For the recordings of relative sunshine duration there were used Campbell-Stokes heliographs. The daily relative sunshine has been established on the basis of all day recordings. The monthly average values have been calculated as an arithmetic mean of daily average values. Meteorological and actinometric data measured in the following two Romanian localities are used in this chapter: Bucharest (latitude 44.5 °N, longitude 26.2 °E, altitude 131 m above sea level) and Jassy (47.2 °N, 27.6 °E, 130 m a.s.l.). The climatic index of continentality is 131.9% at Bucharest and 129.9% at Jassy (Badescu 1991). Thus, the climate of both localities is temperate - continental.

A second database (HOUREAD) consists of measurements in Bucharest and Jassy during about 1200 days in January and July. In Bucharest data were collected in the years 1960–1969 while in Jassy the data were collected during 1964–1973 (RMHI 1974). The data consist of global and diffuse solar radiation, total amount of cloud cover and ambient temperature. They were measured at 6.00, 9.00, 12.00, 15.00 and 18.00 local standard time (LST) in July and at 9.00, 12.00 and 15.00 LST in January. The readings of global and diffuse solar irradiance (G and D, respectively) were performed on Robitzsch actinographs. These third-class pyranometers are generally known to yield measurements of $\pm$ 10% accuracy (see e.g. Coulson (1975), Garg and Garg (1993)). Their maximum relative error was evaluated at 5% by the Romanian Meteorological and Hydrological Institute (Ciocoiu et al. (1974), Neacsa and Susan (1984), Creteanu (1984)) but verification was performed by using second-class thermoelectric pyranometers (Costin 2000). The shading ring method was used to measure the diffuse radiation.

A third database (METEORAR) refers to meteorological measurements during the whole year 1961 in Bucharest and Constanta (IMH 1961). The data are values measured at 1.00, 7.00, 13.00 and 19.00 LST for ambient temperature, air relative humidity and total cloud cover amount. Also, the database contains daily average values for the atmospheric pressure. Data for Bucharest are used in this chapter.

3 Sunshine Number

In this section the sunshine number is introduced and some of its properties are described. Let us consider an observer placed in point P on Earth surface. For that observer, the sunshine number $\xi(t)$ is defined as a time dependent Boolean variable, as follows:

$$\xi(t) = \begin{cases} 0 & \text{if the sun is covered by clouds at time } t \\ 1 & \text{otherwise} \end{cases}. \qquad (13.2)$$

Let us consider a time t during the day-time and a time interval Δt centered on t. We assume the distribution of the clouds over the sky, as well as the dynamics of this distribution, are not known. Then, $\xi(t)$ may be considered as a random variable during the time interval Δt. The probability for the sun being covered by clouds during Δt is denoted $p(\xi = 0, t, \Delta t)$ and the probability that the sun will shine during the same time period is denoted $p(\xi = 1, t, \Delta t)$. Because ξ is a Boolean variable, the two probabilities are related by the following normalization condition:

$$p(\xi = 0, t, \Delta t) + p(\xi = 1, t, \Delta t) = 1. \tag{13.3}$$

Measures for the probabilities $p(\xi = 1, t, \Delta t)$ and $p(\xi = 0, t, \Delta t)$ are now introduced. One denotes by $s(t, \Delta t)$ the total number of time units with the sun shining during the time interval Δt centered on t. Then, the probability $p(\xi = 1, t, \Delta t)$ may be defined as usual by the ratio between $s(t, \Delta t)$ and Δt:

$$p(\xi = 1, t, \Delta t) = \frac{s(t, \Delta t)}{\Delta t} = \sigma(t, \Delta t). \tag{13.4}$$

Here $\sigma(t, \Delta t)$ is the common relative sunshine for the time interval Δt centered on t.

A measure for the probability $p(\xi = 0, t, \Delta t)$ is introduced now by using results of integral geometry and geometrical probabilities (Badescu 2002). Some simplifying hypotheses are necessary. First, the observer in point P sees the Sun and the cloud cover at height h as two convex figures K'_S and K'_C, respectively (Fig. 13.1a). These figures are confined to the celestial vault K'_0. Second, the celestial vault and the two figures are projected onto a horizontal plane (Fig. 13.1b). The projected celestial vault is a plane circle K_0 whose surface area is $A_0 = 2\pi h(R_E + h)$ and perimeter length is $L_0 = 2\pi[h(2R_E + h)]^{1/2}$ (here R_E is Earth radius). The projected cloud and sun become the plane convex figures K_C and K_S, respectively, of surface area A_C and A_S and perimeter length L_C and L_S, respectively. The following treatment applies to these plane projected figures instead of the original figures confined on the surface of a sphere. This is our second hypothesis.

The Sun has a slow apparent movement on the celestial vault. Therefore, for a reasonably short time interval Δt, both K_0 and K_S may be assimilated to fixed figures. The cloud position is not known a priori. Therefore, the figure K_C may be assimilated to a random figure. Theorem 3.3 of Filipescu et al. (1981, p. 48) refers to the probability for the random figure K_C to intersect the fixed figure K_S placed inside the fixed figure K_0. This is precisely the probability $p(\xi = 0, t, \Delta t)$. The argument

Fig. 13.1 (a) Figures K'_s and K'_C obtained by projecting the Sun and the cloud, respectively, on the celestial vault K'_0. (b) Plane figures K_S and K_C obtained by projecting K'_s and K'_C, respectively, on the plane circle K_0

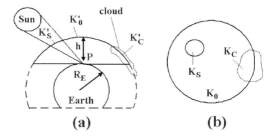

applies in case the state of the sky doesn't change during the time interval Δt. Using the result in Filipescu et al. (1981, p. 48) one finds:

$$p(\xi = 0, t, \Delta t) = \frac{2\pi(A_S + A_C) + L_S L_C}{2\pi(A_0 + Ac) + L_0 L_C}. \tag{13.5}$$

The intersection between the projected cloud and the projected celestial vault (or, in other words, between figures K_C and K_0) constitutes a new figure $K_{0C} \equiv K_C \cap K_0$. This is a random figure and its mean surface area A_{0C} may be evaluated by using a formula due to L. A. Santalo (Santalo 1950; Filipescu et al. 1981, pp. 115–117):

$$A_{0C} = \frac{2\pi A_C}{2\pi(A_0 + A_C) + L_0 L_C} A_0. \tag{13.6}$$

The definition of the total cloud cover amount is $C(t, \Delta t) \equiv A_{0C}/A_0$ and use of Eq. (13.6) yields:

$$C(t, \Delta t) = \frac{A_{0C}}{A_0} = \frac{2\pi A_C}{2\pi(A_0 + A_C) + L_0 L_C}. \tag{13.7}$$

Use of Eqs. (13.5) and (13.7) gives:

$$p(\xi = 0, t, \Delta t) = C \frac{2\pi(A_S + A_C) + L_S L_C}{2\pi A_C}. \tag{13.8}$$

The quantities A_S and L_S may be computed as functions of the Sun zenith angle and the height of the cloud layer. Details are given in Badescu (1992). However, the projected surface of the clouds K_C is usually much more extended than the projected surface of the Sun K_S and the following relations apply:

$$A_S \ll A_C, \qquad L_C L_S \ll 2\pi A_C. \tag{13.9}$$

Use of Eqs. (13.8) and (13.9) yield the following very useful approximation:

$$p(\xi = 0, t, \Delta t) \cong C(t, \Delta t). \tag{13.10}$$

This completes the procedure of defining measures for the probabilities $p(\xi = 1, t, \Delta t)$ and $p(\xi = 0, t, \Delta t)$. Equations (13.4) and (13.10) show that, for appropriate values of t and Δt, these probabilities may be computed by using measurements performed routinely by meteorological stations.

Finally, use of Eqs. (13.3), (13.4) and (13.10) gives:

$$p(\xi = 0, t, \Delta t) \cong C(t, \Delta t) = 1 - \sigma(t, \Delta t). \tag{13.11a,b}$$

Equation (13.11b) is a popular relationship used by many models of computing solar radiation on cloudy sky. The quantity $1 - \sigma$ is sometimes called cloud shade. Thus, Eq. (13.11) simply states that the total cloud cover amount equals the cloud shade. The usual attitude is to postulate Eq. (13.11b). Here we proved rigorously this relationship and the assumptions necessary to derive it were also outlined.

3.1 Statistical Moments and Measures

The statistical moments of order k, $M_k(\xi)$, of a random Boolean variable ξ may be defined by using the following relations:

$$M_k(\xi) = \sum_{\xi=0,1} \xi^k p(\xi) \qquad (k = 1, 2, \cdots). \qquad (13.12)$$

Among these moments, the mean $M \equiv M_1(\xi)$ plays an important role. In case the random Boolean variable is the sunshine number, use of Eqs. (13.4), (3.11) and (13.12) allows writing:

$$M_k(\xi, t, \Delta t) = \sigma(t, \Delta t) = 1 - C(t, \Delta t) \qquad (k = 1, 2, \cdots). \qquad (13.13)$$

The central statistical moments $M_k(\xi - M)$ of a random Boolean variable ξ are defined as usual:

$$M_k(\xi - M) = \sum_{\xi=0,1} (\xi - M)^k p(\xi) \qquad (k = 1, 2, \cdots). \qquad (13.14)$$

The second central moment $M_2(\xi - M)$ and the standard deviation $D \equiv (M_2(\xi - M))^{1/2}$ are frequently used in applications.

The mean is sensitive to extreme values. In normal or symmetrical distributions these extremes balance out. The standard deviation is a measure of data spreading given in the same units as the actual values. The standard deviation is a good unbiased estimate for normal (and unimodal) distributions but can become a highly unreliable estimate if skewness exists in the data.

In case the random Boolean variable is the sunshine number, use of Eqs. (13.4) and (13.14) allows writing the first four central statistical moments:

$M_1(\xi - M, t, \Delta t) = 0,$

$M_2(\xi - M, t, \Delta t) \equiv D(t, \Delta t) = C(t, \Delta t)[1 - C(t, \Delta t)],$

$M_3(\xi - M, t, \Delta t) = C(t, \Delta t)[1 - C(t, \Delta t)][2C(t, \Delta t) - 1],$

$M_4(\xi - M, t, \Delta t) = C(t, \Delta t)[1 - C(t, \Delta t)]\{1 - 3C(t, \Delta t)[1 - C(t, \Delta t)]\}.$ (13.15a-d)

In practice, the moments of order three and four are used within the skewness γ_3 and the kurtosis γ_4, respectively, which are defined by:

$$\gamma_3 \equiv \frac{M_3(\xi - M)}{D^{3/2}},$$

$$\gamma_4 \equiv \frac{M_4(\xi - M)}{D^2} - 3.$$ (13.16a,b)

Skewness measures deviations from symmetry. It will take a value of zero when the distribution is a symmetric bell- shaped curve. A positive value indicates the values are clustered more to the left of the mean with most of the extreme values to the right. Kurtosis is a measure of the relative peakness of the curve defined by the distribution of the values. A normal distribution will have a kurtosis of zero while a positive kurtosis indicates the distribution is more peaked than a normal distribution.

In case the random Boolean variable is the sunshine number, use of Eqs. (13.15) and (13.16) allows writing:

$$\gamma_3(t, \Delta t) \equiv \frac{2C(t, \Delta t) - 1}{\sqrt{C(t, \Delta t)[1 - C(t, \Delta t)]}},$$

$$\gamma_4(t, \Delta t) \equiv \frac{1 - 6C(t, \Delta t)[1 - C(t, \Delta t)]}{C(t, \Delta t)[1 - C(t, \Delta t)]}.$$ (13.17a,b)

Figure 13.2 shows the dependence of the central statistical moments on the total cloud cover amount. At small values of $C(t, \Delta t)$ the skewness is negative and the values of ξ are clustered more to the right of the mean (i.e. $\xi = 1$) with most of the extreme values to the left. The reverse situation applies at large values of total cloud

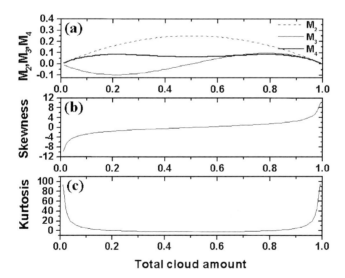

Fig. 13.2 Statistical properties of the sunshine number ξ as a function of the total cloud amount C. (**a**) Central moments $M_2(\xi\text{-M})$, $M_3(\xi\text{-M})$ and $M_4(\xi\text{-M})$ given by Eqs. (13.15b-d). Here M is the mean, given by Eq. (13.15a). (**b**) skewness, given by Eq. (13.17a); (**c**) kurtosis, given by Eq. (13.17b)

amount. The kurtosis is positive for both very small and very large values of $C(t, \Delta t)$ and this means the distribution is more peaked than a normal distribution.

3.2 Time Averaged Statistical Measures

One denotes by $x(t, \Delta t)$ any one of the quantities $p(\xi = 1, t, \Delta t)$, $p(\xi = 0, t, \Delta t)$, $\sigma(t, \Delta t)$, $C(t, \Delta t)$ and the statistical moments defined above. Let us consider a time interval $\Delta t'$ consisting of m non-overlapping time intervals Δt. Of course, $\Delta t' = m\Delta t$. Let be t_i $(i = 1, \ldots, m)$ the moments in the middle of the m time intervals. One defines the average value $x(\Delta t')$ of $x(t, \Delta t)$ on the time interval $\Delta t'$ in the following simple way:

$$x(\Delta t') \equiv \frac{1}{m} \sum_{i=1}^{m} x(t_i, \Delta t). \tag{13.18}$$

Equation (13.11) may be used m times, for t_i $(i = 1, \ldots, m)$. Summation over m and use of Eq. (13.12) yields:

$$\sigma(\Delta t') = 1 - C(\Delta t'). \tag{13.19}$$

Therefore, the simple relation (13.11) keeps its meaning for arbitrary long time intervals.

3.3 Estimation of Statistical Measures

The total cloud cover amount is reported by many meteorological stations on a hourly (or longer time period) basis. We denote by $\tilde{C}(t, \Delta t)$ the observed values of total cloud cover amount. Relative sunshine values are also reported for $\Delta t = 1$ hour and particular moments of time t during the day. We denote these measured values by $\tilde{\sigma}(t, \Delta t)$.

Equation (13.11b) is not strictly fulfilled in practice because both the observed values of total cloud cover amount and the measured values of relative sunshine (or, in other words, the measured values of cloud shade) are affected by errors.

The errors affecting the cloud shade are mainly caused by the relatively inefficiency of the Campbell - Stokes heliographs for low sun (zenith angle exceeding $85°$) (Hay 1979). Also, the Campbell-Stokes recorder is known to prone to over burn the sunshine cards during intermittent strong sunshine producing an overestimation of daily sunshine duration of more than 100% (Painter 1981). Other perturbations can be caused by the formation of frost on the globe of the sunshine recorder or in case of a highly rate of atmospheric humidity, which sometimes make impossible the burning of sunshine cards (Harrison and Coombes 1986). Note however that the following relationship is a good approximation:

$$\sigma \cong \tilde{\sigma}. \tag{13.20}$$

Therefore, the Eq. (13.11b) between C and σ may replaced by the following relationship giving C as a function of measured relative sunshine:

$$C \cong 1 - \tilde{\sigma}.\qquad(13.21)$$

On the other hand, the total cloud amount estimates are subjective by nature and prone to perspective problems faced by the observer. Clouds with a large vertical extent obscure a greater fraction of the background clear sky when viewed near horizon than when viewed overhead. This effect is greater for moderate amount of cloud and vanishes for near overcast skies (Harrison and Coombes 1986).

When the relationship between total cloud cover amount and cloud shade is considered, new complications occur since sunlight can penetrate thin clouds and certain opaque cloud forms sufficiently to operate a sunshine recorder. A theory to relate cloud shade to total cloud cover amount by taking into consideration all these phenomena is missing. Consequently, all the existing formulae relating the two quantities are empirical by nature.

The relationship between cloud shade and total cloud cover amount was studied in Badescu (1991) by using ROMETEO database. In computation we used 1721 pairs of monthly average values $(C, \tilde{C})$. They are shown in Fig. 13.3 in the form $C - \tilde{C}$ vs. C. Note that most of these values are superposed. In general, C is greater than the observed total cloud cover amount by as much as 0.2 and the difference $C - \tilde{C}$ is a maximum for C in the range 0.3 to 0.7. Relatively similar results have been found by other authors. Reddy (1974) obtained for the Indian latitudes ($L = 8° - 36°$N) a yearly variation of $C - \tilde{C}$ between 0.02 in March and 0.17 in August.

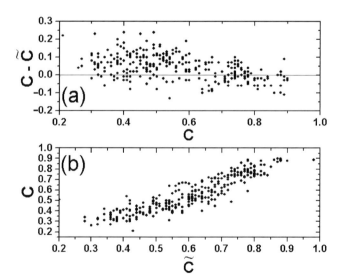

Fig. 13.3 (a) The difference $C - \tilde{C}$ as function of C; (b) C as function of $\tilde{C}$. Here C is the "real" total cloud amount, estimated by Eq. (13.21) from measured data on relative sunshine $\tilde{\sigma}$ while $\tilde{C}$ is the observed total clod amount. Monthly mean values from METEORAR database where used

At the same latitudes Raju and Karuna Kumar (1982) and Rangarajan et al. (1984) find $C - \tilde{C}$ to have maximum values of 0.25 and 0.2, respectively for $\tilde{C}$ in the range 0.4 to 0.7. Harrison and Coombes (1986) find for the latitudes of Canada ($L = 42° - 74°\text{N}$) that $C - \tilde{C}$ can be as high as 0.3 and is a maximum for $\tilde{C}$ in the range 0.4 to 0.7.

Figure 13.3b shows the distribution of observed and measured data in the plane $(C, \tilde{C})$. In previous works (Badescu 1990, 1991) four empirical relationship relating C and $\tilde{C}$ were studied. They were obtained by fitting various expressions to the database ROMETEO by using a least-squares technique. The linear relationship we obtained is $C = 0.004 + 0.92\tilde{C}$. The value 0.92 is in good agreement with the fact that in practice the weather observer will often report $\tilde{C} = 1$ even when $C(\cong 1 - \tilde{\sigma}) < 1$. This anomaly is now well known from previous investigations (Reddy 1974; Hoyt 1977; Raju and Karuna Kumar 1982; Rangarajan et al. 1984; Harrison and Coombes 1986).

Of course, a polynomial fit can usually be improved by adding more terms and doing so without physical justification is a questionable practice. However, in the case we study here there is physical rationale to recommend the use of a non-linear relationship. Indeed, Fig. 13.3a shows that the difference $\tilde{C} - C$ is a maximum for C in the range 0.3 to 0.7. The following quadratic relationship resulted:

$$C = 0.73\tilde{C} + 0.27\tilde{C}^2 \,. \tag{13.22}$$

Note than in deriving this relationship we assumed that a cloudiness sky would be correctly recorded as such by both the sunshine recorder and the weather observer (i.e. $\tilde{C} = 0$ implies $C = 0$). The accuracy of both linear and quadratic models is rather similar: the mean bias error is 0.53% for the linear model and -0.13% for the quadratic one while the root mean square error in the two cases is 12.8% and 12.01%, respectively.

The meteorological database HOUREAD was divided into three smaller databases, as follows:

- (i). A database consisting of those records for which $G < 1.2D$. We assumed that in this case the sun was covered by clouds. Consequently, this databasis is associated to a null value of the sunshine number ($\xi = 0$). Theoretically, if the cloud cover is thick enough, the beam radiation vanishes and global radiation should consist in diffuse radiation only (i.e. $G = D$). However, diffuse radiation was measured with a shadow ring, which normally diminishes the amount of incident radiation. This and the need to cover the measurement errors lead to the above quite arbitrary classification criterion.
- (ii). A database consisting of records for which $G > 3D$. We assumed that in this case the sun was not covered by the sun, because a large beam irradiance exists. Thus, this database is associated to $\xi = 1$.
- (iii). A database with records that do not belong to (i) and (ii) above. These records are usually associated to thin clouds (Suehrcke 2000). Database (iii) is not used in this work.

The two smaller databases (i) and (ii) were stratified according to the observed value of total cloud cover amount $\tilde{C}$. Ten classes of cloudiness were considered,

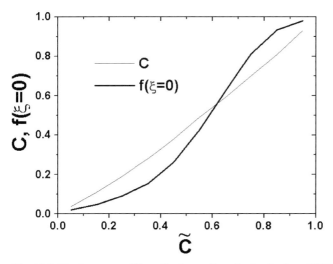

Fig. 13.4 The frequency $f(\xi = 0)$ of recordings in the database HOUREAD associated to the sun covered by clouds as a function of the observed total cloud amount $\tilde{C}$. The "real" total cloud amount $C (\equiv p(\xi = 0, \Delta t))$ predicted by Eq. (13.22) is also shown

i.e. $\tilde{C} = 0 - 0.1, 0.1 - 0.2, \cdots, 0.9 - 1$ for both databases. For any of these cloudiness classes the number of recordings with $\xi = 0$ and $\xi = 1$, respectively, was determined. This allowed to evaluate the frequency of $\xi = 0$ and $\xi = 1$, respectively, as a function of the cloudiness class. The frequency $f(\xi = 0)$ is shown in Fig. 13.4 as a function of the observed total cloud cover amount $\tilde{C}$, together with $C (\equiv p(\xi = 0, t, \Delta t))$ predicted by Eq. (13.22). There is reasonable good agreement between the two quantities.

4 Simple Clear Sky Model

The sunshine number may be used in estimating solar beam, diffuse and global irradiance during days with cloudy sky. However, most models of computing solar radiation in cloudy days are based on models allowing solar irradiance (or irradiation) evaluation on clear sky. In this section we present such a simple clear sky model. However, the choice of this model is arbitrary and does not affect the way of using the sunshine number in case of computing solar irradiance on cloudy skies.

The atmospheric state is quantified by most simple clear sky models through the atmospheric air pressure, temperature and relative humidity, all of them measured at ground level. Here we present a simple technique mainly based on the MAC model for clear sky (Davies and McKay 1982; Davies et al. 1988) which takes into account the main effects that diminish the solar radiation flux during its passage through the atmosphere: the absorption by the ozone layer, the Rayleigh scattering by molecules, the extinction by aerosols and the absorption by water vapor, respectively, as follows.

For a given Julian day n the intensity I^0 of the extraterrestrial solar radiation perpendicular on sun rays is computed with:

$$I^0 = I_{sc}(1.00011 + 0.034221 \cos\theta + 0.00128 \sin\theta$$
$$- 0.000719 \cos 2\theta + 0.000077 \sin 2\theta). \qquad (13.23)$$

Here $\theta \equiv 2\pi(n-1)/365$ and the solar constant is $I_{sc} = 1366.1\,\mathrm{Wm^{-2}}$.

Solar rays of different zenith angles will follow paths of different length in the atmosphere before reaching the ground. Of course, the longer the path, the stronger will be the interaction between the solar radiation and the atmospheric constituents. The usual measure for the path length is the standard air mass m_{st}, which is evaluated by using the following formula (Kasten 1966):

$$m_{st} = \left[\cos z + 0.15\,(93.885 - z)^{-1.253}\right]^{-1}. \qquad (13.24)$$

Here the sun zenith angle z enters in degrees. In case information about the atmospheric pressure $p(h)$ and the air temperature $T(h)$ at altitude h of the site is available, the standard air mass may be affected by a correction procedure originating in the international atmospheric model CIRA 1961 (Badescu 1987). First, the standard ambient temperature at sea level $T_{st}\,(h=0)$ may be evaluated from the measured value $T(h)$ (K) at altitude h (meters) by

$$T_{st}(h=0) = T(h) + 0.0065h. \qquad (13.25)$$

The parameter h here is different of course from the height h of the cloud cover encountered in Section 3. Equation (13.25) takes into account that the temperature of a stable atmosphere decreases by increasing altitude. Second, the standard atmospheric pressure at altitude h, $p_{st}(h)$, is computed by using an adiabatic atmosphere model:

$$p_{st}(h) = p_{st}(h=0) \left[\frac{T(h)}{T_{st}(h=0)}\right]^{5.2561}, \qquad (13.26)$$

where the standard pressure at sea level is $p_{st}(h=0) = 1017.085\,\mathrm{hPa}$. Finally, the air mass m affected by correction may be obtained from:

$$m = \frac{p(h)}{p_{st}(h)} m_{st}. \qquad (13.27)$$

The correction Eq (13.27) takes account that, for a given geometrical distance traveled by a solar ray, the interaction with the atmospheric constituents is stronger when the atmosphere is denser. This results in an effectively longer geometrical distance.

The equivalent depth of the ozone layer in the atmosphere is denoted u_o. In case the solar zenith angle is different from zero, the length x_o of the radiation path through the ozone layer is a function of the air mass:

$$x_o = m u_o. \qquad (13.28)$$

The transmissivity T_o of the ozone layer is computed by:

$$T_o = 1 - a_o,\qquad(13.29)$$

where the absorption coefficient by ozone is given by the following fitted relationship (x_o enters in mm):

$$a_o = \frac{0.1082x_o}{1 + 13.86x_o^{0.805}} + \frac{0.00658x_o}{1 + (10.36x_o)^3}$$

$$+ \frac{0.00218}{1 + 0.0042x_o + 0.00000323x_o^2}.\qquad(13.30)$$

A depth of ozone layer $u_o = 3.5\,\text{mm}$ may be assumed in calculations.

The relationship proposed in Davies et al. (1988, p. 18) for the transmissivity T_r of the atmosphere due to Rayleigh scattering by molecules is in need of correction. Therefore, we used the equation proposed by the Jossefson model (Davies et al. 1988, p. 21):

$$T_r = 0.9768 - 0.0874m + 0.010607552m^2$$
$$- 8.46205 \cdot 10^{-4}m^3 + 3.57246 \cdot 10^{-5}m^4 - 6.0176 \cdot 10^{-7}m^5.\quad(13.31)$$

The transmissivity T_a for aerosols depends on the air mass as follows

$$T_a = k_a^m,\qquad(13.32)$$

where the unit air mass aerosol transmissivity k_a was chosen to be 0.84 to fit the clear sky measurements in database HOUREAD. Note that in a previous work (Badescu 1997) we used $k_a = 0.9$ as an average between the values 0.91, 0.94, 0.87 and 0.90 from the four European localities analyzed in Davies et al. (1988) (i.e. De Bilt, Hamburg, Kew and Zurich).

One of the atmospheric constituents with strong influence on the amount of solar energy reaching the ground is the water vapor. A measure for this quantity is the equivalent thickness u_w of precipitable water layer, which may be evaluated by using the following formula due to Leckner (1978):

$$u_w = \frac{4.93u(h)}{T(h)} \exp\left(26.23 - \frac{5416}{T(h)}\right).\qquad(13.33)$$

Here $u(h)$ (between 0 and 1) is air relative humidity at ground level, while u_w results in mm. In case the solar zenith angle is different from zero, the length x_w of the radiation path through the water precipitable layer is a function of the air mass:

$$x_w = mu_w.\qquad(13.34)$$

The absorptivity a_w by water vapor is given by (here u_w enters in mm):

$$a_w = \frac{0.29 x_w}{(1 + 14.15 x_w)^{0.635} + 0.5925 x_w} . \tag{13.35}$$

The global irradiance perpendicular on sun rays for cloudless sky, $G_{perp,cs}$, is evaluated at ground level as a sum of a direct component ($I_{perp,cs}$) and a diffuse component ($D_{perp,cs}$):

$$G_{perp,cs} = I_{perp,cs} + D_{perp,cs} . \tag{13.36}$$

The direct irradiance $I_{perp,cs}$ is given in terms of the intensity of the extraterrestrial solar radiation perpendicular on sun rays I^0 by the following formula:

$$I_{perp,cs} = I^0 \left(T_o T_r - a_w \right) T_a . \tag{13.37}$$

The diffuse component is given by:

$$D_{perp,cs} = D_{perp,cs,r} + D_{perp,cs,a} , \tag{13.38}$$

where the diffuse irradiance on clear sky due to Rayleigh scaterring ($D_{perp,cs,r}$) and due to aerosol scattering ($D_{perp,cs,a}$), respectively, are given by

$$D_{perp,cs,r} = I^0 \frac{T_o (1 - T_r)}{2} , \tag{13.39}$$

$$D_{perp,cs,a} = I^0 \left(T_o T_r - a_w \right) \left(1 - T_a \right) a_a g . \tag{13.40}$$

Here a_a is the spectrally-averaged single-scattering albedo for aerosols while g is the ratio of forward to total scattering by aerosols, given by

$$g = 0.93 - 0.21 \ln m . \tag{13.41}$$

A value $a_a = 0.75$ may be used in calculations (Davies et al. 1988, p. 36).

The direct, diffuse and global clear sky solar irradiances on a horizontal surface at ground level (I_{cs}, D_{cs} and G_{cs}, respectively) are computed with:

$$I_{cs} = I_{perp,cs} \cos z, \quad D_{cs} = D_{perp,cs} \cos z, \quad G_{cs} = G_{perp,cs} \cos z, \tag{13.42a-c}$$

where the sun zenith angle z may be computed from

$$\cos z = \sin \varphi \sin \delta + \cos \varphi \cos \delta \cos \omega . \tag{13.43}$$

Here φ, δ and ω are the geographical latitude, the solar declination and the solar hour angle, respectively. The solar declination angle is given in radians by:

$$\delta = 0.006918 - 0.399912 \cos \theta + 0.070257 \sin \theta - 0.006759 \cos 2\theta$$
$$+ 0.000907 \sin 2\theta - 0.002697 \cos 3\theta + 0.00148 \sin 3\theta . \tag{13.44}$$

The solar hour angle is given in degrees by

$$\omega = 15° \cdot |12 - LAT|, \qquad (13.45)$$

where LAT is the local apparent (true solar) time, determined (in hours) from the local standard time LST (hours), the equation of time ET (minutes), the geographical longitude of the site LS (degrees) and the standard meridian LSM (degrees) for the time zone:

$$LAT = LST + \frac{ET}{60} \pm \frac{LSM - LS}{15}. \qquad (13.46)$$

The positive sign in the last term of this equation is for places West of Greenwich while the negative sign is for places East of Greenwich. For Romania the negative sign applies and $LSM = 30°E$. The equation of time is given by:

$$ET = 9.87 \sin(2f) - 7.53 \cos(f) - 1.5 \sin(f). \qquad (13.47)$$

Here $f \equiv 2\pi(n - 81)/364$ and the result is in minutes. These complete calculations of clear sky solar irradiances.

5 Solar Irradiance Computation on Cloudy Sky

Figures 13.5a and 13.5b show the ratios $G/G_{perp,cs}$ and $D/D_{perp,cs}$, respectively, for the whole database HOUREAD. Here $G_{perp,cs}$ and $D_{perp,cs}$ were evaluated by using the simple clear sky model described in section 4 while G and D are measured values of global and diffuse instantaneous irradiance. The dependence of both $G/G_{perp,cs}$ and $D/D_{perp,cs}$ on the solar zenith angle z is obviously a function of the observed total cloud cover amount $\tilde{C}$.

Many simple and very simple models were developed to computed solar global irradiation (or irradiance) G on a horizontal surface during days with cloudy sky (see, for instance, chapters 1 and 2 of Festa and Ratto, 1993 and chapters in this book). One of these is the modified Kasten model $G = G_{cs}(1 - 0.72\tilde{C}^{3.2})$ (Kasten 1983; Davies et al. 1988). Note that the regression coefficients 0.72 and 3.2 were obtained by fitting the model to the measured data. Therefore, $\tilde{C}$ (and not C) is the right parameter describing the state of the sky in the modified Kasten model. The model was initially developed in Germany to deal with hourly irradiation values. It proved to be the best among the twelve models tested under the latitudes and climate of Romania (Badescu 1997). In this last work a good accuracy was found in case of computing instantaneous solar global irradiance, too.

Two remarks concerning the modified Kasten model follow. First, its accuracy depends on solar altitude. Second, the model does not allow the cloudy sky global irradiance G to exceed G_{cs} In fact, multiple reflections between ground and clouds make sometimes global irradiance on a cloudy sky to exceed with a few percent the values for clear sky (Suehrcke and McCormick 1992). Finally, results given in Kasten and Czeplak (1980) show that during days with overcast sky, the ratio G/G_{cs} has a weak dependence on zenith angle (see their figure 13.3).

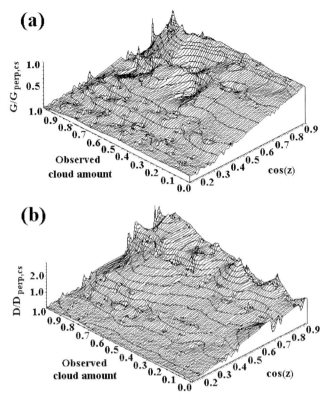

Fig. 13.5 (a) The ratio between global solar irradiance on a horizontal surface, G, and global solar irradiance perpendicular on sun's rays on cloudless sky, $G_{perp,cs}$, as a function of observed total cloud $\tilde{C}$ and the cosine of zenith angle z. **(b)** same as (a) in case of the ratio between diffuse solar irradiance on a horizontal surface, D, and diffuse solar irradiance perpendicular on sun's rays on cloudless sky, $D_{perp,cs}$. All available HOUREAD data were used

The above remarks prompted us to propose the following simple model for computing instantaneous global solar irradiance in days with arbitrary cloudy sky

$$G = G_{perp,cs}(a + b\tilde{C}^c)\cos^d z, \tag{13.48}$$

where a, b, c and d are coefficients to be obtained by fitting Eq. (13.48) to the measured data. In Eq. (13.48) we preferred to use the ratio $G/G_{perp,cs}$ instead of the ratio G/G_H^0 (here $G_H^0 = I^0 \cos z$ is solar irradiance on a horizontal surface at the top of the atmosphere). Both ratios are less dependent on seasons than G itself. The last ratio is specific to Prescott-like models (see Prescott (1940) and p. 102 in Festa and Ratto (1993) and Chapter 5 in this book for a review). Its calculation involves astronomical parameters only. The first ratio was proposed by Angstrom (1924) and was subsequently widely used (see p. 89 of Festa and Ratto (1993) and Chapter 5 in this book for reviews). It makes the influence of the local effects on the right hand side of Eq. (13.48) to decrease. A comparison between Angstrom-like and Prescott-like models was first made in Angstrom (1956). Note, however, that Eq. (13.48) uses the total cloud cover amount as a parameter related to the state of the sky while both Angstrom-like and Prescott-like models use relative sunshine for the same purpose. Equation (13.48) reduces to Kasten's model for particular values of the coefficients (i.e. $a = 1$, $b = -0.72$, $C = 3.2$ and $d = 1$).

The following simple model may be used for computing instantaneous diffuse solar irradiance in days with arbitrary cloudy sky:

$$D = D_{perp,cs}(\alpha + \beta \tilde{C}^{\chi})\cos^{\delta} z, \qquad (13.49)$$

where α, β, χ and δ are coefficients to be obtained by fitting Eq. (13.49) to the measured data.

A few remarks about the physical meaning of the regression coefficients in Eq. (13.48) follow. Similar remarks apply in case of the four regression coefficients in Eq (13.49). First, take account that $G_{cs} = G_{cs,perp}\cos z$ is the clear sky global irradiance on a horizontal surface. Also, $G/G_H^0 (\equiv T^G)$ and $G_{cs}/G_H^0 (\equiv T_{cs}^G)$ is the atmospheric transmittance for global radiation on cloudy sky and clear sky, respectively. These definitions allow re-writing Eq. (13.48) as follows:

$$T^G = T_{cs}^G(a + b\tilde{C}^c)\cos^{d-1} z. \qquad (13.50)$$

For convenience let us assume $z = 0°$ (i.e. the sun is at zenith). Then, another form of Eq. (13.50) is:

$$T^G = T_{cs}^G[(a+b)\tilde{C}^c + a(1 - \tilde{C}^c)]. \qquad (13.51)$$

The meaning of the coefficients a and b in Eq. (13.51) is obvious in case of $c = 1$. The r.h.s. of Eq. (13.51) is given by a weighted superposition of contributions from both parts of the sky (i.e. from the part of the sky covered, and respectively not-covered, by clouds). The coefficient a is the weighting factor for the contribution from the part of the sky free of clouds while $a + b$ is a similar factor for the part of the celestial vault covered by clouds. Both weighting factors can be seen as measures of the brightness/darkness of a given sky region. On clear sky ($\tilde{C} = 0$), $a_{\tilde{C}=0,z=0°}$ should be unity. On overcast sky ($\tilde{C} = 1$) one has:

$$b_{\tilde{C}=1,z=0} = (T^G/T_{cs}^G)_{\tilde{C}=1,z=0°} - 1. \qquad (13.52)$$

When c increases above unity, both $\tilde{C}^c$ and the contribution of the clouds area in the r.h.s. of Eq. (13.51) diminishes. A reverse situation happens for $c < 1$, when $1 - \tilde{C}^c$ decreases and the contribution to the transmittance ratio from the region of the sky free of clouds diminishes, too. Sometimes negative values of c could arise in connection with high values of $\tilde{C}$ (see Tables 13.2 and 13.3). The superposition interpretation makes no sense in these cases as $1 - \tilde{C}^c$ is negative (but Eq. (13.51) is of course still valid from a mathematical view-point).

To make evident the meaning of the coefficient d in Eq. (13.48) one should re-mind that the dependence of the air mass on zenith angle is already incorporated in the atmospheric transmittance. In both extreme cases (i.e. $\tilde{C} = 0$ and $\tilde{C} = 1$), Eq. (13.50) gives $T^G/T_{cs}^G \propto \cos^{d-1} z$. For $d = 1$ the transmittance ratio is constant during the day. Thus, values of d different from unity may be related to temporal changes in the atmospheric structure and properties.

5.1 Sunshine Number not Considered

Equations (13.48) and (13.49) were fitted through a standard least-square procedure to all available data in Bucharest and Jassy (database HOUREAD). Figure 13.6 shows the results obtained for solar global irradiance at Bucharest. We used two common statistical indicators to test the models accuracy. They are the residual average and the residual standard deviation. Results are shown in Table 13.1.

Visual inspection of Fig. 13.6 suggests a better model performance during days with clear and overcast sky (large and small values of the ratio G/G_{cs}, respectively). The models represented by Eqs. (13.48) and (13.49) are relatively bias-free since the residual average is quite small for both global and diffuse irradiance. The residual standard deviation (rsd) shows that diffuse radiation data are more dispersed than global radiation data. Note that the rsd values of the simple models tested by Badescu (1997) by using the same HOUREAD data varied between 0.35 and 0.45. However, the significantly lower rsd values in Table 13.1 are not surprising, because the present models propose best-fit formulas.

In the instance of global irradiance one can perceive that the coefficient a is close to unity, in good concordance with the modified Kasten model. Also, the coefficient d is close to unity. Consequently, Eq. (13.48) shows that G/G_{cs} has a very weak dependence on zenith angle, in agreement with previous results by Kasten and Czeplak (1980). However, the coefficient b is significantly smaller than the value 0.72 from the Kasten model. When an overcast sky is considered ($\tilde{C} = 1$)

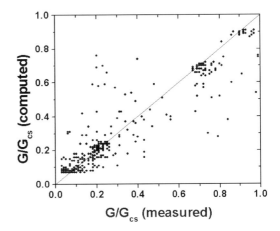

Fig. 13.6 The ratio between global solar irradiance, G, and global solar irradiance on cloudless sky, G_{cs}. Computed versus measured values on a horizontal surface. All available HOUREAD data for Bucharest were used

Table 13.1 Regression coefficients to be used in Eqs. (13.48) and (13.49). All available HOUREAD data were used (January and July). Accuracy indicators for both equations are also shown

Type of Solar Radiation	Locality	a (α)	b (β)	c (χ)	d (δ)	Rezidual average	Rezidual standard deviation
Global (Eq. 13.48)	Bucharest	0.9825	−0.5644	3.4111	1.0639	$3.68 \cdot 10^{-4}$	0.1237
	Jassy	0.9763	−0.5842	4.2094	1.1009	$2.46 \cdot 10^{-4}$	0.1223
Diffuse (Eq. 13.49)	Bucharest	1.5535	1.2507	0.9304	1.508	$4.18 \cdot 10^{-3}$	0.3211
	Jassy	1.6305	0.9736	0.8713	1.5317	$2.13 \cdot 10^{-3}$	0.3559

the Kasten model predicts $G \approx 0.28 G_{cs}$. This is smaller than the value predicted by other models, such as Adnot et al. (1979) where $G \approx 0.36 G_{cs}$ and BCLS (Barbaro et al. 1979; Badescu 1987) where $G \approx 0.34 G_{cs}$. If we take into account that d is close to unity for both Bucharest and Jassy, our model predicts $G \approx 0.41 G_{cs}$. The coefficient c is higher in Jassy than in Bucharest and in both localities it exceeds the value adopted in the Kasten model.

In the instance of clear sky diffuse radiation ($\tilde{C} = 0$) one can observe that $D/D_{cs} \propto \cos^{1/2} z$. This is in reasonable concordance with the BCLS model where $D/D_{cs} \propto \cos^{1/3} z$ (Barbaro et al. 1979). A positive coefficient β means that the diffuse irradiance increases by increasing total cloud amount, as expected. The dependence of the diffuse irradiance D on $\tilde{C}$ is noticeably, but only slightly, nonlinear. We conclude that the strongly nonlinear dependence of the global irradiance G on $\tilde{C}$ is mainly due to the interaction between beam radiation and clouds.

The nonlinear dependence of both global and diffuse solar radiation on $\tilde{C}$ prompted us to stratify the actinometric data according to the observed total cloud cover amount. Three classes of cloudiness were considered. The low cloudiness data include the actinometric data associated to $\tilde{C}$ between 0.0 and 0.4 while the medium and high cloudiness data refer to data associated to $\tilde{C}$ values between 0.4 and 0.7, and between 0.7 and 1.0, respectively. Equations (13.48) and (13.49) were fitted again, this time to these stratified data. Table 13.2 shows the results. Note that the actinometric data associated to $\tilde{C} = 0.4$ for example belong to two cloudiness classes (i.e. 0.0–0.4 and 0.4–0.7). Consequently, there is a difference between the total observation numbers in Table 13.1 and Table 13.2, respectively. The residual standard deviation in Table 13.2 shows that the radiation data on very cloudy skies are more dispersed than those from clear skies, in agreement with previous knowledge.

Global radiation is considered first. For low cloudiness ($\tilde{C}$ between 0.0 and 0.4) the coefficient a is close to unity. The low negative value of b and the high value of c show a weak dependence on $\tilde{C}$. Also, d is close to unity. Consequently, from Eq (13.48) one learns that $G/G_{cs} \approx$ const. and global solar irradiance follows approximately a cosine law (i.e. $G \propto G_{cs} = G_{cs,perp} \cos z$). This is easily to accept, because at low cloudiness global radiation mainly consists in beam radiation. The cosine law is verified for medium and high cloudiness classes. However, the factor

Table 13.2 Regression coefficients to be used in Eqs. (13.48) and (13.49) for various cloudiness classes. All available HOUREAD data were used (Bucharest and Jassy, January and July). Accuracy indicators for both equations are also shown

Type of Solar Radiation	Cloudiness class	a (α)	b (β)	c (χ)	d (δ)	Rezidual average	Rezidual standard deviation
Global	0.0–0.4	0.9756	−0.1113	4.9376	1.0827	$7.18 \cdot 10^{-4}$	0.0656
(Eq. 13.48)	0.4–0.7	0.9843	−0.8869	4.5137	1.0899	$2.25 \cdot 10^{-5}$	0.1594
Diffuse	0.7–1.0	0.9050	−0.5051	4.7158	1.0643	$2.44 \cdot 10^{-4}$	0.1437
(Eq. 13.49)	0.0–0.4	1.5264	6.6979	3.3176	1.2433	$-1.13 \cdot 10^{-4}$	0.2136

in front of the cosine depends on the total cloud cover amount $\tilde{C}$ and on the coefficient b. On overcast sky ($\tilde{C} = 1$) one can see that $G \propto 0.4 G_{cs}$, in good agreement with the models already quoted.

Next, diffuse radiation is considered. For low and medium cloudiness ($\tilde{C}$ between 0.0 and 0.7), the coefficient δ in Eq. (13.49) varies between 1.24 and 1.36. Consequently, from Eq. (13.49) one can see that $D \propto \cos^{0.24 \cdots 0.36} z$. This is in good agreement with the relationship $D \propto \cos^{1/3} z$ considered by Barbaro et al. (1979). At higher cloudiness ($\tilde{C} = 0.7 - 1.0$) the diffuse irradiance decreases by increasing $\tilde{C}$, because the coefficient χ is negative. This is not emphasized by the non-stratified actinometric data (see Table 13.1).

5.2 Sunshine Number Considered

The instantaneous values of both direct and diffuse irradiance at ground level depend on the fact that the sun is (or is not) covered by clouds. In other words, they depend on the values of the sunshine number ξ defined in section 3. We already showed that the statistical properties of the sunshine number depends on the total cloud cover amount C.

The databases (i) and (ii) described in section 3 correspond to $\xi = 0$ and $\xi = 1$, respectively. Figure 13.7 shows the global and diffuse irradiance in databases (i) and (ii) as a function of the solar zenith angle z. Separate "clouds" of data can be seen as the available data are associated to readings at a three hour interval. The data in both subsets (i) and (ii) cover well enough the range of all solar zenith angles. This encouraged us to fit again the Eqs. (13.48) and (13.49), this time to the smaller subsets of data (i) and (ii). Figure 13.8 and Table 13.3 show the results. Figure 13.8 has to be compared with Fig. 13.6, where no stratification of data according to the sunshine number ξ was used. The accuracy of both Eqs. (13.48) and (13.49) strongly increases when the new set of regression coefficients is used (compare the residual standard deviation from Tables 13.3a and 13.3b, on one hand, and from Tables 13.1 and 13.2, on the other hand). This proves that the sunshine number ξ is an important (key) parameter as far as instantaneous irradiance data are concerned. The random

Fig. 13.7 Global (*G*) and diffuse (*D*) solar irradiance on a horizontal surface as a function of the cosine of zenith angle *z*. (**a**) Recordings for which *G* > 3*D*; (**b**) Recording for which *G* < 1.2*D* – in this case only the global solar irradiance is represented. Subsets of HOUREAD database were used

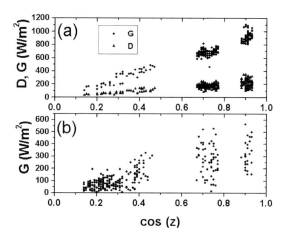

deviation is larger for diffuse than for global radiation, whatever the sun is, or is not, covered by clouds. The random deviation increases by increasing the cloudiness class, for both global and diffuse radiation and in case of both covered and uncovered sun.

First, let us consider solar global radiation. When data is stratified on cloudiness classes the random deviation is, surprisingly, smaller when the sun is covered by clouds (compare residual standard deviation in Tables 13.3a and 13.3b, respectively). If non-stratified data are used (i.e. $\tilde{C} = 0 - 1$) the best accuracy corresponds to a sun *not* covered by clouds. When diffuse radiation is considered, the random deviation is smaller when the sun is *not* covered by clouds, in case of both stratified and un-stratified data upon cloudiness class (compare Tables 13.3a and 13.3b, respectively).

Fig. 13.8 The ratio between global solar irradiance, *G*, and global solar irradiance on cloudless sky, G_{cs}. Computed versus measured values on a horizontal surface. (**a**) $\xi = 0$ (i.e. sun covered by clouds); (**b**) $\xi = 1$ (i.e. sun not covered by clouds). Subsets of HOUREAD data base were used

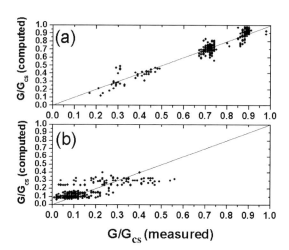

Table 13.3 Regression coefficients to be used in Eqs. (13.48) and (13.49) for various cloudiness classes as a function of the shine sun number ξ. **(a)** $\xi = 0$ (i.e. the sun is covered by clouds); **(b)** $\xi = 1$ (i.e. the sun is not covered by clouds). All available HOUREAD data were used (Bucharest and Jassy, January and July). Accuracy indicators for both equations are also shown

(a)

Type of Solar radiation	Cloudiness class	a (α)	b (β)	c (χ)	d (δ)	Rezidual average	Rezidual standard deviation
Global (Eq. 13.48)	0.0–0.4	0.1734	0.1603	0.1796	0.4926	$2.10 \cdot 10^{-4}$	$3.2 \cdot 10^{-4}$
	0.4–0.7	0.5088	−0.1351	−0.2894	0.6740	$-5.67 \cdot 10^{-4}$	$4.30 \cdot 10^{-3}$
	0.7–1.0	0.0200	0.3025	−0.8491	0.8233	$-5.30 \cdot 10^{-4}$	$5.51 \cdot 10^{-3}$
	0.0–1.0	0.3871	−0.0920	123.99	0.7851	$-1.98 \cdot 10^{-4}$	$5.15 \cdot 10^{-3}$
Diffuse (Eq. 13.49)	0.0–0.4	0.9227	1.4705	0.6460	1.1668	$-9.10 \cdot 10^{-4}$	0.037
	0.4–0.7	1.1801	1.0045	0.5706	1.3561	$-3.68 \cdot 10^{-3}$	0.070
	0.7–1.0	0.3272	1.4871	−0.8051	1.5142	$-2.82 \cdot 10^{-3}$	0.085
	0.0–1.0	2.5602	−0.6976	1.0000	1.5173	$-2.26 \cdot 10^{-3}$	0.085

(b)

Type of Solar Radiation	Cloudiness class	a (α)	b (β)	c (χ)	d (δ)	Rezidual average	Rezidual standard deviation
Global (Eq. 13.48)	0.0–0.4	0.9609	0.0909	0.9963	0.8943	$1.09 \cdot 10^{-4}$	$3.18 \cdot 10^{-3}$
	0.4–0.7	0.7803	0.2702	0.2594	0.8583	$1.21 \cdot 10^{-3}$	$5.47 \cdot 10^{-3}$
	0.7–1.0	0.6372	0.3859	0.0065	0.2142	$1.44 \cdot 10^{-3}$	0.0845
	0.0–1.0	0.9588	0.1037	1.025	0.8885	$1.09 \cdot 10^{-3}$	$3.78 \cdot 10^{-3}$
Diffuse (Eq. (2))	0.0–0.4	1.1592	1.3461	2.3561	1.3503	$1.38 \cdot 10^{-4}$	0.030
	0.4–0.7	1.1229	0.6995	1.2216	1.4792	$-2.43 \cdot 10^{-3}$	0.040
	0.7–1.0	0.8711	0.8068	−0.0370	1.7597	$3.36 \cdot 10^{-3}$	0.047
	0.0–1.0	1.1648	0.5699	1.3168	1.4187	$6.91 \cdot 10^{-4}$	0.033

Almost all the values of the regression coefficients are significantly different from those when the shine sun number ξ is not taken into account (compare Tables 13.3a and 13.3b, on one side, and Table 13.2, on the other side). An except is the coefficient δ from the diffuse radiation formula Eq. (13.49), which ranges in about the same domain as that of Table 13.2, for all cloudiness classes and for both values of ξ.

The coefficients a and α in Table 13.3 have a maximum for the cloudiness class 0.4–0.7 for covered sun. This is different from Table 13.2, where a and α are rather weakly dependent on the cloudiness class. When global radiation is considered and the data is not stratified upon cloudiness (i.e. $\tilde{C} = 0 - 1$), the values of the coefficient a in case of $\xi = 1$ (i.e sun uncovered by clouds) are close to those from Table 13.1. The difference between the values of a (and α) in Tables 13.3 and 13.2, respectively, is larger for diffuse radiation at $\xi = 1$, and even much larger when the sun is covered by clouds ($\xi = 0$).

It seems no similarities can be found between the shape of the dependence of coefficients b (and β) on the cloudiness class in Tables 13.2 and 13.3, respectively.

When the data is not stratified upon cloudiness $(\tilde{C} = 0 - 1)$ the values of coefficient χ for $\xi = 0$ (Table 13.3a) are close to those of Table 13.1. In the same case $\tilde{C} = 0 - 1$, the coefficient c for global radiation depends significantly on the sunshine number ξ (see Tables 13.3a and 13.3b) and differs from the values in Table 13.1 that do not take into account ξ. The values of c and χ decrease by increasing the cloudiness class, whatever the value of ξ is. When ξ is not included in the analysis a similar situation occurs in case of the coefficient χ (see Table 13.2). However, in case of global radiation, the coefficient c depends weakly on the cloudiness class (Table 13.2). Except values for two cloudiness classes at $\xi = 1$, the other values of c and χ are significantly under-unitary and even negative in Table 13.3. This means that including ξ as a model parameter makes the significance of the parameter $\tilde{C}$ to diminish (compare the over-unitary or slightly under-unitary values of c and χ in Tables 13.1 and 13.2).

6 Solar Radiation Data Generation

Using the present model to generate time-series of solar radiation data depends on the procedure of simulating the sequences of sunshine number ξ. One expects these sequences to show autocorrelation features, especially in case of short time intervals. To our knowledge there is very few information concerning the sequential properties of ξ. This is quite surprising, taking into account the easy way to performing the observations, that is (perhaps) simpler than total cloud cover estimation. Researchers preferred to focus on the statistical features of solar irradiance/irradiation (see, e.g. Skartveit and Olseth (1992), Jurado et al. (1995), Tovar et al. (1998), Suehrcke (2000) and chapters in this book). Studying the sequential properties of ξ is beyond the scope of this chapter. One expects these properties to depend on the time interval between consecutive observations, on the total cloud cover amount, on the type and shape of clouds, on the wind speed and on other factors.

Equations (13.48) and (13.49) were used during the following ad-hoc procedure of solar radiation data generation. A random number (say $\tilde{p}(\xi = 0)$) with a priori assumed uniform distribution between 0 and 1 was generated each time when a new solar irradiance value was computed. This random number is used in connection with the statement $p(\xi = 0, t, \Delta t) = C(t, \Delta t)$ giving the probability for the sun being covered by clouds at time t (see Eq. (13.10)). If $\tilde{p}(\xi = 0) \leq p(\xi = 0, t, \Delta t) = C$, then one assumes that the sun is covered by clouds. Consequently, $\xi = 0$. If $\tilde{p}(\xi = 0) > p(\xi = 0, t, \Delta t) = C$, then one assumes that the sun is not covered by clouds and $\xi = 1$. The "real" total cloud amount C was evaluated as a function of the observed total cloud amount $\tilde{C}$ by using Eq. (13.22). A more appropriate treatment, base on the knowledge of the sequential properties of ξ, would allow the procedure to generate the random number $\tilde{p}(\xi = 0)$ as a function of the time t and previous values of sunshine number ξ.

Figure 13.9 shows time-series of measured and synthetic solar radiation data. Input meteorological data from database METEORAR for July 1961 in Bucharest

Fig. 13.9 Solar radiation time-series on a horizontal surface at Bucharest (July). **(a)** Measured global solar irradiance in July 1961; **(b)** Synthetic global solar irradiance. Input data from METE-ORAR database was used in this last case. The time interval used in computation was three hours. The first four statistical moments and the first three values of the k-"three hour" lag auto-correlation coefficient $r(k)$ $(k = 1, 2, 3)$ are also shown

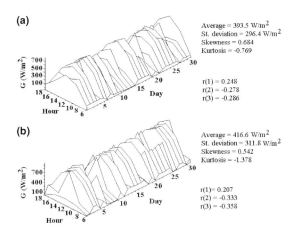

(a)

Average = 393.5 W/m^2
St. deviation = 296.4 W/m^2
Skewness = 0.684
Kurtosis = -0.769

$r(1) = 0.248$
$r(2) = -0.278$
$r(3) = -0.286$

(b)

Average = 416.6 W/m^2
St. deviation = 311.8 W/m^2
Skewness = 0.542
Kurtosis = -1.378

$r(1) = 0.207$
$r(2) = -0.333$
$r(3) = -0.358$

were used. The time interval between two consecutive measurements or computations was three hours. In case of synthesized data, the sunshine number ξ was computed according to the above simple procedure. Thus, the sunshine number autocorrelation properties were neglected. This could be a rather realistic assumption in case of the large time interval adopted here. Visual inspection shows reasonably good similarity between the sequential features of measured and synthetic time-series, with a slightly more abrupt time-variation in the second case.

Complete comparison of the measured and synthetic time-series requires analysis of both sequential and distribution features (see e.g. Badescu (2001) and references therein). Facts about the sequential characteristics can be obtained from the auto-correlation and/or partial auto-correlation plots. Here the indicator of auto-correlation is the k-"three hour" lag auto-correlation coefficient $r(k)$. The first three values of $r(k)$ are reported in Fig. 13.9. There is a reasonable concordance between generated and measured data. Both time series are weakly auto-correlated, as expected. Note that the sign of $r(2)$ and $r(1)$ is different. Consequently, a first order differencing is sufficient to obtain stationarity. This is not surprising, taking into account the daily periodicity of solar irradiance.

Several ways of comparing the distributions of measured and synthesized time-series are now in use. Here we shall simply compare their first statistical moments. Section 3 gives some information about the meaning of these moments. Here a reasonable agreement exists between the first two moments (i.e. average and standard deviation) of measured and synthetic time-series (see Fig. 13.9). For most technical solar energy applications, this sort of agreement is required. The concordance is weaker in case of the skewness and kurtosis.

Figure 13.10 shows time-series of synthetic global solar radiation data computed under the same conditions as those of Figure 13.9b, except the time interval between two consecutive generated values is shorter. Decreasing the time interval from three to one hour has not spectacular consequences (compare Fig 13.9b and Fig 13.10a, respectively).

Fig. 13.10 Same as
Fig 13.9(b) except the time
interval used in computation
is shorter. (**a**) One hour time
interval; (**b**) ten minutes time
interval

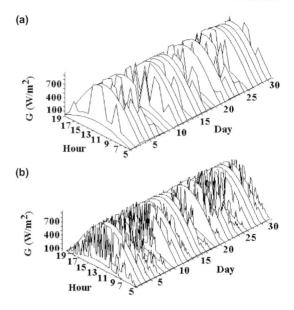

However, the sequential features of the time-series change significantly when the time interval is in the range of a few minutes (Figs 13.10b). Additional work is of course necessary to include the expected autocorrelation of the sunshine number ξ at these short time intervals.

6.1 Computer Program

The computer program g_main.for on the CD-ROM attached to the book (written in Fortran 77) may be used to generate solar radiation time series by using the procedure described above. The regression coefficients in this computer program were derived by using Romanian meteorological and actinometric data as shown in section 5.

7 Conclusions

A new parameter related to the state of the sky, called the sunshine number, is fully defined in this chapter. The sunshine number is 1 or 0 if the sun is not, or is covered, by clouds, respectively. Elementary statistical properties of this parameter are presented here.

The new parameter was used to develop a new kind of simple solar radiation computing model. This model is based on two parameters describing the state of the sky, i.e. the common total cloud cover amount and the sunshine number. So far, most simple models use only one parameter describing the state of the sky, i.e. relative sunshine *or* total cloud amount. The development of the new simple model was performed in two stages. First, a regression formula to compute instantaneous cloudy sky global irradiance on a horizontal surface was derived. Second, this formula was adapted to be used in association with the sunshine number.

The regression formula Eq. (13.48) to compute instantaneous cloudy sky global irradiance was suggested by previous testing of the modified Kasten model (Kasten 1983). Our formula takes into account both the solar zenith angle and the total cloud cover amount. The physical meaning of the four regression coefficients in Eq. (13.48) was briefly explained. They are related to the dependence of the atmospheric transmittance for solar radiation on various parameters, such as the brightness/darkness of the cloud cover and the temporal changes in the atmosphere structure and properties. Another, somewhat similar, formula was proposed to compute diffuse solar irradiance (Eq. (13.49)).

Fitting Eqs. (13.48) and (13.49), to Romanian data shows expected results as low bias errors and larger errors in case of diffuse radiation as compared to global radiation. A detailed discussion of the regression coefficient values shows generally a good concordance with existing models. Fitting the same two formulas to actinometric data stratified upon total cloud amount generally leads to an increased computation accuracy. The residual standard deviation shows that the radiation data on very cloudy skies are more dispersed than those from clear skies (see Table 13.2).

A second stratification of the actinometric data was performed according to the sunshine number. Fitting the formulas (13.48) and (13.49) to these double-stratified data leads to a significant accuracy improvement.

There is very few information concerning the distribution and sequential properties of the sunshine number ξ. However, to give perspective for our findings the model was applied to generate time-series of solar radiation data. The expected autocorrelation of ξ was neglected. Visual inspection and a brief statistical analysis showed a reasonably good similarity between the sequential and distribution features of measured and synthetic data when the time interval between successive calculations is three hours. Decreasing the time interval from three to one hour doesn't change significantly the results. Almost all the values of the regression coefficients are significantly different from those corresponding to the case when the sunshine number is not taken into account. We have proved that the sunshine number is a key parameter, important insofar as the computation of "instantaneous" radiation data is concerned.

However, the sequential features of the generated time-series changes significantly when the time interval is in the range of a few minutes. Thus, additional work is necessary to include the expected autocorrelation of ξ at these short time intervals.

Future work should also focus on the dependence of the regression coefficients in Eqs. (13.48) and (13.49) on site. The good performance of the "parent" Kasten model in various European localities suggests a weak dependence. This conjecture has to be checked, of course, for other countries with climates similar to Romania.

References

Angstrom A (1924) Solar and terrestrial radiation. Q J R Meteorol Soc 50: 121

Angstrom A (1956) On the computation of global radiation from records of sunshine. Arkiv Fur Geofysik 2: 471

Atlas, Romania (1972–1979) Ed. Academiei Bucharest

Badescu V, Popa C, Popescu B, Motoiu I, Duduruz M (1984) Elemente pentru un model climatologic al Romaniei util in calculul radiatiei solare (in Romanian). Sesiune ICEMENERG, paper II.1, Bucharest 27–28 Nov

Badescu V, Popa C (1986) Criteriu climatic de stabilire a regiunilor omogene din punct de vedere meteorologic si a anului tipic de determinare a disponibilului de radiatie solara (in Romanian). Hidrotehnica 31(9): 265–268

Badescu V (1987) Can the model proposed by Barbaro et al be used to compute global solar radiation on the Romanian territory ? Solar Energy 38: 247–254

Badescu V (1990) Observations concerning the empirical relationship of cloud shape to point cloudiness (Romania). J Appl Meteor 29: 1358–1360

Badescu V (1991) Studies concerning the empirical relationship of cloud shade to point cloudiness (Romania). Theor Appl Climatol 44:187–200

Badescu V (1992) Over and under estimation of cloud amount: theory and Romanian observations. Int J Solar Energy 11: 201–209

Badescu V (1997) Verification of some very simple clear sky and cloudy sky models to evaluate global solar irradiance. Solar Energy 61: 251–264

Badescu V (2001) Synthesis of short-term time-series of daily averaged surface pressure on Mars. Environmental Modelling & Software, 16: 283–295

Badescu V (2002) A new kind of cloudy sky model to compute instantaneous values of diffuse and global solar irradiance. Theor Appl Climatol 72: 127–136

Barbaro S, Coppolino S., Leone C, Sinagra E (1979) An atmospheric model for computing direct and diffuse solar radiation. Solar Energy 22: 225–228

Bazac GhC (1983) Influenta reliefului asupra principalelor caracteristici ale climei Romaniei (in Romanian). Ed. Academiei, Bucharest

Bener P (1984) Survey and comments on various methods to compute the components of solar irradiance on horizontal and inclined surfaces. In Handbook of methods of estimating solar radiation, Swedish Council for Building Research, Stockholm, Sweden, pp 47–77

Ciocoiu I, Elekes I, Glodeanu F (1974) Corelatia dintre radiatia globala si durata de stralucire a soarelui (in Romanian). In: Culegerea lucrarilor RMHI pe anul 1972, RMHI, Bucharest, pp 265–275

Creteanu V (1984) Cadastrul energiei solare destinat nevoilor energetice (in Romanian). In: St Cerc Meteorol, RMHI, Bucharest, pp 33–41

Costin V (2000) Private communication

Coulson KL (1975) Solar and terrestrial radiation, Academic Press, New York

Davies JA, McKay DC (1982) Estimating solar irradiance and components. Solar Energy 29:55–64

Davies JA, McKay DC, Luciani G, Abdel-Wahab M (1988).Validation of models for estimating solar radiation on horizontal surfaces, volume 1, IEA Task IX, Final Report, Atmospheric Environment Service of Canada, Downsview, Ontario, Canada

Festa R, Ratto CF (1993) Solar radiation statistical properties. Technical Report for IEA Task IX, University of Genova.

Filipescu D, Trandafir R, Zorilescu D (1981) Probabilitati geometrice si aplicatii (in Romanian). Editura Dacia, Cluj-Napoca

Garg HP, Garg SN (1993) Measurement of solar radiation. I. Radiation instruments. Renewable Energy 3 (4/5): 321–333

Harrison AW, Coombes CA (1986) Empirical relationship of cloud shade to point cloudiness (Canada). Solar Energy 37(6): 417–421

Hay JE (1979) Calculation of monthly mean solar radiation for horizontal and inclined surfaces. Solar Energy 23: 301–307

Hoyt DV (1978) Percent of possible sunshine and total cloud cover. Mon Wea Rev. 105: 648–652

IMH (1961) Anuarul meteorologic (in Romanian). Institutul de Meteorologie si Hidrologie, Bucuresti

IMHR (1964–1972) Meteorological Annual (in Romanian). Romanian Meteorological and Hydrological Institute, Bucharest

Jurado M, Caridad JM, Ruiz V (1995) Statistical distribution of the clearness index with solar radiation data integrated over five minute intervals. Solar Energy 55: 469

Kasten H (1966) A new table and approximation formula for the relative optical air mass. Archiv fur Meteorol Geophys und Bioklim B: 206–223

Kasten F (1983) Parametrisierung der globalstrahlung durch bedeckungsgrad und trubungsfactor. Ann Met 20: 49–50

Kasten F, Czeplak G (1980) Solar and terrestrial radiation dependent on the amount and type of clouds. Solar Energy 24: 177–189

Leckner B (1978) The spectral distribution of solar radiation at the Earth's surface – elements of a model.Solar Energy 20: 143–150

May BR, Collingbourne RH, McKay DC (1984) Catalogue of estimating methods. In Handbook of methods of estimating solar radiation, Swedish Council for Building Research, Stockholm, Sweden, pp 4–32

Neacsa O, Susan V (1984) Unele caracteristici ale duratei stralucirii soarelui deasupra teritoriului Romaniei (in Romanian). St Cerc Meteorol, RIMH, Bucharest pp. 99–113

Painter, HE (1981) The performance of a Campbell-Stokes sunshine recorder compared with a simultaneous record of normal incidence radiance. Meteor Mag 110–102

Prescott JA (1940) Evaporation from a water surface in relation to solar radiation. Trans R Soc Sci Aust 64: 114

Raju ASN, Karuna Kumar K (1982) Comparison of point cloudiness and sunshine duration derived cloud cover in India. Pageoph 120 495

Rangarajan S., Swaminsthan MS, Mani A (1984) Computation of solar radiation from observations of cloud cover. Solar Energy 32: 553

Reddy SJ (1974) An empirical method for estimating sunshine from total cloud amount. Solar Energy 15: 281

Santalo LA (1950) Sobre unas formulas integrales i valores medios referentes a figuras covexas moviles en el plano, Contr Cientificas Univ De Buenos Aires A 1(2):284–294

Skartveit A, Olseth JA (1992) The probability density and the auto correlation of short-term global and beam irradiance. Solar Energy 49: 477

Suehrcke H, McCormick PG (1992) A performance prediction method for solar energy systems. Solar Energy 48: 169

Suehrcke H (2000) On the relationship between duration of sunshine and solar radiation on the earth's surface. Angstrom's equation revisited. Solar Energy 68: 417–425

Tovar J, Olmo FJ, Alados-Arboledas L (1998) One minute global irradiance probability density distributions conditioned to the optical air mass. Solar Energy 62: 387–393

Chapter 14
The Meteorological Radiation Model (MRM): Advancements and Applications

Harry D. Kambezidis and Basil E. Psiloglou

1 Introduction

The estimation of hourly and daily solar radiation on inclined surfaces starts with the determination of the corresponding hourly values on horizontal plane. For this reason the *Atmospheric Research Team* (ART) at the *National Observatory of Athens* (NOA) initially developed the so-called MRM (*Meteorological Radiation Model*; Kambezidis and Papanikolaou 1989; Kambezidis and Papanikolaou 1990a; Kambezidis et al. 1993a,b; Kambezidis et al. 1997). The goal of the development of MRM was to derive solar radiation data at places where these are not available. To do that the implementation in the algorithm of the more widely available meteorological data (viz. air temperature, relative humidity, barometric pressure and sunshine duration) was considered. A solar code with such characteristics is particularly useful for the generation of Solar Atlases in areas with a moderately dense meteorological network.

The original form of MRM version one (MRM v1) worked efficiently under clear-sky conditions, but it could not work under partly cloudy or overcast skies. MRM v2 introduced new analytical transmittance equations and, therefore, became more efficient than its predecessor. Nevertheless, this version still worked well under clear sky conditions only. These deficiencies were resolved via the development of the third version of MRM (MRM v3), derived by T. Muneer's research group at Napier University, Edinburgh (Muneer et al. 1996; Muneer 1997; Muneer et al. 1997; Muneer et al. 1998) after successful co-operation between ART and his group. MRM v3 was included in the book edited by Muneer (1997). Through the EU JOULE III project on *Climatic Synthetic Time Series for the Mediterranean Belt* (CliMed), a further development of the MRM was achieved, which is referred to

Harry D. Kambezidis
National Observatory of Athens, Greece, e-mail: harry@meteo.noa.gr

Basil E. Psiloglou
National Observatory of Athens, Greece, e-mail: bill@meteo.noa.gr

as version four (MRM v4), providing further improvement in relation with partly cloudy and overcast skies. Prof. Hassid, Technion University of Israel, used MRM v4 to make simulations and comparison with Israeli solar radiation data. In using the code, he found some errors mainly in the calculation of the course of the sun in the sky, which were corrected by him. On the other hand, Gueymard (2003) in an inter-comparison study employing various broadband models used MRM v4 and found it not to be performing well in relation to others. Further elaboration of MRM v4 by ART for the purpose of this book resulted in discovering more severe errors in the transmittance and solar geometry equations, which were corrected concluding to a new version of MRM (MRM v5). MRM was successfully used by the *Chartered Institution of Building Service Engineers* (CIBSE) of UK in 1994 under the *Solar Data Task Group* (Muneer 1997). Apart from that specific task, MRM can be used in a variety of applications, among of which the most important nowadays are:

- to estimate solar irradiance on horizontal plane to be used as input parameter to codes calculating solar irradiance on inclined surfaces with arbitrary orientation,
- to estimate solar irradiance on horizontal plane with the use of available meteorological data to derive the solar climatology at a location,
- to fill gaps of missing solar radiation values in a historic series from corresponding observations of available meteorological parameters,
- to provide algorithms for engineering purposes, such as solar energy applications, photovoltaic efficiency, energy efficient buildings and daylight applications, with needed (simulated) solar radiation data.

The objectives of this chapter are (i) to describe the origin, the algorithms, the recent improvements and the test results of MRM; (ii) to show how a detailed statistical analysis of the estimates produced by radiation models can be done to evaluate in depth the performance of solar algorithms such as MRM.

2 Stages of the MRM Development

This section provides a detailed description of all five versions of MRM, which are currently available, and also establishes the relation of the newly developed MRM v5 to the previous versions. MRM v1 refers to clear sky conditions only. The cloudy conditions were introduced in MRM v2 and later. MRM can calculate in either Local Standard Time (LST) or Local Apparent Time (LAT) depending on the declaration upon starting the programme run. MRM v1 was completed in the period 1987–1992. MRM v2 was improved in the period 1994–1996, MRM v3 in 1998–2000, MRM v4 2002–2004 and the current MRM v5 in late 2006 - beginning of 2007. Recently the MRM v5 was used to simulate the radiation levels during the solar eclipse of 29 March 2006 (Psiloglou and Kambezidis, 2007).

2.1 MRM Version 1

Direct Beam Irradiance for Cloudless Skies

MRM is a broadband empirical algorithm for simulation and estimation of solar irradiance on horizontal surface. According to Bird and Hulstrom (1981a,b), the direct beam component normal to the horizontal plane under clear sky and natural (without anthropogenic influence) atmosphere is given by the formula:

$$I_b = 0.975\, I_{ex}\, \sinh T_a\, T_r\, T_o\, T_w\, T_{mg} \tag{14.1}$$

where h is the solar elevation angle (in radians), I_{ex} is the extraterrestrial solar irradiance normal to the solar rays on any day number of the year, DN (DN=1-365), which is given, according to the European Solar Radiation Atlas (1989), by:

$$I_{ex} = I_o\left[1 + 0.035\cos\left[2\pi(DN-4)/366\right]\right] \tag{14.2}$$

where $I_o = 1353\,\mathrm{Wm^{-2}}$ (the solar constant value introduced in 1971), T_a is the optical transmittance of aerosols due to Mie scattering, T_r the optical transmittance of molecules due to Rayleigh scattering, T_o the optical transmittance due to ozone absorption, T_w the optical transmittance due to water vapour absorption and T_{mg} the optical transmittance due to mixed gases absorption (i.e. CO_2, O_2).

For the explicit calculation of the above optical transmittances, the following equations are applied. According to Iqbal (1983), the optical transmittance due to Mie (aerosol) scattering is calculated by:

$$T_a = \exp\left[-\delta_a^{0.873}\left(1 + \delta_a - \delta_a^{0.7088}\right)m'^{0.9108}\right] \tag{14.3}$$

where δ_a is the aerosol optical thickness, m is the optical air mass (Kasten 1966);

$$M = \left[\sin h + 0.15\left(93.885 - \theta_z\right)^{-1.253}\right]^{-1} \tag{14.4}$$

where θ_z is the solar zenith angle, in degrees. The pressure-corrected m for sites with P other than P_o=1013.25 hPa (sea level pressure) is $m'=m(P/P_o)$.

The aerosol optical thickness (AOT), according to Shettle and Fenn (1975), can be calculated by the formula:

$$\delta_a = 0.2758\,\delta_{a,\lambda=0.38} + 0.3500\,\delta_{a,\lambda=0.50} \tag{14.5}$$

where the values of $\delta_{a,\lambda=0.38}$ and $\delta_{a,\lambda=0.50}$ vary. For UK (Muneer et al. 1996) and Athens, Greece (Pisimanis et al. 1987) these are 0.72 and 0.56, while for the rural areas of Greece 0.35 and 0.27 (Pisimanis et al. 1987). For the US Standard Atmosphere of 1976, $\delta_{a,\lambda=0.38}=0.3538$ and $\delta_{a,\lambda=0.50}=0.2661$. The relation between δ_a and λ is:

$$\delta_{a,\lambda} = \beta\,\lambda^{-\alpha} \tag{14.6a}$$

known as the Ångström's equation (Ångström 1929; 1930); Eq. (14.6a) expresses δ_a as a function of wavelength, λ, the turbidity coefficient, β, and a coefficient related to the size distribution of the aerosol particles, α. If α and β are not known for a location, typical values can be obtained from various studies published in the international literature under different atmospheric conditions (e.g. rural, urban, marine, etc). A typical value for α is 1.3. It is worth mentioning that the values of AOT are only weakly influenced by the geographic variation. In cases that visibility observations are available from a near-by meteorological station, these can contribute to the estimation of β through the empirical relation (McClatchey and Selby 1972):

$$\beta = 0.55^{\alpha} \, (3.912/V - 0.01162)[0.02472(V\text{-}5) + 1.132] \tag{14.6b}$$

where V denotes the horizontal visibility, in km.

According to Iqbal (1983), the optical transmittance due to Rayleigh (molecular) scattering can be calculated by:

$$T_r = \exp[-0.0903 \, m'^{0.84}(1 + m' - m'^{1.01})] \tag{14.7}$$

and the optical transmittance due to ozone absorption is:

$$T_o = 1 - [0.1611x_o \, (1 + 139.48x_o)^{-0.3035} - 0.002715x_o \\ (1 + 0.0440x_o + 0.0003x_o{}^2)^{-1}] \tag{14.8}$$

where $x_o = l_o \, m$, and l_o is the total column of ozone in the atmosphere, measured in units of atm-cm ($1 \, \text{atm-cm} = 1 \times 10^{-3}$ DU). This is approximated for the Northern Hemisphere, as in Van Heuklon (1979), by the formula (a similar formula is also available for locations in the Southern Hemisphere):

$$l_o = d_1 + \{d_2 + d_3 \, \sin[d_4 \, (N - 30)] + d_5 \, \sin[d_6 \, (\theta + \Delta)]\}[\sin^2(d_7 \, \phi)] \tag{14.9}$$

where $d_1 = 0.235$, $d_2 = 0.150$, $d_3 = 0.040$, $d_4 = 0.9865$, $d_5 = 0.020$, $d_6 = \pi^2/10800$ and $d_7 = 1.28 \, \pi/180$. In Eq. (14.9), θ is the geographic longitude, in degrees, positive to the East of Greenwich, ϕ is the geographic latitude, in degrees, positive in the Northern Hemisphere; Δ is a correction factor taking the value $20°$ for eastern or $0°$ for western longitudes.

According to Lacis and Hansen (1974), the optical transmittance due to water vapour absorption can be calculated by:

$$T_w = 1 - 2.4959 \, x_w \, [(1 + 79.034 \, x_w)^{0.6828} + 6.385 \, x_w)]^{-1} \tag{14.10}$$

where $x_w = l_w \, m$, and l_w is the total column of precipitable water, in cm, approximated by (Gates 1962):

$$l_w = 0.23 \, e_m \, 10^{-H/22000} \tag{14.11}$$

where e_m is the partial water vapour pressure, in mmHg, at the station's height, H, in metres. This expression for l_w was preferred in MRM v1 instead of that used by Bird and Hulstrom (1981a,b) because it gave better results at locations with Mediterranean climate (Pissimanis et al. 1987).

Finally, according to Bird and Hulstrom (1981a,b), the optical transmittance due to the mixed gases in the atmosphere can be approximated by:

$$T_{mg} = \exp\left(-0.0127\, m'^{0.26}\right) \tag{14.12}$$

Diffuse Irradiance for Cloudless Skies

The diffuse horizontal component, I_d, based on Bird and Hulstrom (1981a,b) under clear skies is given by the following relationship:

$$I_d = 0.79\, I_{ex}\, \sinh T_o\, T_w\, T_{mg}\, T_{aa}\left[0.5(1-T_r)+0.84(1-T_{as})\right]/ \tag{14.13a}$$
$$(1-m+m^{1.02})$$

where

$$T_{aa} = 1 - 0.6(1-T_a)\left(1-m+m^{1.06}\right) \tag{14.13b}$$

$$T_{as} = 10^{-0.045\cdot m'^{0.7}} \tag{14.13c}$$

where T_a is given by Eq. (14.3), T_{aa} is the broadband transmittance due to the absorption by aerosols and T_{as} is the attenuation from aerosol scattering alone (Watt 1978). The coefficient of 0.6 in Eq. (14.13b) was preferred to that proposed by Bird and Hulstrom (1981a,b), i.e. 0.1, as giving better results at locations with Mediterranean climate.

Global Irradiance for Cloudless Skies

The global horizontal irradiance, I_g, under clear skies, is, therefore, given by the sum of the direct beam in Eq. (14.1) and the diffuse components of Eq. (14.13a), so that:

$$I_g = I_b + I_d \tag{14.14a}$$

However, the effect of multiple ground-atmosphere reflections can be accounted for, scaling Eq. (14.14a) by the adequate factor $(1-\rho_g\rho_a)^{-1}$:

$$I_g = (I_b + I_d)/(1-\rho_g\rho_a) \tag{14.14b}$$

where ρ_g is the ground albedo (usually given the value of 0.2) and ρ_a is the albedo of the cloudless sky. The latter can be computed from the relation:

$$\rho_a = 0.0685 + 0.16\,(1-T_{as}) \tag{14.15}$$

For the estimation of solar position in the sky on any day of the year, DN, the algorithm SUNAE (Walraven 1978) and its subsequent corrections (Wilkinson 1981; Muir 1983; Kambezidis and Papanikolaou 1990b; Kambezidis and Tsangrassoulis

Fig. 14.1 Comparison of
hourly values for the global
horizontal radiation between
MRM simulations (solid and
dashed lines) and measurements (circles and crosses)
for Larissa, Greece (solid line,
circles) and Thessaloniki,
Greece (dashed line, crosses)
on 1/1/1981 and 25/12/1988,
respectively. The x-axis is
LAT, in hours, while the
y-axis is solar irradiation in
MJm^{-2}. Source: Kambezidis
and Papanikolaou (1990a)

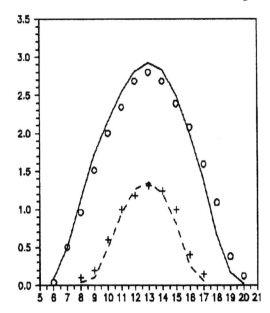

1993) have been used. Estimates of the diffuse and direct beam irradiance components obtained from MRM on any particular date, say on a minute-by-minute
basis, can easily lead to integrated hourly, daily, monthly and annual irradiation
values.

Based on this version, Kambezidis and Papanikolaou (1990a) compared the
MRM-simulated irradiance values with real ones on specific clear days a various
locations in Greece. Below a reproduction of two of their Figures is given (see
Figs. 14.1, 14.2). The agreement seems to be fairly good in all cases. Both figures
are drawn in LAT.

2.2 MRM Version 2

Version 2 of MRM was derived in order to incorporate new knowledge in the radiative transfer field. First, I_o was set equal to $1367\,Wm^{-2}$. Then, new analytical
formulae for the transmittance functions, found to perform better, were introduced.
A description of MRM v2 is given below.

Direct Beam Irradiance for Cloudless Sky

This is slightly different from Eq. (14.1), i.e.:

$$I_b = I_{ex} \sinh T_a\, T_r\, T_o\, T_w\, T_{mg} \tag{14.16a}$$

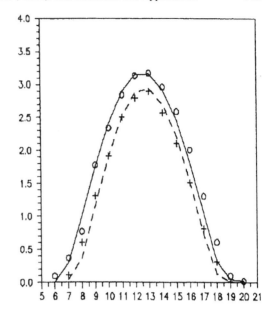

where I_{ex} is given by Eq. (14.2) and I_0 has an updated value. The factor of I_{ex} in Eq. (14.2) is replaced here by a more exact formula for deriving the correction for the Sun-Earth distance, S, according to the day number of the year. The relationship is (Duffie and Beckman 1980):

$$S = 1.00011 + 0.034221\cos D + 0.00128\sin D +$$

$$+0.000719\cos 2D + 0.000077\sin 2D \qquad (14.16b)$$

where D is called the day angle, in radians, expressed as:

$$D = 2\pi(DN - 1)/365 \qquad (14.16c)$$

The transmittance function for the aerosols is the same with that in Eq. (14.3) with the difference that the air mass, m_j, is now calculated for each atmospheric process j, according to Gueymard (1995), i.e.:

$$m_j = \left[\sinh + a_{j1}\, h\,(a_{j2} + h)^{a_{j3}}\right] \qquad (14.17)$$

and Table 14.1; the pressure-corrected air mass, according to its previous definition, is:

$$m_j' = m_j(P/P_0) \qquad (14.18)$$

The transmittance function for the Rayleigh scattering is given (Davies et al., 1975) by:

Table 14.1 Coefficients a_j for the optical air masses m_j in Eq. (14.17)

Process (j)	a_{j1}	a_{j2}	a_{j3}
Aerosols (a)	0.031141	0.10	92.4710
Mixed gases (mg)	0.456650	0.07	96.4836
Ozone (o)	268.4500	0.50	115.420
Rayleigh (r)	0.456650	0.07	96.4836
Water vapour (w)	0.031141	0.10	92.4710

$$T_r = 0.972 - 0.08262\,m_r + 0.00933\,m_r{}^2 - 0.00095\,m_r{}^3 + \\ + 0.000437\,m_r{}^4 \tag{14.19}$$

The transmittance functions for the ozone, water vapour and mixed gases absorptions are exactly the same with those in Eqs. (14.8), (14.10) and (14.12), respectively, the only differences being for $x_o = l_o m_o$ in Eq. (14.8), $x_w = l_w m_w$ in Eq. (14.10) and m'_{mg} in Eq. (14.12). The total column of the precipitable water is given by Leckner (1978):

$$l_w = 0.493\,e_m/T_d \tag{14.20a}$$

with:

$$e_m = e_s\,(RH/100) \tag{14.20b}$$

In Eq. (14.20d) RH is the relative humidity at the station's height, in %, and e_s is the saturation water vapour pressure, in hPa, given by Gueymard (1993):

$$e_s = \exp(22.329699 - 49.140396\ T_{do}{}^{-1} - 10.921853\ T_{do}{}^{-2} - \\ - 0.39015156\,T_{do}) \tag{14.20c}$$

with

$$T_{do} = T_d/100 \tag{14.20d}$$

where T_d is the air temperature at the station's height, in K.

Diffuse Irradiance for Cloudless Sky

The diffuse horizontal component, I_d, based on the works of Bird and Hulstrom (1979) and Dave (1979) under clear skies, is given by Eq. (14.13a), but in MRM v2 the coefficient 0.79 was dropped:

$$I_d = I_{ex}\sin h\,T_o\,T_w\,T_{mg}\,T_{aa}[0.5(1 - T_r) + 0.84(1 - T_{as})]/(1 - m + m^{1.02}) \tag{14.21a}$$

$$T_{aa} = 1 - 0.1(1 - T'_a)(1 - m + m^{1.06}) \tag{14.21b}$$

T_{as} is given by Eq. (14.13c), and:

$$T'_a = \exp[-\delta_a{}^{0.873}(1+\delta_a-\delta_a{}^{0.7088})\,1.58663] \tag{14.21c}$$

where m is given by Kasten's and Young's (1989) expression:

$$m = [\sinh+0.50572\,(h+6.07995)^{-1.6364}]^{-1} \tag{14.21d}$$

Eq. (14.21c) comes from Eq. (14.13d) after setting m=1.66.

Global Irradiance for Cloudless Sky

This is given by Eq. (14.14b) in which ρ_a is calculated as follows:

$$\rho_a = 0.0685 + 0.17\,(1-T'_a) \tag{14.22}$$

Solar Irradiance for Cloudy Skies

MRM v2 introduced for the first time an option for calculating the solar radiation components under cloudy sky conditions. That was achieved by taking into account the daily sunshine duration, SD, in hours. Therefore, the various solar radiation components were calculated according to the ratio of measured, SD_m, to theoretical, SD_t, sunshine duration:

$$I'_b = I_b\,(SD_m/SD_t) \tag{14.23a}$$

$$I'_d = I_d\,(SD_m/SD_t) + K\,(1-SD_m/SD_t)\,(I_b+I_d) \tag{14.23b}$$

$$I'_g = (I'_b+I'_d)/(1-\rho_g\rho_a) \tag{14.23c}$$

$$SD_t = 2\omega/15 \tag{14.23d}$$

$$\omega = \cos^{-1}(-\tan\phi\,\tan\gamma) \tag{14.23e}$$

$$\gamma = (0.006918 - 0.399912\cos D + 0.070257\sin D - 0.006758\cos 2D +$$
$$+0.000907\sin 2D - 0.002697\cos 3D + 0.00148\sin 3D)(180/\pi) \tag{14.23f}$$

where ω is the hour angle of either sunrise or sunset, in degrees, γ is the solar declination, in degrees and K an empirical coefficient dependent upon ϕ (Berland and Danilchenco (1961). Typical values of K for various geographic latitudes are given in Table 14.2. I'_d in Eq. (14.23b) is considered to consist of two parameters: (i) the diffuse component under cloudless sky multiplied by the sunshine fraction, i.e., $I_d(SD_m/SD_t)$, in order to derive a new diffuse component for the first time that the sky becomes cloudless, and (ii) the global radiation in cloudless conditions multiplied by the non-sunshine fraction, i.e., $(1-SD_m/SD_t)$, in order to give contribution to another diffuse component for the time that the sky is cloudy. I'_g in Eq. (14.23c) has the same form with I_g in Eq. (14.14b).

Table 14.2 Typical values of the empirical parameter K for various latitudes, ϕ, in the Northern Hemisphere (source: Berland and Danilchenco 1961)

K	ϕ (degrees)
0.32	30
0.32	35
0.33	40
0.34	45

Figure 14.3 shows a comparison between measured global horizontal irradiation and the MRM-derived for NOA site (Athens, $\phi = 314.97°$ N, $\theta = 23.72°$ E, 107 m a.m.s.l.) in the whole year of 1990. The comparison is extremely good above the level of $12\,\text{MJm}^{-2}$ approximately.

2.3 MRM Version 3

Version 3 of MRM treats the cases of clear and overcast skies exactly as v2 does, but deals more efficiently with the cases of partly cloudy sky conditions. Muneer et al. (1996) recognised a log-log linear relationship between the hourly diffuse-to-direct beam ratio (DBR $= I'_d/I'_b$) and the direct transmittance ($k_b = I'_b/I_{ex}$) in the form:

$$DBR = A\, k_b^{\,C} \tag{14.24}$$

where A and C are parameters to be estimated via a linear least-squares method applied to observed data of DBR and k_b.

The practical merit of Eq. (14.24) is that it can be used to substitute Eq. (14.23b) in order to estimate values of diffuse irradiance based on values of direct beam and extraterrestrial irradiance and without acquiring sunshine data:

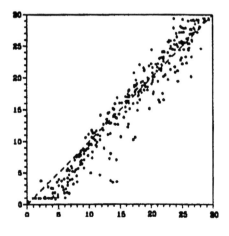

Fig. 14.3 Comparison between MRM simulations (y axis) with measurements (x axis) for Athens within 1990. The unit of solar radiation in both axes is MJm^{-2}. The dashed line represents the equation y=x. Source: Kambezidis (1998)

$$I'_d = I_b A k_b{}^C \tag{14.25}$$

Nevertheless, the very estimation of A and C presumes some initial data of concurrent direct beam and diffuse irradiance values under mixed sky scenarios, a case that constitutes a "drawback" of the method, since few solar radiation stations exist worldwide. For the UK a generalised expression of Eq. (14.24) is in the form $DBR = 0.285211 K_b{}^{-1.00648}$ (Gul et al. 1998; Muneer et al. 1998). For Romania, Kambezidis and Badescu (2000) derived the relationships $DBR = 0.2262 k_b{}^{-0.8542}$ for Iasio and $DBR = 0.1457 k_b{}^{-1.048}$ for Bucharest.

The global horizontal irradiance in MRM v3 is, therefore, estimated as:

$$I'_g = I'_b + A k_b{}^C I_b = I_b(1 + A k_b{}^C) SD_m/SD_t \tag{14.26}$$

Muneer (1997) presents some results regarding evaluations of MRM v3 against real hourly data at several locations in the UK, ignoring the enhancement due to multiple ground-atmosphere reflections introduced by Eq. (14.14b). The evaluation of MRM v3 was very good: an overall correlation coefficient of 92% was obtained. This striking performance of MRM v3 for UK is significantly influenced by the very detailed input information of sunshine duration in terms of hourly fractions in the ratio SD_m/SD_t, instead of the much less informative daily totals of sunshine duration, which was used in MRM v2.

Figure 14.4 presents a comparison between measured hourly global horizontal irradiance values in London with MRM-derived for a period of four years (1992–1995). The agreement seems much better than that of Fig. 14.3 in the sense that the tuft of the data points is concentrated around the y = x line (not shown) in contrast to that in Fig. 14.3 where the data points show a different slope from that of the y = x line. A similar improvement is shown with daily irradiation values from the Stornoway station for the same period as for London (see Fig. 14.5). It is worth noticing that both cases refer to all sky conditions.

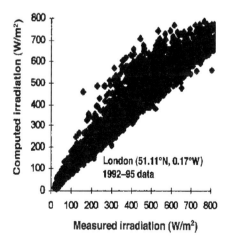

Fig. 14.4 Comparison of MRM-derived and measured hourly values of global horizontal irradiance (Wm^{-2}) under all sky conditions for London in the period 1992-1995. Source: Muneer et al. (1998)

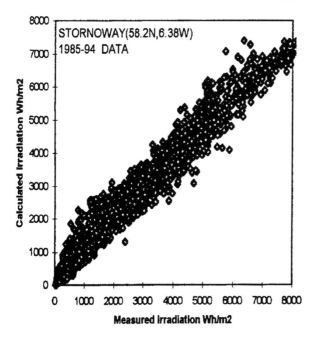

Fig. 14.5 As in Fig. 14.4, but for Stornoway in the period 1992-1995. Source: Muneer et al. (1998)

2.4 MRM Version 4

This new version of MRM is actually a hybrid version obtained from combining in a certain manner some ramifications of the previous two versions. In particular, MRM v4 discerns between four similar but different sub-versions denoted by v4.1, v4.2, v4.3, and v4.4, defined explicitly hereafter. These developments were accomplished within the framework of the EC JOULE III project CliMed.

MRM v4.1 estimates global horizontal irradiance via Eq. (14.21a) under over-cast skies ($I'_b=0$), via Eq. (14.26) under partly cloudy skies (as in v3), and via Eq. (14.14a) under clear skies (as in v2). It is worth noting here that the diffuse irra-diance expressed by Eq. (14.21a) does not coincide with that given by Eq. (14.25) in the case of overcast skies; this is what makes the difference between v3 and v4.1.

MRM v4.2 estimates global horizontal irradiance exactly as v4.1, and then scales it by the factor $(1 - \rho_g \, \rho_a)^{-1}$ in order to account for multiple ground and atmo-spheric reflections, too. The diffuse values are calculated as in v4.1.

MRM v4.3 estimates global horizontal irradiance for overcast and partially cloudy skies as v4.1 does, but uses Eq. (14.26) instead of Eq. (14.14a) under clear skies (i.e. with $SD_m = SD_t$).

MRM v4.4 estimates global horizontal irradiance as in v4.3, and then scales it by the factor $(1 - \rho_g \, \rho_a)^{-1}$. The diffuse values are calculated as in v4.1.

All four sub-versions of MRM v4 coincide with MRM v3 for the cases of partly cloudy skies. However, each one differs from both v2 and v3 in the case of overcast

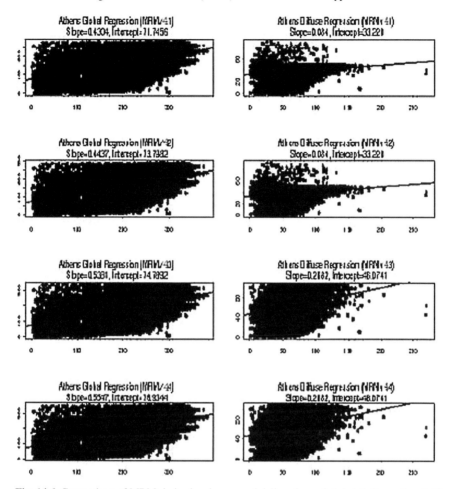

Fig. 14.6 Comparison of MRM-derived and measured daily values of global (left panel) and diffuse (right panel) horizontal irradiation (Whm^{-2}) under all sky conditions for Athens in the period May 1989 - September 1995. Source: Final CliMed report

skies. In the case of clear skies, the innovation is made only in v4.3 and v4.4, so that Eqs. (14.14a) or (14.14b) being used in the previous versions have been replaced by Eq. (14.26) or its scaled version, respectively. By comparing the simulated values derived from these four innovations of MRM with real data, it becomes possible to investigate the influence of multiple ground-atmosphere reflections on the performance of MRM. Similarly, one may also investigate how the above-described modifications of the MRM algorithm (regarding exclusively the cases of overcast and clear skies) influence its performance. MRM v4.4 gives the best performance of all sub-versions, as shown in the Figs. 14.6 and 14.7 taken from the CliMed final report.

During the CliMed project there was collected a rather large volume of data from sites in France, Greece, Italy, Portugal and Spain. A first filtering of the data was

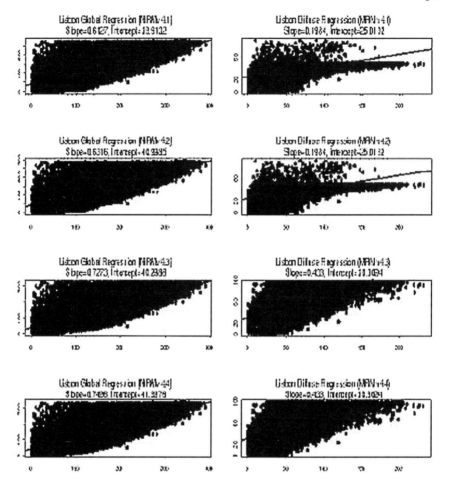

Fig. 14.7 As in Fig. 14.6, but for Lisbon in the period January 1985 - December 1989. Source: Final CliMed report

done by imposing criteria referring to climatic zones, proximity of stations and completeness of the data series. This procedure reduced the number of the acceptable data sets to 614.

All data that passed the above criteria were further passed through quality requirements regarding the span of air temperature, relative humidity, irradiance (global and diffuse) and sunshine duration. The data sets from Athens (314.97° N, 23.72° E, 107 m a.m.s.l.) and Lisbon (38.7° N, 9.1° W, 70 m a.m.s.l.) were the only ones to include hourly values of both global and diffuse horizontal irradiance along with all other meteorological parameters needed for the MRM algorithm. Even so, the values of barometric pressure were missing from both data sets, and, therefore, this parameter was set at 1013.25 hPa throughout. In the case of the Athens' atmosphere, the values of α and β in the Ångström's turbidity formula have been estimated by Kambezidis et al. (1993a) so as to give the mean diurnal variation for each season of the year in the period 1975–1993. This yielded an average annual value for AOT, namely $\delta_a = 0.394$, which was also used in the case of Lisbon because of absence of similar estimates. The period covered by the Athens' data ranged from May 5, 1989 through September 28, 1993, corresponding to 14528 hourly data. The period covered by the Lisbon data ranged from January 1, 1985 through December 31, 1989, corresponding to 22838 hourly data.

Finally, the sample size of the Athens data was reduced to 11848, and that of Lisbon to 20495; the rejected data corresponded to the computed extraterrestrial irradiance values on horizontal plane and found to be smaller than the corresponding observed hourly values of global irradiance. This is a common problem in observed data sets, which may be attributed partly to limitations of the algorithm simulating the extraterrestrial irradiance and partly to inaccuracies in measuring small values of irradiance mainly for $\theta_z \leq 85$ degrees, or yet to the existence of some sky diffuse irradiation before sunrise or after sunset.

Figure 14.6 shows the comparison between the measured global (left) and diffuse (right) horizontal irradiations in Athens (May 1989 – September 1993) in the first row and the MRM-simulated ones from versions 4.1 to 4.4. The x-axis is time, in days, in the whole experimental period.

Figure 14.7 shows the comparison between the measured global (left) and diffuse (right) horizontal irradiations in Lisbon (January 1985 – December 1989) in the first row and the MRM-simulated ones from versions 4.1 to 4.4. The x-axis is time, in days, in the whole experimental period.

From both figures it is seen that the diffuse radiation is not simulated well by all four sub-versions of MRM v4.

2.5 MRM Version 5

This version of the MRM code was completely considered from the beginning for the purpose of this book. The reason was that preliminary runs of MRM v4.4 with available meteorological data (air temperature, relative humidity, barometric pressure and sunshine duration) for various locations in Greece showed a strange behaviour of the global horizontal radiation throughout the year, an observation that had not been made systematically before with the use of the previous MRM versions. Gueymard (2003) in an inter-comparison study employing various broadband models used MRM v4 and found it not to be performing well in relation to other codes. The peculiarity was mainly attributed to the solar geometry routine of the code and the analytical expressions of some transmittance functions (particularly the Rayleigh one). This made ART to try to improve MRM further. The effort resulted in a new version of MRM. Recent knowledge on the subject was successfully incorporated in the model, including the new solar constant of $1366.1 \, \mathrm{Wm^{-2}}$. The description of MRM v5 is given below in a different format than that for the previous versions. Whichever analytical expression for any parameter of the model is not given in this Section, it is implied that MRM v5 uses the corresponding of MRM v4.4. In the latest version, the user can either feed the model with a measured value of l_o (from ozone-sondes, for example) or use the relationships developed for

Greece and Israel in this work or even the original Van Heuklon expression given by Eq. (14.9) above.

Transmittance Functions

The general transmittance function, T_i, for seven atmospheric gases (H_2O, O_3, CO_2, CO, N_2O, CH_4 and O_2) is (Psiloglou et al. 1994, 1995a, 1995b, 1996, 2000):

$$T_i = 1 - \{A' m\, l_i / [(1 + B' m\, l_i)^{C'} + D' m\, l_i]\} \qquad (14.27)$$

where l_i is the vertical column for each of the gases and A', B', C', D' coefficients. The l_i's are used in the analytical expressions of the corresponding transmittance functions (see below). The values of the coefficients are given in Table 14.3.

The only difference in the T_i's between v5 and v4.4 regards the transmittance function for the mixed gases, T_{mg}, which is given by:

$$T_{mg} = T_{CO2}\, T_{CO}\, T_{N2O}\, T_{CH4}\, T_{O2} \qquad (14.28a)$$

where all T's in the right-hand side of Eq. (14.28a) are the transmittance functions of the specific gases; the values of the corresponding l_i's have been considered to be 350, 0.075, 0.28, 1.6 and 2.095×10^5 atm-cm (Psiloglou et al. 1995a, 2000).

It should be noticed that for the estimation of T_o, if l_o, in atm-cm or DU, is not available from in-situ measurements, the Van Heuklon (1979) approximation is used as in Eq. (14.9) for the Northern Hemisphere. Also, for the estimation of T_w, Eqs. (14.20a,b,c,d) are used:

$$l_w = 0.00493\, e_s\, RH / T_d \qquad (14.28b)$$

Section 4 gives an inter-comparison between four methodologies for the calculation of l_w, i.e., by Gates (1962), Paltridge and Platt (1976), Leckner (1978), and Perez et al. (1990). From the results, the choice in using the third expression in MRM v5 is justified.

The Rayleigh scattering transmittance function is (Psiloglou et al. 1995b):

$$T_r = \exp[-0.1128\, m'^{0.8346} (0.9341 - m'^{0.9868} + 0.9391\, m')] \qquad (14.29)$$

Table 14.3 Values of the coefficients A', B', C' and D', in Eq. (14.27)

Gas	A'	B'	C'	D'
H_2O	3.0140	119.300	0.6440	5.8140
O_3	0.2554	6107.260	0.2040	0.4710
CO_2	0.7210	377.890	0.5855	3.1709
CO	0.0062	243.670	0.4246	1.7222
N_2O	0.0326	107.413	0.5501	0.9093
CH_4	0.0192	166.095	0.4221	0.7186
O_2	0.0003	476.934	0.4892	0.1261

T_r in the above equation has a completely different behaviour to that given by Eq. (14.19). This has been one of the main improvements in MRM v5, the others being the use of a new solar geometry routine, new expressions for the absorption and scattering of solar radiation by gases and aerosols and use of the Bird-Hulstrom expression for the estimation of the diffuse horizontal component.

The Mie scattering transmittance function is (Yang et al. 2001):

$$T_a = \exp\left\{-m\beta\left[0.6777 + 0.1464\,m\beta - 0.00626\,(m\beta)^2\right]^{-1.3}\right\} \tag{14.30}$$

where the Ångström's turbidity parameter, β, is in the range 0.05–0.4 for low to high aerosol concentrations. Some indicative values of β are given in Table 14.4 from Iqbal (1983).

Another way of estimating β, if it is not known from measurements, is by using Yang et al.'s (2001) expression, which relates β to the geographical latitude, ϕ, and the altitude of the station, H. This expression is:

$$\beta = \beta' + \Delta\beta \tag{14.31a}$$

$$\beta' = (0.025 + 0.1\cos\phi)\exp(-0.7H/1000) \tag{14.31b}$$

$$\Delta\beta = \pm(0.02 \sim 0.06) \tag{14.31c}$$

where β' represents the annual mean value of turbidity and $\Delta\beta$ the seasonal deviation from the mean, i.e., low values in winter, high values in summer.

The aerosol absorption function, T_{aa}, is (Bird and Hulstrom 1980, 1981a, 1981b):

$$T_{aa} = 1 - 0.1\,(1 - m + m^{1.06})\,(1 - T_a) \tag{14.32a}$$

The expression for the aerosol scattering, T_{as}, is:

$$T_{as} = T_a/T_{aa} \tag{14.32b}$$

Solar Irradiance for Clear Sky (Clear Sky MRM)

From Bird and Hulstrom (1981b), I_b is estimated as in MRM v2, i.e., through Eq. (14.16a). The reason for adopting the expression of Bird and Hulstrom again

Table 14.4 Indicative values of Ångström's turbidity parameter β representing various atmospheric conditions, for different ranges of visibility, V.

Atmospheric condition	β	V [km]
Clean	0.05	340
Clear	0.1	28
Turbid	0.2	11
Very turbid	0.4 - 0.5	< 5

is that the method developed by Muneer in v3 does not seem to be very helpful in the sense that one needs to have long-term series of diffuse and direct beam irradiance measurements in order to compute the coefficients A and C in Eq. (14.24).

As for calculating I_d, this is taken by the relevant expressions developed by Atwater and Brown (1974) (see also Psiloglou et al. 2000):

$$I_{ds} = I_{ex} \cos\theta_z \, T_{aa} \, T_o \, T_w \, T_{mg} \, 0.5 \, (1 - T_{as} \, T_r) \tag{14.33a}$$

$$I_{dm} = (I_b + I_{ds}) \left[\rho_g \, \rho_a / (1 - \rho_g \, \rho_a) \right] \tag{14.33b}$$

$$I_d = I_{ds} + I_{dm} \tag{14.33c}$$

where I_{ds} is the circumsolar diffuse radiation produced by a single-scattering mode of molecules and aerosols, and I_{dm} the diffuse component reflected by the ground and backscattered by the atmosphere, and

$$\rho_a = 0.0685 + 0.16 \, (1 - T_{a1.66}) \tag{14.33d}$$

where $T_{a1.66}$ implies the value of T_a at $m = 1.66$ (or $\theta_z = 53°$).

The global horizontal irradiance is then given (Psiloglou et al. 2000):

$$I_g = I_b + I_d \tag{14.34}$$

Solar Irradiance for Cloudy Skies (Cloudy Sky MRM)

The direct beam component is given in relevance to MRM v2, Eq. (14.23a), by the expression:

$$I'_b = I_b \, T_c \tag{14.35a}$$

$$T_c = k \, (SD_m / SD_t) \tag{14.35b}$$

where k is a coefficient taking values in the range 0.75–1.0. When setting $k = 1$ (which is the usual case) Eq. (14.35a) identifies itself with Eq. (14.23a). Here, SD_t is given by Eq. (14.23d).

The diffuse (I'_d) and global (I'_g) components are then given by (Psiloglou et al. 2000):

$$I'_{ds} = I_{ds} \, T_c + K \, [1 - T_c] \, (I_b + I_{ds}) \tag{14.35c}$$

$$I'_{dm} = (I'_b + I'_{ds}) \left[\rho_g \, \rho'_a / (1 - \rho_g \, \rho'_a) \right] \tag{14.35d}$$

$$I'_d = I'_{ds} + I'_{dm} \tag{14.35e}$$

$$I'_g = I'_b + I'_d \tag{14.35f}$$

$$\rho'_a = 0.0685 + 0.16 \, (1 - T_{a1.66}) + V \, (1 - SD_m / SD_t) \tag{14.35g}$$

Table 14.5 Efficiency of the MRM versions. The grading scale is poor, sufficient, good and excellent

Version	Clear sky	Cloudy Sky	Remarks
1	Sufficient	Poor	Essentially Bird's and Hulstrom's model.
2	Good	Poor	Modified Bird's and Hulstrom's model.
3	Good	Good	New concept in estimating diffuse component.
4.1	Good	Poor	Mixture of v2 and v3.
4.2	Good	Poor	As v4.1, but with introduction of the reflections from ground and atmosphere.
4.3	Good	Poor	Uses the calculations for I_g of v4.1 under cloudy skies and of v4.2 for clear ones.
4.4	Good	Sufficient	As v4.1, but with introduction of the reflections from ground and atmosphere.
5	Excellent	Good	Completely new approach in the design of the algorithm.

where ρ'_a is the albedo of the cloudy sky, K an empirical coefficient given in Table 14.2, for various geographic latitudes (Berland and Danilchenco 1961) and V is a parameter varying between 0.3 and 0.6.

Table 14.5 summarises the advantages and disadvantages of all versions of MRM and serves the purpose of their inter-comparison.

3 Results and Discussion

This section is devoted to the evaluation of the performance of the newly introduced v5 of the MRM algorithm. The means used for such an evaluation are mostly of statistical nature.

Therefore, in order to investigate the accuracy of the algorithm, the Typical Meteorological Years (TMYs) for 3 sites were used, i.e., for Eilat (29.5° N, 34.9° E) and Bed-Dagan (32.0° N, 34.8° E) in Israel, and for Athens (37.97° N, 23.72° E) in Greece. Hourly values of dry bulb air temperature, relative humidity, global and diffuse radiation on horizontal surface (for model comparison) as well as daily sunshine values were available for all three stations. (The sunshine duration has been measured with the classical Campbell-Stokes sunshine recorder or heliograph.) In addition, hourly values of barometric pressure for Athens were available, while for the Israeli sites P = 1000 hPa was set in the input data. For the ozone total column daily values in DU, from TOMS satellite for Athens and Jerusalem (31.78° N, 35.22° E) were used covering the period between July 1996 and September 2004. With these data, corrected Van Heuklon-type equations were developed for the estimation of the total column of ozone in the atmosphere for both areas (see Eq. (14.6)). The area of Jerusalem was selected as the closest site to Eilat and Bed-Dagan with available TOMS ozone total column data for the MRM validation. This validation was performed for all three sites under two scenarios: summertime and whole TMY;

in each scenario the clear sky MRM routine and the cloudy sky one ran for the clear and cloudy days, respectively. It should be mentioned here that a TMY consists of real data covering a number of years. The statistical procedure for the derivation of a TMY selects the most representative months (January to December) from the available data set to make up the typical year.

The Root Mean Square Error (RMSE) and the Mean Bias Error (MBE), both expressed in Wm^{-2} and in percent of the measured mean values, were used as indicators for the MRM performance, i.e.:

$$RMSE \text{ (in } Wm^{-2}) = [\Sigma (I_{gm} - I_{gc})^2/N]^{1/2} \qquad (14.36a)$$

$$RMSE \text{ (in \%)} = [RMSE \text{ (in } Wm^{-2})/(\Sigma I_{gm}/N)] \times 100 \qquad (14.36b)$$

$$MBE \text{ (in } Wm^{-2}) = \Sigma (I_{gm} - I_{gc})/N \qquad (14.37a)$$

$$MBE \text{ (in \%)} = [MBE \text{ (in } Wm^{-2})/(\Sigma I_{gm}/N)] \times 100 \qquad (14.37b)$$

where I_{gm} and I_{gc} are measured and model-estimated values of global radiation, N is the number of hourly data points during the examined period and the summation is performed N times. The same statistical estimators apply to the diffuse component, if in the above expressions I_{gm} and I_{gc} are replaced by I_{dm} and I_{dc}, respectively.

In Fig. 14.8, a comparison of the new expressions for the estimation of the ozone total column daily values, in DU, with the original Van Heuklon one, for Athens and Jerusalem, are given. The data points are measurements from the TOMS spectroradiometer on-board the Earth Probe satellite, the bold solid lines are the new expressions and the grey solid ones the originally derived by Van Heuklon. A remarkable difference in amplitude and a smaller difference in phase are observed. This is probably due to the fact that the Van Heuklon general formula was derived in late '70s, when the ozone depletion had just started. Nowadays, this phenomenon is well monitored and the "ozone hole" is found at other parts of the world than the original locations over the Antarctica and later the Arctic. A preliminary sensitivity analysis showed a 3% improvement to the MRM performance by using the modified Van Heuklon expression.

3.1 Clear Sky MRM

From the available data (one TMY for each site), the "clear" or almost "clear days" in each month of the TMY were selected for the validation of the model. A day was characterised as "clear" if the measured sunshine duration (SD_m) was greater than or equal to the 95% of the maximum sunshine duration (SD_t) of that day. For this sub-dataset the MRM ran using its clear sky routine. This was achieved by equating the daily sunshine durations to the theoretical ones at the site and using these values in the input data file. Apart from validating MRM in each month of the TMY, the algorithm ran for the clear days of the summer (June to September) as well as those

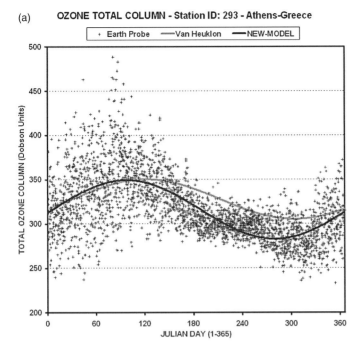

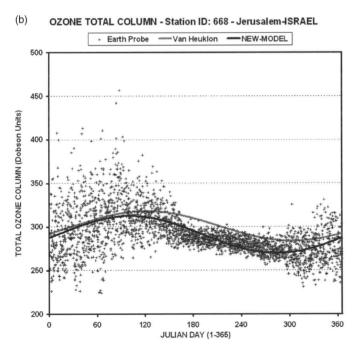

Fig. 14.8 Comparison of the proposed equations (black solid line) for the estimation of total ozone column with the originally derived by Van Heuklon (grey solid line) for (a) Athens, Greece and (b) Jerusalem, Israel

in the whole TMY. These scenarios were adopted for just an inter-comparison of the MRM efficiency on clear days over different time spans (hours, days, months, years).

The RSME and MBE results, in Wm^{-2}, for the global and diffuse horizontal radiation components along with their mean monthly measured values are given in Tables 14.6–14.8. The brackets in the RMSE and MBE columns indicate their values in %.

From the Tables 14.6–14.8 it is easily seen that the MBE (%), a measure of the overestimation $(-)$ or underestimation $(+)$ of the computed values with respect to the measured ones, lies between -6.3% and $+10.7\%$ for the global component and between -31.7% and $+14.6\%$ for the diffuse one on the clear sky summer days for the 3 sites considered. These figures become -9.3% and $+16.6\%$ for the global and -65.4% and $+20.2\%$ for the diffuse on the clear sky TMY days. The bias of

Table 14.6 RSMEs and MBEs for the mean hourly global and diffuse horizontal radiation components and their mean monthly measured values for "clear days" in Eilat, Israel

Month	Mean diffuse rad. $[Wm^{-2}]$	Mean global rad. $[Wm^{-2}]$	RMSE $[Wm^{-2}$ (%)]		MBE $[Wm^{-2}$ (%)]	
			Diffuse rad.	Global rad.	Diffuse rad.	Global rad.
1	65.8	384.7	24.9	22.8	-21.9	-0.9
			(314.9)	(5.9)	(-33.3)	(-0.2)
2	514.2	424.5	35.9	16.7	-31.0	-14.4
			(62.8)	(3.9)	(-54.3)	(-1.7)
3	78.7	496.6	22.4	26.6	-16.6	-16.3
			(28.5)	(5.4)	(-21.1)	(-3.3)
4	88.1	554.6	16.4	36.5	-12.3	-29.4
			(18.6)	(6.6)	(-13.9)	(-5.3)
6	79.6	562.2	22.7	39.9	-114.7	-15.8
			(28.5)	(14.1)	(-22.3)	(-2.8)
7	89.1	543.1	19.8	41.7	-14.5	-29.2
			(22.3)	(14.7)	(-8.4)	(-5.4)
8	111.3	5714.8	22.3	34.0	15.0	16.2
			(20.0)	(5.9)	(13.5)	(2.8)
9	90.5	530.2	18.3	42.4	-1.2	13.7
			(20.2)	(8.0)	(-1.3)	(2.6)
10	72.1	458.0	26.8	214.3	-21.7	-8.9
			(314.2)	(6.0)	(-30.1)	(-1.9)
11	72.3	396.3	15.9	19.4	-13.1	-5.9
			(22.0)	(4.9)	(-18.1)	(-1.5)
12	52.6	349.6	33.3	21.4	-29.6	-2.9
			(63.3)	(6.1)	(-56.3)	(-0.8)
Summer	92.6	553.3	20.8	39.5	-2.9	-3.8
			(22.4)	(14.1)	(-3.1)	(-0.7)
Year	814.0	518.4	23.0	35.3	-14.0	-4.7
			(26.5)	(6.8)	(-8.0)	(-0.9)

Table 14.7 As in Table 14.6, but for Bed-Dagan, Israel

Month	Mean diffuse rad. [Wm^{-2}]	Mean global rad. [Wm^{-2}]	RMSE [Wm^{-2} (%)]		MBE [Wm^{-2} (%)]	
			Diffuse rad.	Global rad.	Diffuse rad.	Global rad.
1	60.6	375.3	23.3 (38.5)	26.0 (6.9)	−16.9 (−27.9)	8.0 (2.1)
2	69.2	443.7	17.4 (25.1)	22.2 (5.0)	−11.9 (−17.1)	11.8 (2.7)
3	67.4	489.1	28.7 (42.6)	29.1 (6.0)	−18.4 (−27.2)	−114.5 (−3.6)
5	77.2	560.0	24.9 (32.3)	52.3 (9.3)	−10.2 (−13.9)	−14.1 (−2.5)
6	70.8	566.2	24.6 (34.7)	45.3 (8.0)	−16.9 (−23.8)	−12.1 (−2.1)
7	88.9	538.5	22.3 (25.0)	42.5 (7.9)	2.9 (3.3)	−27.4 (−5.1)
8	76.1	542.3	23.5 (30.9)	26.0 (4.8)	−8.8 (−11.5)	−2.9 (−0.5)
9	84.0	471.1	19.1 (22.7)	60.8 (12.9)	3.1 (3.7)	−29.8 (−6.3)
10	72.8	411.4	21.3 (29.3)	80.0 (19.5)	−9.5 (−13.1)	−38.5 (−9.3)
11	67.1	338.6	30.9 (46.0)	89.3 (26.4)	−5.6 (−8.4)	−21.7 (−6.4)
12	65.4	313.1	14.8 (22.7)	55.9 (17.9)	−9.7 (−14.8)	−29.3 (−9.3)
Summer	80.0	529.5	22.4 (27.9)	43.7 (8.2)	−19.7 (−24.6)	−18.1 (−3.4)
Year	75.3	479.1	23.6 (31.3)	53.7 (11.2)	−7.5 (−10.0)	−17.3 (−3.6)

the global horizontal radiation is in totally acceptable limits for energy oriented applications and stands on firm grounds even for research studies, wherever only this component is of importance. The higher bias for the diffuse horizontal radiation is, in every broadband model, unavoidable as this strongly depends on the accurate determination of the atmospheric composition (and, therefore, turbidity) into all directions of the sky vault.

The RMSE estimator, in %, a measure of the power contained in the estimated values in excess to that possessed by the real ones, lies between 4.8% and 17.9% for the global component and between 14.3% and 39.5% for the diffuse on the clear sky summer days for the three sites. In the case of the clear TMY days the above figures become 14.3% and 63.3% for the diffuse and 3.9% and 26.4% for the global component. Similar conclusions for the MBE to those for the summer clear days can be drawn here for the three sites.

Figures 14.9–14.11 present the estimated vs. measured values of the global radiation on "clear days" for all three sites (a) during the summer period, and (b) the whole TMY. The bulk of data points along the y=x line displays the effectiveness of the MRM code.

Table 14.8 As in Table 14.6, but for Athens, Greece

Month	Mean diffuse rad. [Wm^{-2}]	Mean global rad. [Wm^{-2}]	RMSE [Wm^{-2} (%)]		MBE [Wm^{-2} (%)]	
			Diffuse rad.	Global rad.	Diffuse rad.	Global rad.
1	132.6	294.2	35.3 (26.6)	61.5 (20.9)	23.8 (17.9)	23.7 (8.0)
2	89.9	348.9	32.5 (36.1)	85.8 (24.6)	−22.6 (−25.1)	58.0 (16.6)
4	108.2	509.3	44.8 (41.4)	31.2 (6.1)	−35.1 (−32.5)	12.5 (2.5)
5	129.8	532.7	19.6 (15.1)	79.1 (14.8)	−3.7 (−2.9)	42.3 (7.9)
6	103.0	531.4	40.7 (39.5)	51.5 (9.69)	−32.6 (−31.7)	27.8 (5.2)
7	148.2	498.0	21.1 (14.3)	44.8 (9.0)	15.3 (10.4)	12.0 (2.4)
8	162.8	545.0	31.3 (19.2)	97.3 (17.9)	23.7 (14.6)	58.3 (10.7)
9	151.5	452.6	26.5 (17.5)	58.7 (13.0)	19.3 (12.7)	13.9 (3.1)
10	147.2	378.8	32.6 (22.1)	59.6 (15.7)	29.7 (20.2)	49.6 (13.1)
11	137.0	314.3	30.4 (22.2)	54.6 (17.4)	25.6 (18.7)	33.0 (10.5)
12	64.2	270.8	45.9 (71.5)	45.6 (16.9)	−42.0 (−65.4)	27.7 (10.2)
Summer	141.4	506.8	29.9 (21.1)	63.1 (12.4)	6.4 (4.5)	28.0 (5.5)
Year	133.3	470.4	32.5 (24.3)	61.4 (13.0)	2.3 (1.7)	26.9 (5.7)

3.2 Cloudy Sky MRM

From the available data (one TMY for each site), all days (clear and cloudy) were selected for the validation of the model under the same scenarios as in the case of the clear sky MRM, i.e., summertime and whole TMY. For these datasets the MRM ran using its cloudy sky routine. This was achieved by entering the observed daily sunshine durations in the input data file. Figures 14.12–14.14 show the performance of MRM in all sky conditions. The tuft of the data points in the diagrams for all the TMY days displays a dispersion, which is accounted for the greater RMSE of the diffuse component in comparison to the corresponding one for the clear days scenario.

The RMSE and MBE values for the entire TMY of Eilat are:

- RMSE: 17.8 % (global radiation) and 41.9 % (diffuse radiation).
- MBE: +1.2 % (global radiation) and +42.2 % (diffuse radiation).
- Mean: 456.6 Wm^{-2} (global radiation) and 111.8 Wm^{-2} (diffuse radiation).

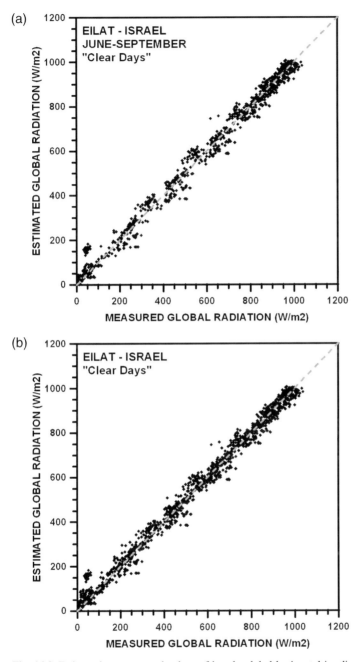

Fig. 14.9 Estimated vs. measured values of hourly global horizontal irradiance for "clear days" (a) during the summer period, and (b) in the whole TMY for Eilat, Israel. The y = x (dashed) line shows the degree of correlation between the MRM-estimated and the measured values

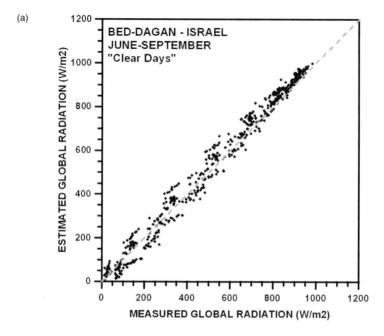

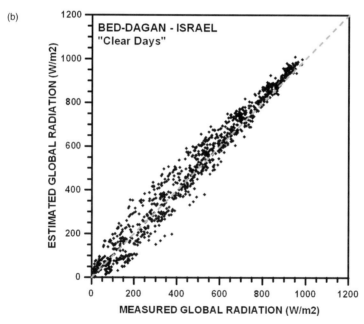

Fig. 14.10 As in Fig. 14.9, but for Bed-Dagan, Israel

(a)

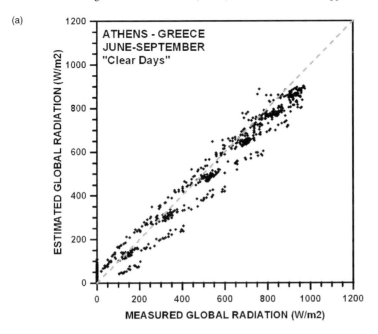

(b)

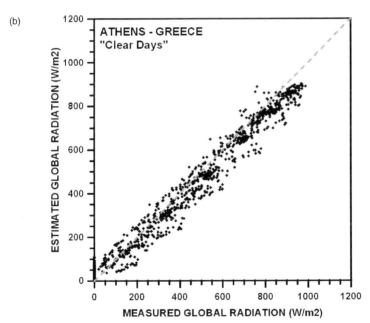

Fig. 14.11 As in Fig. 14.9, but for Athens, Greece

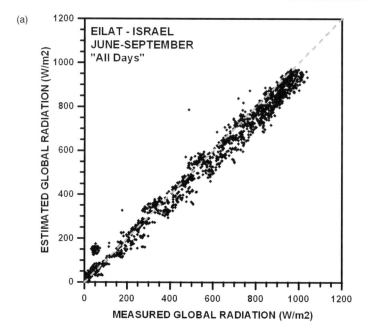

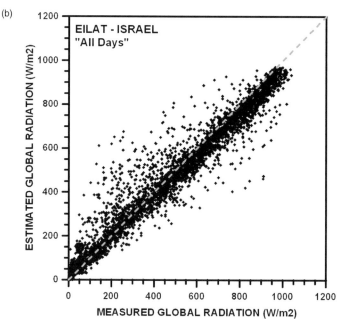

Fig. 14.12 Estimated vs. measured values of global horizontal irradiance for "all days" (a) during the summer period, and (b) in the whole TMY for Eilat, Israel. The y = x (dashed) line shows the degree of correlation between the MRM-estimated and the measured values

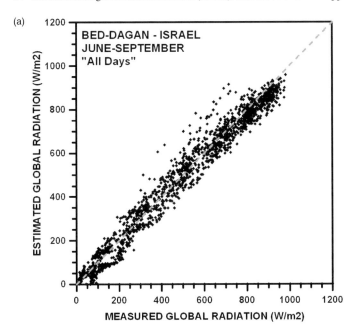

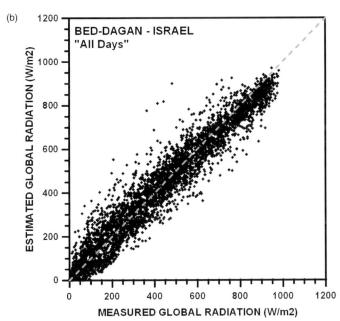

Fig. 14.13 As in Fig. 14.12, but for Bed-Dagan, Israel

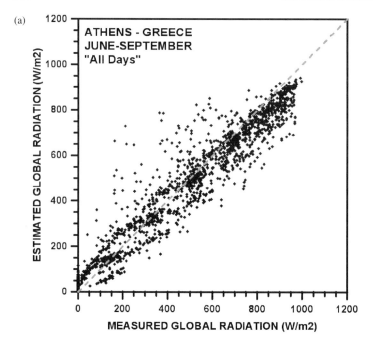

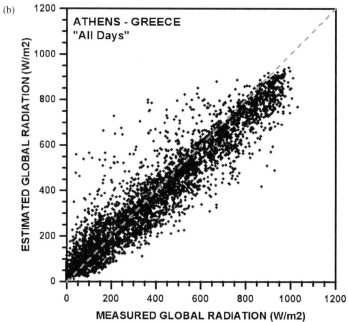

Fig. 14.14 As in Fig. 14.12, but for Athens, Greece

The above values of the statistical estimators for Bed-Dagan are:

- RMSE: 16.1 % (global radiation) and 39.5 % (diffuse radiation).
- MBE: +1.0 % (global radiation) and −0.9 % (diffuse radiation).
- Mean: $420.6\,\mathrm{Wm}^{-2}$ (global radiation) and $106.2\,\mathrm{Wm}^{-2}$ (diffuse radiation).

The same statistical estimators for Athens are as follows:

- RMSE: 39.2 % (global radiation) and 28.3 % (diffuse radiation).
- MBE: +11.7 % (global radiation) and +5.5 % (diffuse radiation).
- Mean: $368.7\,\mathrm{Wm}^{-2}$ (global radiation) and $144.2\,\mathrm{Wm}^{-2}$ (diffuse radiation).

4 Inter-Comparison of Precipitable Water Expressions

An inter-comparison is given here between four different approximate relationships that estimate the amount of precipitable water, l_w, in cm. These are:

4.1 Gates' Formula

Gates (1962) gave Eq. (14.11) for estimating l_{wG}, i.e.,:

$$l_{wG} = 0.23\,e_m\,10^{-H/22000} \tag{14.38a}$$

where e_m is the partial water vapour pressure, in mmHg, at the station's height, H, in metres. By combining (14.38a) and (14.20d) and converting hPa of e_s into mmHg, one gets:

$$l_{wG} = 0.17\,e_s\,(RH/100\,T_d)\,10^{-H/22000} \tag{14.38b}$$

where e_s is the saturation water vapour pressure, in hPa, T_d the dry bulb air temperature, in K, and RH the relative humidity, in %; e_s is estimated through Eq. (14.20c). For the calculations to follow, H is taken as 107 m a.m.s.l. This site corresponds to the Actinometric Station of NOA (ASNOA).

4.2 Paltridge's and Platt's Relationship

These authors in their 1976 book suggest the following general equation, which brings the calculated precipitable water, l'_w, at the station's height to that at the reference conditions of 1013.25 hPa (sea level pressure) and 273.15 K, i.e.,:

$$l_{wPP} = l'_w\,(P/1013.25)^{0.75}\,(273.15/T_d)^{0.5} \tag{14.39}$$

In Eq. (14.39) the expression for l'_w is taken here from Leckner's (1978), following expression.

4.3 Leckner's Regression

Leckner (1978) used the following relationship:

$$l_{wL} = 0.493\,e_s\,RH/T_d \qquad (14.40)$$

where RH is the relative humidity, in %, and T_d is the air temperature, in K, at the station's height.

4.4 Perez's Relationship

Perez et al. (1990) gave the following approximate formula for l_{wP}:

$$l_{wP} = \exp(0.07\,t_d - 0.075) \qquad (14.41)$$

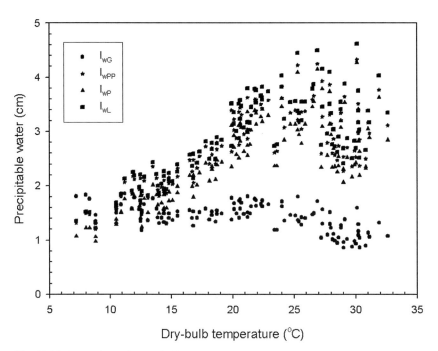

Fig. 14.15 Comparison of precipitable water estimated by various expressions vs. air temperature at ASNOA

where t_d is the dry bulb air temperature, in degrees Celcius.

To perform an inter-comparison between the above-mentioned equations, data of dry bulb temperature, relative humidity and barometric pressure have taken from the records of ASNOA. Figure 14.15 shows the various l_w values from Eqs. (14.38b), (14.39), (14.40) and (14.41).

It is seen from Fig. 14.15 that the Gates' formula gives very low values, while the other three methodologies present comparable estimations of l_w. MRM v5 chose to use the Leckner's equation, as this has been widely used in the international literature and its data points in Fig. 14.15 lie within the bundle of data points of l_{wPP} and l_{wP}.

5 Conclusions

A new version of the Meteorological Radiation Model (MRM v5) has been derived, which is able to estimate solar irradiation values from most commonly available meteorological parameters, viz. barometric pressure, relative humidity, ambient temperature and sunshine duration. This version is completely new as regards its predecessors (v1-v4) in the sense that a new approach to building it was followed. The international literature was extensively searched for finding the best performing expressions in the various parts of the model (transmittance, solar geometry, air mass, gas absorption and aerosol scattering optical depths), while some expressions derived by ART (for the total ozone column, the aerosol transmittance and other gases) have been incorporated.

The MRM v5 was tested with Athens, Greece, and Eilat and Bed-Dagan, Israel, data (Typical Meteorological Years consisting of hourly values of dry bulb temperature, barometric pressure, humidity and daily values of sunshine duration). From the deployed statistics, it was found that this version of MRM performs better than its predecessors (v1-v4) in both global and diffuse radiation components.

The experience with MRM v5 seems to propose further improvement to any similar type of models – i.e. models that rely on common meteorological data for estimating solar irradiation values. The improvement is emphasised on the development of a better (regional) description of the diffuse (or direct beam) irradiation characteristics, as well as to be able to break down daily sunshine values into hourly values correctly.

From all the above, it seems that the MRM v5 is a promising broadband model that can easily be used at places where no solar radiation stations exist or to derive solar radiation data series with increased reliability at locations where meteorological stations operate. In this sense, a region can be filled with "artificial" solar data, which can result in a Solar Atlas for the area for energy applications. The other advantage of the MRM is that it can be used as a tool to fill solar radiation data gaps in a series of measurements. That was the case for the Solar Data Task Group of CIBSE for various solar radiation stations around UK.

Acknowledgements The CliMed (*Climatic Synthetic Series for the Mediterranean Belt*) project was partially funded by the European Commission (DG-XII), under the JOULE-III project, contract no. JOR3CT960042, in which project ART participated.

Accompanying Material

The Chapter is accompanied by a CD-ROM, where the MRM code is included in an executable form together with some guide for the reader. For the sake of easiness a working example (input data and results) has been prepared.

References

Ångström A (1929) On the atmospheric transmission of sun radiation and on dust in the air. Geografis. Annal. 2: 156–166.

Ångström A (1930) On the atmospheric transmission of sun radiation. Geografis. Annal. 2 & 3: 130–159.

Atwater MA and Brown PS (1974) Numerical computations of the latitudinal variations of solar radiation for an atmosphere of varying opacity. J. Appl. Meteorol. 13: 289–2914.

Berland G and Danilchenco VY (1961) The continental distribution of solar radiation. Gidrometeorzdat, St. Peterburg.

Bird RE and Hulstrom RL (1979) Application of Monte Carlo technique to insolation characterization and prediction. US SERI Tech. Report TR-642-761: 38.

Bird RE and Hulstrom RL (1980) Direct insolation models. US SERI Tech. Report TR-335–344.

Bird RE and Hulstrom RL (1981a) Review, evaluation and improvement of direct irradiance models. Trans. ASME, J. Sol. Energy Eng. 103: 182–192.

Bird RE and Hulstrom RL (1981b) A simplified clear-sky model for the direct and diffuse insolation on horizontal surfaces US SERI Tech. Report TR-642-761: 38.

Dave JV (1979) Extensive data sets of the diffuse radiation in realistic atmospheric models with aerosols and common absorbing gases. Solar Energy 21: 361–369.

Davies JA, Schertzer W and Nunez M (1975) Estimating global solar radiation. Boundary-Layer Meteorol. 9: 33–52.

Duffie JA and Beckman WA (1980) Solar engineering of thermal processes. J. Wiley, New York.

ESRA-European Solar Radiation Atlas (1989) Commission of the European Communities, version I.

Gates DM (1962) Energy exchange in the biosphere. Harper & Row, New York.

Gueymard C (1993) Assessment of the accuracy and computing speed of simplified saturation vapour equations using a new reference dataset. J. Appl. Meteorol. 32: 1294–1300.

Gueymard C (1995) SMARTS2, a simple model of the atmospheric radiative transfer of sunshine: algorithms and performance assessment. Rep. FSEC-PF-270-95, Florida Solar Energy Center, Cocoa, USA.

Gueymard C (2003) Direct solar transmittance and irradiance predictions with broadband model. Part I: detailed theoretical performance assessment. Solar Energy 74: 355–379.

Gul MS, Muneer T and Kambezidis HD (1998) Models for obtaining solar radiation from other meteorological data. Solar Energy 64: 99–108.

Iqbal M (1983) An introduction to solar radiation. Academic Press, New York.

Kambezidis HD and Papanikolaou NS (1989) Total solar irradiance flux through inclined surfaces with arbitrary orientation in Greece: comparison between measurements and models. In: XIV Assembly of EGS, pp. 13–17, Barcelona, Spain.

Kambezidis HD and Papanikolaou NS (1990a) Total solar irradiance on tilted planes in Greece. Technika Chronika B 10: 55-70 (in Greek).

Kambezidis HD and Papanikolaou NS (1990b) Solar position and atmospheric refraction. Solar Energy 44: 143–144.

Kambezidis HD and Tsangrassoulis AE (1993) Solar position and right ascension. Solar Energy 50: 415–416.

Kambezidis HD, Founda DH and Papanikolaou NS (1993a) Linke and Unsworth-Monteith turbidity parameters in Athens. Q.J. Roy. Meteorol. Soc. 119: 367–374.

Kambezidis HD, Psiloglou BE, Tsangrassoulis AE, Logothetis MA, Sakellariou NK and Balaras CA (1993b) A meteorology to give solar radiation on titled plane from meteorological data. In: ISES World Congress, Farkas J. (ed), Budapest, Hungary, pp. 99–104.

Kambezidis HD, Psiloglou BE and Synodinou BM (1997) Comparison between measurements and models of daily total irradiation on tilted surfaces in Athens, Greece. Renew. Energy 10: 505–518.

Kambezidis HD (1997) Estimation of sunrise and sunset hours for location on flat and complex terrain: review and advancement. Renew. Energy 11: 485–494.

Kambezidis HD (1998) The "Meteorological Radiation Model". Bull. Hell. Assoc. Chart. Mech.-Electr. Engineers 3/2: 38–42 (in Greek).

Kambezidis HD and Badescu V (2000) MRM: a new solar radiation computing model. Application to Romania. In: VII Conf. "Efficiency, comfort, energy preservation and environmental protection", CONSPRESS (publ.), pp. 195–199.

Kasten F (1966) A new table and approximate formula for relative optical air mass. Arch. Meteorol. Geophys. Bioklimatol. B 14: 206–223.

Kasten F and Young AT (1989) Revised optical air mass tables and approximation formula. Appl. Optics 28: 124–1214.

Lacis AA and Hansen JE (1974) A parameterization for the absorption of solar radiation in the earth's atmosphere. J. Atmos. Sci. 31: 118–132.

Leckner B (1978) Spectral distribution of solar radiation at the Earth's surface-elements of a model. Solar Energy 20: 443–450.

McClatchey RA and Selby JE (1972) Atmospheric transmittance from 0.25 to 38.5 μm: computer code LOWTRAN-2. Air Force Cambridge Laboratories, AFCRL-72-0745, Environment Research Paper, pp. 4214.

Muir LR (1983) Comments on "The effect of the atmospheric refraction in the solar azimuth". Solar Energy 30: 295.

Muneer T, Gul M, Kambezidis HD and Alwinkle S (1996) An all-sky solar meteorological radiation model for the United Kingdom. In: CIBSE/ASHRAE Joint National Conf., CIBSE/ASHRAE (Eds), Harrogate, UK, pp. 271–280.

Muneer T (1997) Solar radiation and daylight models for the energy efficient design of buildings. 1st edn, Architectural Press, pp. 65–70.

Muneer T, Gul MS and Kambezidis HD (1997) Solar radiation models based on meteorological data. In Proc. ISES World Congress, Taegon, Korea.

Muneer T, Gul M and Kambezidis HD (1998) Evaluation of all-sky meteorological model against long-term measured hourly data. Energy Conv. & Manag. 39: 303–3114.

Paltridge GW and Platt CMR (1976) Radiative processes in meteorology and climatology. American Elsevier, New York.

Perez R, Ineichen P, Seals R, Michalsky J and Stewart R (1990) Modeling daylight availability and irradiance components from direct and global irradiance. Solar Energy 44: 271–289.

Pisimanis DK, Notaridou VA and Lalas DP (1987) Estimating direct, diffuse and global solar radiation on an arbitrarily. Solar Energy 39: 159–172.

Psiloglou BE and Kambezidis HD (2007) Performance of the meteorological radiation model during the solar eclipse of 29 March 2006. Special issue on: The total solar eclipse of 2006 and its effects on the environment. In: Zerefos C, Mihalopoulos N, Monks P (eds). Atmospheric Chemistry and Physics 7: 6047–6059.

Psiloglou BE, Santamouris M and Asimakopoulos DN (1994) On the atmospheric water vapour transmission function for solar radiation models. Solar Energy 53: 445–453.

Psiloglou BE, Santamouris M and Asimakopoulos DN (1995a) Predicting the broadband transmittance of the uniformly-mixed gases (CO_2, CO, N_2O, CH_4 and O_2) in the atmosphere for solar radiation models. Renewable Energy 6: 63–70.

Psiloglou BE, Santamouris M and Asimakopoulos DN (1995b) On broadband Rayleigh scattering in the atmosphere for solar radiation modelling. Renewable Energy 6: 429–433.

Psiloglou BE, Santamouris M, Varotsos C and Asimakopoulos DN (1996) A new parameterisation of the integral ozone transmission. Solar Energy 56: 573–581.

Psiloglou BE, Santamouris M and Asimakopoulos DN (2000) Atmospheric broadband model for computation of solar radiation at the Earth's surface. Application to Mediterranean climate. Pure Appl. Geophys. 157: 829–860.

Settle EP and Fenn RW (1975) Models of the atmospheric aerosol and their optical properties. In Proc. AGARD Conf. No. 183 on ''Optical propagation in the atmosphere'', 2.1–2.16.

Van Heuklon TK (1979) Estimating atmospheric ozone for solar radiation models. Solar Energy 22: 63–68.

Walraven R (1978) Calculating the position of the sun. Solar Energy 20: 393–3914.

Watt D (1978) On the nature and distribution of solar radiation. H CP/T2552-01 US Dept. of Energy.

Wilkinson BJ (1981) An improved FORTRAN program for the rapid calculation of the solar position. Solar Energy 27: 67–68.

Yang K, Huang GW and Tamai N (2001) A hybrid model for estimation of global solar radiation. Solar Energy 70: 13–22.

Chapter 15
Chain of Algorithms to Compute Hourly Radiation Data on Inclined Planes used in Meteonorm

Jan Remund

1 Introduction

Meteonorm Version 6.0 (Edition 2007) (Remund et al. 2007) is a global climatological database. It's especially designed (but not only) for planners of active solar systems like PV plants or solar thermal systems. These planners do generally use statistical values of meteorological data. Planning is based mainly on a long term forecast of irradiance and other parameters. Nowadays many planners use simulation software (especially for more complex projects).

Most of the software need as input so-called typical years. Such typical years contain hourly data for one year or more (8760 lines of input per year) of global radiation and temperature and sometimes further parameters like wind speed. There are three main types of such time-series: measured values, typical meteorological years (TMY) or Design Reference Years (DRY) and mathematically generated time series. Each of them have their advantages and disadvantages:

- Measured values (ground and satellite data). They give the most precise information, but have several disadvantages: there are only few ground stations existing worldwide with the whole set of information, at least 10 years of data is needed and it's expensive and time consuming to get them. From satellite data only radiation parameters can be derived.
- TMY types of data include variations of several years in one year and so only a data set of one year is needed for simulation. Depending on the application, they are more or less adapted (Argiriou et al. 1999). They are site dependent. Generally TMY reproduce well the 10-years statistical distributions of the original data; in particular, extreme values are included. In mountainous regions with high horizons solar radiation parameters are only representative for small areas due to strong variations of horizon line from place to place.

Jan Remund

Meteotest, Bern, Switzerland, e-mail: remund@meteotest.ch

- Mathematically generated time series. There are two subgroups: stochastic methods (Markov chains, autoregressive processes) and Fourier analyses. The advantage of the mathematically generated time series is that they are in principle site independent and can be used for any place. The statistical distribution and the auto- and cross-correlations are designed to fit measured values as much as possible. One year includes information of several ones as in TMYs. Generally mean monthly distributions are achieved (mean extreme values).

Meteonorm belongs to the group of mathematically generated time series (first subgroup). The reason for choosing this type of model was that it could provide data for any site in the world (together with interpolation of monthly means) and the time series correspond to typical values.

The chain of algorithms has been developed over the last decade and enhanced step by step. Earlier versions have been described in Remund et al. (1998) and Remund and Kunz (2003c).

2 Chain of Algorithms

2.1 Aim of the Chain

Algorithms linked to form focused chains of algorithms are needed to fill gaps between generally available data resources and the parameters requested by solar applications. Based on measurements and/or interpolation worldwide monthly mean values of global radiation are available. But most applications need hourly time series of at least global radiation. This information gap can be filled with an algorithmic chains resource. The missing parameters are mainly values that are not stored (too many values), not available (not public) or not measured (requiring too complicated equipment). The main missing parameter is the global radiation on inclined planes. The diffuse radiation part is needed to calculate this value. Frequently only monthly values of climatic data are available only for certain parts of the world, so techniques to generate hourly values world wide are required.

2.2 Definition of the Meteonorm Chain

The basic inputs to the chain are monthly mean values of the Linke turbidity factor and global radiation. The outputs of the chain are hourly values of global radiation on inclined planes. This is achieved via stochastic generation of daily and hourly values of global radiation by splitting global radiation into beam and diffuse radiation and finally calculating the radiation on inclined planes (Table 15.1 and Fig. 15.1). The stochastic process leads to an hourly dataset of a statistically average year with average mean, minimum and maximum values.

Table 15.1 Models used in the chain of algorithms

Model	Parameters	Model reference
Solar geometry	div.	Bourges 1985
Clear sky radiation	Beam	Rigollier et al. 2000; Remund et al. 2003a
	Diffuse	Rigollier et al. 2000; Remund et al. 2003a
	Global	Global + diffuse
Daily profile	Global	Remund et al. 2003a
Stochastic generation	Global day	Aguiar et al. 1988; Remund et al. 2003a
	Global hour	Aguiar and Collares-Pereira 1992; Remund et al. 2003a
Rad. Separation	Diffuse/beam	Perez et al. 1991
Tilted planes	Global/diffuse	Perez et al. 1986
High horizons	Global/diffuse	Remund et al. 1998

The following description is concentrated on new or corrected models. When using a chain of algorithms, it is an important fact, that one part of the chain can influence other parts (further down the line). In the test this fact was looked at in particular. Additional parameters like temperature, dewpoint temperature and wind speed also available in Meteonorm are not described in this paper. In the accompanying CD-ROM the reader can find the demo version of Meteonorm as well as the description of the software and the theory.

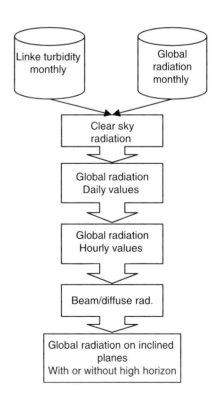

Fig. 15.1 Models used in chain of algorithms

3 Clear Sky Radiation

In Meteonorm Version 6.0 a slightly modified European Solar Radiation Atlas (ESRA) clear sky radiation model is used (Rigollier et al. 2000). The changes have been presented in Remund et al. (2003b).

Linke turbidity (TL) is used for input of the ESRA clear sky model. For version 6.0 a new turbidity climatology has been included. It's based on ground measurements of Aeronet (Holben et al. 2001) and satellite measurements of MISR and MODIS (NASA 2007) for the years 2000–2006.

During validation of downstream models, it was detected, that the obtained values of TL are too high and therefore the clear sky radiation too low. The reason for this was not examined in detail, but similar observations have been reported by Duerr and Ineichen (2007). The monthly mean Linke turbidity (TL_m) is lowered with Eq. (15.1):

$$TL'_m = TL_m \cdot (1.133 - 0.0667 \cdot TL_m) \qquad (15.1)$$

High turbidity values are reduced more than lower values. For mean conditions at mid latitudes and industrialized regions like Europe with Linke turbidity of about 5, the value is lowered by 20% to a value of 4.

Additionally it was detected, that with varied turbidity values the observed distribution of clear sky conditions could be matched better. Also models producing beam radiation gave better results, when using varied turbidities. By default the daily Linke turbidity (TL_d) values are varied stochastically (optionally it can be set constant) (Eq. 15.2).

$$TL_d(d) = \phi_1 \cdot TL_d(d-1) + r$$
$$\phi_1 = 0.7$$
$$\sigma(TL'_m) = 0.1 \cdot TL'_m$$
$$\sigma' = \sigma \cdot \left(1 - \phi_1{}^2\right)^{0.5}$$
$$r = N(0, \sigma')$$
$$TL'_m \cdot 0.75 < TL_d < TL'_m \cdot 1.2 \qquad (15.2)$$

where ϕ_1 ist the first order autocorrelation, $\sigma(TL_m)$ the standard deviation of TL_m perturbations depending on monthly means of TL_m, σ' standard deviation of the normally distributed random function and r the normally distributed random variable with expected value 0 and standard deviation σ'.

4 Monthly Means of Global Radiation

Meteonorm Version 6.0 contains monthly mean values of global radiation (G_h) of several databases. The most important and extensive database is the Global Energy

Balance Archive (GEBA) (Gilgen et al. 1998). A total of 452 sites are included. The main time period is 1981–2000.

With help of spatial interpolation models the monthly means are calculated for any place. Two different models are used for this: If the radiation network is dense enough (the nearest site is less than 50 km away) a kind of Shepard's gravity interpolation (Lefèvre et al. 2001) is used (Eq. 15.3).

$$G_h(x) = \sum w_i G_h(x_i)$$
$$w_i = \left[(1-\delta_i)/\delta_i^2\right]/\sum w_k \text{ with}$$
$$\delta_i = d_i/R \text{ for } d_i < R$$
$$w_i = 0 \text{ otherwise}$$
$$d_i^2 = f_{NS}^2 \cdot \left\{s^2 + [v \cdot (z_2 - z_1)]^2\right\}$$
$$\text{for } z_2 - z_1 < 1600\,m$$
$$f_{NS} = 1 + 0.3 \cdot |\Phi_2 - \Phi_1| \cdot \left[1 + (\sin\Phi_1 + \sin\Phi_2)/2\right] \tag{15.3}$$

where w_i is the weight i, R the search radius (max. 2000 km), v the vertical scale factor (150), s the horizontal (geodetic) distance [km], z_1 and z_2 the altitudes of the sites [km], i the number of sites (maximum 6) and Φ_1 and Φ_2 the latitudes of the two points.

If the nearest site is more than 300 km away, a precalculated map with a grid resolution of $1/3°$ (37×37 km at the equator) based on ground stations and geostationary satellites is used (Remund et al. 2003c). If the distance of the nearest site is between 50 and 300 km a mixture of both informations is used.

4.1 Validation

The accuracy of this interpolation comes to a global mean of $15\,W/m^2$ (8%) for monthly values and $11\,W/m^2$ (6%) for yearly values (estimated by calculating the crosscorrelation of a control sample).

5 Daily Values of Global Radiation

The model of Aguiar et al. (1988) provided the starting point for the used methodology. It calculates daily values of global radiation with monthly mean values of global radiation as input. A change in this model was implemented in the chain of algorithms:

The original model gives one single distribution of daily clearness index values for any one monthly mean value. The model does not take into account any local factors like site altitude above sea level (higher maximum irradiation values at higher altitude) or different turbidity situations.

The whole system of the matrices was therefore changed from a clearness index basis to clear sky clearness index basis.

Formulated like this, the maximum values do correspond automatically to the clear sky model predictions used.

The monthly mean Linke turbidity factors are used to drive the clear sky model to obtain the required monthly mean daily values of global clear sky radiation needed to calculate daily clear sky clearness index values in any selected month for any point. This change required the daily Markov transition matrices tables to be completely revised to match the new formulation. The description of the methodology can be found at Aguiar et al. (1998). For version 6.0 only a few corrections of the matrices have been made to get better results for extreme low or high radiation conditions. The current matrices are listed in Remund et al. (2007).

5.1 Validation

The calculated mean values are adapted to the measured, so there is no difference at this level. The distribution has been tested at 5 stations of the Baseline Surface Radiation Network (BSRN) (WRCP 2001) (Table 15.2) with Kolmogorov-Smirnov (KS) test (Massey 1951). The stations were chosen because of their global distribution, the top standard of measuremnts and their availability. The KS test was chosen, as stochastically generated values can't be compared value by value to measured data. Distribution tests are suited best for this kind of data.

The interval distance p is defined as

$$p = \frac{x_{max} - x_{min}}{m}, \quad m = 100 \tag{15.4}$$

where x_{min} and x_{max} are the extreme values of the independent variable. Then, the distances between the cumulative distribution function are defined, for each interval, as

Table 15.2 Kolmogorov-Smirnov (KSI over %) test for daily global radiation

Site	Latitude [°]	Longitude [°]	Altitude [m]	KSI over %
Payerne (Switzerland)	46.82	6.95	490	0.0%
Camborne (United Kindom)	50.22	−5.32	88	0.0%
Boulder (CO, USA)	40.13	−105.23	1689	0.0%
Alice Springs (Australia)	−23.80	133.88	545	0.0%
Ilorin (Nigeria)	8.53	4.57	398	8.4%

$$D_n = \max |F(x_i) - R(x_i)|, \; x_i \in [x_{\min} + (n+1)p, x_{\min} + np] \tag{15.5}$$

$F(x_i)$ is the cumulative distribution function of the measured and $R(x_i)$ of the modelled data and $n = 1 \ldots m$ levels.

If at any of the intervals considered, this distance as given in equation (Eq. 15.5) is above a critical value V_c (which depends on the population size N) the null hypothesis that the sets are statistically the same must be rejected. The critical value is calculated for 99.9 % level of confidence (Eq. 15.6)

$$V_c = \frac{1.63}{\sqrt{N}}, N \geq 35 \tag{15.6}$$

A special test (KSI over) (Espinar et al. 2007) was used to estimate the proportion of the distribution, where the critical value is overshot:

$$aux = \begin{cases} D_n - V_c & \text{if } D_n > V_c \\ 0 & \text{if } D_n \leq V_c \end{cases} \tag{15.7}$$

The KSI over % parameters are then calculated as the trapezoidal integral of that auxiliary vector and its corresponding normalization to the critical area:

$$KSI \text{ over} \% = \frac{\int aux \, dx}{a_{critical}} \cdot 100 \tag{15.8}$$

where $a_{critical}$ is calculated as

$$a_{critical} = V_c \cdot (x_{\max} - x_{\min}) \tag{15.9}$$

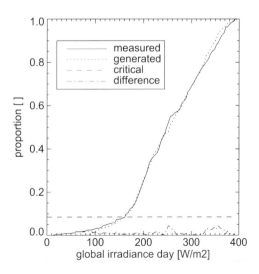

Fig. 15.2 Cumulative distribution functions of daily values of global irradiance for Alice Springs (Australia)

Generally a good agreement is achieved. At 4 of the 5 sites the distributions are statistically the same (Table 15.2). Figure 15.2 shows a typical cumulative distribution function for Camborne.

6 Hourly Values of Global Radiation

6.1 Generation of Hourly Values

The stochastic generation of hourly irradiance values from the daily mean profile is based on the model of Aguiar and Collares-Pereira (1992) (TAG model: **T**ime dependent, **A**uto-regressive, **G**aussian model). This model consists of two parts: the first part calculates an average daily profile $G_h{}^a$ while the second part computes the hourly variations $y(h)$ (Eq. 15.10). In Meteonorm Version 6.0 both parts have been used in a changed mode.

$$G_h = G_h{}^a + y(h) \tag{15.10}$$

6.1.1 Average Daily Profile

Based on the hypothesis that the average day global irradiance mean profile should exactly mirror the clear day global profile in form a model has been introduced in EU project SoDa (Wald et al. 2002). The proposed method for mean daily irradiance profile generation is therefore based on the use of the global radiation clear sky profile to calculate the global irradiance profile for all days in the time series (Eq. 15.11).

$$G_h{}^a = G_d \cdot \frac{G_c}{G_{c,d}} \tag{15.11}$$

Where G_d is the daily global horizontal irradiance, G_c the clear sky hourly global irradiance, $G_{c,d}$ the daily clear sky global irradiance. Other authors like Gruter et al. (1986) have used this approach as well.

The advantage of this model is that daily values of beam or diffuse radiation do not have to be known in advance. The model fits perfectly to the upper edge of the distribution, i.e. the clear sky profile, which is needed as a first step in the chain of algorithms. A short validation can be found in Remund et al. (2003a).

6.1.2 Hourly Variations

The generation of hourly values with the TAG model is governed by the autocorrelation and the standard deviation function.

The autocorrelation function has been adapted to 5 BSRN/Surfrad sites in the USA (Table Mountain, Fort Peck, Bonville, Penn State Univ., Sioux Falls) (NOAA, 2007). This subset was chosen, as it showed the best results. The standard deviation model has been modeled by hand. Adapted models showed less good results (Eq. 15.12).

$$y(h) = \phi_1 \cdot y(h-1) + r$$
$$\phi_1 = 0.148 + 2.356 \cdot K_t - 5.195 \cdot K_t^2 + 3.758 \cdot K_t^3$$
$$\sigma(K_t) = 0.32 \cdot \exp\left[-50 \cdot (K_t - 0.4)^2\right] + 0.002$$
$$\sigma' = \sigma \cdot \left(1 - \phi_1^2\right)^{0.5}$$
$$r = N\left(0, \sigma'\right) \tag{15.12}$$

where ϕ_1 is the first order autocorrelation, $\sigma(Kt)$ is the standard deviation of y perturbations and K_t is daily values of clearness index.

During the generation process the values of $y(h)$ have to be limited, as no negative irradiance or irradiance higher than the clearness irradiance can occur. This leads to a distortion of the theorethical distribution and the autocorrelation of $y(h)$. The problem of the non-Gaussian distribution of the intermittent hourly values was accounted for in Graham and Hollands's (1990) model using a function that maps the Gaussian distribution to a beta distribution.

A simpler procedure was chosen in the present model. The distortion of the first order correlation is corrected using a multiplication factor, k (Eq. 15.13). In this procedure, the value of the standard deviation, which is well reproduced by the model, is retained. Thus, in calculating the standard deviation, the uncorrected first order auto-correlation value must be used. The effect of including the factor k would be to increase the standard deviation. Since, however, ϕ_1 is reduced again during data generation, the standard deviation defined by the model can be used.

$$y(h) = k \cdot \phi_1 \cdot y(h-1) + r \tag{15.13}$$
$$\text{correction factor:} k = 2.0$$

6.1.3 Validation

The calculated mean values are adapted to the measured, so there is no difference at this level. The distribution has been tested at 5 BSRN sites (Tables 15.2 and 15.3)

Table 15.3 Kolmogorov-Smirnov test (KSI over %) for hourly global radiation

Site	KSI over %
Payerne	0.9%
Camborne	12.5%
Boulder	12.6%
Alice Springs	17.5%
Ilorin	10.3%

Fig. 15.3 Cumulative distribution functions of hourly values of global irradiance for Camborne

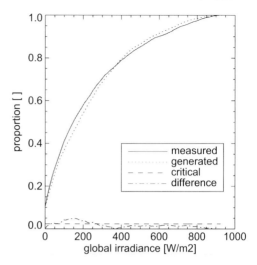

with Kolmogorov-Smirnov (KS) test: Generally a good agreement is achieved. Nevertheless at all sites there are areas, where the critical value is overshoot (Tab. 3). Most of the sites show biggest differences at 50–300 W/m². Figure 15.3 shows a typical cumulative distribution function for Camborne. Figure 15.4 shows the histograms of the same site.

The results for Ilorin may astonish at first sight, showing here the second best result whereas in the test of daily values having clearly the worst result. The errors of earlier steps of the chain often induce larger errors on next steps. In the case of the two steps daily and hourly generation of global radiation this is not necessarily the case. The distribution of hourly values of global radiation is not totally dependent

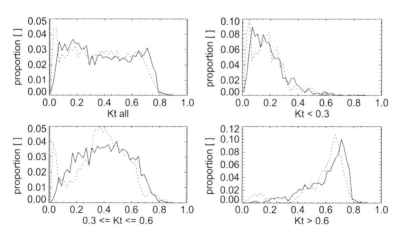

Fig. 15.4 Histograms of hourly global irradiance for Camborne depending on the daily clearness index (Kt). Measured values: full line, generated: dotted line

Table 15.4 Comparison between yearly means of measured and generated diffuse and beam values. BSRN sites are marked with a [B]

Station	Years	Diffuse meas. [W/m^2]	Diffuse gen. [W/m^2]	Diffuse difference [W/m^2]	Beam meas. [W/m^2]	Beam gen. [W/m^2]	Beam difference [W/m^2]
Uccle (Belgium)	1981–90	65.9	64.2	−1.7	81.6	88.5	6.9
Trier (Germany)	1981–90	67.1	63.7	−3.4	92.9	106.7	13.8
Dresden (Germany)	1981–90	61	65.3	4.3	100.3	96.8	−3.5
Hamburg (Germany)	1981–90	61.5	60.7	−0.8	84.5	91.9	7.4
Braunschweig (Germany)	1981–90	67.5	64.3	−3.2	85.3	97.7	12.4
Würzburg (Germany)	1981–90	69.2	63.3	−5.9	96	115.7	19.7
Weihenstephan (Germany)	1981–90	65.6	65.8	0.2	98.5	105.7	7.2
Payerne [B] (Switzerland)	2005	63.7	66.3	2.6	154.5	148.1	−6.4
Alice Springs [B] (Australia)	2005	56.5	69	12.5	315.4	295.2	−20.2
Camborne [B] (England)	2005	69.3	68.4	−0.9	105.7	109.6	3.9
Bonville [B] (USA)	2005	73.5	78.7	5.2	174.8	173.4	−1.4
Goodwin [B] Creek (USA)	2005	75.1	83.8	8.7	189.3	165.9	−23.4
Penn State [B] Univ. (USA)	2005	74.9	74.3	−0.6	135.4	144.2	8.8
Desert Rock [B] (USA)	2005	51.4	53.3	1.9	290.5	295.2	4.7
Sioux Falls [B] (USA)	2005	69.5	69	−0.5	183.7	182.3	−1.4
Table Mountain [B] (USA)	2003	66.4	68.8	2.4	229.8	222.1	−7.7
Fort Peck [B] (USA)	2005	66.1	62.5	−3.6	187.6	189.8	2.2
Mean value		66.1		1.0	153.3		1.4
RMSE (%)				4.7 (7.1 %)			11.3 (7.4 %)

from daily distribution. So errors may be compensated. Additionally the KSI over % test exaggerates often the differences as it show the percentage of values over a certain threshold.

The autocorrelation was examined for 17 sites (Table 15.4). The first autocorrelation value (ac(1) – which is the measured equivalent to ϕ_1 in Eq. 15.11) and the standard deviation (sd – which is the measured equivalent of σ in Eq. 15.11) depending on the daily clearness index (K_t) were compared graphically (Fig. 15.5). The autocorrelation ac(1) is underestimated on average by 14%, the standard

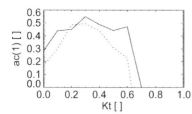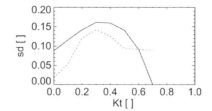

Fig. 15.5 Comparison of measured values (full line) and generated (dotted line) autocorrelation (ac(1)) and standard deviation sd at Cambourne

deviation (sd) is underestimated on average by 23%. Tests with enhanced values showed better results in this test, but did lead to much less accurate beam and diffuse separation. This was the reason to leave the values at this level.

6.2 Splitting the Global Radiation to Diffuse and Beam

The models of Perez for the splitting of global radiation have been examined briefly. The models are very similar after the correction of Skartveit (1998). The disadvantage of using a model like Perez or Skartveit is that the hourly diffuse values can not be known without (stochastic) generation of hourly global values. Therefore the beam and diffuse values depend to a certain extent on random numbers. The use of mean daily profiles to calculate the beam and diffuse profile is not reliable. Both the Skartveit and Perez models depend on the hourly variations from one hour to the next. Mean profiles and hourly values with variations do not give the same result.

6.2.1 Validation

The model performance has been tested at 17 sites by looking at the yearly means of generated diffuse and beam irradiances (Table 15.4). As only yearly means are validated here, the results don't have to be mixed with validation based on hourly values, which show much bigger errors.

The calculated yearly means of beam radiation have a mean bias error (mbe) of 1.4 and a root mean squared error (rmse) of $11.3 \, \text{W/m}^2$ (7.2 %) (definition e.g. in Argiriou, 1999). For diffuse radiation an mbe of 1.0 and a rmse of 4.7 (7.0%) was estimated. The validation shows different results for BSRN and the other sites. For BSRN sites the beam estimation has an mbe of $-4 \, \text{W/m}^2$ and an rmse of $10.6 \, \text{W/m}^2$. The reason for this was not examined. The accuracy could be enhanced compared to older versions by introducing changes in the standard deviation and the autocorrelation function of the stochastic generation (Eq. 15.11) and a new turbidity climatology with daily variations. On a global scale, the error in calculated diffuse radiation does not show regional patterns. The error distribution shows a sligth yearly pattern. In winter the rmse's are registered somewhat bigger.

Table 15.5 Kolmogorov-Smirnov test (KSI over %) for hourly beam radiation

Site	KSI over %
Payerne	22.8%
Camborne	81.6%
Boulder	59.9%
Alice Springs	106%
Ilorin	183%

The distributions of generated and measured diffuse radiation are similar, but do differ statistically. In Table 15.5 the KSI over% test at 5 sites (Table 15.2) for hourly beam radiation are listed. Figure 15.6 shows the distributions at Boulder (CO, USA).

Figure 15.6 shows a clear difference between measured and generated values at lower beam values ($< 500\,\mathrm{W/m^2}$). This effect can be seen at most other test sites as well.

6.3 Radiation on Inclined Planes

The Perez model (Perez et al. 1986) enables global and diffuse radiation to be calculated on an inclined surface using two input values, hourly global horizontal and diffuse horizontal irradiance. Additionally models from Hay (1978), Gueymard (1987) and Skartveit and Olseth (1986) are included. Hay's model distinguishes from the 3 other models by it's simple structure and the small number of input values.

As input for the calculation global radiation, diffuse radiation and albedo is needed.

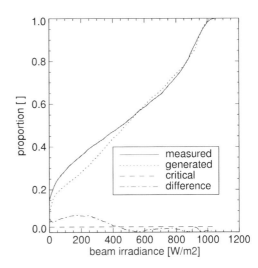

Fig. 15.6 Cumulative distribution functions of hourly beam irradiance for Boulder (CO, USA)

6.3.1 Albedo Model

The albedo is calculated with a model that calculates daily albedo as a function of temperature of the last weeks (Eq. 15.14):

For areas with great amount of snow:

$$\rho = 0.618 - 0.044 \cdot Ta, 0.2 \leq \rho \leq 0.8$$

For the rest of the world:

$$\rho = 0.423 - 0.042 \cdot Ta, 0.2 \leq \rho \leq 0.8 \tag{15.14}$$

where Ta is mean temperature of the last 14 days and $\rho =$ surface albedo.

A distinct separation for areas with a great amount of snow and areas with less snow was found:

- Sum of precipitation for 3 winter months > 150 mm
- Mean temperature for 3 winter months < $-3°C$

A great amount of snow enhances the possibility of fresh snow and therefore higher albedoes and prolongates the time of snow cover in spring. For temperatures below $-10°C$ and above $10°C$ there is no difference between the models.

6.3.2 Validation of Radiation on Inclined planes

The monthly and yearly means are examined. The tests were carried out for 14 sites in many different climate zones and different inclinations (18° to 90°) (PVPS 2007). The measurements were made with pyranometers for horizontal and inclined planes. The sites have been filtered to choose only reliable data.

The following sites where chosen depending on the length of the records, the tilt angle and the region: Locarno-Magadino (Switzerland), Burgdorf (Switzerland), Bern (Switzerland), Mt. Soleil (Switzerland), Liestal (Switzerland), Gontenschwil (Switzerland), Akakuma (Japan), Akamatsu/Kobe (Japan), Toyooka (Japan), Kanan-Town (Japan), Cloppenburg (Germany), Holzkirchen (Germany), Mexicali (Mexico), Huvudsta (Sweden).

The four different models did show similar results (Table 15.6). Due to uncertainties based on the measurements the differences are too small to rank the models seriously. Nevertheless for monthly values Perez model showed the best results, followed by Gueymard's, Hay's and Skartveit's models, respectively. The average mbe error of Perez' model was $0 \, W/m^2$ and the rmse standard deviation $8 \, W/m^2$.

At some sites the differences have a distinct yearly pattern with an overestimation in winter (e.g. Switzerland). In other regions these effects are not visible.

For yearly average values, the mbe of Perez model was $-1 \, W/m^2$ and the rmse $6 \, W/m^2$ (4%). For inclination above 50° the calculation is partly better. For facades Hay's and Skartveit's model are the best, followed by Gueymard's and Perez' models.

Table 15.6 Accuracy of monthly and yearly values of radiation on tilted planes

	Perez	Hay	Gueymard	Skartveit
Monthly mbe $[\text{W}/\text{m}^2]$	0.4	−2.7	−1.7	−3.4
Monthly rmse $[\text{W}/\text{m}^2]$	8.0	9.9	8.8	8.6
Yearly mbe $[\text{W}/\text{m}^2]$	−0.9	−5.2	−3.8	−6.2
Yearly rmse $[\text{W}/\text{m}^2]$	5.5	7.2	6.5	7.6
Yearly rmse $[\%]$	4.0	5.3	4.7	5.5
Yearly rmse 0–50° $[\text{W}/\text{m}^2]$	5.0	7.7	6.7	8.1
Yearly rmse $> 50°$ $[\text{W}/\text{m}^2]$	8.2	3.1	5.0	3.2

This validation doesn't have to be misinterpreted as tilt model validation based on measured values, as the stochastically generated values used here as input have an important influence on the results.

6.4 Modification of Irradiance Due to Horizon

The aim of the modification method described here is to calculate the radiation at sites with raised (distant) horizons. It is clear that direct radiation is affected by a raised (i.e. non-horizontal) horizon in such a way that when the sun is occluded by the horizon, no direct radiation can impinge on the inclined surface. In other words, the surface in question receives less direct radiation than it would with a horizontal skyline. In calculating hourly values, a check has therefore to be made whether the sun is above or below the skyline. If occluded by the skyline, the direct radiation on the inclined surface is zero.

The hourly direct radiation on an inclined surface is set to zero, if the the sun has not yet risen or has already set, if the sun is behind the surface or if the sun is behind the skyline.

The diffuse radiation components of the Perez model are processed as follows:

- Circumsolar component: this is treated in the same way as direct radiation.
- The horizontal ribbon: this part of the diffuse radiation remains unchanged, i.e. it retains its original value independently of skyline profile. This is assumed for the reason that the sky immediately above the horizon is often brighter than the rest of the sky. This applies not only in regions with practically level horizons but also in mountainous regions. In mountainous regions in summer, this is often caused by the bright convective clouds that tend to form above ridges and peaks.
- Diffuse isotropic and reflected irradiance are calculated as follows:
  If the skyline is not horizontal, a larger proportion of ground and smaller proportion of sky is visible to the surface. This implies that the view factors must be modified when a raised skyline is present. The skyline profile is normally given as a closed polygon whose points are specified in terms of azimuth and altitude. The proportion of the sky which, despite the existence of a skyline profile is still seen by the inclined surface may be calculated by numerical integration.

7 Conclusions

The models constructed are based as much as possible on published work. Nevertheless some important new validated chain links were introduced.

Together with a spatial interpolation model, time series can be obtained for any site.

Testing against observed data shows that the stochastic generation of the hourly short wave parameters is possible and that the quality is acceptable. The biases are generally small. The rmse's for beam irradiance come to 7% and for irradiation on inclined planes to 4%.

Together with the accuracy of the spatial interpolation – wich we assume in a first approximation to be independent – the accuracy for the calculation of yearly means of global radiation on inclined planes on sites without monthly measurements comes to 9%.

The distributions of daily global irradiance values are similar to the measured at all test sites. For hourly values the discrepancies are bigger and for beam irradiance the distributions don't pass the nullhypothesis of KS test. Nevertheless the distributions are similar at most sites for the biggest part of possible values (especially for higher values above $500\,\mathrm{W/m^2}$).

References

Aguiar R, Collares-Pereira M, Conde JP (1988) A simple procedure for generating sequences of daily radiation values using a library of markov transition matrices. Solar Energy 40: 269–279

Aguiar R, Collares-Pereira M (1992) TAG: A time-dependent auto-regressive, Gaussian model. Solar Energy 49: 167–174

Argiriou A, Lykoudis S, Kontoyiannidis C, Balaras A, Asimakopoulos D, Petrakis M, Kassomenos P (1999) Comparison of methodologies for tmy generation using 20 years data for Athens, Greece. Solar Energy 66: 33–45

Bourges B (1985) Improvement in solar declination calculations. Solar Energy, Vol. 35: 367–369

Collares-Pereira M, Rabl A (1979) The average distribution of solar radiation: Correlations between diffuse and hemispherical values. Solar Energy 22: 155–164

Dürr B, Ineichen P (2007): private communication (IEA SHC Task 36).

Espinar B, Ramirez L, Drews A, Beyer HG, Zarzelejo LF, Polo J, Martin L (2007): Analysis of different error parameters applied to solar radiation data from satellite and german radiometric stations. Internal paper of IEA SHC Task 36.

Gilgen H, Wild M, Ohmura MA (1998). Means and trends of shortwave incoming radiation at the surface estimated from Global Energy Balance Archive data. Journal of Climate 11: 2042–2061.

Graham V, Hollands K (1990) A method to generate synthetic hourly solar radiation globally. Solar Energy 44: 333–341.

Gruter W, Guillard H, Möser W, Monget JM, Palz W, Raschke E, Reinhardt RE, Schwarzmann P, Wald L (1986) Solar Radiation Data from Satellite Images. Determination of Solar Radiation at Ground Level from Images of the Earth Meteorological Satellites - An Assessment Study. D. Reidel Publishing Company.

Gueymard C (1987) An anisotropic solar irradiance model for tilted surfaces and its comparison with selected engineering algorithms. Solar Energy 38: 367–386

Hay JE (1978) Calculation of monthly mean solar radiation for horizontal and inclined surfaces. Solar Energy 23: 301–307

Holben BN, Tanre D, Smirnov A, Eck TF, Slutsker I, Abuhassan N, Newcomb WW, Schafr J, Chatenet B, Lavenue F, Kaufman YF, Vande Castle J, Setzer A, Markham B, Clark D, Frouin R, Halthore R, Karnieli A, O'Neill NT, Pietras C, Pinker RT, Voss K, Zibordi G (2001) An emerging ground-based aerosol climatology: Aerosol Optical Depth from AERONET. J Geophys. Res. 106: 12 067–12 097

Lefèvre M, Remund J, Albuisson M, Wald L (2002) Study of effective distances for interpolation schemes in meteorology. Annual Assembly, European Geophysical Society, Nice, April 2002. Geophysical Research Abstracts 4: EGS02-A-03429

Massey FJ (1951) The Kolmogorov-Smirnov test for goodness of fit. Journal of American Statistical Association 46: 68–78

NASA (2007) http://eosweb.larc.nasa.gov/ and http://modis-atmos.gsfc.nasa.gov/index.html

NOAA (2007): http://www.srrb.noaa.gov/surfrad/

Perez R, Stewart R, Arbogast C, Seals R, Scott J (1986) An anisotropic hourly diffuse radiation model for sloping surfaces: Description, performance validation, site dependency evaluation. Solar Energy 36: 481–497

Perez R, Seals R, Ineichen P, Stewart R, Menicucci D (1987) A new simplified version of the Perez Diffuse Irradiance Model for tilted surfaces. Solar Energy 39: 221–231

Perez R, Ineichen P, Seals R, Michalsky J; Stewart R (1990) Modeling daylight availability and irradiance components from direct and global irradiance. Solar Energy 44: 271–289

Perez R, Ineichen P, Maxwell E, Seals R, Zelenka A (1991) Dynamic Models for hourly global-to-direct irradiance conversion. Edited in: Solar World Congress 1991. Volume 1, Part II. Proceedings of the Biennial Congress of the International Solar Energy Society, Denver, Colorado, USA, 19–23 August 1991.

PVPS (2007) http://www.iea-pvps.org/

Remund J, Salvisberg E, Kunz S (1998) Generation of hourly shortwave radiation data on tilted surfaces at any desired location. Solar Energy 62: 331–334

Remund J, Wald L. Page J (2003a) Chain of algorithms to calculate advanced radiation parameters. Proceedings of the ISES solar world congress 2003, Göteborg Sweden. CD-ROM Paper P6 38.

Remund J, Wald L, Lefèvre M, Ranchin T, Page J (2003b) Worldwide Linke turbidity information. Proceedings of the ISES solar world congress 2003, Göteborg Sweden. CD-ROM Paper O6 18.

Remund J, Kunz S (2003c) The new version of the worldwide climatological database METEONORM. Proceedings of the ISES solar world congress 2003, Göteborg Sweden. CD-ROM Paper P6 39.

Remund J, Kunz S, Schilter C (2007) METEONORM Version 6.0. Meteotest, Fabrikstrasse 14, 3012 Bern, Switzerland. www.meteonorm.com

Rigollier C, Bauer O, Wald L (2000) On the clear sky model of the ESRA with respect to the heliosat method. Solar Energy 68: 33–48

Skartveit A, Olseth JA (1985) Modelling slope irradiance at high latitudes. Solar Energy 36: 333–344

Wald L, Albuisson M, Best C, Delamare C, Dumortier D, Gaboardi E, Hammer A, Heinemann D, Kift R, Kunz S, Lefèvre M, Leroy S, Martinoli M, Ménard L, Page J, Prager T, Ratto C, Reise C, Remund J, Rimoczi-Paal A, Van der Goot E, Vanroy F, Webb A (2002) SoDa: a project for the integration and exploitation of networked solar radiation databases. In: Environmental Communication in the Information Society, W. Pillmann, K. Tochtermann Eds, Part 2, pp 713–720. Published by the International Society for Environmental Protection, Vienna, Austria

WCRP (2001) Baseline Surface Radiation Network (BSRN). Sixth BSRN Science and Review Workshop (Melbourne, Australia, 1 - 5 May 2000). WCRP, Informal Report No. 17/2001, World Meteorological Organization, Geneva.

Chapter 16
Modelling UV–B Irradiance in Canada

John Davies and Jacqueline Binyamin

1 Introduction

Ultra-violet B (UV–B) radiation (defined here as the 290–325 nm waveband) constitutes less than 1% of the total irradiance reaching the ground but has important adverse biological effects. Life is shielded partially from this radiation by the ozone in the stratosphere which is equivalent to a depth of about 3 mm or 300 DU (Dobson units). This thin shield has been significantly attenuated by chlorine damage from halocarbons. Systematic ozone reductions of more than 5% have been found globally. Man-made chlorofluorocarbons are being phased out following the Montreal Protocol in 1987 and the ozone layer is expected to recover within about 60 years. Until then stratospheric ozone concentrations will be less than normal and UV–B irradiance reaching the earth's surface will be larger than normal.

However, the UV–B radiation band does not pose biological problems only during periods of reduced stratospheric ozone. Biological concerns for UV–B damage existed before depletion of stratospheric ozone. This radiation damages DNA, immune systems, skin (erythema, skin cancers) plant growth and phytoplankton. Photobiologists have determined experimentally the spectral variation in effects in the form of action spectra or spectral weights which are applied to spectral irradiance measurements. However, spectral irradiance measurements are rare and must be supplemented with model calculations. In this chapter, we describe a model developed for Canada and show how it performs in calculating climatological estimates of spectral irradiance and biologically-weighted irradiance. Since there is little information on the vertical distribution of ozone and aerosol which Zeng et al. (1994) found important in successfully modelling short-term cloudless sky irradiances, we adopt a more climatological approach by seeking to model monthly averaged

John Davies
McMaster University, Hamilton, Ontario, Canada, e-mail: davies.j@sympatico.ca

Jacqueline Binyamin
University of Winnipeg, Canada, e-mail: binyamin@winnipeg.ca

daily irradiances using readily available data. The model described may be used for shorter periods. It has been shown to produce instantaneous broadband irradiance estimates which compare with surface measurements as well as estimates from the inversion of satellite measurements of reflected UV–B irradiances (Binyamin et al., submitted).

2 The Model

The model (Davies et al. 2000) calculates spectral irradiance G_λ at 1 nm wavelength λ intervals as the sum of cloudless sky and overcast sky components:

$$G_\lambda = (1 - C)G_{\lambda,0} + CG_{\lambda,\otimes}, \qquad (16.1)$$

where $G_{\lambda,0}$ is cloudless sky irradiance, $G_{\lambda,\otimes}$ overcast irradiance and C the fraction of the sky covered by cloud. This simple formulation is commensurate with the hourly surface-based cloud observations that are used. A biological dose D is obtained from spectral irradiances from

$$D = \iint w_\lambda G_\lambda d\lambda dt, \qquad (16.2)$$

where w_λ is a spectral biological weighting. Cloudless and overcast spectral irradiances are calculated from solutions of the radiative transfer equation. The delta-Eddington (DE) method (Joseph et al. 1976) has been used mainly but comparisons have been made with more rigourous solutions from the discrete ordinates (DO) method (Stamnes et al. 1988). The latter is potentially an exact solution to the radiative transfer equation since it allows the scattering phase function to be expanded in any number of terms. The exactness of the solution increases with the number of terms. In practice, an 8–term expansion has been found adequate for the UV–B spectrum (Wang and Lenoble 1994; Zeng et al. 1994). The DE method combines the Eddington approximation (Shettle and Weinman 1970) with a Dirac delta function to approximate the large forward peak in the phase function for asymmetric scattering in aerosols and clouds. Madronich (1993) has used this method to calculate UV–A (320–400 nm) and UV–B irradiances. This method is computationally much faster than the DO method but Forster and Shine (1995) found for clear, aerosol-free skies and at high solar zenith angles that the DE compares poorly with 8-stream and 16-stream DO calculations in the UV–B band at wavelengths below 305 nm. They attributed this difference to the inadequacy of the phase function approximation at wavelengths smaller than 305 nm where ozone absorption is strong. Differences between the two methods were much smaller under overcast skies.

In this chapter irradiance estimates from both algorithms are compared with measurements from single monochromator Brewer spectrophotometers (Bais et al. 1996) in Canada. This instrument measures in the 290–325 nm range rather than the 280 to 315 or 320 nm range which usually defines the UV–B waveband.

The CD-ROM which accompanies this volume contains programs which implement this model using the delta-Eddington or the discrete ordinates methods for solving the radiative transfer equation and sample data input files.

2.1 Extraterrestrial Spectral Irradiance

The model uses extraterrestrial irradiance measurements sampled approximately every 0.05 nm from the Solar Spectral Irradiance Monitor (SUSIM) instrument on the third Atmospheric Laboratory for Applications and Science (ATLAS-3) space shuttle mission. The Brewer instrument measures irradiance through a triangular filter with a full width at half maximum of 0.55 nm. The effect of the Brewer instrument's triangular filter on measurements at a given wavelength was mimicked numerically by averaging weighted irradiances within ± 0.55 nm of each nanometer. The weighting increased linearly from 0 at a distance of 0.55 nm from the centre wavelength to 1 at that wavelength. Strictly speaking surface irradiance should be calculated at fine spectral intervals and then weighted appropriately to mimic the Brewer instrument's filter. Davies et al. (2000) showed that the two averaging procedures yielded virtually identical broadband surface irradiances.

2.2 Atmospheric Optical Properties

Since there are few measured atmospheric vertical profiles of ozone, temperature T, pressure p and humidity, summer and winter midlatitude and subarctic model atmospheres containing these variables for 50 atmospheric levels from the surface to 120 km (Kneizys et al. 1988) were used (Table 16.1). Radiative transfer calculations need spectral values of optical depth τ, single scattering albedo ω_0 and asymmetry factor g for each atmospheric layer l. Layer values are calculated from components for ozone absorption, molecular scattering and aerosol extinction from

$$\tau(\lambda,l) = \tau_o(\lambda,l) + \tau_m(\lambda,l) + \tau_a(\lambda,l), \tag{16.3}$$

$$\omega(\lambda,l) = \frac{\tau_m(\lambda,l) + \omega_a(\lambda,l)\tau_a(\lambda,l)}{\tau(\lambda,l)}, \tag{16.4}$$

and

$$g(\lambda,l) = \frac{g_a(\lambda,l)\,\omega_a(\lambda,l)\,\tau_a(\lambda,l)}{\tau(\lambda,l)\,\omega(\lambda,l)}. \tag{16.5}$$

Unless stated otherwise, an optical property q for each nanometer was calculated as a weighted mean using

$$\bar{q} = \frac{\sum w_\lambda q_\lambda S_\lambda}{\sum w_\lambda S_\lambda}. \tag{16.6}$$

In the presence of cloud, cloud optical properties replace those calculated from Eqs. (16.3)–(16.5).

Table 16.1 Model atmosphere

Layer	From (km)	To (km)	Levels
Upper atmosphere	120	30	50–27
Stratosphere	30	10	27–12
Troposphere	10	2	12–4
Cloud	3	2	4–3
Boundary	2	Surface	3–1

Figure 16.1 shows the spectral variation of optical depth components at a 0.05 nm resolution for a midlatitude summer model atmosphere (332 DU). Ozone absorption dominates at wavelengths below 300 nm, scattering dominates above 310 nm and all attenuants are important between 300 nm and 310 nm. Vertical profiles of cloudless sky transmittance (calculated from time-integrated and spectrally-integrated irradiances) show that the stratosphere reduces irradiance by about 75% while the underlying atmosphere attenuates by less than 15% (Davies et al. 2000).

2.2.1 Ozone Absorption

The climate model's ozone profile was adjusted each day by multiplying each level by the ratio of daily Brewer measurements of ozone depth to the depth for the climate profile. Ozone spectral optical depth for each atmospheric layer was

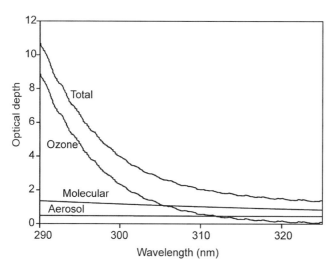

Fig. 16.1 Spectral variation of total optical depth and its component parts due to ozone, molecules and aerosol

determined as the product of weighted mean, temperature-dependent ozone absorption cross-sections (Paur and Bass 1985) and the ozone molecular number density calculated from the adjusted measurements.

2.2.2 Molecular Scattering

The spectral optical depth for molecular scattering in each layer was calculated as the product of the weighted mean optical cross-section calculated following Elterman (1968) and the molecular number density at standard air pressure

2.2.3 Aerosol Optical Properties

Since there are no aerosol data for the UV–B band for Canada, aerosol optical properties were interpolated from the data of Shettle and Fenn (1979). Weighted means of optical cross-sections were unnecessary because the optical properties vary slowly with wavelength in this band. These data consist of spectral extinction and absorption coefficients and asymmetry factors for the boundary layer, troposphere, stratosphere and upper atmosphere. Boundary layer aerosols are classified into rural, urban and maritime with a further breakdown according to visibility. Tropospheric aerosols are classified by season (cooler, i.e. fall/winter, and warmer, i.e. spring/summer) and visibility. Stratospheric aerosols are classified into background and different levels of volcanic contamination for the same two seasons. Upper atmosphere aerosols are classified into normal and several volcanic contamination states. Values of optical parameters were interpolated for each wavelength for each atmospheric level and averaged for each layer

2.2.4 Cloud Optical Properties

Cloud optical properties are unknown for this waveband. Mie theory was applied to calculate single scattering albedo ω_c and asymmetry factor g_c using the refractive index data of Hale and Querry (1973) for water. Calculations were made for two cloud droplet equivalent radii: $7\,\mu m$ for arctic regions (Leontyeva and Stamnes, 1994) and $10\,\mu m$ for temperate regions (Han et al. 1994). The results (Table 16.2) are

Table 16.2 Co-albedo and asymmetry factor for two equivalent radii r_e

	$r_e = 7\,\mu m$	$r_e = 10\,\mu m$
$(1 - \omega_c)$	3×10^{-6}	5×10^{-6}
g_c	0.8709	0.8587

Table 16.3 Median cloud optical depths from the DE and DO methods

Station and year	DE	DO
Resolute (1995)	12.2	11.2
Churchill (1993)	13.6	13.5
Winnipeg (1993)	17.1	17.0
Toronto (1993)	17.3	16.4

similar to those presented by Slingo and Schrecker (1982) and Hu and Stamnes (1994). Cloud optical depth was calculated (Binyamin et al, 2007) from overcast sky irradiance measurements by the method described by Leontyeva and Stamnes (1994) for the total solar spectrum. Mie calculations for the UV-B band show that the extinction efficiency factor is approximately constant with wavelength and approaches an asymptotic value of 2 for both equivalent cloud drop radii. Therefore, cloud optical depth can be considered constant in this waveband. Cloud optical depth was iterated until model spectrally integrated irradiance converged with measurement. Overcast data for snow free conditions were used since snow albedo is difficult to specify because it varies greatly with surface contamination and state of snow. Initially, optical depths were calculated for four stations from both the DE nd DO methods. Median optical depths from both for a one year are presented in Table 16.3.

Agreement between the two sets of optical depth is excellent and we conclude that the DE method is appropriate for this purpose. Cloud optical depths were then calculated by the DE method for nine stations, most for several years. Median optical depths for each station are given in Table 16.4. With the exception of Alert and Resolute, the two Arctic stations, optical depth varies between 16.6 and 23 around an overall median of 18.7. Since differences in irradiances calculated with the value of 18.7 and individual station values were small, the former is used for midlatitude stations. Station values were used for arctic stations.

Table 16.4 Median cloud optical depths from the DE method

Station and year(s)	Latitude (deg N)	Optical depth
Alert (1995)	82.52	5.0
Resolute (1993–1996)	74.70	10.2
Churchill (1993–1996)	58.74	15.0
Edmonton (1993–1996)	53.54	19.6
Regina (1994–1995)	50.50	17.4
Winnipeg (1993)	49.88	17.1
Montreal (1993–1994)	45.47	20.0
Halifax (1993–1996)	44.62	16.6
Toronto (1993–1996)	43.67	23.0

2.2.5 Surface Albedo

Albedo was calculated as a linear function of daily snow depth between 0.05 for a snow-free ground (Bowker et al. 1985) and 0.75 for a snow cover of 30 cm or greater. The value of 0.75 was inferred from Stony Plain data (approximately 37 km west of Edmonton) where measurements were made with extensive snow cover. Albedo is independent of wavelength and any effects of melting and snow contamination are ignored.

3 Measurements

Daily ozone depths and UV–B irradiances once or twice an hour were made by single-monochromator Brewer ozone spectrophotometers in the national network maintained by the Meteorological Service of Canada (MSC). The instrument provides measurements of very good accuracy and precision (Bais et al. 1993). However, the single monochromator version used in Canada suffers from instrumental stray light contamination at wavelengths less than about 305 nm (Bais et al. 1996). Kerr and McElroy (1993) argued that below 290.5 nm there is no detectable radiation signal and that the mean value for this spectral region is the stray light and can be subtracted from the whole spectrum. Bais et al. (1996) showed that this is an effective correction and the MSC has applied it to the Brewer irradiance measurements used in this study. We have applied an additional correction. On the basis of work by Krotkov et al. (1998) and Wang et al. (2000) measurements were increased by 6% to compensate for cosine error.

MSC also provided daily snow depth and hourly meteorological observations including total cloud opacity which measures the effective cloud cover on a scale from 0 to 1. Total cloud amount records were unsuitable because they were not recorded on a continuous linear scale. Meteorological data were linearly interpolated for the times of Brewer irradiance measurements from hourly meteorological.

4 Model Estimates of Spectral Irradiance

4.1 Comparison of Irradiances from the DE and DO Methods

The spectral variation of DE/DO (the ratio of the spectral irradiance calculated using the delta-Eddington model to that from the discrete ordinatesmodel) with solar zenith is plotted for a cloudless and overcast mid-latitude summer atmosphere (302 DU) in Fig. 16.2. For zenith angles $\leq 40°$, the ratio exceeds unity by less than 4% for cloudless skies and by less than 2% for overcast. At larger zenith angles the ratio decreases rapidly and becomes smaller than unity at wavelengths below

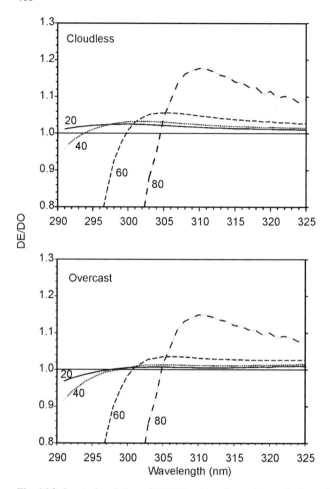

Fig. 16.2 Spectral variation of DE/DO with solar zenith angle for a cloudless and overcast mid-latitude summer atmosphere (302DU)

305 nm, the start of the decrease shifting to shorter wavelengths with decreasing zenith angle. This confirms the findings of Forster and Shine (1995) and casts doubt on the suitability of DE for wavelengths below 300 nm in sub-arctic and arctic regions. The method should perform satisfactorily in mid-latitudes for wavelengths above 300 nm.

To show the net effect of regional differences in cloudiness and sun angle we calculated irradiances for seven wavelengths (295, 300, 305, 310, 315, 320 and 325 nm) for both models for one year and from these the difference DE-DO as a fraction of DO for one arctic, one subarctic and two midlatitude stations. Results for Resolute, Churchill, Winnipeg and Toronto are summarized in Table 16.5.

Table 16.5 Spectral values of (DE-DO)/DO for Resolute (1995), Churchill, Winnipeg and Toronto in 1993

Station	295 nm	300 nm	305 nm	310 nm	315 nm	320 nm	325 nm
Resolute	−0.301	−0.041	−0.001	0.009	0.0	0.002	0.0
Churchill	−0.030	0.014	0.029	0.027	0.024	0.022	0.021
Winnipeg	−0.005	0.027	0.024	0.024	0.013	0.013	−0.031
Toronto	−0.012	0.016	0.012	0.015	0.006	0.008	0.011

Except for Resolute, for wavelengths less than 305 nm, spectral irradiances from the DE method exceed those from DO by less than 3%. Overestimation of this magnitude can be anticipated from Fig. 16.2. Underestimation at Resolute at 295 nm arises from larger solar zenith angles while at the other stations smaller zenith angles than at Resolute preserve the values for the other wavelengths at 295 nm.

4.2 Comparison of Calculated with Measured Irradiances

Model performance is summarized for seven wavelengths in Table 16.6 as the average daily fractional difference for one year between model calculations C and measurements M, i.e. (C-M)/M. For wavelengths larger than 300 nm differences for both models are mainly within 5% and both models tend to underestimate irradiances. This is apparent in scatter diagrams of calculated against measured irradiances at 295 nm for Resolute and Winnipeg (Fig. 16.3). At this wavelength, the scatters for both DE and DO at Resolute show no trend and DE performs better than DO at Winnipeg. Results for Resolute do not support the underestimation of irradiance from the DE method below 305 nm expected from Fig. 16.2. For both DE and DO the largest differences occur at these shorter wavelengths but the DO results are not clearly superior to those for DE. We believe that this indicates problems in measuring low light.

Table 16.6 Spectral values of (DE-M)/M (bold) and (DO-M)/M for Resolute (1995) and Churchill, Winnipeg and Toronto in 1993

Station	295 nm	300 nm	305 nm	310 nm	315 nm	320 nm	325 nm
Resolute	**−0.013**	**0.204**	**−0.077**	**0.068**	**−0.019**	**−0.021**	**−0.030**
	0.245	0.274	−0.079	0.058	−0.018	−0.019	−0.030
Churchill	**−0.043**	**0.208**	**0.020**	**0.060**	**−0.073**	**−0.081**	**−0.050**
	−0.010	0.183	−0.024	−0.023	−0.091	−0.098	−0.069
Winnipeg	**−0.009**	**0.025**	**−0.065**	**−0.012**	**−0.038**	**−0.023**	**0.014**
	−0.042	−0.002	−0.087	−0.035	−0.050	−0.036	−0.002
Toronto	**0.081**	**0.130**	**−0.027**	**0.019**	**−0.056**	**−0.041**	**−0.019**
	0.094	0.113	−0.039	0.004	−0.061	−0.048	−0.029

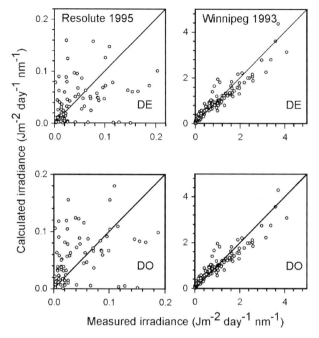

Fig. 16.3 Correlation between calculated (DE or DO) and measured daily spectral irradiance at 295 nm for Resolute in 1995 and Winnipeg in 1993

Figure 16.4 shows examples of the agreement between mean daily model calculations and measurements for seasonally contrasting months for Resolute (April and June, 1995), Churchill (March and June, 1993), Winnipeg (June and November, 1993) and Toronto (June and December, 1993). 1993). Model calculations agree well with measurements at wavelengths greater than 300 nm. Poorer agreement below 300 nm is attributed to the difficulty of measuring such low irradiance levels and to light leakage even though a correction has been applied to irradiances at wavelengths less than 305 nm. The Winnipeg and Toronto data still show evidence of stray light leakage in the corrected measurements. We conclude that the model performs well using both DE and DO in the arctic to temperate range of climatic conditions that are represented.

4.3 Comparison of Calculated with Measured Biological Doses

The significance of differences between measured and calculated irradiances is examined after applying action spectra for biological effects. For this purpose we selected action spectra for DNA damage (Setlow, 1974) and erythema (McKinlay and Diffey, 1987). These spectra normalized to unity are plotted in Fig. 16.5.

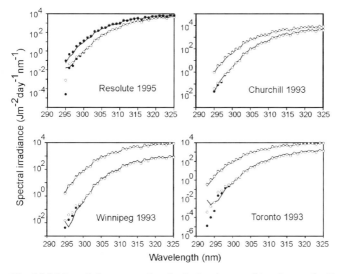

Fig. 16.4 Mean daily measured and calculated spectral irradiances for Resolute (April and June, 1995), Churchill (March and June, 1995), Winnipeg (June and November, 1993) and Toronto (June and December, 1993). Measurements are shown with a solid line, DE calculations by an open circle and DO calculations by a dot

Below 305 nm the relative response is at least one order of magnitude larger than at longer wavelengths, thus amplifying any measurement and model deficiencies below 305 nm. Nevertheless, average spectral doses calculated by multiplying the measured and modeled irradiances in Fig. 16.4 by action spectra still agree well.

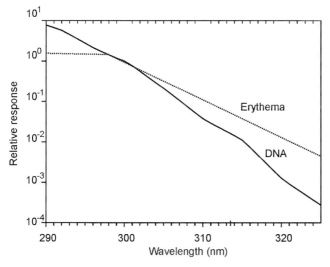

Fig. 16.5 Action spectra for DNA damage and erythema

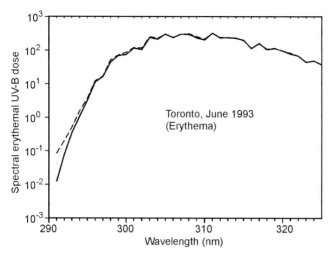

Fig. 16.6 Mean daily spectral erythemal UV–B dose for June 1993 at Toronto. Measured irradiance is shown by a solid line, calculated irradiances from the DE method by a dotted line and from the DO method by a dashed line

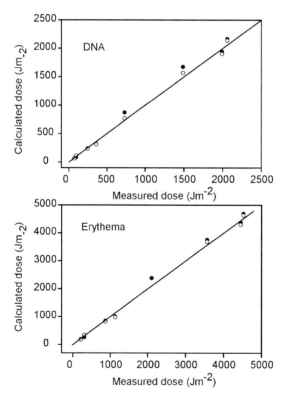

Fig. 16.7 Correlation between doses using calculated and measured spectral irradiances for DNA and erythema. Dots represent DE and open circles represent DO. The 1:1 lines are shown

An example is shown for the mean daily erythemal dose at Toronto in June 1993 (Fig. 16.6). Doses for DNA and erythema calculated from the measured and calculated data in Fig. 16.4 are plotted in Fig. 16.7.

Results from the DO method compare very well with those from measurements. The 1:1 lines on these plots coincide with plots of regressions with no intercept between doses calculated by the DO method and measured doses. The DE deficiency in calculating ozone absorption at wavelengths below 300 nm proves not to be important since dose calculations are strongly weighted by irradiances at longer wavelengths. This must be true for dose calculations with other action spectra except those which might terminate close to 300 nm. The effect of irradiance overestimation by DE (Fig. 16.2) is apparent in Fig. 16.7 but is probably of little importance. The average dose overestimation, determined as the slope of a regression fit with no intercept, is 3% for all three action spectra compared with zero for DO.

5 Conclusions

The DE method performed satisfactorily in Canadian mid-latitude, subarctic and arctic atmospheres to calculate broadband cloud optical depth and spectral irradiances at wavelengths above 305 nm. It should not perform similarly in other climates. Inadequacies at wavelengths smaller than 305 nm do not affect broadband irradiances and doses. Its fast computing time makes it an attractive choice. Doses from the DE method were consistently overestimated by about 3% for the two selected action spectra. This is probably an acceptable error. The DO method is recommended where greater accuracy is required for irradiances at wavelengths below 305 nm.

Although Binyamin, Davies and McArthur (submitted) showed that the model, using both DE and DO methods, provides instantaneous broadband UV–B irradiances which compare favourably with measured irradiances and those calculated from the Canada Centre for Remote Sensing satellite model (Li et al. 2000), the model is best suited for climatological estimates of mean daily spectral irradiance for individual months since it uses averaged atmospheric profiles of ozone, cloud optical depth and aerosol properties.

The model's most important variable inputs are daily ozone depth and hourly cloud cover. Surface-based measurements of ozone depth can be replaced by satellite estimates such as those from the Total Ozone Mapping Spectrometer. Although surface observations of cloud cover are made hourly in Canada many countries observe it once every three hours. This is not a serious restriction since earlier work with global solar irradiance models (Davies and McKay, 1989) showed that satisfactory results can be obtained using linearly interpolated hourly cloud cover from such observations. Cloudiness can also be estimated from sunshine measurements.

References

Bais AF, Zerefos CS, Meleti C, Ziomas IC, Tourpali K (1993) Spectral measurements of solar UVB radiation and its relation to total ozone, SO_2 and clouds. J Geophys Res 98, 5199–5204

Bais AF, Zerefos CS, McElroy CT (1996) Solar UVB measurements with the double and single-monochromator Brewer ozone spectrophotometers. Geophys Res Lett 23: 833–836

Barker HW, Curtis TJ, Leontieva E and Stamnes K (1998) Optical depth of overcast cloud across Canada: Estimates based on surface pyranometer and satellite measurements. J Climate 11: 2980–2994

Binyamin J, Davies JA, McArthur LJB (2007) UV–B Cloud optical properties for Canada (submitted)

Binyamin J, Davies JA, McArthur LJB (2007) Comparison of UV–B broadband Brewer measurements with irradiances from surface-based and satellite-based models (submitted)

Bowker DE, Davis RE, Myrick DI, Stacy K, Jones WT (1985) Spectral reflectance of natural targets for use in remote sensing studies. NASA reference publication 1139, NASA, Langley Research Center, Hampton, Va

Davies JA and McKay DC(1989) Evaluation of selected models for estimating solar radiation on horizontal surfaces. Solar Energy 43:153–168.

Davies JA, Kuhn P, Duhamel G, Binyamin J, Runnalls K (2000) An ultraviolet (290–325 nm) irradiance model for southern Canadian conditions. Physical Geography 21: 327–344

Elterman L (1968) UV, visible and IR attenuationfor altitudes to 50 km. Air Force Cambridge Research Laboratories EnvironmentalResearch Paper No 285,59p

Forster PM de F and Shine KH (1995). A comparison of two radiation schemes for calculating ultraviolet radiation. Quart J Roy Meteor Soc 121: 1113–1131

Hale GM, Querry MR (1973) Optical constants of water in the 200 nm to $200 \mu m$ wavelength region. Appl Opt 12: 555–562

Han Q, Rossow WB, Lacis AA (1994) Near-global survey of effective droplet radii in liquid water clouds using ISCCP data. J Climate 7, 465–497

Joseph JH, Wiscombe WJ, Weinman JA (1976) The delta-Eddington approximation for radiative flux transfer. J Atmos Sci 33: 2452–2459

Kerr JB, McElroy CT (1993) Evidence for large upward trends of ultraviolet-B radiation linked to ozone depletion. Science 262: 1032–1035

Kneizys FX, Shettle EP, Abreu LW, Chetwynd JH, Anderson GP, Gallery WO, Selby JEA, Clough SA (1988) Users guide to LOWTRAN 7, Technical Report 88–0177, Air Force Geophysics Laboratory, Bedford, Mass

Krotkov NA, Bhartia PK, Herman JR, Fioletov V, Kerr J (1998) Satellite estimation of spectral surface UV irradiance in the presence of tropospheric aerosols: 1 Cloud-free case. J Geophys Res 103: 8779–8793

Leontieva E, Stamnes K (1994) Estimations of cloud optical thickness from ground-based measurements of incoming solar radiation in the Arctic. J Climate 7: 566–578

Li Z, Wang P and Cihlar J (2000) A simple and efficient method for retrieving surface UV radiation dose rate from satellite. J Geophys Res 105: 5027–5036

Madronich S (1993) UV radiation in the natural and perturbed atmosphere. In: Tevini M (ed) UV–B radiation and ozone depletion. Lewis, Boca Raton, pp. 17–69

McKinlay AF, Diffey BL (1987) A reference action spectrum for ultraviolet-induced erythema in human skin. In: Passchler WR, Bosnajokovic BFM (eds) Human exposure to ultraviolet radiation: risks and regulations. Elsevier, Amsterdam

Pauer RJ, Bass AM (1985) The ultraviolet cross-sections of ozone:II Results and temperature dependence. In Zerefos C, Ghaz A (eds) Atmospheric Ozone Proceedings of the Quadrennial Ozone Symposium, Kalkidiki, Greece, Reidel, pp. 606–616

Setlow RB (1974) The wavelengths in sunlight effective in producing skin cancer: a theoretical analysis. Proc Natl Acad Sci USA 71: 3363–3366

Shettle EP, Fenn RW (1979) Models for the aerosols of the lower atmosphere and the effects of humidity variations on their optical properties. AFGL Technical Report 79–0214, Air Force Geophysics Laboratory, Environmental Research Papers, No 676, Bedford, Mass

Shettle EP, Weinman JA (1970) The transfer of solar irradiance through inhomogeneous turbid atmospheres evaluated by Eddington's approximation. J Atmos Sci 27: 1048–1055

Stamnes K, Tsay S-C, Wiscombe WJ and Jayawerra K (1988) Numerically stable algorithm for discrete ordinate method radiative transfer in multiple scattering and emitting layered media. Appl Opt 27: 2503–2509

Slingo A, Schrecker HM (1982) On the shortwave radiative properties of stratiform water clouds. Quart J Roy Meteor Soc 108: 407–426

Wang P, Lenoble J (1994) Comparison between measurements and modeling of UV–B irradiance for clear sky: a case study. Appl Opt 33: 3964–3971

Wang P, Li Z, Cihlar J, Wardle DI, Kerr J (2000) Validation of an UV inversion algorithm using satellite and surface measurements. J Geophys Res 105: 5037–5048

Zeng JR, McKenzie R, Stamnes K, Wineland M, Rosen J (1994) Measured UV spectra compared with discrete ordinate method simulations. J Geophys Res 99: 23019–23030

Chapter 17
Angular Distribution of Sky Diffuse Radiance and Luminance

José Luis Torres and Luis Miguel Torres

1 Introduction

As a rule, available global irradiance data in weather stations are referred to the horizontal plane. On the other hand, the different solar systems are generally placed on sloping surfaces. Therefore, it is necessary to establish procedures for calculating the existing irradiance on tilted planes in which direct, diffuse and ground reflected irradiances are evaluated in separated ways. The importance of the diffuse fraction of the global radiation is commonly underestimated. However, it must be pointed that in latitudes from 40 to $60° \text{N}$, this fraction may become 40% to 60% of the yearly radiation received on a horizontal plane. Although this fraction can be lower on a sloping plane, it still supposes an important percentage that should be calculated in the most exact way, especially, in climates with frequent covered skies.

Estimation of the diffuse component has been tackled in most cases by means of models that calculate the radiation on a tilted plane from radiation data on the horizontal plane. Existing models basically differ in the treatment each of them makes of the sky diffuse radiation. In fact, diffuse radiation is caused by a number of complex processes due to the interaction of solar radiation with the molecules and particles of the atmosphere, i.e., simple and multiple scattering and absorption phenomena, which may take place simultaneously or not at every wavelength, modify the intensity and the spectrum of the incident radiation in the high part of the atmosphere and redistribute the energy in different directions until it reaches the Earth surface. The physical bases of the aforementioned phenomena have been known for a long time since J.W. Strutt (later known as Lord Rayleigh) in 1871 set the physical laws that govern the light dispersion for very small particles and later in 1908, Mie proposed his theory for bigger spherical particles. All the described phenomena cause that the

José Luis Torres
Public University of Navarre, Pamplona, Spain, e-mail: jlte@unavarra.es

Luis Miguel Torres
Public University of Navarre, Pamplona, Spain, e-mail: lmtorresgarcia@gmail.com

diffuse radiation has an anisotropic nature, highly non-uniform, and that the energy received from the different areas of the sky vault may be different. The newest models for calculating radiation on the tilted plane from radiation on the horizontal plane provide approximations for this physical reality. Three areas of brightness more or less differentiated depending on the insolation conditions are considered.

An alternative to this procedure consists of carrying out the calculation of the diffuse radiance on a sloping plane by integrating the sky radiance distribution coming from the part of the sky that is "seen" by the said plane. This alternative is particularly interesting for estimating the radiation on planes placed in urban environments or complex terrains. In these situations, the presence of obstacles more or less close to the plane of collection means that some parts of the sky vault may not be seen by the said plane at given moments of the day and, as a consequence, it is necessary to know the radiance corresponding to the hidden area in order to achieve a exact estimation of the available energy. The application of the described alternative is clear for the design of active and passive solar systems as well as for determining the thermal and energetic performance of buildings. In addition, the present interest for integrating photovoltaic systems gives importance to the calculation of the existing radiation on vertical surfaces due to the fact that modules are usually placed on the vertical walls of large buildings exhibiting a relatively small surface for solar exploitation on their roofs.

Moreover, the sky radiance and luminance have the same origin and nature (Luminance is directly related to radiance through the luminous efficacy), and the *relative* distributions of both magnitudes are almost identical, being different their absolute values. As a consequence, there is a parallelism between the angular distribution models of radiance and luminance in the sky vault. Luminance models can also be used for designing illumination systems. In fact, different computer tools, as *Radiance*, take into consideration the distribution of luminance in their calculations of both outside and inside illumination.

The best way of knowing the radiance or luminance distribution in the sky is by measuring it. Nevertheless, there are very few places with the said existing measurements (or even registrations of illuminance on the horizontal plane) whereas, generally, there are irradiance data available. Therefore, the use of mathematical models of angular distribution of radiance or luminance becomes essential in most places and there is where their importance lies.

2 Definitions

- **Radiant flux** (ϕ): is the total energy of the electromagnetic radiation emitted, received or carried for unit of time.
- **Radiance** (L): the radiant flux, in a direction, transmitted by an elemental beam that goes through a given point and propagates according to a solid angle $d\Omega$ that contains the given direction, divided by the product between the value of the solid angle, the area of a section of this beam that contains the point and the

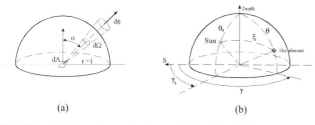

Fig. 17.1 (a) Radiant flux and radiance. (b) Angles defining the position of the sun and of the sky element

cosine of the angle between the normal to the section and the direction of the beam. This magnitude is also called intensity of radiation and is defined by the following formula (see Fig. 17.1.a):

$$L = \frac{d\phi}{dA \cdot \cos\alpha \cdot d\Omega} \quad (\text{Wm}^{-2}\text{sr}^{-1}) \tag{17.1}$$

- **Luminous flux** (ϕ_v): it results of the weighting of the radiation of the radiant flux according to the fotopic curve. The fotopic curve takes into account the effect of radiation on a patron fotometric observer.
- **Luminance** (L_v): conceptually, it is equal to the radiance as the only difference is that the luminous flux is considered instead of the radiant one. Its mathematical expression is as follows:

$$L_v = \frac{d\phi_v}{dA \cdot \cos\alpha \cdot d\Omega} \quad (\text{cdm}^{-2}) \tag{17.2}$$

- **Illuminance** (D): the total luminous flux incident on a surface, per unit area.
- **Relative scattering indicatrix function** $(f(\xi))$: it represents the radiance/luminance in a determined direction produced by the dispersion of the sunbeams, divided by that corresponding to the perpendicular direction to those beams. It is a function of the angular distance (ξ) between the direction of the sunbeams and the direction in question (see Fig. 17.1.b).
- **Gradation function** $g(\theta)$: it represents the drop or rise in sky radiance/luminance from the zenith towards the horizon.

3 Models of Angular Distribution of Diffuse Radiance and Luminance

Most proposed models for calculating the distribution of the radiance or the luminance in the sky have an empirical nature and are referred to average conditions. Many pieces of work on these models have dealt with skies whose characteristics may simplify the analysis, such as clear or overcast skies. Moreover, it is possible that these conditions are actually the ones of interest. For instance, the overcast sky situation is the most unfavorable for the design of illumination installations, and, consequently, the reference for the calculus.

Amongst all the different models that can be found in related literature, it has been decided to describe only those that exhibit some of the characteristics listed below: to be defined themselves as valid for all types of skies, to present different mathematical formulations, or to be still employed for comparison with the results of new proposals. The general nature of these models forces them to establish internal procedures that take into account the type of sky. This means that the sky conditions must be parameterized by means of a series of indexes which not always agree and which allow to distinguish between different categories of sky that, in most cases, do not exactly agree either. Anyway, in order to make the applicability of the models easier, these indexes have been chosen so that they are easily calculated from normally available data in any weather station.

Even if most models have an empirical nature and are referred to average conditions, some others have been proposed that tackle the problem from a rational basis. There are some based on the resolution of the equation of radiative transference in the atmosphere (semi-empirical model) and others that take into consideration the statistical nature of clouds in order to approximate calculations of radiance values in a certain place of the sky and in short periods of time (as Three Discrete Radiance Components model).

Anyone, by simple observation, can see that radiance and luminance in the sky are not uniform and have a distributed nature. In contrast to this fact, the first models for irradiance calculation on tilted planes established very strong simplifications. On the one hand, the isotropic model considered a uniform and isotropic radiance all over the entire sky vault. In this case, there is a direct relation between radiance and the diffuse irradiance (G_d) given by:

$$L = \frac{G_d}{\pi} \tag{17.3}$$

On the other hand, the heliocentric model assumed that all the sky radiance came from the same direction than the direct radiation. This is mathematically expressed by:

$$L = \frac{G_d}{\cos \theta_s} \delta(\cos \theta - \cos \theta_s)\delta(\gamma - \gamma_s) \tag{17.4}$$

Where $\delta(x)$ is a delta of Dirac function. Also, θ and γ are the zenith and azimuth angles of the sky element and θ_s and γ_s are the same quantities but corresponding to the sun.

A combination of the previous proposals was presented by Hay (1978) in his *Fixed Combinational Model,* where he assigned an isotropic nature to one half of the diffuse radiance and a directional from the sun nature to the other half.

Of course, it has been well-known for a long time that radiance/luminance have not, in general, an isotropic and uniform nature. Lambert (1760) (cited by Kittler et al 1997), Schramm (1901) (cited by Kittler and Valko 1993) and, subsequently,

Moon and Spencer (1942) made already clear that luminance from different spots was, in fact, different. The latter pointed out that the isotropic case could not be even maintained in the overcast sky situation. A function (later known as gradiation function) was proposed that explained the variation of radiance/luminance in this type of skies, as the observer moved from the zenith to the horizon. Overcast skies were considered to have a distribution of radiance/luminance practically independent of the sun's position and which only depended on the zenith angle of the point where the radiance/luminance was to be calculated. Actually, the isotropy situation of radiance could only happen in a bright overcast sky ideal.

Hay himself improved his *Fixed Combinational Model* incorporating an anisotropy index that was a function of the sky conditions (in particular, of the direct and extraterrestrial solar radiation and of the optical air mass). With this index, it was possible to change the part of the diffuse radiance corresponding to the isotropic and heliocentric components, resulting a model called *Variable Combinational Model*.

After these very simple first approximations to the problem of radiance/luminance, other complex models were developed. They will be described below, taking into account preceding considerations. Models to be presented are classified as follows: we will initially distinguish between the ones specifically proposed for luminance and the ones for radiance. Among the latter, three subgroups will be made: semi-empirical, stochastic and empirical.

3.1 Models Proposed for the Angular Distribution of Luminance

The first advances in modeling the angular distribution of luminance in the sky were carried out in skies presenting some continuity, such as overcast or clear skies. On the one hand, Moon and Spencer (1942) suggested the first CIE standard with non-uniform nature for covered skies that was put into practice in the classical expression that follows:

$$\frac{L_v(\theta)}{L_{v,z}} = \frac{1 + 2 \cdot \cos \theta}{3} \tag{17.5}$$

On the other hand, Kittler (1967) established the expression of luminance for clear skies, which was later included in the ISO/CIE standard (1996):

$$\frac{L_v(\theta, \gamma)}{L_{v,z}} = \frac{(1 - \exp(-0.32/\cos \theta))(0.91 + 10 \cdot \exp(-3\xi) + 0.45 \cdot \cos^2 \xi)}{0.274 (0.91 + 10 \cdot \exp(-3\xi) + 0.45 \cdot \cos^2 \xi)} \tag{17.6}$$

In the expression above, the scattering indicatrix appears along with the gradation function.

Between the two described extreme situations many other types of sky can be found. A number of works have been developed trying to propose expressions valid for every type of sky, including the aforementioned extremes. There are proposals in which the luminance distribution in a given sky is calculated by linear combination of the ones corresponding to some skies of reference. Others postulate general expressions that include coefficients whose variation takes into account the type of sky.

3.1.1 Models for Every Type of Sky Considering a Linear Combination of Skies of Reference

Perez's model (Perez et al. 1990). These authors formulated an operational model based on the combination of the proposals that the CIE had previously made for the distribution of luminance in four types of sky: clear, clear turbid, intermediate and overcast. In the said model, the luminance in a point become the product between the zenith luminance and a geometrical factor, ψ:

$$L_v(\theta, \gamma) = L_{v,z} \cdot \psi \tag{17.7}$$

For each of the cited type of sky, the geometrical factor follows a specific and different expression although, in every case, it depends on the position of both the point and the sun. For the rest of skies, the geometrical factor to be considered is obtained by linear combination of two of the ones corresponding to the four specified skies of reference. It is assumed that any type of sky is characterized by two parameters: the sky brightness (Δ) and the sky clearness (ε). The said parameters define the types of sky of reference to be considered as well as the weight corresponding to each of the types when performing the linear interpolation to obtain the geometrical factor.

A modification of Perez's model is known as ASRC-CIE-combination. In it, the luminance in a point is calculated as a linear combination of the expressions corresponding to the four types of sky of reference being the weights, again, dependant on the sky clearness ε.

Matsuura's model (Matsuura and Iwata 1990). The structure of this model is similar to the one of Perez, although there are two differences:

1. Only three skies of reference are considered for the lineal combination: clear, intermediate and overcast. The expressions used for them are the same as those of Perez's model.
2. The coefficients of the contributions of each sky of reference depend on the ratio between diffuse and global illuminance.

3.1.2 Luminance Models with a General Formulation Including Coefficients Dependant on the Type of Sky

Perradeau's model (1988). According to this model, the luminance in a point of the sky vault relative to the horizontal diffuse irradiance is calculated by the product of three functions: a gradation function, an indicatrix function and a third one dependant on the zenith angle of the sun. The coefficients of the corresponding expressions are tabulated as a function of the five types of sky under consideration (overcast, intermediate overcast, intermediate mean, intermediate blue and blue). The nebulosity index (see Eq. 17.23a) is the parameter for determining the type of sky to be considered among the five mentioned.

Perez all-weather model (Perez et al. 1993). In this model, the proposed expression for the luminance in a point relative to another point of reference $(l_v(\theta,\gamma))$ is as follows:

$$l_v(\theta,\gamma) = \frac{L_v(\theta,\gamma)}{L_{v,ref}} = [1+a\cdot\exp(b/\cos\theta)]\left[1+c\exp(d\xi)+e\cos^2\xi\right] \quad (17.8)$$

The zenith $(L_{v,ref}=L_{v,z})$ may be used as reference direction, for which $\theta=0$ and $\xi=\theta_s$. As a consequence, luminance in a point may be determined by knowing the one of the zenith and the value of the coefficients a,b,c,d and e that depend on the conditions of the sky parameterized by Δ, ε and θ_s.

CIE standard general sky (CIE 2004). In this standard, the relative luminance of a point with respect of the zenith luminance is obtained by the product of the gradation and indicatrix functions, both of them applied to the direction of the point in question, divided by the same product corresponding to the zenith direction. In brief:

$$\frac{L_v(\theta,\gamma)}{L_{v,z}} = \frac{f(\xi)\cdot g(\theta)}{f(\theta_s)g(0)} \quad (17.9)$$

$$\frac{L_v(\theta,\gamma)}{L_{v,z}} = \frac{\{1+c\left[\exp(d\xi)-\exp(d\pi/2)\right]+e\cos^2\xi\}\cdot\{1+a\cdot\exp(b/\cos\theta)\}}{\{1+c\left[\exp(d\theta_s)-\exp(d\pi/2)\right]+e\cos^2\theta_s\}\cdot\{1+a\cdot\exp(b)\}}$$

It can be appreciated that the formulation is similar to the one proposed by Perez in his all-weather model. Nevertheless, some differences must be emphasized. On the one hand, a new term $(\exp(d\pi/2))$ appears in the scattering indicatrix function. It was introduced by Kittler (1994) as a necessary correction according to the concept of relative scattering indicatrix.

On the other hand, six groups of a and b coefficients and other six groups of c, d and e coefficients are considered instead of employing continuous parameters for establishing the type of sky and the value of the coefficients of Eq. (17.9). Subsequently, six different gradation and indicatrix functions exist whose combination may produce up to thirty six different types of sky. Notwithstanding, the standard only includes those fifteen considered to be of more interest, although some work in low latitudes seems to make clear the need of increasing the current type of skies under consideration (Wittkopf and Soon 2007).

In order to choose the type of sky to be used in each case, one of the following procedures may be employed:

1. By comparison of the theoretical functions of gradation, $g(\theta)$, and indicatrix, $f(\xi)$, with their observed counterparts. The observed gradation function may be determined from luminance measurements in different points of the sky located on the plane of the solar meridian and on another one perpendicular to it. In contrast, for establishing the observed indicatrix function, it is necessary to know the luminance in different points of almucantars of different altitudes.
2. By the analysis of the ratio between the zenith's luminance and the diffuse illuminance (D_v) on a horizontal plane. In this respect, Kittler et al (1997) graphically

represented $L_{v,z}/D_v = f(\theta_s)$ and observed how the said ratio could be used for identifying the type of sky (at least, for solar elevation angles lower than $30°$).
3. By the analysis of the relative luminances with respect to the diffuse illuminance in a large number of points, according to Tergenza's (1999) proposal.

Anyway, it must be remembered that the luminance distributions given by this model and by the previous ones are characterized by:

1. Being symmetrical with respect to the solar meridian plane.
2. Adapting well to clear or homogeneously overcast skies (this is a consequence of the distributions being continuous). Nevertheless, they can also provide a good approximation for skies with scattered clouds in many practical calculations.

3.2 Models Proposed for Radiance Angular Distribution

3.2.1 Radiance Distribution Model of Semi-Empirical Nature

These models are based on the fact that the radiance/luminance distribution in the sky vault may be calculated by solving the equation of radiative transfer in the atmosphere, which establishes that:

$$\frac{dL(\theta,\gamma)_\lambda}{d\tau_\lambda} = -L(\theta,\gamma)_\lambda + J(\theta,\gamma)_\lambda \tag{17.10}$$

where $J(\theta,\gamma)_\lambda$ is a source function with which the contribution of the emission and scattering to the radiation in the atmosphere is considered. τ is the optical path and λ is the wavelength.

In this respect, Siala and Hooper (1989) developed a model where both the case of a simple dispersion and the realistic one of multiple dispersions are considered. In the latter, the approximation of successive dispersion orders was adopted for solving the equation of radiative transfer.

Other simplifications are also included. Most notoriously, the equation of radiative transfer is applied to the whole bandwidth of the sky radiation (without paying attention to the monochromatic nature of the said equation) and the emission contribution is disregarded, which reduces the source function to the dispersion component. For the calculation of the source function, Henyey-Greenstein phase function is used as it has been proved to fulfil the necessary requisites (Petty 2004) and facilitates the solution of the equation of radiative transfer.

The mathematical formulation of the normalized radiance with respect to the solar constant (G_{sc}) (obtained by solving the equation of radiative transfer) for a sole dispersion is as follows:

$$\frac{L}{G_{sc}} = \frac{h_1 \cos\theta_s}{4\pi(\cos\theta - \cos\theta_s)} \frac{1-g^2}{(1+g^2-2g\cos\xi)^{3/2}} [\exp(-h_2 \sec\theta) - \exp(-h_2 \sec\theta_s)]$$
$$for \quad \theta \neq \theta_s \tag{17.11a}$$

and

$$\frac{L}{G_{sc}} = \frac{h_1 \cdot h_2}{4\pi \cos \theta_s} \cdot \frac{1 - g^2}{(1 + g^2 - 2g \cos \xi)^{3/2}} \exp(-h_2 \sec \theta_s) \; for \; \theta = \theta_s \quad (17.11b)$$

Parameters h_1, h_2 and g were obtained from the observed data by a non-linear least squares fitting and g is the asymmetry factor of the scattering phase function.

For bigger orders of dispersion, the previous expressions become more difficult due to the appearance of integrals of the phase function. The small improvement in the accuracy achieved by including two orders of scattering instead of one does not seem to justify the added complexity of the model.

The calibration of the model, for which data collected in Canada during a year was averaged over all sky conditions, provided values of 1.287, 0.46 and 0.438 for the parameters $h_1, h_2 \, y \, g$.

3.2.2 Empirical Models of Radiance Distribution

Three component continuous distribution model or TCCD (Hooper and Brunger 1980). In this model, the absolute radiance of a point on the sky vault is obtained by the sum of the contributions of three differentiated components: the isotropic, the circumsolar and the horizon brightness, according to the following expression:

$$L(\theta, \gamma) = G_d \left[A_0 + A_1 \left(\frac{\theta}{\pi/2} \right)^2 + A_2 \exp(-c \cdot \xi \cdot \exp(d \cdot \theta_s)) \right] \quad (17.12)$$

The first term inside the square brackets represents the isotropic contribution, which is constant through the sky dome. The second one corresponds to the horizon brightness which is due to the bigger optical air mass that scatters the radiation towards the observer from that direction. As it can be observed, this term follows the square of the zenith angle and is independent of the azimuth. The third term, and last, is the circumsolar term, which is a consequence of the anisotropic scattering of the radiation by the elements of the atmosphere, with a bias towards the smaller dispersion angles, i.e, closer to the direction of the sun beams.

The expression already follows the general rule which establishes that radiance/luminance in a clear sky element depends on (ξ) and (θ).

In Eq. (17.12) c and d are two positive constants whose values, in the calibration carried out with the radiance in clear skies data collected by Steven (1979) in Canada, were 0.0145423 and 0.0231798 respectively.

The anisotropy factors A_i, on their part, depend on the atmospheric conditions and let the model consider any kind of sky. In this way, Rosen et al (1989) collected the values corresponding to skies with different cloud covers and different angles of the solar zenith.

Brunger's model. Brunger (1987) decided to keep the structure of the three components corresponding to TCCD model but replaced the parabolic function that represented the horizon-brightening (or darkening) with a cosine function. He did so after confirming from radiance measurements that the cosine formulation, proposed by Moon and Spencer (1942), was a good fit to the horizon-brightening. Furthermore, he opted for determining the normalized radiance with respect to the horizontal diffuse irradiance. He argued that, with this procedure, the variance of the values correspondding the areas of high zenith angles was reduced as opposed to the other normalization alternative which was to use the zenith radiance.

$$\frac{L(\theta,\xi)}{G_d} = \frac{a_0 + a_1\cos\theta + a_2\exp(-a_3\xi)}{\pi(a_0 + 2a_1/a_3) + 2a_2 \cdot I(\theta_s,a_3)} \tag{17.13}$$

The denominator of the second member of Eq. (17.3) comes from the condition that the integral of the radiance extended through the entire sky dome must be equal to the horizontal diffuse irradiance and, as a consequence, it happens that for the normalized radiance:

$$\int_0^{2\pi}\int_0^{\pi/2}[a_0 + a_1\cos\theta + a_2\exp(-a_3\xi)]\cos\theta \cdot \sin\theta \cdot d\theta \cdot d\gamma = \tag{17.14}$$
$$= \pi(a_0 + 2a_1/a_3) + 2a_2 I(\theta_s,a_3) = 1$$

with

$$I(\theta_s,a_3) = \frac{1+\exp(-a_3\pi/2)}{a_3^2+4}\left[\pi - \left(1 - \frac{2(1-\exp(-a_3\pi))}{\pi a_3(1+\exp(-a_3\pi/2))}\right)(2\theta_s\sin\theta_s - 0.02\pi\sin(2\theta_s))\right]$$
$$\tag{17.15}$$

The parameters a_i are calculated from two indexes that characterize the several sky conditions and which, in this model, are atmospheric clearness index $(k_t = G/G_0)$ and the diffuse fraction or cloud ratio $(Ce = G_d/G)$. They are collected in Table 6.1 of Brunger (1987).

Three discrete radiance components model (TDRC). The previous models, as some others that will be described subsequently, take into account average distributions of the radiance in the sky and use continuous functions for it. Nevertheless, in a given instant, partially covered skies do not show that kind of behaviour. In order to model the real time radiance in each area of the sky and, as a consequence, introduce time as a variable, Rosen (1983) and Rosen and Hooper (1987a) proposed a model in which the radiance of partially covered skies comes from three components. The first one, the clear component, corresponds to clear skies, i.e. it represents the radiation received from cloud-free regions of sky. The second is the scattered component; it is the received radiation from the part of a cloud that is not exposed to direct radiation and from the base of the same cloud. The last term is the reflected component, the one coming from the part of the cloud exposed to direct radiation.

Furthermore, the model assumes that each of the mentioned components, which mean discrete contributions to the total radiance, has a continuous distribution through the sky and can be modelled by TCCD model.

For the real time case, the base equation of the model is:

$$L(\theta,\gamma,t) = L_i(\theta,\gamma,t) \qquad (17.16)$$

Where the subscript i indicates the kind of sky element: clear, scattered or reflected. This model can also be used for determining the radiance distribution averaged throughout a period of time with the following formulation:

$$\overline{L}(\theta,\gamma) = \sum_{i=1}^{i=3} L_i(\theta,\gamma) \cdot x_i(\theta,\gamma) \qquad (17.17)$$

being x_i a geometrical factor that represents the fraction of time when the component of the radiance L_i is in the position of the sky given by (θ,γ).

For the calculation of the aforementioned factors, the use of the Cloudy Sky Geometry, that describes the geometrical configuration of partially clouded skies, has been proposed. In this configuration, clouds are considered as vertical cylinders randomly distributed through the atmosphere. Rosen and Hooper (1988) developed analytical expressions for the said factors, valid for the case of only one layer of clouds. For the application of TCCD model to each of the previous components (clear, scattered or reflected), Rosen and Hooper (1987b) establish expressions for the anisotropy factors (A_i). In this way, for the clear sky component (L_1), factors for this kind of skies are to be used (as the ones that can be found in Rosen et al (1989)). For the scattered component, the coefficients are a function of an empirical parameter (P) related to the fractions of clear and covered sky as well as of the factors for clear and opaque skies, as the following expression shows:

$$A_i^{non-opaque} = (1-P)A_i^{opaque} + (P)A_i^{clear} \ for \ i = 0,1,2 \qquad (17.18)$$

Lastly, for the reflected component (L_3) a term is added to L_1 in order to consider the reflected direct radiation, which takes into account the attenuation of the solar radiation in the atmosphere and the directional albedo of clouds.

A comparison with other models let conclude that the described one seems to be able to represent the instantaneous distribution of the radiance as well as the time averaged.

Igawa's model (Igawa et al 2004). In this model the formulation for obtaining the radiance and luminance distributions is the same for both magnitudes, being similar to that proposed in the CIE 2004 standard, that is, the one collected in Eq. (17.9).

The coefficients for radiance can be obtained from the kind of sky that is determined by the sky index (Si) as it is shown in the following expressions:

$$a = 4.5/[1+0.15 \cdot \exp(3.4 \cdot Si)] - 1.04 \qquad (17.19)$$
$$b = -1/[1+0.17 \cdot \exp(1.3 \cdot Si)] - 0.05$$
$$c = 1.77 \cdot (1.22 \cdot Si)^{3.56} \cdot \exp(0.2 \cdot Si) \cdot (2.1 - Si)^{0.8}$$
$$d = -3.05/[1+10.6 \cdot \exp(-3.4 \cdot Si)]$$
$$e = 0.48/[1+245 \cdot \exp(-4.13 \cdot Si)]$$

Table 17.1 Classification of kinds of sky according to the sky index

Sky index range	Classification of sky conditions
$Si \geq 1.7$	Clear Sky
$1.7 \geq Si > 1.5$	Near Clear Sky
$1.5 \geq Si > 0.6$	Intermediate Sky
$0.6 \geq Si > 0.3$	Near Overcast Sky
$Si \leq 0.3$	Overcast Sky

The sky index obeys the equation:

$$Si = \frac{G}{sG} + Cle^{0.5} \tag{17.20}$$

Five kinds of sky are distinguished according to the value of Si, as shown in Table 17.1.

sG is the standard global irradiance obtained from Kasten's global irradiance of clear sky, considering that Linke's turbidity factor is equal to 2.5.

$$sG = 0.84 \cdot \frac{G_{sc}}{m} \cdot \exp(-0.0675 \cdot m) \tag{17.21}$$

and the optical air mass is:

$$m = \left[\cos\theta_s + 0.50572 \cdot (96.07995 - \theta_s)^{-1.6364} \right]^{-1} \text{ with } \theta_s \text{ in deg} \tag{17.22}$$

Cle is the cloudless index, previously called nebulosity index by Perradeau.

$$Cle = \frac{1 - Ce}{1 - Ces}; \ Ce = \frac{G_d}{G} \tag{17.23a,b}$$

Ce is the cloud ratio and Ces is the standard cloud ratio, given by:

$$Ces = 0.01299 + 0.07698m - 0.003857m^2 + 0.0001054m^3 - 0.000001031m^4 \tag{17.24}$$

As a consequence, once known the position of the point in the sky vault where the radiance is to be determined, the moment when the calculation is to be carried out and the global and diffuse irradiance on horizontal plane, it is possible to determine the radiance relative to the zenith with Eq. (17.9) and Eqs. (17.19) to (17.24).

In order to obtain the absolute radiance of a point, it is necessary to know the zenith radiance. This can be measured or calculated taking into account that the integration of radiances through the entire seen hemisphere is equal to the diffuse irradiance on the horizontal plane, which leads to:

$$G_d = \int_0^{2\pi} \int_0^{\pi/2} L(\theta,\gamma) \cdot \cos\theta \cdot \sin\theta \cdot d\theta \cdot d\gamma = \int_0^{2\pi} \int_0^{\pi/2} l_r(\theta,\gamma) \cdot L_z \cdot \cos\theta \cdot \sin\theta \cdot d\theta \cdot d\gamma \tag{17.25}$$

and, therefore:

$$L_z = \frac{G_d}{\int_0^{2\pi} \int_0^{\pi/2} l_r(\theta, \gamma) \cdot \cos\theta \cdot \sin\theta \cdot d\theta \cdot d\gamma} \tag{17.26}$$

Once the distribution of $l_r(\theta, \gamma)$ is known, the integral of the denominator of Eq. (17.26) can be carried out thereby calculating the value of the zenith radiance. Another option is to calculate L_z/G_d from the following equation:

$$L_z/G_d = \sum_{k=0}^{k=4} A(k) \cdot K_c^k \tag{17.27}$$

with

$$A(k) = \sum_{j=0}^{j=6} \left[B(j,k) \cdot Cle^{0.5j} \right] \ and \ B(j,k) = \sum_{i=0}^{i=5} C(i,j,k) \cdot (\pi/2 - \theta_s)^i \tag{17.28}$$

Coefficients $C(i, j, k)$ appear in Table 17.2 and Kc (clear sky index) results from the ratio between the global irradiance (G) and the standard global irradiance (sG).

Example 1. Calculate the diffuse radiance in a point of the sky given by a zenith angle of 30° and an azimuth of 15° west when:

$$\theta_s = 35.88°$$
$$\gamma_s = -28.32° \ (east \ negative)$$

Global and diffuse irradiances are 700 and 80 Wm^{-2}, respectively.

Solution 1. First of all, the angle (ξ) between the direction of the sun and that of the point where the radiance is being evaluated is calculated by means of the expression:

$$\cos\xi = \cos\theta_s \cdot \cos\theta + sen\theta_s \cdot sen\theta \cdot \cos(\gamma_s - \gamma)$$

By introducing the value of θ_s in Eq. (17.22) an air mass of 1.232 is obtained.

Once m has been calculated, by means of Eqs. (17.21) and (17.24) sG and Ces can be obtained, their values becoming 857.1 and 0.1022 respectively.

The cloud ratio (Ce) is calculated by dividing the diffuse irradiance by the global radiance, according to Eq. (17.23b), resulting 0.1143.

Now the cloudless index can be calculated with Eq. (17.23a). $Cle = 0.9865$.

The sky index is:

$$Si = \frac{700}{857.1} + 0.9865^{0.5} = 1.8099 \approx 1.81$$

As a consequence, it can be said that, according to Igawa classification, this is a clear sky.

The use of the set of Eq. (17.19) results in the following values for a, b, c, d and e parameters:

$$a = -0.97714; \; b = -0.408697; \; c = 15.845; \; d = -2.9828; \; e = 0.42145$$

All the values necessary for the calculation of the radiance relative to the zenith have already been obtained. Introducing them in the following equation results in:

$$l_r(\theta, \gamma) = \frac{\{1 + c\,[\exp(d\xi) - \exp(d\pi/2)] + e\cos^2\xi\} \cdot \{1 + a \cdot \exp(b/\cos\theta)\}}{\{1 + c\,[\exp(d\theta_s) - \exp(d\pi/2)] + e\cos^2\theta_s\} \cdot \{1 + a \cdot \exp(b)\}} = 1.803$$

Determination of the zenith radiance by means of Eq. (17.27) requires the previous calculation of the clear sky index and of the coefficients A_k.

$$K_c = \frac{G}{sG} = 0.817$$

Coefficients $B(j, k)$ can be calculated taking into account the coefficients $C(i, j, k)$ of the Table 17.2 and θ_s. They are shown in Table 17.3.

Lastly, the radiance in the specified point of the sky is obtained by multiplying $l_r(\theta, \gamma)$ and L_z and it results in $46\,\mathrm{Wm^{-2}sr^{-1}}$.

In the case of the luminance distribution, the expression is the same as the one for the radiance. The same data can be used for characterizing the sky and, as a consequence, for determining the coefficients of the model. The only difference is that, when calculating the zenith luminance, the radiance calculated for the same point (following the previous procedure), must be multiplied by the diffuse luminous efficacy (η_d):

$$L_{vz} = \eta_d \cdot G_d \sum_{k=0}^{k=4} A(k) \cdot K_c^k \tag{17.29}$$

and

$$L_v(\theta, \gamma) = \frac{f(\xi) \cdot g(\theta)}{f(\theta_s)g(0)} \cdot L_{v,z} \tag{17.30}$$

Igawa et al (2004) validated different models of angular radiance and luminance distribution using data corresponding to a period of time longer than a year, in Japan.

With respect to irradiance, the two models showing a better behaviour were the one proposed by these authors (20.4% RMSE) and the All Weather Model of Perez (22.8% RMSE), in spite of the fact that this second model was not specifically developed for determining radiance. Regarding the luminance, the two cited models were again the ones showing lower RMSE (15.8% and 19%, respectively). Results obtained by Ineichen (2005) also reveal that the two said models are among the three most exact for the luminance. In this case Perez's model exhibits a 36% RMSE whereas the one of Igawa reaches a value of 37%. It must be pointed that the study of Ineichen was performed considering ten weeks during the months of February, March and April, with data of Geneva; the average luminance of these data being lower than that observed by Igawa.

Table 17.2 Coefficients $C(i, j, k)$

k=0	j\i	5	4	3	2	1	0
	6	−2.3236	3.8397	1.3678	−2.8773	0.5302	−0.9167
	5	5.9466	−8.437	−8.4637	11.3017	−0.896	2.7842
	4	−4.8599	4.1516	15.9467	−16.8828	0.093	−2.8711
	3	1.1362	1.8522	−12.3903	11.6859	0.3772	1.0626
	2	0.2833	−1.9825	4.3546	−3.7111	−0.0829	−0.146
	1	−0.0817	0.33	−0.4699	0.4062	−0.0316	−0.0564
	0	−0.0068	0.0177	−0.02	0.009	0.0044	0.4015
k=1	j\i	5	4	3	2	1	0
	6	41.7667	−101.3222	83.3628	−44.0612	8.0616	7.835
	5	−128.3895	313.3116	−245.8816	131.5008	−38.0729	−26.2167
	4	140.6794	−344.9908	254.23	−137.6762	64.8618	31.5218
	3	−61.5252	150.9138	−100.4209	54.5553	−48.3263	−15.2722
	2	7.8043	−18.4959	9.7897	−3.9888	14.3852	1.6549
	1	0.199	−0.7237	0.3678	−0.806	−0.8417	0.2509
	0	0.0888	−0.2173	0.2446	−0.0842	−0.0767	−0.0509
k=2	j\i	5	4	3	2	1	0
	6	−140.165	329.5323	−283.3008	142.5802	0.0588	−24.2309
	5	450.5763	−1067.6426	872.1058	−442.3932	38.0638	85.2745
	4	−530.7575	1266.6167	−971.174	496.1507	−112.2037	−109.9849
	3	258.314	−617.6864	428.3022	−220.7694	112.7302	61.4569
	2	−37.2452	86.0313	−44.2429	20.6861	−41.5866	−11.1184
	1	−1.1026	3.9671	−2.6865	3.9616	2.8287	−1.2699
	0	−0.3572	0.8603	−1.097	0.3359	0.3435	0.0863
k=3	j\i	5	4	3	2	1	0
	6	178.9761	−414.3978	382.3517	−189.0674	−17.1604	26.7019
	5	−585.76778	1369.3651	−1197.9089	597.8211	18.6492	−98.6168
	4	717.1565	−1698.494	1398.0054	−701.8434	57.7891	133.1647
	3	−377.9488	899.2027	−681.3328	345.6662	−99.9641	−79.6806
	2	63.7066	−149.2886	93.5278	−47.5262	47.1414	17.6094
	1	2.2956	−7.2571	6.1345	−5.5006	−4.1306	1.0003
	0	0.4479	−1.032	1.7185	−0.3234	−0.5077	−0.3754
k=4	j\i	5	4	3	2	1	0
	6	−79.2551	181.5249	−178.8391	86.4222	13.7469	−9.3016
	5	259.0233	−599.2154	558.7982	−273.5933	−28.3222	36.8154
	4	−323.93	758.4764	−665.1001	327.9505	2.5328	−52.2028
	3	178.5947	−422.7656	343.9805	−171.1919	27.4527	32.8606
	2	−34.0204	80.2519	−57.1731	28.9888	−17.6413	−8.0731
	1	−0.9299	2.9337	−2.739	2.0469	1.9924	−0.1519
	0	−0.0673	0.12	−0.5003	−0.0077	0.2274	0.1944

3.2.3 Stochastic Model

It is a fact that the radiance in a point of the sky vault shows a stochastic nature
derived from the statistical behaviour of the different attenuator elements of the atmosphere. Trying to configure models that adjust as much as possible to the physical

Table 17.3 Coefficients $B(j,k)$

k\j	0	1	2	3	4	5	6
4	0.026	2.884	−8.752	−6.307	42.484	−45.346	14.981
3	−0.180	−6.690	27.617	−8.276	−71.592	92.460	−33.158
2	0.202	5.001	−28.750	34.446	17.435	−49.600	21.212
1	−0.099	−1.380	11.079	−23.003	15.358	0.800	−2.859
0	0.406	0.081	−1.231	3.732	−4.756	2.644	−0.521

reality, Siala and Hooper (1987) proposed a stochastic model for the angular distribution of the radiance in the sky vault. In this model, the radiance in a point results from the sum of a deterministic component and a stochastic one. The former can be calculated by means of the preceding models and it provides the average values in the said point. When it comes to dealing with the stochastic component, the starting point is the direct relation between the variance and the means of the radiance in a point of the sky vault (Siala et al 1987).

The basic expression of the stochastic model is the following:

$$L(\theta, \gamma, t) = \overline{L}(\theta, \gamma, t) \cdot (1 + S_t(\theta, \gamma)) \tag{17.31}$$

Where $\overline{L}(\theta, \gamma, t)$ corresponds to the determinist part and S_t is a random variable that can be considered as the normalized deviation from the means of the named apparent scattering optical thickness.

In the study that was carried out, it was noticed that the variable S_t was not Gaussian, so another normally distributed variable was generated, $W_t = f^t(S_t)$ by means of the application of a suitable transformation function. For determining this transformation function, the radiance was considered to follow a LogNormal distribution. The study of the time series of W_t corresponding to several sectors of the sky vault let set the most suitable ARMA(p,q) models. These models resulted to be mainly of the kind ARMA(2,1). As a consequence, the value of W_t was related to the ones of the same variable in two previous moments, as the following expression shows:

$$W_t = a_1 \cdot W_{t-1} + a_2 \cdot W_{t-2} + e_t - b_1 \cdot e_{t-1} \tag{17.32}$$

In which a_i, b_i are the basic parameters of the autoregressive part and of the moving average part of the model and the e_t, e_{t-1} are random shocks. Once the ARMA model parameters are estimated, the stochastic model let generate synthetic radiance series in a given position of the sky dome by applying the inverse process to that followed for elaborating the model. Steps to perform are described below.

1. Generation of random shocks
2. Application of ARMA model in order to obtain the time series of Wt
3. Inverse transformation in order to obtain St
4. Introduction of the average radiance and application of Eq. (17.31) in order to obtain the radiance in a given point and instant.

By means of the described process, radiance time series can be generated for every point in the sky. It must be remembered that these series, being different, are undistinguishable from the statistical point of view and equally probable as they show the same means, variances and autocorrelations. Therefore, the model cannot be expected to simulate a particular occurrence of sky conditions but rather to describe the average of a great number of said events with their intact statistical characteristics.

4 Equipment for Observing the Radiance/Luminance Distribution from the Sky Dome

Practical models for establishing the radiance or luminance distribution in the sky have an empirical nature, according to the remarks of the previous sections. Measuring equipment has been used for obtaining data of the said distribution, necessary for establishing the models. Some work on radiance/luminance distribution includes both the proposal of an angular distribution model and the designing of specific measuring equipment. For instance, Brunger (1987) designed and made an instrument for obtaining diffuse radiance measurements in different points of the sky. The said instrument was composed of a pyroelectric radiometer with a limited vision angle, a robot for pointing to different points of the sky and a shadow ball for hiding the sun.

One of the problems the first measuring instruments presented was the long time they spent on scanning the full sky dome. Nowadays, current equipment obtains data in considerable shorter time and there are even some technologies available to get information in real time.

Present equipment for measuring the radiance/luminance in different points of the sky can be classified according to different criteria. In this case, two groups have been considered: instruments with mobile sensors and with static sensors. The first ones are generally called sky scanners. They show a similar technology, are extensively diffused and have been used for a long time. In contrast, the instruments with static sensors are characterized by different technologies and a more restricted and recent use.

Instruments above obtain ground data; notwithstanding, some studies have been performed that try to relate the angular distribution of the luminance in the sky with the images coming from satellite sensors (Ineichen 1997).

4.1 Sky Scanners

According to CIE 108–1994, these instruments measure the luminance of 145 points of the sky dome. The two more commonly used are described below.

Sky scanner PRC Krochmann. In this instrument, a silicon photovoltaic sensor measures the luminance coming from 145 points of the sky dome. The measurement in the sun direction is left in blank in order to avoid damaging the Si- photovoltaic device. The sensor vision angle is 10°, as specified by the manufacturer. Changes in azimuth and elevation are obtained by combination of two different movements: a horizontal support rotates around a vertical axe and the sensor itself rotates around a horizontal axe (see Fig. 17.2.a). It is a quick instrument as 150 measurements are taken in every scan process in about 35 seconds (zenith luminance is recorded six times). The optical system is thermostatised at 35 °C in order to avoid the problems derived from the dependence of measurements with the temperature and the ice formation in the window sensor. Table 17.4 shows some specifications of interest of this scanner.

Sky Scanner EKO. Different models of this instrument measure luminance, radiance or both. In particular, model LR is designed to measure luminance and radiance of 145 points of sky hemisphere. It has two sensors (the one for luminance includes a $V(\lambda)$ filter) with an aperture angle of 11° and made of silicon photodiodes. Both sensors are integrated in a mobile head. A two-axis tracker let the head sweep the said 145 points. Table 17.4 contains the basic specifications of this instrument. Figure 17.2.b contains an image of model MS-321LR. In addition, Fig. 17.3 shows the graph of radiance distribution obtained with the software associated to this sky scanner. As it can be appreciated, the sky sphere is divided into 145 rectangular patches, with similar subtended solid angle and with known directions and limits. In every patch radiance/luminance is considered to be uniform.

4.2 Measuring Instruments of Radiance/Luminance Angular Distribution with Static Sensors

Solar-Igel. This instrument includes 135 silicon sensors (Gruffke et al 1998), distributed considering a radial orientation, according to the configuration based on the concept developed by Appelbaum (1987) that minimises overlapping and missed areas of the sky.

Each sensor is located in the bottom of a tube in order to get an aperture angle of 14.2°. Unlike sky scanners, Solar-Igel is an static instrument with no need of tracking, as shown in Fig. 17.4.a.

Table 17.4 Sky Scanner specifications

	PRC Krochmann	EKO
$V(\lambda)$ match	$F_1 < 3\%$	$F_1 < 2.5\%$
Measuring field angle	10°	11°
Measuring range:		
-Luminance	$0 - 65000$ (cd m^{-2})	$0–50000$ (cd m^{-2})
-Radiance		$0–300$ Wm^{-2} sr^{-1}
Total measuring points	145	145
Total measuring time (aprox.)	35"	3'
Operating temperature	-20°-+40°	0-+40°

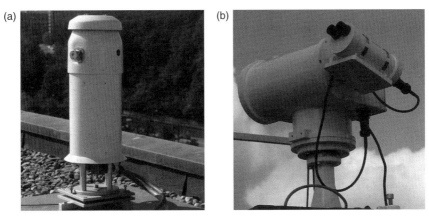

Fig. 17.2 (**a**) Image of Sky Scanner Krochmann (photo courtesy of Dr. Ineichen). (**b**) Image of Sky Scanner EKO

Each sensor registers the incident irradiance on its active surface coming from the portion of sky seen by it. The 135 measurements taken simultaneously let know the radiance distribution in the sky (Fig. 17.4.b). Apart from the described sensors, the instrument has an additional silicon sensor in the zenith, placed horizontally, with a range of angular response of 180°. This last sensor is used for avoiding possible ambiguities in the determination of global, diffuse and direct irradiance on the horizontal plane when the measurements of the 135 sensors are considered. These ambiguities may be caused by partial overlapping among the areas seen by the different sensors as well as by the existence of areas in the sky dome that are not observed by any of them.

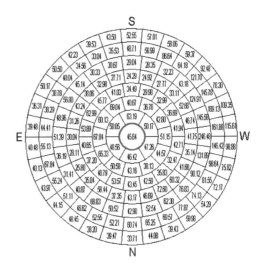

Fig. 17.3 Graph of radiance distribution obtained by Sky Scanner EKO

(a) 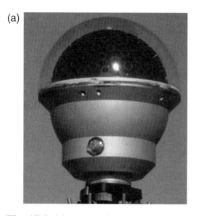   (b)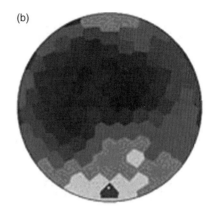

Fig. 17.4 (**a**) Image of SolarIgel. (**b**) Graphical representation of the radiance registred by 135 sensors. Darker patches show lower radiance values. The patch with a white point correspond to the sun position

Measuring instruments using charged coupled device (CCD). The Whole Sky Imager (Shields et al. 1998) can be cited as one of this kind of instruments. A digital image of the whole sky is obtained with it, with a field of view of 180° in the visible and near infrared bands. A later image processing let determine the radiance distribution in the said bands. The sensor is composed of a digital CCD camera that receives radiation through a fisheye in order to cover the 2π stereoradians of the sky dome. The sun is covered by a circular monitorized occultor with the aim of avoiding the camera saturation as well as the distortion of the field of radiance.

A number of works have also been performed using monochromatic CCD cameras. In these cases, in order to determine the radiance in the whole bandwidth of the diffuse radiation, it is necessary to correlate the amount of energy corresponding to the limited region of the sensor with that of the whole spectrum.

Acknowledgements We would like to thank professors Marian de Blas and Almudena García for their help with text revisions and with the figures of Chapter 17.

References

Appelbaum J (1987) A solar radiation distribution sensor. Solar Energy 39: 1–10
Brunger A (1987) The magnitude variability and angular characteristics of the shortwave sky radiance at Toronto. Ph.D. thesis, Toronto University
CIE S 011/E (2004). Spatial distribution of daylight – CIE Standard General Sky.
Gruffke M, Heisterkamp N, Matzak S, Otorjohann E, Vob J (1998) Solar irradiation measurement with silicon sensors. In: Schmid J (ed) 2nd World conference and exhibition on photovoltaic solar energy conversion. Vienna
Hay JE (1978) Measurement and modelling of shortwave radiation of inclined surfaces. In: Third Conference on atmospheric radiation. Amer. Meteor. Soc., Davis, pp 150–153

Hooper FC, Brunger AP (1980) A model for the angular distribution of sky radiance. J. Solar Energy Engineering 102:196–202

Igawa N, Koga Y, Matsuzawa T, Nakamura H (2004) Models of sky radiance distribution and sky luminance distribution. Solar Energy 77:137–157

Ineichen P (1997) Sky luminance distribution from Meteosat images. 3rd Satellight meeting. (Working paper), Les Marécottes

Ineichen P (2005) Angular distribution of the diffuse illuminance. Université de Genève. www.heliosat3.de/documents

Ineichen P, Molineaux B (1993) Characterisation and comparison of two sky scanners: PRC Krochmann and EKO instruments. IEA Task XVII expertmeeting, Geneva

Kittler R (1967) Standarisation of the outdoor conditions for the calculation of the daylight factor with clear skies. In: Hopkinson RG (ed) Conference sunlight in buildings, Rotterdam, pp 273–286

Kittler R (1994) Some qualities of scattering functions defining sky radiance distributions. Solar Energy 53:511–516

Kittler R, Valko P (1993) Radiance distribution on densely overcast skies: comparison with CIE luminance standard. Solar Energy 51:349–355

Kittler R, Perez R, Darula S (1997) A new generation of Sky Standards. In: 8th European lighting conference, Amsterdam, pp 359–373

Matsuura K, and Iwata TA (1990) Model of daylight source for the daylight illuminance calculations on the all weather conditions. In: Spiridonov A (ed) 3rd International daylighting conference, Moscow

Moon P, Spencer D (1942) Illumination from a non-uniform sky. Illuminating Engineering 37:707–726

Perez R, Ineichen P, Seals R, Michalsky J, Stewart R (1990) Modeling daylight availability and irradiance components from direct an global irradiance. Solar Energy 44:271–289

Perez R, Seals R, Michalsky J (1993) All weather model for sky luminance distribution. Preliminary configuration and validation. Solar Energy 50:235–245

Perraudeau M.(1988) Luminance models. In: National lighting conference and daylighting colloquium, Cambridge, pp 291–292

Petty GW (2004) A first course in atmospheric radiation. Sundog, Madison, Wisconsin

Rosen MA (1983) The characterization and modelling of the angular distribution of diffuse sky radiance. M.A. Sc thesis, Toronto University

Rosen MA, Hooper FC (1987a) A model for the instantaneous distribution of diffuse sky radiance. In: 11th Canadian congress of applied mechanics, Edmonton, pp E104–E105

Rosen MA, Hooper FC (1987b) A calibration of the three discrete radiance components model. In: Bloss WH, Pfisterer F (eds) 10th Biennal congress of the Int. Solar Energy Soc., Hamburg, pp 3747–3752

Rosen MA, Hooper FC (1988) The development of a model for the geometric description of clouds and cloudy skies. Solar Energy 41:361–369

Rosen MA, Hooper FC, Brunger AP (1989) Characterization and modelling of the diffuse radiance distribution. Solar Energy. 43:281–290

Shields JE, Johnson RW, Karr ME, Wertz JL (1998) Automated day/night whole sky imagers for field assessment of cloud cover distributions and radiance distributions. In: American Meteorological Society (ed) Tenth symposium on meteorological observations and instrumentation, Phoenix

Siala FMF, Hooper FC (1987) Stochastic modelling of the angular distribution. Proc. In: Bloss WH, Pfisterer F (eds) 10th Biennal congress of the Int. Solar Energy Soc., Hamburg, pp 3760–3767

Siala FMF, Hooper FC (1989) A semi-empirical model for the directional distribution of the diffuse sky radiance. In: Renewables- a clean energy solution, Solar Energy Soc. of Can., Ottawa, pp 322–326

Siala FMF, Hooper FC, Rosen MA (1987) An investigation of the statistical nature of sky radiance: The relation between mean and standard deviation. In: 10th Biennal congress of the Int. Solar Energy Soc., Hamburg, pp 3753–3759

Steven MD, Unsworth MH (1979) The diffuse solar irradiance of slopes under cloudless skies. Quarterly Journal of the Royal Meteorological Society 105: 593–602

Tregenza PR (1999) Standard skies for maritime climates. Lighting Research and Technology 31:97–106

Wittkopf SK, Soon LK (2007) Analysing sky luminance scans and predicting frequent sky patterns in Singapore. Lighting Research and Technology 39:31–51

Chapter 18
Solar Radiation Derived from Satellite Images

Jesús Polo, Luis F. Zarzalejo and Lourdes Ramírez

1 Introduction

The accurate knowledge of solar radiation at the earth's surface is of great interest in solar energy, meteorology, and many climatic applications. Ground solar irradiance data is the most accurate method for characterising the solar resource of a given site. However, despite the availability of ground databases is growing up through different measuring networks, its spatial density is usually far too low. In consequence, satellite-derived solar radiation has become a valuable tool for quantifying the solar irradiance at ground level for a large area. Thus, derived hourly values have proven to be at least as good as the accuracy of interpolation from ground stations at a distance of 25 km (Zelenka et al. 1999).

Several algorithms and models have been developed during the last two decades for estimating the solar irradiance at the earth surface from satellite images (Gautier et al. 1980; Tarpley 1979; Hay 1993). They can be generally grouped into physical and pure empirical or statistical models (Noia et al. 1993a, 1993b). Statistical models are simpler, since they do not need extensive and precise information on the composition of the atmosphere, and rely on simple statistical regression between satellite information and solar ground measurements. On the contrary, the physical models require as input the information of the atmospheric parameters that model the solar radiation attenuation through the earth's atmosphere. On the other hand, the statistical approach needs ground solar data and the models suffer of lack of generality.

Jesús Polo
CIEMAT, Solar Platform of Almería, Spain, e-mail: jesus.polo@ciemat.es

Luis F. Zarzalejo
CIEMAT, Solar Platform of Almería, Spain, e-mail: lf.zarzalejo@ciemat.es

Lourdes Ramírez
CIEMAT, Solar Platform of Almería, Spain, e-mail: lourdes.ramirez@ciemat.es

Satellites observing the earth can be grouped, according to its orbit, in polar orbiting and geostationary satellites. The former, with an orbit of about 800 km have high spatial resolution but a limited temporal coverage. The geostationary satellites, orbiting at about 36000 km, can offer a temporal resolution of up to 15 minutes and a spatial resolution of up to 1 km. Most of the methods for deriving solar radiation from satellite information make use of geostationary satellite images.

2 Fundamentals

2.1 Observing the Earth-Atmosphere System

Solar radiation traversing through the atmosphere interacts with the atmospheric constituents before reaching the surface. A part of this radiation is backscattered toward the space, a part is absorbed, and the remainder reaches the ground. The ground absorbs a part of the radiation reaching the earth's surface, while the remainder is again reflected toward the space. Therefore, the radiation emerging from the atmosphere is composed of the solar radiation backscattered by the atmosphere and the radiation reflected by the ground (Fig. 18.1). This is the information received by the sensor of a satellite observing the earth.

Considering the conservation of energy for the earth-atmosphere system, the following expression can be written.

$$I_0 = I_S + E_a + E_t \tag{18.1}$$

Here I_0, I_S, E_a, and E_t are respectively the extraterrestrial solar irradiation, the solar irradiation reflected by the earth-atmosphere system, the solar irradiation absorbed at the atmosphere, and the solar irradiation absorbed at the earth surface.

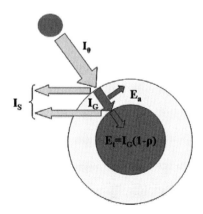

Fig. 18.1 Balance of short-wave radiation the earth-atmosphere system

The solar radiation absorbed at the earth surface may be expressed as a function of surface albedo (ρ) and incident solar irradiation (I_G).

$$E_t = I_G(1 - \rho) \tag{18.2}$$

Therefore, the solar radiation on the earth surface may be expressed as:

$$I_G = \frac{1}{1 - \rho} [I_0 - I_s - E_a] \tag{18.3}$$

In consequence, since the extraterrestrial irradiation (I_0) is well known, by the knowledge of the absorbed energy (E_a) and of the surface albedo, Eq. (18.3) could be used to derive the global radiation (I_G) from the radiation measured by the satellite's radiometer (I_S). Equation (18.3) is thus the fundamental equation for all the models aiming at deriving solar radiation from satellite images.

The use of satellite images to estimate the solar radiation has, in fact, noticeable advantages, in particular the following are worth to mention:

- Satellites collect information for large extensions of ground at the same time, which allows to identify the spatial variability of solar radiation at ground level.
- When the information available (satellite images) can be superimposed, that is, corresponds to the same area, it is possible to study the time evolution of values in an image pixel or in a certain geographic area.
- Satellites images allow the analysis of the solar resource in a potential emplacement that has no previous ground measuments.

2.2 The Cloud Index Concept

Basically, a satellite image used in solar radiation retrieval is a measure of the earth's radiance in the visible channel at a specific time and over a spatial window. The radiance values recorded by the radiometer of the satellite can vary according to the state of the atmosphere, from clear sky situations to complete overcast, and depending also on the reflectance of the ground surface. In this sense, satellite images give information of the cloudiness at a given time and site. In consequence, a normalized parameter describing the cloudiness can be defined from the radiance measurements of the satellite radiometer. This parameter is denoted as the cloud index, which can be mathematically defined as (Cano et al. 1986; Diabaté et al. 1989),

$$n = \frac{\rho - \rho_g}{\rho_c - \rho_g} \tag{18.4}$$

where ρ is the reflectance viewed by the satellite radiometer (denoted as the instantaneous planetary albedo), ρ_c is the reflectance of the clouds (i.e., cloud albedo), and ρ_g is the reflectance of the ground (i.e., ground albedo).

The reflectance can be obtained from the radiance, which is the actual magnitude measured by the satellite radiometer, by the next expression assuming isotropy,

$$\rho = \frac{\pi L_{\Delta}}{E_{\Delta}} \qquad (18.5)$$

where L_{Δ} is the radiance and E_{Δ} denotes the incident solar irradiance within the spectral band of the satellite radiometer.

The ground albedo corresponds to a clear, clean and dry sky, which can be associated with the minimum satellite count. On the other hand, cloud albedo, representing a heavily overcast sky, is associated to the maximum satellite count. Satellite counts must be properly normalized for avoid sun and/or satellite geometric effects (Ineichen and Perez 1999). From Eq. (18.4) is clear that under complete clear sky situations the instantaneous planetary albedo would be close to the ground albedo, and the cloud index would tend to zero; likewise, under complete overcast skies the instantaneous planetary albedo would approach to the cloud albedo and the cloud index tend to the unity.

The importance of the cloud index concept bases on the fact that satellite information (basically cloud cover information) can be related with the solar irradiance incoming to the earth surface. Consequently, most empirical/statistical methodologies to retrieve solar irradiance from satellite images rely on the assumption of linear relationship between the atmospheric transmittance and the cloud index (Cano et al. 1986; Diabaté et al. 1988; Schmetz. 1989; Diabaté et al. 1989; Noia et al. 1993a; Ineichen and Perez. 1999; Zelenka et al. 1999; Perez 2002; Rigollier et al. 2004; Zarzalejo et al. 2005).

3 Geostationary Meteorological Satellite Images

Geostationary satellites are positioned at an exact height above the Earth (about 36000 km), and they are rotating around the earth axis at the same speed as the earth does. Since they remain practically stationary, they are facing and collecting images of the same portion of the earth disc. Geostationary satellites are always positioned above equator, having thus a limited use in latitudes beyond 60–70 degrees north or south. This is also the reason for their decreasing spatial resolution with the latitude.

Geostationary meteorological satellites make use of instruments (mainly radiometers) on board to scan the images. According to the specific characteristics of the radiometers the following levels of resolution can be stated:

- The spatial resolution is the geographical size of the pixel in the image, which depends on the angular section detected by the sensor, denoted as instantaneous field of view (IFOV).
- The time resolution indicates the time interval between two consecutives images.
- The spectral resolution is associated to the spectral range and the number of channels (spectral bands) of the sensor.

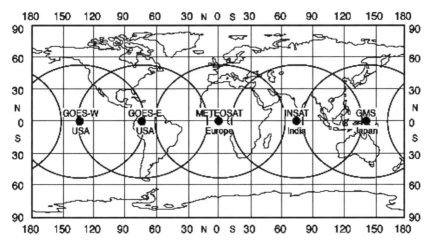

Fig. 18.2 Geostationary satellite coverage (Hauschild et al. 1992)

- The radiometric resolution characterizes the sensitivity of the radiometer and it is expressed as the number of bits needed for registering the pixel information.

Most of geostationary satellites include at least three main channels: a visible channel ($\sim$0.5–1.1 μm), thermal infrared channel ($\sim$10.5–12.5 μm), and a water vapour channel ($\sim$5.7–7.1 μm). Nevertheless, last generations of geostationary satellites are adding more numbers of spectral channels.

The whole earth is covered by about seven geostationary satellites positioned at regular intervals around the equator. Figure 18.2 illustrates the coverage of some geostationary satellites; two additional satellites, Elektro (Russia) and FY-2 (China), complete the list.

4 Satellite-Based Models for Deriving Solar Radiation

Several computational methods have been developed in the past two decades for estimating the downward solar irradiance from satellite information. Most of them have been evolved with improvements in many aspects of the modelling. In consequence, the former distinction between purely empirical or physical models has been diluted towards a more hybrid character. Nowadays most of currently used models for deriving solar radiation from satellite images contain both empirical and physical information. A review of current models is briefly presented here. The description of the Janjai et al. model has been intentionally removed from this compilation, since it is presented in another chapter of this book (Janjai et al. 2005).

4.1 Heliosat Model

Heliosat model was originally proposed by Cano (Cano et al. 1986) and later it has been modified and improved through different versions (Diabaté et al. 1988, 1989; Beyer et al. 1996). Heliosat models have been particularly used with Meteosat satellite images. Heliosat 1 was initially a pure empirical model and it evolved incorporating physical atmospheric parameters, such as the Linke turbidity factor in Heliosat 2 (Rigollier et al. 2004), and aerosol and other atmospheric absorbers parameters in the future Heliosat-3 version focused on Second Generation Meteosat (MGS) (Hammer et al. 2003).

Heliosat 1 model consist basically on proposing a linear relationship between the cloud index (Eq. 18.4) and the clearness index (k_T), defined as the global hourly irradiance normalised by the extraterrestrial irradiance,

$$k_T = an + b \qquad (18.6)$$

where a and b parameters have to be fitted with ground data.

Heliosat 2 deals with atmospheric and cloud extinction separately. As a first step the irradiance under clear skies is calculated by using the ESRA clear sky model (Rigollier et al. 2000), where the Linke turbidity factor is the only parameter required for the atmosphere composition. The Linke turbidity factor is defined as the number of clear and dry atmospheres (Rayleigh) that would yield the radiation extinction observed. The downward (sun to ground, T_d) and upward, Tu, (Happ et al. 1989) transmittances of the clear atmosphere, estimated by the ESRA clear sky model, are used to correct the estimation of the ground and cloud reflectance.

$$\rho_g = \frac{\rho - \rho_{atm}}{T_d T_u}$$
$$\rho_c = \frac{\rho_{eff} - \rho_{atm}}{T_d T_u} \qquad (18.7)$$

Here the intrinsic reflectance of the atmosphere, ρ_{atm}, can be estimated from the diffuse irradiance under clear sky, and the effective cloud albedo, ρ_{eff}, is defined by (Rigollier et al. 2004),

$$\rho_{eff} = 0.78 - 0.13[1 - \exp(-4\sin^4 \alpha)] \qquad (18.8)$$

being α the solar elevation angle.

Finally, the clear sky index, k_c, defined as the global irradiance normalised to the clear sky global irradiance, is related with the cloud index by the following empirical fit,

$$
\begin{array}{ll}
n < -0.2 , & k_C = 1.2 \\
-0.2 \leq n < 0.8 , & k_C = 1 - n \\
0.8 \leq n < 1.1 , & k_C = 2.0667 - 3.6667\,n + 1.6667\,n^2 \\
1.1 \leq n , & k_C = 0.05
\end{array}
\qquad (18.9)
$$

Some authors make use of the Heliosat method philosophy but they have modified particular aspects of the modelling. For instance, it can be mentioned the modification of the k_c-n relationship by including moments of the cloud index distribution (Zarzalejo. 2005), or the study of the influence of the three dimensional characteristics of cloud in the cloud index determination (Girodo et al. 2006). Dagestad et al. developed some corrections in the cloud index estimation for non-lambertian reflectivity and for the backscattered radiation from air molecules (Dagestad and Olseth 2007).

As a consequence of the enhanced information on spatial structure and spectral channels of MSG, Heliosat model is being improved towards a more physical model. Heliosat-3 model uses a new type of calculation scheme based on radiative transfer modelling, the SOLIS scheme (Mueller et al. 2004). MSG provides the potential for the retrieval of atmospheric parameters, and other platforms like GOME/ATSR-2 can be used also for the aerosols extinction. Therefore, Heliosat-3 calculation scheme does not rely on Linke turbidity factor as the only atmospheric parameter for modelling the solar radiation attenuation (Hammer et al. 2003).

4.2 The Operational Model of Perez et al.

The operational model of Perez et al. is an evolution of the original Cano model (Cano et al. 1986). It has been developed in the ASRC (Atmospheric Sciences Research Centre) of the University of Albany and applied to the GOES satellite images (Perez et al. 2002).

The model proposes to estimate the global hourly irradiance (I_G) from the cloud index (n) and the global irradiance for the clear sky (I_{CS}) by the following expression,

$$I_G = I_{CS} f(n) [0.0001 I_{CS} f(n) + 0.9]$$

$$f(n) = 2.36 n^5 - 6.2 n^4 + 6.22 n^3 - 2.36 n^2 - 0.58 n + 1$$

(18.10)

The clear sky global irradiance determination is based on the Kasten model with some modifications introduced by the authors in order to taking into account a revised formulation of the Linke turbidity coefficient which allows to remove its dependence on solar geometry (Ineichen and Perez 2002).

The model also makes use of external information concerning snow cover (provided for USA and Canada by the National Operational Hydrologic Remote Sensing Center) that allows the dynamic modification of the algorithm for determining the cloud index. Moreover, the model accounts of sun-satellite angle effects individually for each pixel.

Finally, the authors propose a methodology based upon the DIRINT model (Perez et al. 1992) and the new Linke turbidity coefficient formulation (Ineichen and Perez 2002) for estimating the direct normal irradiance (DNI) from global hourly irradiance. New methodologies developed by the authors permit to correct the retrieved irradiance values for applying to complex terrain, where high reflectance surface or juxtaposition of high and low reflectance surfaces occurs (Perez et al. 2004).

4.3 BRASIL-SR Model

The BRASIL-SR model is an evolution of the physical model developed by Möser and Raschke during the 80's (Moser and Raschke 1983). It provides solar irradiation maps using the two-stream approach to solve radiative transfer equation, the GOES satellite images and ground data on temperature, surface albedo, relative humidity and visibility (Pereira et al. 2000; Martins et al. 2007). The surface global irradiation is determined by the cloud index, and the sky transmittances for clear (T_{clear}) and cloudy (T_{cloud}) conditions by,

$$I_G = I_0 [n\, T_{cloud} + (1-n)\, T_{clear}] \qquad (18.11)$$

The boundary values for the sky transmittance are estimated by using a two-stream radiative transfer scheme that accounts for absorption and scattering by gases and aerosols assuming realistic atmospheres, that need of climatic data (temperature, relative humidity, surface albedo, visibility and cloud properties) as input. The model also estimates the DNI as function of the clear sky and cloud transmittance for beam solar radiation. The former is obtained by the radiative transfer scheme and the latter from the cloud index.

4.4 The DLR-SOLEMI Method for DNI

The DLR-SOLEMI model is focused exclusively on direct normal irradiance derived from Meteosat satellite images (Schillings et al. 2004a, 2004b). The model is based on Bird's clear sky model for estimating the DNI under clear sky conditions (Bird and Hulstrom 1983),

$$DNI_{clear} = I_{0n} T_R T_g T_o T_w T_a \qquad (18.12)$$

where T_i represents the transmittances for the attenuation by Rayleigh scattering, uniformly mixed gases, ozone, water vapour and aerosols, respectively, and I_{0n} is the extraterrestrial normal irradiance.

The atmospheric parameters used in the Bird's clear sky model are taken from different meteorological satellites and other sources, such as TOMS (Total Ozone Mapping Spectrometer), CDC-NOAA (National Oceanic and Atmospheric Administration-Climate Diagnostic Center) for water vapour, and NASA-GACP (Global Aerosol Climatology Project).

For cloudy conditions DNI is estimated by multiplying the DNI_{clear} by an additional transmission coefficient for clouds. The latter is obtained from the cloud index (in percentage),

$$T_{cloud} = \frac{100 - n}{-100} \qquad (18.13)$$

The cloud index is derived from the IR and VIS channel images of Meteosat satellite by linear interpolation between the expected cloud-free value and a threshold for a fully cloudy pixel, and it varies in the range 0–100.

5 Assessment of Solar Radiation Derived from Satellite

The different methods described here for estimating the solar radiation from satellite images have been used and tested by several authors. The assessment of these models have been made using qualified ground data covering different time periods (from a couple of months to several years) and ground areas (from a couple of locations to large areas covered by about a hundred of ground stations). Most authors use MBE (Mean Bias Error) and RMSE (Root Mean Squared Error) as parameters to quantify the accuracy of their results. Table 18.1 summarises the results of several authors concerning the accuracy of the satellite models in terms of MBE and RMSE averaged for the whole region and time period of the different estimations.

According to the results showed in the Table 18.1 the general accuracy of satellite models for global hourly radiation lies usually around 17–25% in RMSE. Greater errors are found for the normal beam irradiance as it can be expected due to the strong dynamics associated to this component of the solar radiation. Nevertheless, according to (Zelenka et al. 1999) a substantial portion of these errors can be attributed to measurement errors by the surface instruments and, more importantly, to the genuine micro-variability of the irradiation field. In the case of daily estimations the accuracy is better than for hourly values, being in the range of 10–15%. On the other hand, the use of RMSE as quality measure in daily values is questionable, since it must be properly normalised in order to be expressed as percentage

Table 18.1 Review of satellite models assessment

Author	Component	RMSE %	RMSE %	MBE %	MBE %
		Hourly	Daily	Hourly	Daily
(Rigollier et al. 2004)	Global	19.93	12.90	−2.56	−0.50
(Lefévre et al. 2007)	Global		18.13		−0.65
(Pielke. 1984)	Global		16.00		
(Zelenka et al. 1999)	Global	23.00			
(Pereira et al. 2003)	Global	25.57	13.02	2.67	2.86
(Zarzalejo. 2005)	Global	18.25		0.27	
(Schillings et al. 2004b)	Beam	36.1		4.3	

and the daily radiation has large absolute variations during the year and the mean daily radiation could not be the best choice for normalisation. In this sense, a great effort is being made by the Task 36 of Solar Heating and Cooling - International Energy Agency (http://www.iea-shc.org/task36/index.html) concerning standardisation of procedures for benchmarking and for quality measures of satellite estimations.

6 Availability of Data on the Web

The needs of industry and research for information on solar radiation parameters have encouraged the development of web-based systems and databases offering solar radiation information from satellite data. The information available is mainly solar radiation maps, but also some webs acts as data servers. On the next a general review of best known web sites concerning solar radiation derived from satellite is made.

Satel-light (www.satel-light.com) was probably one of the first web sites that provide solar radiation data. This service, denoted also as the European database of daylight and solar radiation, was the result of a project funded by the European Union from 1996 to 1998. The methodology is based on heliosat model and it covers Europe and a small region of the North Africa. The information provided concerns solar irradiance and illuminances statistical data in terms of monthly means of hourly and daily radiation fro the period from 1996 to 2000.

The project SoDa (Solar Data) is an effort to consolidate different databases through a WWW server (www.soda-is.com) containing solar radiation parameters and other relevant information (Wald et al. 2002, 2004): long-term time series of daily irradiation, climatological data and derived quantities, simulation of radiation under clear skies, and simulation of different solar systems are, among others, the main kind of data offered. One of the main improvements of SoDa regards the access to the information coming from networked information sources that are geographically dispersed. Concerning radiation data SoDa makes use of helioclim databases that are based upon heliosat model and Meteosat satellite images. Helioclim-2 database offers hourly values of solar radiation computed from Meteosat second generation (MSG) images since February 2004. Finally, a worldwide map of averaged solar radiation, as well as some other maps, can be obtained from the web.

The Solar and Wind Energy Resource Assessment (SWERA) is a multinational project financed by the UNEP-GEF (United Nations Environment Programme – Global Environment Facility) aimed at performing a detailed survey of solar and wind energy resources of various developing countries (swera.unep.net). Solar radiation maps are available for several regions in Africa, South and Central America and East Asia. The methodology includes the BRAZIL-SR, Perez et al. and Heliosat satellite models.

PVGIS (Geographical Assessment of Solar Energy Resource and Photovoltaic Technology) integrates solar radiation data derived from satellite with a GIS (Geographical Information System) for offering data and radiation maps for Europe, Africa and South-West Asia (re.jrc.ec.europa.eu/pvgis/). The integration of solar radiation data with GIS is particularly helpful in areas with complex terrain. Solar radiation models incorporate physically-based and empirical equations to provide rapid and accurate estimates of radiation over large regions, while considering also surface inclination, orientation and shadowing effects.

The NASA Surface Meteorology and Solar Energy is big archive of over 200 satellite-derived and solar energy parameters, globally available at a resolution of $1° \times 1°$ (eosweb.larc.nasa.gov/sse/). In general, meteorology and solar insolation are obtained from the NASA Earth Science Enterprise (ESE) program's satellite and reanalysis research data.

Finally, the Australian Bureau of Meteorology runs a computer model for estimating solar radiation from images of the geostationary satellite MTSAT-1R and offers data and solar radiation maps for Australia (www.bom.gov.au/sat/solradinfo .shtml).

7 Conclusions

Solar radiation derived from satellite images has become a highly valuable method for analysing the solar resource of a given site. Since meteorological satellites can now cover most of the earth, the satellite images can be used to estimate solar radiation where no measured data exists. The models, formerly classified as physical or pure statistical, have evolved recently towards a hybrid concept. Thus statistical models are now including physical parameters for characterising the solar radiation attenuation under clear skies conditions. An important effort to ease the access to the satellite derived information has been made during the last years by incorporating the model to web services that offer radiation data almost worldwide.

Information on the CD-ROM

The accompanying CD-ROM includes a set of simple functions and subroutines concerning solar geometry calculations and the clear sky model of ESRA as well. The source code is written in MatLab language, but it is easy to translate it to any other computer language. These functions are of general purpose and they are useful for many applications, not only for those concerning solar radiation derived from satellite images. In addition, a pdf file is also included. It shows a set of maps of solar global irradiation for Spain. The document contains monthly means and annual mean of global solar irradiation on horizontal surface as a result of applying a satellite model based upon heliosat method to Meteosat images from 1994 to 1996.

References

Beyer HG, Costanzo C and Heinemann D (1996.) Modifications of the Heliosat procedure for irradiance estimates from satellite images. Solar Energy 56: 207–212

Bird RE and Hulstrom RL (1983.) Review, evaluation and improvement of direct irradiance models. J. Sol. Energy Eng. 103: 182–192

Cano D, Monget JM, Albuisson M, Guillard H, Regas N and Wald L (1986.) A method for the determination of the global solar radiation from meteorological satellite data. Solar Energy 37: 31–39

Dagestad KF and Olseth JA (2007.) A modified algorithm for calculating the cloud index. Solar Energy 81: 280–289

Diabaté L, Demarcq H, Michaud-Regas N and Wald L (1988.) Estimating incident solar radiation at the surface from images of the Earth transmitted by geostationary satellites: the Heliosat Project. International Journal of Solar Energy 5: 261–278

Diabaté L, Moussu G and Wald L (1989.) Description of an operational tool for determining global solar radiation at ground using geostationary satellite images. Solar Energy 42: 201–207

Gautier C, Diak G and Masse S (1980.) A simple phisical model to estimate incident solar radiation at the surface from GOES satellite data. Journal of Applied Meteorology 19: 1005–1012

Girodo M, Mueller RW and Heinemann D (2006.) Influence of three-dimensional cloud effects on satellite derived solar irradiance estimation–First approaches to improve the Heliosat method. Solar Energy 80: 1145–1159

Hammer A, Heinemann D, Hoyer C, Kuhlemann R, Lorenz E, Muller R and Beyer HG (2003.) Solar energy assessment using remote sensing technologies. Remote Sensing of Environment 86: 423–432

Happ, H, Lin, W. H., Raschke, E., Rieland, M. and Stuhlmann, R. Solar radiation atlas of Africa. Total and diffuse fluxes at ground level measured by geostationary satellites. Personal communication sent: 1989.

Hauschild H, Reiss M, Rudulf B and Schneider U (1992.) Die verwendung von satellitendaten im WZN. Met. Zeitschrift 1: 58–66

Hay JE (1993.) Satellite based estimates of solar irradiance at the earth's surface–I. Modelling approaches. Renewable Energy 3: 381–393

Ineichen P and Perez R (1999.) Derivation of cloud index from geostationary satellites and application to the production of solar irradiance and daylight illuminance data. Theoretical and Applied Climatology 64: 119–130

Ineichen P and Perez R (2002.) A new airmass independent formulation for the Linke turbidity coefficient. Solar Energy 73: 151–157

Janjai S, Laksanaboonsong J, Nunez M and Thongsathitya A (2005.) Development of a method for generating operational solar radiation maps from satellite data for a tropical environment. Solar Energy 78: 739–751

Lefévre M, Wald L and Diabaté L (2007.) Using reduced data sets ISCCP-B2 from the Meteosat satellites to assess surface solar irradiance. Solar Energy 81: 240–253

Martins FR, Pereira EB and Abreu SL (2007.) Satellite-derived solar resource maps for Brazil under SWERA project. Solar Energy 81: 517–528

Moser W and Raschke E (1983.) Mapping of global radiation and of cloudiness from METEOSAT image data. Theory and ground truth comparisons. Meteorologische Rundschau 36: 33–41

Mueller RW, Dagestad KF, Ineichen P, Schroedter-Homscheidt M, Cros S, Dumortier D, Kuhlemann R, Olseth JA, Piernavieja G and Reise C (2004.) Rethinking satellite-based solar irradiance modelling: The SOLIS clear-sky module. Remote Sensing of Environment 91: 160–174

Noia M, Ratto CF and Festa R (1993a.) Solar irradiance estimation from geostationary satellite data: I. Statistical models. Solar Energy 51: 449–456

Noia M, Ratto CF and Festa R (1993b.) Solar irradiance estimation from geostationary satellite data: II. Physical models. Solar Energy 51: 457–465

Pereira, E. B., Martins, F. R., Abreu, S. L., Beyer, H. G., Colle, S., Perez, R. and Heinemann, D.,
  2003. Cross validation of satellite radiation models during SWERA project in Brazil. Proceedings of: ISES Solar World Congress, Göteborg (Sweden)
Pereira EB, Martins FR, Abreu SL, Couto P, Stuhlmann R and Colle S (2000.) Effects of burning
  of biomass on satellite estimations of solar irradiation in Brazil. Solar Energy 68: 91–107
Perez R, Ineichen P, Maxwell E, Seals R and Zelenka A (1992.) Dynamic global-to-direct irradiance conversion models. ASHRAE Transactions 98: 354–369
Perez R, Ineichen P, Moore K, Kmiecik M, Chain C, George R and Vignola F (2002.) A new
  operational model for satellite-derived irradiances: description and validation. Solar Energy
  73: 307–317
Perez R (2002.) Time Specific Irradiances Derived from Geostationary Satellite Images. J. Sol.
  Energy Eng. 124: 1
Perez R, Ineichen P, Kmiecik M, Moore K, Renne D and George R (2004.) Producing satellite-
  derived irradiances in complex arid terrain. Solar Energy 77: 367–371
Pielke, R. A., 1984. Mesoscale meteorological modeling. Academic Press, Orlando (USA)
Rigollier C, Bauer O and Wald L (2000.) On the clear sky model of the ESRA – European Solar
  Radiation Atlas – with respect to the heliosat method. Solar Energy 68: 33–48
Rigollier C, Lefèvre M and Wald L (2004.) The method Heliosat-2 for deriving shortwave solar
  radiation from satellite images. Solar Energy 77: 159–169
Schillings C, Mannstein H and Meyer R (2004a.) Operational method for deriving high resolution
  direct normal irradiance from satellite data. Solar Energy 76: 475–484
Schillings C, Meyer R and Mannstein H (2004b.) Validation of a method for deriving high reso-
  lution direct normal irradiance from satellite data and application for the Arabian Peninsula.
  Solar Energy 76: 485–497
Schmetz J (1989.) Towards a surface radiation climatology: Retrieval of downward irradiances
  from satellites. Atmospheric Res. 23: 287–321
Tarpley JD (1979.) Estimating incident solar radiation at the surface from geostationary satellite
  data. Journal of Applied Meteorology 18: 1172–1183
Wald, L., Albuisson, M., Best, C., Delamare, C., Dumortier, D., Gaboardi, E., Hammer, A., Heine-
  mann, D., Kift, R., Kunz, S., Lefèvre, M., Leroy, S., Martinoli, M., Menard, L., Page, J., Prager,
  T., Ratto, C., Reise, C., Remund, J., Rimoczi-Pall, A., van der Goot, E., Vanroy, F. and Webb,
  A., 2004. SoDa: a web service on solar radiation. Proceedings of: The 2004 Eurosun Congress,
  Freiburg (Germany). pp. 921–927.
Wald, L., Albuisson, M., Best, C., Delamare, C., Gaboardi, E., Hammer, A., Heinemann, D., Kift,
  R., Kunz, S., Lefèvre, M., Leroy, S., Martinoli, M., Menard, L., Page, J., Prager, T., Ratto,
  C., Reise, C., Remund, J., Rimoczi-Pall, A., van der Goot, E., Vanroy, F. and Webb, A., 2002.
  SoDa: a project for the integration and exploitation of networked solar radiation databases. Pro-
  ceedings of: 16th International Conference "Informatics for Environmental Protection", Vienna
  (Austria). pp. 713–720
Zarzalejo LF, (2005). Estimación de la irradiancia global horaria a partir de imágenes de satélite.
  Desarrollo de modelos empíricos. Tesis Doctoral en el programa: Física Atómica y Nuclear y
  Energías Renovables, Universidad Complutense de Madrid (España)
Zarzalejo LF, Ramírez L and Polo J (2005.) Artificial intelligence techniques applied to hourly
  global irradiance estimation from satellite-derived cloud index. Energy 30: 1685–1697
Zelenka A, Perez R, Seals R and Renne D (1999.) Effective accuracy of satellite-derived hourly
  irradiances. Theoretical and Applied Climatology 62: 199–207

Chapter 19
Generation of Solar Radiation Maps from Long-Term Satellite Data

Serm Janjai

1 Introduction

A solar radiation map is an illustration revealing the geographical distribution of solar radiation covering an area of interest. It demonstrates the solar energy potentials of that area. With this information, it is a useful tool for optimum site-selection of a solar energy system. It has also applications in agriculture, climatology and environmental studies. A solar radiation map can be generated by using solar radiation data obtained from a network of solar radiation measurement stations (Suwantrakul et al. 1984; Palz and Greig 1996). However, such a method is not applicable to most parts of the world due to insufficiency or lack of the measurement stations. One of alternative solutions to this problem is to use satellite-derived solar radiation data to create a solar radiation map.

In the past 30 years, a number of methods for estimating solar radiation from satellite data have been developed (Vonder Haar 1973, Tarpley 1979, Gautier et al. 1980, Exell 1984, Moser and Rachke 1984, Zelenka 1986, Cano et al. 1986, Dedieu et al. 1987, Sorapipatana et al. 1988, Sorapipatana and Exell 1989, Diabate et al. 1989, Perez et al. 1990, Czeplak et al. 1991, Nunez 1993, Pinker and Laszlo 1995, Beyer et al. 1996, Hirunlabh et al. 1997, Zelenka et al. 1999, Ineichen and Perez 1999, Hammer et al. 1999, Wyser et al. 2002, Perez et al. 2002, Schillings et al. 2004, Perez et al. 2004, Janjai et al. 2005, Vignola et al. 2006, Martins et al. 2006). However, only a limited number of investigations have been carried out for the tropics with long-term satellite data. In this chapter, a method for generating solar radiation maps in a tropical environment with high aerosols loads using long-term satellite data is presented. The method is applied to Lao People's Democratic Republic (Lao PDR), an Asian tropical country, as a case study. Only the maps of global horizontal solar radiation are emphasized in this work.

Serm Janjai
Silpakorn University, Nakhon Pathom, Thailand, e-mail: serm@su.ac.th

2 Description of the Method

This method was first developed for mapping solar radiation over the tropical Western Pacific Ocean (Nunez 1993). Then it was improved and used to generate solar radiation maps in a tropical environment with a case study of Thailand (Janjai et al. 2005). In this work, this method was further improved and employed to generate a solar radiation map for Lao People's Democratic Republic, a neighboring country of Thailand. The improvements are as follows: a) Absorption of aerosols in the upwelling radiation is accounted for in the satellite model; b) Geographical distribution of precipitable water over a country is obtained from the interpolation of the precipitable water at the network of meteorological stations, instead of using latitude dependent functions; c) Geographical distribution of the total ozone column from TOMS/EP is used, instead of employing the fixed values from ground-based measurements.; d) Surface elevation is included in the calculation of air mass.

It is noted that our method does not emphasize on the calculation of solar irradiance for each hourly satellite image because at a time scale of a fraction of an hour, the cloud field is strongly random, especially in the tropics. This imposes serious constraints on the ability of satellites to map irradiance with a once or twice hourly scan images. By contrast, the regional cloud structure emerges after daily averaging. Therefore, our method is aimed at the calculation of long-term average of daily radiation using long-term satellite data. All parameters involved in the satellite model are calculated on the monthly average basis. The outcome of the calculation is solar radiation climatology which is usually required for generating a solar map for solar energy applications.

Our improved method still consists of 5 steps: preparation of satellite data, modelling, calculation of model parameters, model validation and radiation mapping. The details of each step are described in the following sections.

2.1 Preparation of Satellite Data

The satellite data used in this work are the digital image data from 4 geostationary satellites: GMS 4, GMS 5, GOES 9 and MTSAT. They were recorded from the visible channel of these satellites. The data periods of GMS 4, GMS 5, GOES 9 and MTSAT are: January-September, 1995, October, 1995-May, 2003, June, 2003-July, 2005 and August, 2005-December, 2006, respectively.

The nine hourly images per day (8:30 am-4.30 pm) for the total period of 12-year (1995–2006) with approximately 35,000 images from these satellites were used in this work. Each image consists of a matrix of pixels which records solar radiation reflected from the earth-atmospheric system in the form of gray levels.

A program computer written in IDL (Interactive Data Language) was developed to read and display these digital image data. A displayed image is in the satellite projection showing the spherical surface of the earth (Fig. 19.1)

Fig. 19.1 Example of satellite data displayed as an image in satellite projection

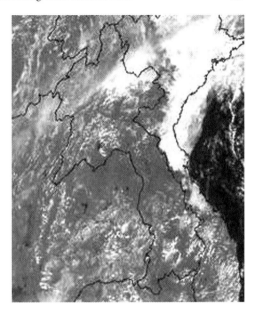

As the distance in the satellite projection images is not proportional to the distance on the ground, the images are transformed into a cylindrical projection, thus making the image distance linear in latitude and longitude. To further rectify the images, a coastline map is superimposed on each image and control points common to both the image and the coastline are adjusted. These rectification and navigation processes are carried out using an IDL computer program developed in this work. Each navigated image, with the resolution of approximately $3 \times 3\,\text{km}^2$, has 450×600 pixels covering an area from the latitude of $13.0\,^\circ\text{N}$ to $23.7\,^\circ\text{N}$ and the longitude of $100.0\,^\circ\text{W}$ to $108.0\,^\circ\text{E}$ as an example shown in Fig. 19.2. The coordinates of every pixel in the rectified image are identified from the coordinates of the image and then used for the further steps of the mapping process.

2.2 Satellite Model

The model used in this work is modified from our previous work (Janjai et al. 2005) by accounting for the aerosols absorption of the upwelling path. According to the modified model, the incident solar radiation which enters the earth's atmosphere is scattered back to the outer space by air molecules and clouds with the cloud-atmospheric albedo of ρ'_A and by atmospheric aerosols with the albedo of ρ'_{aer}. The rest of the radiation continues to travel downwards and is absorbed by ozone, gases, water vapour and aerosols with the absorption coefficients of α'_o, α'_g, α'_w and α'_{aer}, respectively. Upon reaching the surface, the radiation flux is reflected back by the ground with the albedo of ρ'_G. As it travels upwards through the atmosphere, it is

Fig. 19.2 A rectified image in cylindrical map projection

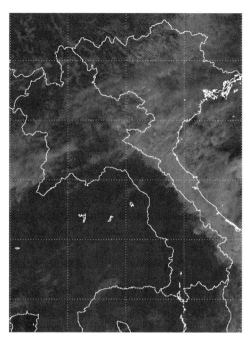

further depleted by aerosol scattering (ρ'_{aer}), aerosols absorption (α'_{aer}) and cloud-atmospheric scattering (ρ'_A). No further absorption due to water vapour, ozone and gases is estimated as it is assumed that the spectral irradiance in the absorption band of these atmospheric constituents has all been absorbed during the downward travel of the solar flux. These processes are schematically shown in Fig.19.3.

The albedo of the earth-atmospheric system (ρ'_{EA}) as detected by the satellite can be written as

$$\rho'_{EA} = \rho'_A + \rho'_{aer} + (1 - \rho'_A - \rho'_{aer})^2(1 - \alpha'_o - \alpha'_w - \alpha'_g - \alpha'_{aer})(1 - \alpha'_{aer})\rho'_G \quad (19.1)$$

Although, ρ'_{EA} is a function of several reflectivities and absorptivities of atmospheric constituents, the most dominant term is ρ'_A, the cloud-atmospheric albedo. This rapidly-varying and unknown term is obtained by re-arranging Eq. (19.1) to gives:

$$\rho'_A = \frac{-(1 - 2CB) \pm \sqrt{(1 - 2CB)^2 - 4C(A + CB^2)}}{2C} \quad (19.2)$$

where

$$\begin{aligned} A &= \rho'_{aer} - \rho'_{EA} \\ B &= 1 - \rho'_{aer} \\ C &= (1 - \alpha'_o - \alpha'_g - \alpha'_w - \alpha'_{aer})(1 - \alpha'_{aer})\rho'_G \end{aligned} \quad (19.3)$$

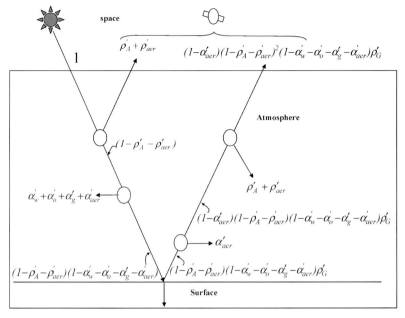

Fig. 19.3 A schematic diagram showing the radiation budget as seen by the satellite ($\rho_A' =$ scattering by gases and cloud, $\rho_{aer}' =$ scattering by aerosols, $\alpha_w' =$ absorption by water vapour, $\alpha_o' =$ absorption by ozone, $\alpha_g' =$ absorption by gas, $\alpha_{aer}' =$ absorption by aerosols, $\rho_G' =$ surface albedo)

Only the $+$ sign in Eq. (19.2) is used. All parameters in Eqs. (19.1), (19.2) are in satellite band: 0.50–0.75 μm for GMS 4, 0.55–0.90 μm for GMS 5, 0.55–0.72 μm for GOES 9 and 0.55–0.80 μm for MTSAT.

The parameters in the right hand side of Eq. (19.2) can be obtained from ground-based measurements and ρ_{EA}' is obtained from the satellites, as will be explained in the next section. As a result, ρ_A' can be calculated from Eq. (19.2). ρ_A' is needed for the calculation of the surface solar radiation. However, ρ_A' cannot be directly used because its values are in the satellite bands. It is necessary to convert ρ_A' into the broadband cloud–atmospheric albedo, $\rho_A(\lambda: 0.3\text{-}3.0\,\mu m)$ using solar radiation measured from pyranometer stations. The details of the conversion will also be described in the next section. The broadband cloud-atmospheric albedo (ρ_A) will be used to calculate the atmospheric transmittance (τ) in the next step.

Keeping the same symbols as stated in Eq. (19.1), but using unprimed notation for broadband parameters, the daily broadband atmospheric transmittance (τ) can be expressed as follows:

$$\tau = \frac{(1 - \rho_A - \rho_{aer})(1 - \alpha_w - \alpha_o - \alpha_g - \alpha_{aer})}{1 - (\rho_A + \rho_{aer})\rho_G} \tag{19.4}$$

The denominator of Eq. (19.4) represents the effect of multiple scattering between the ground and the atmosphere. In Eq. (19.4), ρ_A will be calculated from ρ_A'

and the other parameters are obtained from ground-based data. This allows the values of τ to be obtained. Finally, global radiation (H) is calculated from the daily broadband atmospheric transmittance (τ) and daily extraterrestrial radiation (H_0) as follows:

$$H = \tau\, H_0 \tag{19.5}$$

2.3 Determination of Model Coefficients

2.3.1 Earth-Atmospheric Albedo (ρ'_{EA})

Each pixel of all rectified satellite images is converted into earth-atmospheric albedo using a calibration tables provided by the satellite data agencies. As the calculation of global solar radiation is on the daily average basis, the values of earth-atmospheric albedo are averaged over a day to obtain the daily mean values. These mean values are again averaged over a month to get a monthly mean of daily earth-atmospheric albedo (ρ'_{EA}).

2.3.2 Absorption Coefficients of Water Vapour, Ozone and Gases

These absorption coefficients are all determined from ground-based measurements. In its general form, the absorption coefficients α_i for the atmospheric constituent i is calculated from:

$$\alpha'_i = 1 - \frac{\int_{\lambda_1}^{\lambda_2} I_{0\lambda}\, \tau_{w\lambda}\, d\lambda}{\int_{\lambda_1}^{\lambda_2} I_{0\lambda}\, d\lambda}, \quad \alpha_i = 1 - \frac{\int_{0.3\,\mu m}^{3.0\,\mu m} I_{0\lambda}\, \tau_{w\lambda}\, d\lambda}{\int_{0.3\,\mu m}^{3.0\,\mu m} I_{0\lambda}\, d\lambda} \tag{19.6,7}$$

where α'_i and α_i are the absorption coefficients of constituent i in satellite band (λ_1, λ_2) and broadband, respectively. $I_{0\lambda}$ is the spectral extraterrestrial radiation, $\tau_{i\lambda}$ is the spectral transmittance for constituent i. the calculation of spectral transmittance for each atmospheric constituent is explained as follows.

Water Vapour

The spectral transmittance of water vapour was computed from the relationship given in Iqbal (1983):

$$\tau_{w\lambda} = exp\left[-0.238 k_{w\lambda}\, w m_r / (1 + 20.07 k_{w\lambda}\, w m_r)^{0.45}\right] \tag{19.8}$$

where $k_{w\lambda}$ is the spectral extinction coefficient for water vapour, w is the monthly average pricipitable water and m_r is the relative air mass.

As the measurement of precipitable water in Lao PDR is not available the monthly average pricipitable water (w in cm) is estimated from the formula relating w to the ambient relative humidity (rh in decimal) and the temperature (T in K) developed in Janjai et al. (2005). This formula is as follows:

$$w = 0.8933 \, exp(0.1715 rh P_s / T) \tag{19.9}$$

where Ps is the saturated vapour pressure in mb. This formula was applied to 17 meteorological stations in Lao PDR to obtain the values of w at these stations. These values are again interpolated to all areas corresponding to the satellite pixels. With these precipitable water data, the absorption coefficient of water vapour are finally obtained for all pixels covering the entire country.

Ozone

The calculation of the spectral transmittance of ozone $\tau_{o\lambda}$ is based on the formula described in Iqbal (1983)

$$\tau_{o\lambda} = exp(-k_{o\lambda} l m_r) \tag{19.10}$$

where $k_{o\lambda}$ is a spectral attenuation coefficient for ozone absorption; m_r is a relative air mass and l is the total column ozone in cm.

The daily total column ozone data from TOM/EP satellite is acquired for the region. These data are averaged over individual months to obtain the monthly averaged total column ozone for all pixels of TOMS/EP with a resolution of 1.0 degree (latitude) $\times$ 1.25 degree (longitude). Then these ozone data are interpolated to arrive at the monthly average ozone for each pixel of the geostationary satellites used in this work.

Atmospheric Gases

The spectral transmittance of gases ($\tau_{g\lambda}$) described in Iqbal (1983) was used. It is written as

$$\tau_{g\lambda} = exp[-1.41 k_{g\lambda} m_a / (1 + 118.93 k_{g\lambda} m_a)^{0.45}] \tag{19.11}$$

where $k_{g\lambda}$ and m_a are the gas extinction coefficients and the air mass, respectively.

2.3.3 Surface Albedo

The surface albedo is estimated using the ρ'_{EA} from the images collected at 12:30 h local time. These images are examined for a given month and pixels are selected with the lowest value to create the cloud-free composite image for that month. The effect of cloud shadows are assumed to be negligible because the shadows of the clouds are almost underneath the clouds at 12:30 h in Lao PDR throughout the year. The composite image is converted to the surface albedo using parameterization developed from the 5S radiative transfer model (Tanre' et al. 1986; Janjai et al. 2006).

2.3.4 Absorption and Scattering of Aerosols

Southeast Asia is well-known for the its high atmospheric aerosols loads due to intensive biomass burning both from agricultural activities and forest fire (Von Hoyningen-Huene et al. 1999). These atmospheric aerosols play a very important role in the depletion of the incoming solar radiation. In the calculation of solar radiation from satellite data, it is necessary to know the amount of solar radiation depleted by aerosols. Ideally, this depletion should be obtained from a network of sunphotometer stations. Unfortunately, this is not possible in developing countries. As atmospheric aerosol loads are closely related to visibility data which are generally observed in most meteorological stations. In this work, the relation between the solar radiation depletion by aerosols (D_{aer}) and the visibility (VIS, in km) developed in Janjai et al. 2005 is used. This relation is written as:

$$D_{aer} = 0.3631 - 0.0222VIS + 0.0002VIS^2 \qquad (19.12)$$

The visibility data observed at 17 meteorological stations in Lao PDR (Fig. 19.4) is employed to estimate the depletion, D_{aer} and the results are interpolated to all areas corresponding to the satellite pixels. Since the solar radiation depletion by aerosols results from both scattering and absorption processes, it is necessary to partition the depletion into scattering and absorption. The partition process is conducted by using a 5S radiative transfer model (Tanre' et al. 1986; Janjai et al. 2005)

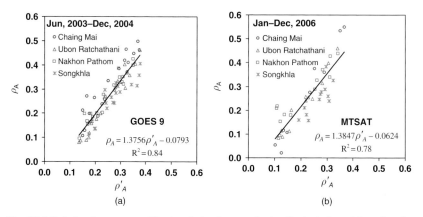

(a)   (b)

Fig. 19.4 Relation between satellite band cloud-atmospheric albedo (ρ'_A) and broadband atmospheric albedo (ρ_A) for GOES 9 and MTSAT

2.3.5 Conversion of the Cloud-Atmospheric Albedo from Satellite Band (ρ'_A) to Broadband (ρ_A)

The conversion of the satellite band (ρ'_A) to broadband (ρ_A) cloud-atmospheric albedo needs solar radiation data from pyranometer stations and the satellite data of the same period. As such solar radiation data are not available in Lao PDR, the solar radiation measured at 4 pyranometer stations in Thailand, namely Chiang Mai (19.78°N, 98.98°E), Ubon Ratchatani (15.25°N, 104.87°E), Songkhla (7.20°N, 100.60°E) and Nakhon Pathom (13.82°N, 100.04°E) is used.

The mathematical expression for the broadband cloud-atmospheric albedo (ρ_A) is obtained from rearranging Eq. (19.4) to give:

$$\rho_A = \frac{(1 - \alpha_w - \alpha_0 - \alpha_g - \alpha_{aer}) - \rho_{aer}(1 - \alpha_w - \alpha_0 - \alpha_g - \alpha_{aer}) - \tau(1 - \rho_{aer}\rho_G)}{1 - \alpha_w - \alpha_0 - \alpha_g - \alpha_{aer} - \tau\rho_G}$$

(19.13)

In Eq. (19.13), the atmospheric transmittance τ was calculated by using Eq. (19.5) with the daily global radiation (H) measured at the four stations. The other parameters of Eq. (19.13) are estimated using the method described in the previous section with the input data measured at the four stations.

The values of the satellite band cloud-atmospheric albedo (ρ'_A) are calculated from Eq. (19.2) using the satellite-derived earth-atmospheric albedo (ρ'_{EA}) and the other parameters at four stations.

The two data sets, ρ'_A and ρ_A, are plotted together for each set of satellite data from GOES 9 and MTSAT as shown in Fig. 19.4 (a)-(b). The two albedos exhibit linear relationships and the resultant regressions are as follows:

$$\rho_A = -0.0793 + 1.3756 \, \rho'_A \quad \text{for GOES9} \tag{19.14}$$

$$\rho_A = -0.0624 + 1.3847 \, \rho'_A \quad \text{for MTSAT} \tag{19.15}$$

For GMS 4 and GMS 5 satellites, the existing relations developed in Janjai et al. (2005) are used. The relations are as follows:

$$\rho_A = -0.0046 + 1.001 \, \rho'_A \quad \text{for GMS 4} \tag{19.16}$$

$$\rho_A = -0.0768 + 1.4846 \, \rho'_A \quad \text{for GMS 5} \tag{19.17}$$

These equations are used to convert ρ'_A and ρ_A for the calculation of the surface radiation

2.4 Validation of the Model

As routine measurement of solar radiation in Lao PDR was not available prior to this project, five new pyranometer stations (Fig. 19.5) were established in different

Laotian cities at: Vientiane Capital, Laung Prabang, Xamnua, Thakhak and Pakxe. For each station, a pyranometer at Kipp & Zonen (model CM 11) is used to measure the global solar radiation. The voltage signal from the pyranometer is recorded by a data logger of Yokogawa (DC100) at a frequency of one second, which is averaged every 10 minutes. All pyranometers were newly calibrated from Kipp & Zonen. The global radiation data for the period of 6-8 months from these stations were collected during the project and used to test the model for the case of MTSAT data. Since radiation data in Lao PDR are not available for testing the model for the period of data collection from GMS5 and GOES9, the global radiation measured in Thailand at 3 stations situated near Thai-Lao border were employed for the tests. These stations are Nongkhai (17.87 °N, 102.72 °E) Nakhon Panom (16.97 °N, 104.73 °E) and Ubon Ratchathani-DEDE (15.25 °N, 104.87 °E). The station of Ubon Ratchathani-

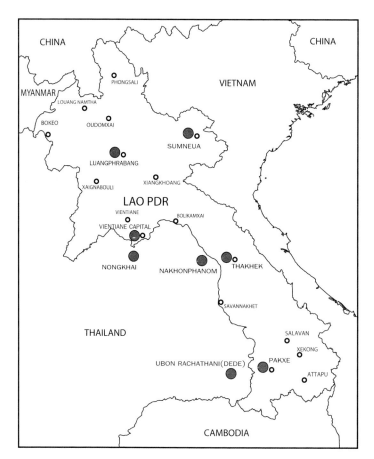

Fig. 19.5 Positions of pyranometer stations and meteorological stations whose data are used in this work. • Pyranometer stations; ○ Meteorological stations for the measurements of relative humidity, temperature and visibility

DEDE is located approximately 10 km away from Ubon Ratchathani station. The positions of all stations used for the tests are shown in Fig. 19.5.

As these solar radiation data were not used in the model development, they were considered to be an independent data set. The monthly average of daily global radiation from the measurement was compared to that obtained from the model. The results are showed in Fig. 19.6 (a)-(d) for 3 satellite data sets. It is observed that the global radiation from the measurements and the calculation are in good agreement with a root mean square difference (RMSD) of 6.6 %, 6.3 %, 8.8 % for GMS 5, GOES 9 and MTSAT, respectively. The total data set gives RMSD of 7.2 %.

As solar radiation data in Lao PDR and in Thailand near Thai-Lao border are not available for the data collection period of GMS 4, the validation of the model for GMS 4 was not carried out in this work. However, the GMS 4 validation has already been done in Janjai et al. (2005).

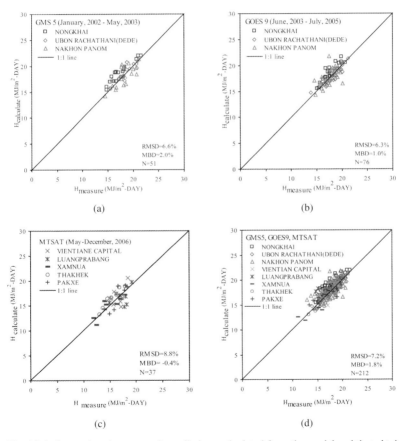

Fig. 19.6 Comparison between solar radiations calculated from the model and that obtained from measurement

2.5 *Mapping of Global Radiation Over Lao PDR*

The monthly averaged values for all model parameters are calculated at each pixel of the satellite image covering Lao PDR. Then the model is used to compute the monthly averaged daily global radiation for every month over a period of 12 years (1995–2006). The calculation of the air mass in the absorption coefficients of the model also accounts for the surface elevation. For each month, the monthly average of daily global radiation was again averaged over the 12 years to obtain the long-term average global radiation. The results are displayed as monthly solar radiation maps and a yearly map.

3 Results and Discussions

The monthly radiation maps and the yearly radiation maps obtained from the above-mentioned process are shown in Figs. 19.7 and 19.8, respectively. As expected, the

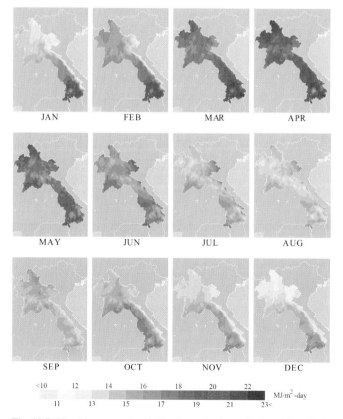

Fig. 19.7 Monthly maps of global horizontal solar radiation of Lao PDR

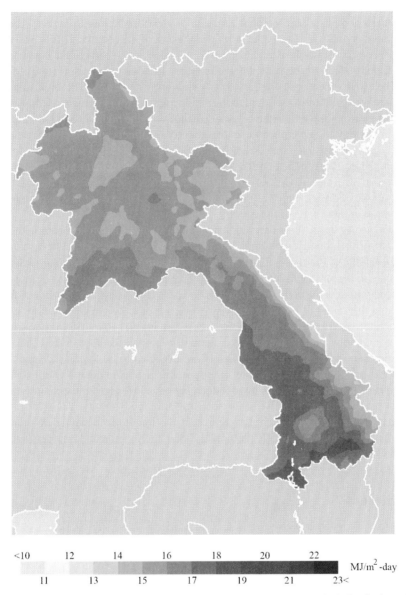

Fig. 19.8 A solar radiation map of Lao PDR, showing the geographical distribution of yearly average of daily global solar radiation over the country

monthly maps demonstrate a seasonal variation of global radiation. In January, the solar radiation is relatively low in the north and the east of the country. This is likely due to the fact that the northeast monsoon still influences these parts of the country, bringing cloudy skies to these mountainous areas. In addition, the sun is still in southern celestial sphere, causing less solar radiation in the north. The high solar radiation areas expand from the south to the entire country from February to May as the sun moves from the south to the celestial equator with the highest solar radiation in April. From May to September, the entire country is influenced by the southwest monsoon, causing rain and cloud cover, which reduces the incident solar radiation. In October, the northeast monsoon starts influencing the eastern part of the country, bringing moist air from the Gulf of Tonkin, which causes cloudy skies in this part. The effect of the northeast monsoon continues to the end of the year, thus decreasing solar radiation in most parts of the country.

From the yearly basis, it is observed that the high solar radiation areas are in the western part of the country with the level of 17–18 MJ/m^2-day. The highest solar radiation level of 18–19 MJ/m^2-day is found in the south, while the low solar radiation areas are in the east and scattering areas of high mountains. The yearly average of daily radiation over the country is found to be 15.8 MJ/m^2-day. The yearly radiation map and the values of monthly and yearly average of daily global radiation for all provinces of Lao PDR are shown in the accompanying CD-ROM of this book.

It must be emphasized that we do not attempt to predict the short term behavior of solar radiation at hourly scales. Errors are expected to be large as (Vignola et al. 2006) In addition, present satellite systems in this region do not have sufficient temporal resolution to attempt this task. Rather, we have investigated the long-term average of daily radiation to arrive at the solar radiation climatology. The solar radiation maps generated with such data represent the long-term behavior of solar radiation patterns which are usually required by solar energy project developers and climatologists.

4 Conclusions

A solar radiation map of Lao PDR has been generated using a physical satellite model and long-term satellite data. The model was modified from the original version by accounting for the aerosol absorption of upwelling radiation and the effect of terrain elevations. The model is based on the radiation budget which traces solar radiation as it is scattered, absorbed and reflected back to the space. This model employs satellite data to estimate cloud-atmospheric albedo in satellite band, which in turn is converted into a broadband cloud-atmospheric albedo using global radiation measured at pyranometer stations. Then this broadband albedo and surface ancillary data were used to map daily global radiation. When tested against the total independent data sets, the model performed satisfactorily, with the RMSD of 7.2 % and MBD of 1.8 %.

Acknowledgements The author would like to thank the Department of Alternative Energy Development and Efficiency, Ministry of Energy of Thailand and Department of Electricity, Lao PDR for supporting this project. The meteorological data used in this project were obtained from the Department of Meteorology and Hydrology of Lao PDR and Thai Meteorological Department. The author would like to thank these organizations. The author gratefully acknowledges Dr. Manuel Nunez for valuable advice and Dr. Jarungsaeng Laksanaboonsong for the development of image navigation program and technical support. The assistance in the data analysis of Mr. Prasan Pankaew and the preparation of the manuscript of Miss Rungrat Wattan are gratefully acknowledged.

References

Beyer HG, Costanzo C, Heinemann D (1996) Modifications of the heliosat procedure for irradiance estimated from satellite images. Solar Energy 56: 207–212.

Cano D, Monget J.M, Albuisson M, Guillard H, Regas N, Wald L (1986) A method for the determination of the global solar radiation from meteorological satellite data. Solar Energy 37: 31–39.

Czeplak G, Noia M, Ratto CF (1991) An assessment of a statistical method to estimate solar irradiance at the earth's surface from geostationary satellite data. Renewable Energy 1(5/6): 737–743.

Dedieu G, Deschamps PY, Kerr YH (1987) Satellite estimation of solar irradiance at the surface of the Earth and Surface albedo using a physical model applied to METEOSAT data. Journal of Climate and Applied Meteorology 26: 79–87.

Diabate L, Moussu G, Wald L (1989) Description of an operation tool for determining global solar radiation at ground using geostationary satellite images. Solar Energy 42(3): 201–207.

Exell RHB (1984) Mapping solar radiation by meteorological satellite. Renewable Energy Review Journal 6(1).

Gautier C, Diak G, Masse S (1980) A simple physical model to estimate incident solar radiation at the surface from GOES satellite data. Journal Applied Meteorology 36: 1005–1012.

Hammer A, Heinemann D, Lorenz E, Luckehe B (1999) Short-term forecasting of solar radiation: a statistical approach using satellite data, Solar Energy 67: 139–150.

Hirunlabh J, Sarachitti R, Namprakai P (1997) Estimating solar radiation at the earth's surface from satellite data, Thammasat International Journal of Science and Technology 1: 69–79.

Ineichen P, Perez R (1999) Derivative of cloud index from geostationary satellites and application to the production of solar irradiance and daylight illuminance data, Theoretical Applications of Climatology 64: 119–130.

Iqbal M (1983) An Introduction to Solar Radiation, Academic Press, New York.

Janjai S, Laksanaboonsong J, Nunez M, Thongsathitya A (2005) Development of a method generating operational solar radiation maps from satellite data for a tropical environment, Solar Energy 78: 739–751.

Janjai, S., Wanvong, W., Laksanaboonsong, J., 2006. The determination of surface albedo of Thailand using satellite data, In Proceeding of the 2nd Joint International Conference on Sustainable Energy and Environment (SEE2006) Bangkok, Thailand, pp. 156–161.

Martins FR, Pereira EB, Abreu SL (2007) Satellite-derived solar resource maps for Brazil under SWERA project, Solar Energy 81: 517–528.

Moser W, Rachke E (1984) Incident solar radiation over Europe from METEOSAT data. Journal of Climate and Applied Meteorology 23: 166–170.

Nunez M (1993) The development of a satellite-based insolation model for the tropical western Pacific ocean, Int. J. Climatol. 13: 607–627.

Palz W, Greig J (1996) European Solar Radiation Atlas, Springer, Heidelberg.

Perez R, Ineichen P, Seals R, Michalsky J, Stewart R (1990) Modeling daylight availability and irradiance components from direct and global irradiance. Solar Energy 40(5): 271–289.

Perez R, Ineichen P, Moore K, Kmiecik M, Chain C, George R, Vignola F (2002) A new operational for satellite-derived irradiances: description and validation. Solar Energy 73(5): 307–317.

Perez R, Ineichen P, Kmiecik M, Moore K, Renne D, George R (2004) Producing satellite–derived irradiances in complex arid terrain. Solar Energy 77: 367–371.

Pinker RT, Laszlo I (1995) Modelling surface solar irradiance for satellite applications on a global scale. J. Appl. Meteorol. 31: 194–211.

Schillings C, Mannstein H, Meyer R (2004) Operational method for deriving high resolution direct normal irradiance from satellite data. Solar Energy 76: 475–484.

Sorapipatana C, Exell RHB, Borel D (1988) A bispectral method for determining global solar radiation from meteorological satellite data. Solar and Wind Technology 5(3): 321–327.

Sorapipatana C, Exell RBH (1989) Mesoscale mapping of daily insolation over Southeast Asia from satellite data. Solar and Wind Technology 6(1): 59–69.

Suwantrakul B, Watabutr W, Tia Sitathani V, Namprakai P (1984) Solar and Wind Potential Assessment of Thailand. Renewable Nonconventional Energy Project, USAID project No. 493-0304, Meteorological Department and King Mongkut's Institute of Technology Thonburi, Bangkok, Thailand.

Tanre' D, Deroo C, Duhaut P, Herman N, Morcrette JJ, Perbos J, Deschamps PY (1986) Simulation of the Satellite Signal in the Solar Spectrum. Technical Report, Laboratoire d' Optique Atmospherique, Universite' des Science et Technique de Lille, 59655 Villeneuve d' Ascq Cedex, France.

Tarpley JD(1979) Estimating incident solar radiation at the surface from geostationary satellite data. Journal of Applied Meteorology 18, 1172–1181.

Vignola F, Harlan P, Perez R, Kmiecik M (2007) Analysis of satellite derived beam and global solar radiation data. Solar Energy 81: 768–772.

Vonder Haar T, Raschke E, Bandeen W, Pasternak M (1973) Measurements of solar energy reflected by the earth and atmosphere from meteorological satellites. Solar Energy 14: 175–184.

Von Hoyningen-Huene W, Schmidt T, Schienbein S, Kee CA, Tick LJ (1999) Climate-relevant aerosol parameters of South-East-Asian forest fire haze. Atmospheric Environment 33: 3183–3190.

Wyser K, O'Hirox W, Gautier C, Jones C (2002) Remote sensing of surface solar irradiance with corrections for 3-D cloud effects. Remote Sensing of Environment 80: 272–284.

Zelenka A (1986) Satellite versus ground observation based model for global irradiation. In Proceedings of the INTERSOL 85 Congress, Montreal, Canada, 2513–2517.

Zelenka A, Perez R, Seals R, Renne D (1999) Effective accuracy of the satellite-derived hourly irradiance. Theoretical and Applied Climatology 62: 199–207.

Chapter 20
Validation and Ranking Methodologies for Solar Radiation Models

Christian A. Gueymard and Daryl R. Myers

1 Introduction

This chapter provides an overview of the methodologies that can be used to validate different types of solar radiation models currently in use in various applications, with a focus on solar energy applications.

2 Types of Models

Different types of models have been developed to provide the community with predictions of solar radiation when or where it is not measured appropriately or at all. An accepted typology of solar radiation models does not currently exist; hence, what is proposed below should be considered tentative. From an exhaustive review of the literature over the past four decades, it is clear that radiation models can be categorized in different ways. The previous chapters concentrated on a few specific types of model. More types do exist, so that nine classification criteria have been identified, as follows.

- Criterion #1—*Type of output data*
  Outputs ideally consist of direct, diffuse and global irradiance, but frequently only one component is necessary (e.g., direct normal irradiance, or global irradiance on a tilted plane). Furthermore, many models try to derive the direct and/or diffuse component from global irradiance used as input (Maxwell 1987; Perez et al. 1992).

Christian A. Gueymard
Solar Consulting Services, Colebrook NH, USA, e-mail: chris@solarconsultingservices.com

Daryl R. Myers
National Renewable Energy Laboratory, Golden CO, USA, e-mail: daryl_myers@nrel.gov

- Criterion #2—*Type of input data*
  Inputs may consist of meteorological variables, climatological data, or irradiance components. These inputs can be from ground sites, or remote sensed from airborne or spaceborne sensors (Maxwell 1998; Muneer et al. 2000; Perez et al. 2002).
- Criterion #3—*Spatial resolution*
  Some models provide predictions for a specific location (generally where the input data come from), others provide gridded results (generally when using satellite-based inputs). The former models have greater spatial resolution, whereas the latter have greater spatial coverage, e.g., a large part of the world (Perez et al. 2002).
- Criterion #4—*Time resolution*
  Irradiance outputs can be of high-resolution (every minute or less), standard resolution (hourly), average resolution (specific day), low-resolution (average hour or average day for a specific month), or climatological (average hour or day over a long-term period, such as 10–30 years). Typically, high-resolution data are needed for solar concentrators, hourly data are used for solar system or building energy simulations, daily data are used in agricultural meteorology, and long-term average data are used in system design and climatology. Some useful models convert mean daily irradiation data into mean hourly data (Collares-Pereira and Rabl 1979; Gueymard 2000).
- Criterion #5—*Spectral resolution*
  Most models considered in this book evaluate the shortwave radiation transmitted by the atmosphere as if the solar spectrum was constituted only of one band, typically 300–4000 nm. Some models, however consider two or more distinct bands for more resolution. For instance, there are models limited to the ultraviolet (below 400 nm) or the photosynthetic waveband (400–700 nm). For specific applications, such as atmospheric physics, remote sensing or prediction of the performance of spectrally-selective devices such as photovoltaic systems or coated glazings, spectral models are necessary. These are reviewed elsewhere (Gueymard and Kambezidis 2004).
- Criterion #6—*Type of methodology*
  The model's methodology can be either deterministic or stochastic (also called "statistical"). A deterministic algorithm tries to determine irradiance for a specific time, which can be in the past, present or future. A stochastic algorithm "invents" solar radiation without this requirement, in a virtual way, but primarily tries to respect some statistical properties of the irradiance time series, such as variance, cumulative frequency distribution, persistence, etc. The interested reader should consult (Gordon 2001) for a short review of stochastic models. It is also possible to combine deterministic and statistical features, but such models are extremely rare; an example is METSTAT (Maxwell 1998).
- Criterion #7—*Type of algorithm*
  Another methodological distinction can be made between physical or semi-physical models, which are derived more or less directly from physical principles (e.g., Gueymard 2008), and empirical models, which are based only/mostly on

measured irradiance data obtained for a specific location and period to predict irradiance at other locations and/or periods (e.g., Perez et al. 1990b).

- Criterion #8—*Surface geometry*

  Irradiance data may be needed for horizontal planes, tilted surfaces, or tracking surfaces that permanently point to the sun. Most solar energy and building applications involve receivers that are either fixed on a tilt or tracking the sun. Similarly, ecological applications often involve the modeling of topographic solar radiation over complex mountainous terrain (e.g., Wang et al. 2006). Solar radiation on such tilted surfaces can be obtained from horizontal radiation data by using so-called "transposition models" (Gueymard 1987; Hay 1993b; Loutzenhiser et al. 2007; Perez et al. 1990a).

- Criterion #9—*Type of sky*

  Most models consider the effect of clouds, which is of primary importance. For some applications however, such as building energy load calculations or solar concentrator resource assessment, irradiance predictions are normally limited to clear-sky conditions, which may require a specific model (Bird and Hulstrom 1981b; Gueymard 1989, 2008; Ianetz et al. 2007; Ineichen 2006; Power 2001).

It is obvious that the number of combinations resulting from this nine-dimensional typology can be considerable. Fortunately, not all combinations are necessary or useful in practice. The desired end-results may also require the use of successive models whose results are linked. For instance, suppose that the monthly-average irradiance on a vertical surface must be evaluated, and that only sunshine information is available. Typically, a first model will be used to derive the global horizontal irradiation. A second model will be necessary to separate direct and diffuse radiation, and a third model will predict the global tilted irradiation from these two components. Sometimes a model is used in a different way than it was originally intended, which complicates the matter. Finally, it must be noticed that some models can also be reversed. This can be done when solar radiation is actually measured, and the model is inverted to evaluate one or more atmospheric characteristic, usually aerosol turbidity, which is normally an input to the model (e.g., Gueymard 1998; Louche et al. 1987).

3 Model Validation Principles

A perfect model does not exist. Even if it existed, this would be impossible to ascertain because the "true" solar irradiance cannot be determined theoretically or measured experimentally with perfect certainty. Chapter 1 described the various sources of uncertainty in experimental radiation measurement. These must always be bore in mind when evaluating the performance of any model against measured data (Gueymard and Myers 2007). Normally, when a new model is proposed in the literature, it should be also tested so that potential users can be sure of its validity under such or such conditions. This, however, is not done systematically by all model authors. Moreover, many models are developed empirically from data measured at one or a few specific sites, and their "universality" must be verified by testing them against data from a variety of other sites. All these studies take time, so that it is usually difficult to recommend a new or recent model for widespread application. Furthermore, changes in radiometric calibration and measurement procedures (as described in Chap.1) can alter the performance of a model, or the performance ranking of similar models (Gueymard and Myers 2007).

Comparison between the irradiance predictions of a model and corresponding measurements—the uncertainty of which should be reported as part of the validation results—is the usual way of assessing its validity and performance (see Sect. 5). In some cases, it is also possible to compare these predictions to "reference" predictions from a validated and more sophisticated model. As has been demonstrated in such a study (Gueymard 2003b), the latter approach has the advantage of not being limited by the particular atmospheric or climatological conditions of a specific experimental site, or the quality of its data. Finally, when experimental data of exceptional quality, with well-characterized and documented uncertainties are available, they can be grouped into a benchmark dataset (Gueymard 2008), against which various models can be tested with confidence. This approach is developed further in Sect. 6.3.

4 Model Sensitivity to Input Errors and Error Analysis

A radiation model is driven by input data that are directly or indirectly related to the optical characteristics of the atmosphere for the location and period considered. The most sophisticated radiative transfer models used in atmospheric physics require a wealth of information about various atmospheric constituents, such as gases and aerosols, and their vertical distribution. In engineering and other disciplines, the input requirements may be vastly different, and usually simpler.

For models whose Criteria 2 and 9 above call for meteorological inputs and clear-sky results, the main factor will be aerosol turbidity, particularly for the direct and diffuse radiation components (Gueymard 2005a), and the second factor will be water vapor (Fig. 20.1). (Note that the total atmospheric water vapor amount is statistically related to the dew-point temperature measured near ground level.) If all-sky results are rather sought, cloudiness becomes generally the critical factor, with greater impact than the two previous effects. Error in one of these inputs will yield significant error in the predicted irradiance. This is the main problem in solar radiation modeling because these atmospheric characteristics are highly variable over both time and space, and are not precisely measured with the desired resolution. Because of this lack of fundamental data, many models attempt to simplify or even ignore these requirements, and use empirical algorithms rather than a physical approach (Criterion #7). This results in a simpler model, albeit with compromised universality due to its reliance on location- or climate-specific information.

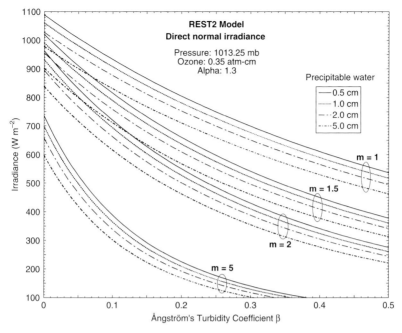

Fig. 20.1 Dependence of direct normal irradiance on turbidity, water vapor and air mass, according to the REST2 model

An appropriate error analysis should be the first step into validating radiation models, but is rarely provided by their authors. A simple example follows, using the case of the Ångström-Prescott equation (Gueymard et al. 1995; Martinez-Lozano et al. 1984)—one of the simplest radiation models ever proposed:

$$H/H_0 = a + bS/S_0 \tag{20.1}$$

where H and H_0 are respectively the terrestrial and extraterrestrial global mean-daily horizontal irradiations, a and b are empirical coefficients, and S and S_0 are the observed mean daily sunshine duration and daylength, respectively. Simple differentiation of Eq. (20.1) provides the effect (or error) on H, ΔH, due to a variation (or error) in S, ΔS:

$$\Delta H/\Delta S = bH_0/S_0. \tag{20.2}$$

Both H_0 and S_0 are obtained from theory (Iqbal 1983; Muneer 2004), and their error can be considered negligible in comparison of those in H or S. For a north latitude of $45°$ and average days in January and July, S_0 amounts to 9.4 and 15.3 hours, respectively, and H_0 amounts to 12.3 and 40.4 MJ m^{-2} per day, respectively (Iqbal 1983). For a typical $b = 0.60$, Eq. (20.2) yields $\Delta H/\Delta S = 0.785$ in January and 1.58 in July. Assuming a possible error $\Delta S = \pm 1$ hour (due to variations in sunshine recorder threshold, humidity, trace overburning, etc., which are typical of Campbell-Stokes sunshine recorders), ΔH becomes ± 0.79 MJ m^{-2} in January and

$\pm 1.58\,\mathrm{MJ\,m^{-2}}$ in July. Differentiating Eq. (20.1) from its log transform provides a different expression:

$$\frac{\Delta H}{H} = \frac{1}{1 + \frac{a}{b}\frac{S}{S_0}}\frac{\Delta S}{S}. \tag{20.3}$$

Equation (20.3) can be used to evaluate the relative error in H due to a relative error, $\Delta S/S$, in S. For instance, using the conditions as above with $a = 0.2$ and $S/S_0 = 0.3$ in January and 0.6 in July, the relative error in H is obtained as 9.7% in January and 5.4% in July.

Such an error analysis becomes more involved when the modeled irradiance depends on more than one variable, although this is also an efficient way to improve the model's performance in most cases. Note, however, that one must be extremely careful in introducing "independent" variables, which may be highly correlated with one another, for two reasons: (1) introducing more variables in a multivariate analysis (even a random variable) artificially inflates the correlation coefficients, and (2) multicollinearity results in conditions where conventional least-squares procedures do not apply. Examination of the data structure is truly needed to produce correct estimates (Myers 1986). In the practice of solar radiation modeling, multicollinearity is very difficult (if not impossible) to eliminate, is sometimes difficult to identify, and may have a strong, sometimes subtle effect on model performance. With proper statistical analysis these adverse effects can be quantified by the so called "variance inflation factor", i.e., the diagonal elements of the inverse correlation matrix. The modeler should become more familiar with this aspect of all models, and in particular multivariate regression models (see, e.g., Fox 1997).

Let us consider the case of the so-called "meteorological models" (Criterion #2), which predict the instantaneous irradiance on a horizontal surface from n input variables, $v_1, v_2 \ldots v_n$, describing the main optical characteristics of the atmosphere (e.g., water vapor, turbidity, ozone amount...). The direct normal irradiance (DNI), E_{bn}, is ideally obtained as the product of the extraterrestrial irradiance, E_{0n}, and n broadband transmittances (one per atmospheric variable):

$$E_{bn} = E_{0n} T_1 T_2 T_3 \ldots T_n. \tag{20.4}$$

The individual transmittance expressions are generally amenable to a generic form $T_i = \exp(-m\,\tau_i)$, where m is the air mass and τ_i the broadband optical depth for atmospheric variable v_i. The air mass is a function of the solar zenith angle, Z. Various functions $m = \mathrm{f}(Z)$ have been proposed in the literature, and two of them can be recommended. They are based on detailed calculations involving the vertical profile of atmospheric constituents (Gueymard 2003b; Kasten and Young 1989). The overall relative error in DNI, $\Delta E_{bn}/E_{bn}$, resulting from individual errors Δv_i in the n input variables v_i can be derived from the log transform of Eq. (20.4), following the usual method for error combination (Bevington and Robinson 2003):

$$\frac{\Delta E_{bn}}{E_{bn}} = \sqrt{\left(\frac{\Delta E_{0n}}{E_{0n}}\right)^2 + m^2\left[\left(\frac{\partial \tau_1}{\partial v_1}\Delta v_1\right)^2 + \left(\frac{\partial \tau_2}{\partial v_2}\Delta v_2\right)^2 \ldots + \left(\frac{\partial \tau_n}{\partial v_n}\Delta v_n\right)^2\right]}$$

$$\tag{20.5}$$

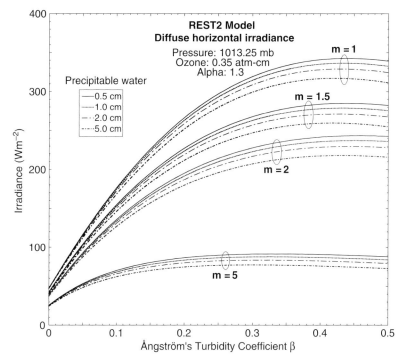

Fig. 20.2 Dependence of diffuse horizontal irradiance on turbidity, water vapor and air mass, according to the REST2 model

The sensitivity of τ_i to an individual error in v_i, $\partial\tau_i/\partial v_i$, must be established. Such an exercise has been done recently (Gueymard 2003a) and provides quantitative information about the predicted DNI's sensitivity to m, turbidity and water vapor—the three major variables under clear skies. With turbidity and precipitable water evaluated in terms of the Ångström turbidity coefficient, β, and precipitable water, w, respectively, the sensitivity of DNI to m, β and w can be represented graphically for some combinations of these input variables (Fig. 20.1), using a model such as REST2 (Gueymard 2008). Similar results, but for diffuse horizontal irradiance, are presented in Fig. 20.2. More sophisticated numerical techniques—such as N-way factorials, Monte Carlo analysis or hypercube sampling—can be used to evaluate the effects of uncertainties in a model's input variables on its predictions (Loutzenhiser et al. 2007).

5 Model Validation and Performance Assessment

An error analysis can only provide general information corresponding to ideal or "worst-case" scenarios. But what is the actual accuracy of a model in practice? Answering this question requires a specialized study called "validation" or "performance assessment". A model can be declared "validated" even if it does not perform very well or better than others. It is only a way of saying that "it works". A true performance assessment consists of a series of tests whose findings are usually summarized by qualitative and/or quantitative results.

5.1 Qualitative Assessment

Most qualitative results appear in the form of scatterplots, which visually indicate the bias (systematic error) and scatter (random error) of predicted *vs* measured values. Results from a perfect model would align on the 1:1 diagonal when compared to their perfectly measured counterparts. As an example, consider the measurement of instantaneous clear-sky irradiance and its prediction from meteorological data. In recent years, progress has been made due to the convergence of significant improvements in radiometry (described in Chap. 1), time resolution (radiation data are now often measured at e.g., 1–6 minutes intervals rather than hourly intervals), measurement of key ancillary data (such as aerosol optical depth or precipitable water), and radiation modeling.

Examples of scatterplots are shown in Fig. 20.3, which compares the predictions of the REST2 (Gueymard 2008) and "Iqbal C" models (Iqbal 1983) to a benchmark dataset defined by 30 carefully-selected one-minute measurements of clear-sky direct and diffuse irradiance at the Southern Great Plains site of the Atmospheric Radiation Measurement program, obtained in May 2003 (Michalsky et al. 2006). (See Sect. 6.3 for more information on this dataset and the methodology used.) A computer program devised to use and compare the predictions of REST2, Iqbal C and other similar clear-sky meteorological models is included on the CD-ROM (filename: 'Models_performance_compar.f').

In most cases, a performance assessment study requires thousands of data points to be displayed, particularly when short-term data and long time periods are considered. This translates into scatterplots that become very difficult to decipher, even more so if their intent is to compare two or more models. An example is given in Fig. 20.4, which compares the predicted global normal irradiance (i.e., incident on a tracking receiver) to pyranometer measurements routinely performed at the Solar Radiation Research Laboratory of the National Renewable Energy Laboratory (NREL) in Golden, Colorado. The clear-sky calculations shown here use the CDRS model (Gueymard 1987) and the more recent Muneer model (Muneer 2004). These are "transposition" models (see Criterion #8 in Sect. 2) that predict the global tilted irradiance from horizontal irradiance. In the present case, measured one-minute ground albedo and direct and diffuse irradiance data are used. Golden (latitude $39.742°$ N, longitude $105.178°$ W, elevation 1829 m) is a site at relatively high altitude enjoying sunny-clean-dry climatic conditions, hence the high clearness index ($K_T = E/E_0$), the low diffuse/global irradiance ratio ($K = E_d/E$) [where E, E_d and E_0 are the global, diffuse and extraterrestrial horizontal irradiances,

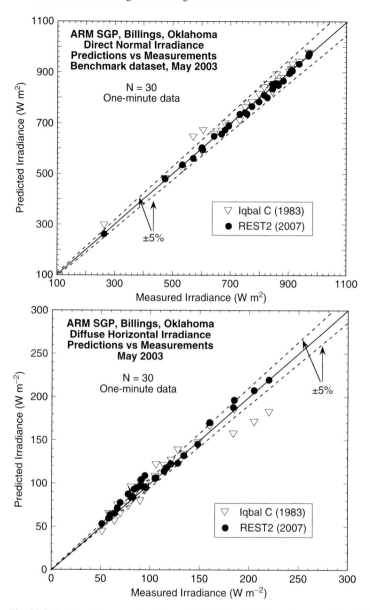

Fig. 20.3 Scatterplots comparing the predicted direct irradiance (top graph) and diffuse irradiance (bottom graph) to measured data for a proposed benchmark dataset (Gueymard 2008) and two clear-sky meteorological models

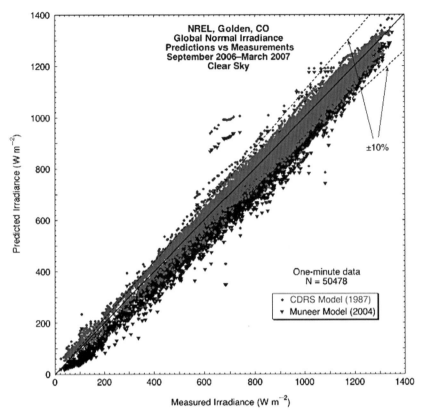

Fig. 20.4 Scatterplot showing the predictions of global normal irradiance with two transposition models compared to one-minute measurements at NREL, Golden, Colorado

respectively], and the large number of data points ($N = 50478$) for the period September 2006–March 2007 used here for demonstration purposes.

A far more informative representation of a model's performance has been introduced in the mid 1980s (Ineichen et al. 1984, 1987). The absolute or relative differences between the predicted and measured irradiances are grouped in bins of an appropriately chosen independent variable, and the mean and standard deviations of these differences are calculated for each bin. This is illustrated in Fig. 20.5, which uses the same exact information as in Fig. 20.4. The independent variables chosen here are K and K_T, because of their prominent importance in characterizing a site's solar climate. In the improved way used here to present these plots, it is possible to legibly display the results of two models (by simply introducing a slight offset in the bin centers), and to superimpose the relative frequency distribution of the independent variable.

Another highly informative diagnostic tool is to plot residuals, or the difference between modeled and measured data as functions of time and/or the various "independent" variables such as air mass, turbidity, etc. Patterns apparent in residual

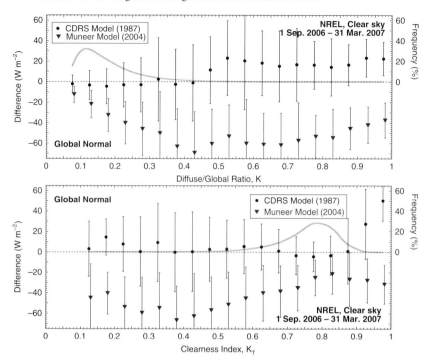

Fig. 20.5 Difference between the predicted and measured global normal irradiance at NREL using two transposition models. Top graph: Distribution of differences in bins of diffuse/global irradiance ratio, K. Bottom graph: Distribution of the same differences in bins of the clearness index, K_T. Bars represent one standard deviation around the mean. The thick gray lines represent the measured relative frequency distribution of K or K_T (right Y-axis)

plots may suggest improvements applied to reduce patterns in these relationships, and move toward a more random distribution of differences, which are preferable to differences that are strong functions of some parameter. (See also the following discussion.)

5.2 Quantitative Assessment

Most generally, the performance of various models is evaluated against a single dataset, in the aim of selecting the best performing model for this particular dataset, and by extrapolation, for the climatic conditions represented by the dataset. Using qualitative information from plots such as Figs. (20.3)–(20.5) would be too difficult or subjective for this task. A statistical analysis of the actual modeling errors must therefore be performed. (Sect. 5.3 goes into more details on how to isolate "modeling errors".) An individual error, e_i, is, by definition, the difference between a predicted value (of radiation, presumably) and the corresponding "true value". It must be emphasized that the true value is never known. Any measured value is only an approximation of the true value, and is therefore uncertain. Unfortunately, the literature generally refers to e_i as an "error", because measured values are normally considered of better quality than modeled values. This is obviously not always the case, and models are often used to test the validity of measurements and, in particular, detect malfunction, miscalibration, etc. (see Sect. 8.3 of Chap. 1). The term "error" should rather be noted "model error", "estimated error", "apparent error", or "observed difference" to avoid confusion with *measurement* "error" or uncertainty. This ambiguity notwithstanding, the two terms "error" and "difference" will both represent e_i in what follows, and will be used interchangeably.

Although true values cannot be measured and true errors cannot be obtained, it is known that random errors do follow statistical laws (Crandall and Seabloom 1970). However, as discussed in Chap. 1, systematic or bias errors are always embedded in measured data, and can only be identified and quantified by calibration and characterization, but *never* totally removed (BIPM 1995). A description of these statistical laws is beyond the scope of this chapter, but essential definitions and tools will be provided, considering the current usage in solar radiation modeling.

The most common bulk performance statistics are the Mean Bias Error (*MBE*), the Root Mean Square Error (*RMSE*) and the Mean Absolute Bias Error (*MABE*), which, for a dataset containing N data points, are defined as

$$MBE = \frac{1}{N}\sum_{i=1}^{N} e_i, \quad RMSE = \sqrt{\frac{1}{N}\sum_{i=1}^{N} e_i^2}, \quad MABE = \frac{1}{N}\sum_{i=1}^{N} |e_i|. \tag{20.6}$$

These formulae provide results in radiation units ($\mathrm{W\,m^{-2}}$ for irradiance and $\mathrm{MJ\,m^{-2}}$ or $\mathrm{kWh\,m^{-2}}$ for irradiation). They are frequently converted into percent values after dividing them by the mean measured irradiance or irradiation. *MBE* is a measure of systematic errors (or bias), whereas *RMSE* is mostly a measure of random errors. *MABE* is more rarely used than the two other statistics. It is worth insisting on the fact that a part of the apparent cumulative error described by *MBE, RMSE* or *MABE* is actually the result of measurement uncertainty. Another part is induced by the uncertainties in the inputs to the model, as discussed in Sect. 4. For these reasons, some authors rather use the nomenclature *MBD, RMSD* and *MABD*, where *D* stands for *difference*. From the discussion just above, this nomenclature is preferable because it does not imply or suggest that the measured values are identical or closer to the true values. For instance, suppose we test three models against a set of measured diffuse irradiance data. The fictitious results are that model A yields an *MBE* or *MBD* of 3.3%, as compared to 0.1% for model B, and −3.2% for model C. Based on these numbers alone, the usual conclusion is that model B performs better since its *MBE* is lower in absolute value. However, if the measured data contained a (typical) systematic error of −2% due to miscalibration, model C would be the actual best performer.

Suppose now that diffuse irradiance is not actually measured, but obtained as the difference between global and direct radiation data, according to Eq. (1.1) of Chap. 1. The −2% systematic error in the diffuse data is now the result of some specific combination of systematic errors in the global and direct data, e.g., 1% in global and 3% in direct. The average random error embedded in the diffuse measurements, e_d, would be estimated from those for global (e_g) and direct (e_b) as

$$e_d = \sqrt{e_g{}^2 + e_b{}^2}. \tag{20.7}$$

Determining the uncertainty of modeled results from all the possible sources of errors, including bias and random errors in the measured data points used for validation, and errors in the model's inputs, is an intricate process. The procedure may be built from the general principles explained in Chap. 1. More details for an actual case of validation involving two models and many stations-years are provided in a recent report (NREL 2007), to which the interested reader is referred.

Contrarily to bias errors, random errors tend to decrease when the data are averaged over some time period. For instance, if the N data points considered so far are averaged over a period of n days, the expected $RMSE$ of this averaged dataset is

$$RMSE_{avg} = RMSE / \sqrt{n}. \tag{20.8}$$

This mathematical fact (BIPM 1995) is relatively well verified in practice when radiation models are used to predict hourly irradiances, which are then averaged over daily to monthly periods (Davies et al. 1975; Davies and McKay 1982). When comparing surface irradiance estimates based on models using gridded data from satellites to measurements from one or more sites in a single cell of the grid, a modified definition of $RMSE$ improves the comparison (Li et al. 1995).

Studies that have relied on these performance statistics alone are numerous (e.g., Badescu 1997; Battles et al. 2000; Davies and McKay 1982; Davies et al. 1988; Davies and McKay 1989; De Miguel et al. 2001; Gopinathan and Soler 1995; Gueymard 2003a; Ianetz and Kudish 1994; Ineichen 2006; Kambezidis et al. 1994; Lopez et al. 2000; Ma and Iqbal 1984; Notton et al. 1996; Perez et al. 1992; Reindl et al. 1990).

MBE and $RMSE$ do not characterize the same aspect of the overall errors' behavior. Therefore, when comparing various models against the same reference dataset, the ranking that is obtained from MBE in ascending order (of absolute value) is frequently different from the ranking obtained from $RMSE$. For a reportedly sounder ranking, other statistical tools have been proposed in the literature. Alados-Arboledas et al. (2000) have used a combination of MBE, $RMSE$, and coefficient of linear correlation, R, between the predicted and measured results. Jeter and Balaras (1986) and Ianetz et al. (2007) have used the coefficient of determination (i.e., the square of the coefficient of linear correlation, R^2) and the Fisher F-statistic (Bevington and Robinson 2003). Similarly, other authors (Jacovides 1998; Jacovides and Kontoyiannis 1995; Jacovides et al. 1996) have used a combination of MBE, $RMSE$, R^2, and t-statistic. Usage of the latter was originally suggested by Stone (1993), who showed that, for $f - 1$ degrees of freedom,

$$t = \sqrt{\frac{(f-1)MBE^2}{RMSE^2 - MBE^2}}. \tag{20.9}$$

With this statistic, the model's performance is inversely related to the value of t. A detailed ranking procedure based on t was later proposed (Stone 1994).

Another convenient ranking tool, the index of agreement, d, was proposed (Willmott 1981, 1982a, b; Willmott et al. 1985) as a measure of the degree to which a model's predictions are error free. The index d varies between 0 and 1, with perfect agreement indicated by the latter value; it has been used in later studies (Alados et al. 2000; González and Calbó 1999; Power 2001).

Muneer et al. (2007) recently proposed an "accuracy score" that appropriately combines six indices: *MBE, RMSE, R^2*, skewness, kurtosis, and the slope of the linear correlation between predicted and reference values. The score's minimum and maximum values are 0 and 6, respectively. A major inconvenience of the method is that each of its individual scores refers to the best performer, so that all calculations need to be redone each time a model is modified or added to the test pool.

Finally, a clever graphical way of summarizing multiple aspects of model performance in a single diagram has been proposed by Taylor (2001).

The diversity of the current performance indicators and ranking tools being used calls for assessment studies with help from statisticians. Expert systems are now being developed, based on, e.g., fuzzy algorithms (Bellocchi et al. 2002). Computerized model evaluation tools are also introduced to simplify the numerical burden associated with extensive statistical calculations (Fila et al. 2003).

The need for more research on the most appropriate and statistically-sound ranking methodologies is confirmed by the results of Sect. 6.3, which presents an example (involving fifteen radiation models of the same type) where the different possible rankings do not agree.

5.3 Performance Assessment Significance

The solar radiation literature is rich in validation reports of new, isolated models, or in performance assessment studies of similar models being intercompared. But, readers or users might ask, what is the significance of all these results? Interestingly, from a philosophical perspective, it has been boldly postulated that "*Verification and validation of numerical models of natural systems is impossible*" and that "*Models can only be evaluated in relative terms, and their predictive value is always open to question*" (Oreskes et al. 1994). These arguments are certainly debatable and can appear of hardly any concern in the context of daily engineering tasks, for instance. Nonetheless, a thorough literature review reveals that, indeed, radiative models are not always "validated" or "verified" convincingly, due to the non-observance of some important rules, which are discussed in what follows.

- Rule #1—*Datasets independence*

The dataset used to validate a model should be as independent as possible from that used to derive it. This can be generally achieved by first randomly selecting, for instance two subsets, one for development, one for validation. The random selection is a better procedure than using, e.g., two years of data for model derivation and one year for model validation, since autocorrelation is likely to exist from one year to the other. Another aspect of this rule is that circular calculations should be absolutely avoided. This would occur, for instance, if the model under scrutiny is used in inverted mode to derive the turbidity data that it needs in normal mode. This rule seems obvious, but is not always followed in practice.

- Rule #2—*Uncertainty analysis*

As indicated in previous sections, an uncertainty analysis of the reference dataset (presumably measured) is essential here. A sensitivity analysis carried out on the model's inputs can then determine how they should be filtered to consider only those conditions that can lead to prediction errors lower than the experimental uncertainty, and the experimental uncertainty should be established and stated.

- Rule #3—*Data filtering*

All available measured data points are not necessarily good for validation purposes. They need to be checked first for inconsistencies, egregious errors, etc. Irradiance data from research-class sites are generally well quality-controlled, but spurious data can still exist. (See the discussion on data quality assessment in Chap. 1, Sect. 8.3.) *A posteriori* tests are recommended with data from any source, using different possible strategies (Claywell et al. 2005; Hay 1993a; Muneer and Fairooz 2002; Muneer et al. 2007). Also, as a result of Rule #2 above, all input data conducive to low accuracy should be discarded. For instance, it is generally observed that under low-sun conditions (high zenith angles) both measured and modeled uncertainties become too high to draw valid conclusions.

- Rule #4—*Unity of time and space*

Ideally, model inputs should have the same time resolution as the validation data, and should be obtained at the same site. This can be rarely achieved due to the frequent constraint that some inputs are not measured on site at the proper frequency, and must therefore be extrapolated, interpolated, or averaged in various ways. When these imperfect data are used for a highly-sensitive input, the model's performance can be significantly degraded. This has been demonstrated in the case of clear-sky meteorological models for instance, in relation with the use of instantaneous *vs* time-averaged turbidity data (Battles et al. 2000; Ineichen 2006; Olmo et al. 2001). This rule cannot be respected either when the model uses gridded input data, such as cloud information, and is tested with site-specific reference data, or "ground truth". This problem is known to introduce significant random errors (Perez et al. 2002). (See also the accuracy discussion at http://eosweb.larc.nasa.gov/cgi-bin/sse/sse.cgi?na+s05#s05 and references therein.)

- Rule #5—*Proper ancillary data*

Only the best possible ancillary data should be used, particularly for the most model-sensitive inputs. This is often critical to the performance of a model. Using low-quality ancillary data results in biased or inconclusive performance assessments (Ineichen 2006).

- Rule #6—*Radiative closure*

Ideally, a model validation should be conducted as a radiative closure experiment. This means that all inputs to the model be measured independently with co-located instruments, at the required frequency, and with sufficiently small uncertainty. If the error bars (estimated uncertainty) in the modeled results overlap the measurement error bars (uncertainty in the measurements) to a significant extent, then the model can be considered validated. If not, either the model's intrinsic performance is deficient, or the input data are of too low quality and do not satisfy the requirements of Rule #5.

- Rule #7—*Validity limits*

In most cases, a model is only validated for specific atmospheric or climatic conditions (see Rule #4). It is important to specify the limits of validity of the model to avoid inadvertent extrapolations from users. It is also common observation that empirically-determined equations using high-order polynomials are subject to divergence if used outside of their intended limits. More efficient mathematical modeling is therefore recommended, using, e.g., polynomial ratios. Finally, the required time scale of the input data must be clearly identified to avoid misinterpretations. For instance, Gueymard et al. (1995) showed that using a sunshine-based global radiation model with empirical coefficients originally developed for yearly-mean sunshine, but incorrectly applied to monthly means for convenience, could be detrimental.

6 Some Performance Assessment Results

Solar radiation model performance is usually carried out from two different perspectives. First, the assessment of models made by their developers is usually referred to as validation, as discussed in previous sections. Secondly, models can be evaluated by testing against independent data sets, usually by authors independent from the model development.

6.1 Performance of Model Elements

One informative segment of model performance is the comparison of model elements or functions to previously-developed similar model components. Examples include so-called "simple" broadband or spectral transmittance models. These types of models were briefly mentioned in Sect. 4, and described in Eq. (20.4). Comparisons of individual transmittance functions from some similar models are detailed elsewhere (Gueymard 1993, 2003b). These model elements may be changed or improved during the model's development or as different, possibly more detailed information on model parameters become available.

For instance, Bird and Hulstrom (1981a,b) compared the transmittance functions developed for their model with those of several other authors (Atwater and Ball 1978; Davies and Hay 1979; Hoyt 1978; Lacis and Hansen 1974; Watt 1978). This pioneering work is regularly updated (Gueymard 1993, 2003b). Of course, for models of this type, the number of parameterized transmission functions need not be identical, nor the input parameters match exactly (e.g., relative humidity and ambient temperature in place of dew point for estimating, or as surrogates for, precipitable water).

A similar example is provided by Thornton and Running (1999) who developed an improved version of a model for solar radiation based solely on ambient temperature developed by Bristow and Campbell (1984). Thornton and Running expanded the model with improved parameterizations of the coefficients, and consideration of a single additional (optional) input variable of dew-point temperature. Their sensitivity study is also remarkable. One site at a time is systematically removed from the set of 40 sites. Model parameters are derived from the remaining 39 sites. The 'new' model is applied to the excluded site, and the *MABE* for that site is computed. The procedure is iterated over all sites, providing data for a multi-way analysis to minimize the pooled *MABE* for the data set. This also provides a means of testing the influence of specific sites (or sets of sites, say, clear *vs* cloudy or desert *vs* continental) on the model's parameters.

These comparisons and modifications of model elements are helpful in developing model improvements and quantifying the causes of relative model biases and differences. The real test of model performance comes from evaluations performed by other authors, using different sets of input data, and especially measured solar radiation data, as discussed above in Sect. 3.

6.2 Independent Model Performance Evaluation

Independent evaluation of models based on input parameters and measured data for different sites is prevalent in the literature. One or more model is evaluated with one or more new and independent validation data sets.

Physical and Empirical Model Evaluation

With physical and empirical model evaluation of single or multiple models, an important problem is that, often, the performance evaluation units and temporal resolution used in the new evaluation are different from those in the original validation. This is a much larger problem for satellite-based model evaluation, discussed briefly in the next subsection.

For example, a model developer may quote *RMSE* and *MBE* errors in terms of percentage errors in monthly-mean totals of radiation, while an independent evaluation may compute the errors in daily irradiation units ($MJ\,m^{-2}$ or $kWh\,m^{-2}$), on the basis of hourly diurnal profiles. Note that, from Sect. 5.3, both the model limits of validity (Rule 7 in section 5.3), and Rule 4 (unity of time and space) should be considered when independent validation is developed. Attempts to extrapolate or interpolate model performance beyond the validation limits should be described in detail, with appropriate caveats. Disparate criteria make it difficult to decide how the variations in data sets (i.e., site dependencies) may be affecting models.

On the other hand, quantitative results in any form still convey an idea of the expected uncertainty in models, as long as the analysis is adequately described. For example, in the original detailed report on his Direct Insolation Simulation Code (DISC) separation model for converting hourly global horizontal to hourly direct beam irradiance, (Maxwell 1987) shows monthly average *MBE* and *RMSE* errors as percent deviations, and shows diurnal profiles of hourly differences between modeled and measured data for specific dates. The performance results are based on data from three sites not used in the model development. In an independent evaluation of the Maxwell model, Perez et al. (1990b) show *MBE* and *RMSE* errors for 13 individual sites (as well as two additional models) sorted by zenith angle and clearness index (K_T) in irradiance units (W m^{-2}). Their analysis gives a more detailed picture of the model performance, but is difficult to compare with the original analysis.

With the increasing interest and need for solar resource information, and reduced availability of ground-based measurements in many countries, an increasing number of authors have carried out independent evaluation studies of multiple solar radiation models. Excellent examples of these review articles include studies considering five models and four data sets (Perez and Stewart 1986); three models and fourteen data sets (Perez et al. 1990b); five models at two sites (Badescu 1997); 38 models of atmospheric (infrared) emission and 15 data sets (Skartveit et al. 1996); a combination of twelve transposition models for tilted surfaces and four albedo submodels at four sites (Psiloglou et al. 1996); two sunshine models, three empirical models, and four datasets (Iziomon and Mayer 2002); seven sunshine models and 77 sites (Soler 1990); six irradiance models at four sites (Battles et al. 2000); 21 irradiance models and six datasets (Gueymard 2003a); and eight models at sixteen sites (Ineichen 2006). Moreover, the Solar Heating and Cooling Programme's Task IX of the International Energy Agency (IEA) supported two important validation studies using many models and international datasets. One was devoted to the prediction of hourly or daily horizontal radiation from meteorological data (Davies and McKay 1982, 1989; Davies et al. 1984, 1988), and the other to the prediction of hourly or daily tilted radiation from horizontal data (Hay and McKay 1985, 1986; Hay 1993b). The latter study included most, if not all, transposition models of the literature then available. The former IEA study only included a small number of meteorological models. A different approach is used in Sect. 6.3, where a sample of all known clear-sky models able to predict both direct and diffuse radiation instantaneously on a horizontal surface is reviewed. This study being for demonstration purposes only, it is geographically limited to only one site, using the small *benchmark* dataset previously mentioned in Sects. 3 and 5.1.

Satellite Model Evaluation

Many of the issues discussed above become much more noticeable and complicated for validation of models based on satellite input data. The questions of temporal and spatial consistency are particularly vexing, as satellite data, while uniform, are usually sparse in time compared to surface observations. Spatial concerns are an even greater problem, since surface observations are 'point' observations, and satellite observations are spatially extended, even if at very high spatial resolution. Perez et al. (1997, 2001) provide a detailed review of these issues. In particular, as one observes the degradation in correlation between solar radiation measurements as ground site-separation increases, one sees the same sort of degradation in the accuracy of satellite model estimates as one moves away from a "ground truth" site. Magnitudes of this degradation start at about $\pm15\%$ for sites within a few kilometers of each other to above 40–60% at distances of several hundred kilometers, for both ground stations and pixels removed from validation sites in satellite estimates (Perez et al. 2001).

As the number of satellite platforms and models evolve, there are also the issues of degradation and 'recalibration' of space-based sensors. These issues, similar to the calibration, degradation, and uncertainty of ground-based sensors discussed in Chap. 1, should be kept in mind when using models using satellite-based input data.

The NASA Surface and Meteorological and Solar Energy (SSE) website (http://eosweb.larc.nasa.gov/sse) provides a great deal of helpful information on both accuracy and methodologies.

The European Community's Helioclim project has links describing the Heliosat model used for the European Solar Radiation Atlas (http://www.helioclim.net/heliosat/index.html), as well as links describing the calibration of MeteoSat instruments (http://www.helioclim.net/calibration/index.html), and solar radiation data quality control (http:// www.helioclim.net/quality/index.html).

Comparisons between results from models using either satellite or ground-based input data are also possible. An example provided here is between the METSTAT meteorological model (Maxwell 1998) for ground-based input data—but modified for use in the 1991–2005 update to the 1961–1990 United States National Solar Radiation data base (http:// rredc.nrel.gov/solar/old_data/nsrdb)—and the recently developed Perez satellite model (Wilcox et al. 2007). Figure 20.6 shows a comparison of the annual average direct beam estimates as gray-scaled background (satellite model estimates) and circles (station-based estimates). The observed differences (up to several $kWh\,m^{-2}$ per day) are due mainly to issues with the quality of the input data for the meteorological model. Most particularly, the move from human observers to automated ceilometer measurements has severely compromised the cloud cover data needed for the modified METSTAT model inputs.

6.3 Model Performance Benchmarking and Ranking

This section provides a concrete example of how a comparative performance assessment study can be conducted when a large number of models is involved. In the present case, a rather exhaustive literature survey provided a list of 35 clear-sky broadband irradiance models that can predict instantaneous or short-term (e.g., hourly) direct and diffuse irradiances from limited information on the optical properties of the atmosphere. A computer program that can use the benchmark dataset

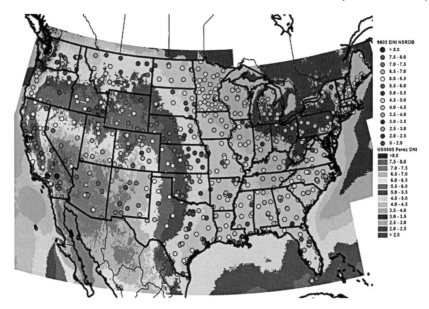

Fig. 20.6 Comparison between meteorological (METSTAT) site-based and (Perez) satellite-based direct normal irradiance (DNI) solar radiation estimates for the 1991–2005 US National Solar Radiation Data Base update (Wilcox et al. 2007). Grayscale contours (originally in color) indicate the average 1998–2005 DNI predictions from the Perez model, and similarly color-coded circles indicate those from the METSTAT model

mentioned in Sect. 5.1 to evaluate all these models is included on the CD-ROM (file 'Models_performance_compar.f'). Only a subset of 15 models is described in what follows, owing to the demonstration purpose of this validation exercise and space limitations. Most of the models have been described and discussed in previous studies (e.g., (Gueymard 1993, 2003b), so that only a summary is provided here, except where more details are necessary. The models are listed in alphabetical order below.

- Model 1—*ASHRAE*

This is the model used by engineers to calculate solar heat gains and cooling loads in buildings. It was first introduced in 1972, but new monthly coefficients have appeared recently (ASHRAE 2005), which are used here. Note that, beyond these empirical coefficients and solar zenith angle, this model does not depend on any atmospheric data.

- Model 2—*Bird*

This is the original Bird model (Bird and Hulstrom 1981a, b), with only one modification, required by changes in turbidity measurement practice. At the time this model was developed, the aerosol optical depth (AOD) was measured by sunphotometers with at most two channels, centered at 380 and 500 nm, hence the model's requirement for the AOD at these two wavelengths. Since the early 1990s, networks of multiwavelength sunphotometers, with typically 5–7 aerosol channels, have expanded worldwide (see, e.g., http://aeronet.gsfc.nasa.gov). Therefore, it is now easier than ever to obtain the turbidity coefficients α and β by fitting the experimental AOD at various wavelengths, $\tau_{a\lambda}$, to Ångström's Law:

$$\tau_{a\lambda} = \beta(\lambda/\lambda_0)^{-\alpha} \qquad (20.10)$$

where $\lambda_0 = 1000$ nm. When α and β are known, the specific AOD at 380 and 500 nm required by Bird's model can be replaced by $\beta 0.38^{-\alpha}$ and $\beta 0.5^{-\alpha}$, respectively. This respects the model's integrity while considerably expanding its applicability.

- Model 3—*CLS*

The Cloud Layer-Sunshine model (Suckling and Hay 1976, 1977) is based on original work by Houghton (1954) and Monteith (1962). The CLS model has been independently validated for average sky conditions during the IEA Task IX mentioned in the previous section, but not for clear skies only—at least outside of Canada. The original expression for the aerosol transmittance, $T_a = 0.95^m$, where m is the air mass, is used here.

- Model 4—*CPCR2*

This two-band model (Gueymard 1989) has already been tested extensively in various studies (Battles et al. 2000; Gueymard 1993, 2003a, 2003b; Ineichen 2006; Olmo et al. 2001). It normally requires separate values of α and β for each of the two wavebands (290–700 nm and 700–4000 nm) considered by the model, i.e., (α_1, β_1) and (α_2, β_2) with the constraint $\beta_1 = \beta_2 0.7^{\alpha_1-\alpha_2}$. This information can be derived from the current sunphotometric data by appropriate application of Eq. (20.10). If not possible, the model can be accommodated with simply $\alpha_1 = \alpha_2 = \alpha$ and $\beta_1 = \beta_2 = \beta$ yielding only a modest degradation of performance.

- Model 5—*ESRA2*

The original version of this model (Rigollier et al. 2000) has been used to derive the latest edition of the European Solar Radiation Atlas (Scharmer and Greif 2000). The Linke turbidity factor, T_L, was then the basis to evaluate the effect of aerosols. However, this factor cannot be measured directly, and therefore needs to be evaluated by inversion of an appropriate irradiance model, using experimental clear-sky direct irradiance as the input. This creates a problem in the context of validation studies since the measured direct irradiance cannot be used both to test the model's predictions and to derive its inputs (see Rule #1 in Sect. 5.3). A new version of the model, which is tested here, rather calculates T_L from air mass, precipitable water, and β (Remund et al. 2003).

- Models 6–8—*Iqbal's Parameterization Models A, B and C*

These models are fully described in the original publication (Iqbal 1983), and have been tested previously, to some extent (Battles et al. 2000; Gueymard 1993). Scatterplots of Iqbal C's model appear in Fig 20.3 for the benchmark dataset considered here.

- Model 9—*Kasten*

This classic model (Kasten 1980, 1983; Kasten and Czeplak 1980) has been expanded to provide direct and global irradiance, both as a function of T_L (Davies and McKay 1989). The latter version is used here. To overcome the difficulty in using T_L while respecting the model's intentional simplicity, a simple linear function of β has been used,

$$T_L = 2.1331 + 19.0204\,\beta \qquad (20.11)$$

where the numerical coefficients have been obtained by combining the empirical determinations of $\beta = \mathrm{f}(T_L)$ proposed by different authors (Abdelrahman et al. 1988; Grenier et al. 1994; Hinzpeter 1950; Katz et al. 1982).

- Model 10—*METSTAT*

As mentioned in Criterion #6 of Sect. 2, this model has a deterministic algorithm that can be combined with statistical features so that correct frequency distributions of hourly irradiances can be obtained despite the use of daily or monthly-average turbidity and cloud input data (Maxwell 1998). The model is used here without these statistical corrections since short-term input data are available for validation, therefore respecting Rule #4 in Sect. 5.3. A modification to the model, however, is necessary since it uses the broadband aerosol optical depth, τ_a, to evaluate the aerosol transmittance. Like T_L, τ_a can only be obtained indirectly from an inverted model and irradiance measurements. To circumvent this problem, a convenient methodology (Molineaux et al. 1998), which uses the concept of equivalent wavelength for broadband turbidity, is used here to derive τ_a from α and β through

$$\tau_a = \beta\,[0.695 + (0.0160 + 0.066\,\beta\,0.7^{-\alpha})m]^{-\alpha}. \qquad (20.12)$$

- Model 11—*MAC*

The McMaster (MAC) model has evolved slightly between its original derivation (Davies et al. 1975) and the latest performance assessment results (Davies and McKay 1989). The version described in the IEA Task IX report (Davies et al. 1988) is used here, with a Rayleigh transmittance formula corrected for its typographic error. In the absence of specific information on the most appropriate aerosol transmittance to be used here, the original formula $T_a = 0.95^m$ (Davies et al. 1975; Davies and Hay 1979), as for the BCLS model, is selected.

- Model 12—*MRM5*

This new, version 5, of the Meteorological Radiation Model, is described in Chap. 14. It contains important changes from the previous version 4 (Muneer 2004), which used incorrect numerical coefficients that considerably affected the model's irradiance predictions (Gueymard 2003a, 2003b).

- Model 13—*REST2*

This two-band model (Gueymard 2008) is based on CPCR2, but incorporates completely revised parameterizations, which have been derived from the SMARTS spectral model (Gueymard 2001, 2005b). REST2 uses the same inputs as CPCR2, with the addition of the amount of nitrogen dioxide in a vertical atmospheric column, which can be defaulted if unknown. See Fig. 20.3 for scatterplots involving this model and the benchmark dataset.

- Model 14—*Santamouris*

This model's algorithm (Santamouris et al. 1999) is in essence similar to that of the Bird model, except that a fixed turbidity is considered.

- Model 15—*Yang*

The direct irradiance predictions of this model (Yang et al. 2001; Yang and Koike 2005) have been shown to perform very well (Gueymard 2003a, b). Its diffuse irradiance predictions have not been evaluated so far.

All radiation models require at least one input variable, namely the solar zenith angle, Z. The simplest models (e.g., ASHRAE) do not need more inputs (besides some empirical coefficients). The most sophisticated models (e.g., REST2) may require as many as 10 more inputs, mostly atmospheric data. For clarity, all the inputs (besides Z) required by the 15 models considered here are compiled in Table 20.1.

Using the specially-developed Fortran program contained on the CD-ROM (file 'Models_performance_compar.f'), all these models (and more) have been run using the 30-point benchmark dataset recently proposed (Gueymard 2008) and previously mentioned in Sects. 3 and 5.1. This dataset is contained in file 'Models_perf_exp. dat.txt'. Of course, to the benefit of the reader, this Fortran program can also

Table 20.1 Inputs required by the 15 clear-sky models under scrutiny

#	Name	m	E_{n0}	ρ_g	p	T	U_o	U_n	w	T_L	τ_a	α	β	ω_a
1	ASHRAE													
2	Bird	✓	✓	✓	✓		✓		✓			✓	✓	
3	CLS	✓	✓	✓	✓				✓					
4	CPCR2	✓	✓	✓	✓		✓		✓			✓	✓	✓
5	ESRA2	✓	✓		✓				✓				✓	
6	Iqbal A	✓	✓	✓	✓	✓	✓		✓			✓	✓	
7	Iqbal B	✓	✓	✓	✓	✓	✓		✓				✓	
8	Iqbal C	✓	✓	✓	✓	✓	✓		✓			✓	✓	
9	Kasten	✓	✓		✓					✓				
10	METSTAT	✓	✓	✓	✓		✓		✓		✓			
11	MAC	✓	✓	✓	✓	✓			✓					
12	MRM5	✓	✓	✓	✓	✓	✓		✓					
13	REST2	✓	✓	✓	✓		✓	✓	✓			✓	✓	✓
14	Santamouris	✓	✓	✓	✓		✓		✓					
15	Yang	✓	✓	✓	✓		✓		✓				✓	

Key: m, air mass; E_{n0}, distance-corrected extraterrestrial irradiance; ρ_g, far-field ground albedo; p, site pressure; T, dry-bulb temperature; U_o, total ozone in the vertical column; U_n, total nitrogen dioxide in the vertical column; w, precipitable water; T_L, Linke turbidity coefficient; τ_a, broadband aerosol optical depth; α, Ångström wavelength exponent; β, Ångström's turbidity coefficient, ω_a, aerosol single-scattering albedo.

Table 20.2 Performance statistics, and their associated ranking (in bold italics), for direct irradiance predicted by the 15 clear-sky models under scrutiny, relative to a benchmark dataset

#	MBE (%)		RMSE (%)		R^2		t		d		AS	
1	3.6	10	12.0	11	0.703	15	1.70	5	0.897	11	2.71	11
2	−0.6	2	2.9	3	0.983	5	1.21	2	0.995	3	4.46	4
3	13.0	14	17.0	14	0.730	11	6.47	11	0.823	14	2.28	14
4	−1.2	5	2.2	2	0.994	3	3.68	8	0.997	2	4.49	3
5	3.1	8	4.4	8	0.979	7	5.50	10	0.989	8	4.38	7
6	2.7	7	3.0	4	0.996	2	10.76	15	0.995	4	4.45	5
7	9.8	12	11.0	10	0.949	10	10.58	14	0.928	10	3.59	10
8	1.2	4	3.1	5	0.982	6	2.31	6	0.995	6	4.40	6
9	1.5	6	5.8	9	0.961	9	1.50	3	0.984	9	3.97	9
10	3.4	9	4.0	7	0.994	4	8.86	12	0.990	7	4.28	8
11	9.9	13	14.9	13	0.725	12	4.87	9	0.846	13	2.35	13
12	−5.8	11	12.7	12	0.718	13	2.82	7	0.897	12	2.66	12
13	0.1	1	1.0	1	0.998	1	0.49	1	0.999	1	4.66	1
14	−20.0	15	23.3	15	0.715	14	9.20	13	0.759	15	1.97	15
15	0.9	3	3.4	6	0.977	8	1.54	4	0.994	6	4.58	2

accommodate other, more voluminous datasets, if prepared with the same format. The present example is purposefully limited to a small dataset, so that the conclusions reached here should not be considered of general or "universal" validity. However, owing to the benchmark status of this dataset, it can be said that a model should not be considered as universal if it does not perform well under the conditions of this dataset.

A statistical analysis has been conducted from the differences between predicted and measured direct, diffuse and global irradiances; the summary performance results appear in Tables 20.2, 20.3 and 20.4, respectively. These tables include the

Table 20.3 Performance statistics, and their associated ranking (in bold italics), for diffuse irradiance predicted by the 15 clear-sky models under scrutiny, relative to a benchmark dataset

#	MBE (%)		RMSE (%)		R^2		t		d		AS	
1	−5.6	5	33.9	12	0.351	14	0.92	4	0.595	14	2.41	10
2	1.0	2	10.8	2	0.933	7	0.49	3	0.980	2	3.94	2
3	−8.8	6	35.3	13	0.289	15	1.41	5	0.583	15	2.13	13
4	14.2	9	16.8	6	0.969	3	8.60	12	0.963	5	3.41	6
5	−19.5	11	24.5	8	0.947	6	7.18	11	0.879	8	2.91	8
6	13.2	8	15.3	5	0.966	4	9.19	14	0.966	4	3.52	4
7	−28.5	15	33.3	10	0.826	9	9.10	13	0.847	9	2.29	11
8	−0.4	1	11.3	3	0.928	8	0.20	1	0.977	3	3.88	3
9	−21.0	12	33.4	11	0.594	10	4.43	8	0.808	10	2.20	12
10	−22.0	13	22.6	7	0.985	2	24.63	15	0.926	7	3.12	7
11	−14.9	10	36.0	14	0.365	12	2.49	6	0.604	13	2.06	14
12	−1.8	3	32.7	9	0.360	13	0.30	2	0.669	12	2.51	9
13	2.7	4	4.4	1	0.993	1	4.28	7	0.997	1	4.16	1
14	28.4	14	43.0	15	0.366	11	4.83	9	0.671	11	1.82	15
15	−10.1	7	15.2	4	0.961	5	4.83	10	0.955	6	3.45	5

Table 20.4 Performance statistics, and their associated ranking (in bold italics), for global irradiance predicted by the 15 clear-sky models under scrutiny, relative to a benchmark dataset

#	MBE (%)		RMSE (%)		R^2		t		d		AS	
1	1.7	*9*	3.4	*8*	0.996	*13*	3.14	*6*	0.999	*1*	4.97	*1*
2	−0.6	*2*	1.8	*4*	0.998	*7*	1.78	*2*	0.843	*9*	3.21	*9*
3	8.5	*14*	9.2	*14*	0.998	*8*	13.45	*14*	0.865	*2*	1.76	*14*
4	1.3	*6*	1.5	*3*	1.000	*3*	9.55	*10*	0.847	*7*	4.08	*3*
5	−0.9	*4*	3.9	*9*	0.994	*15*	1.37	*1*	0.841	*11*	3.90	*6*
6	4.2	*11*	4.6	*11*	1.000	*2*	13.08	*13*	0.859	*3*	2.98	*12*
7	1.3	*7*	2.3	*6*	1.000	*6*	3.84	*7*	0.850	*5*	3.54	*8*
8	0.6	*3*	1.5	*2*	1.000	*5*	2.42	*4*	0.848	*6*	3.93	*4*
9	−1.6	*8*	3.9	*10*	0.995	*14*	2.47	*5*	0.841	*12*	3.17	*10*
10	−1.7	*10*	2.0	*5*	1.000	*4*	9.90	*11*	0.833	*13*	3.92	*5*
11	4.7	*13*	5.3	*13*	0.998	*9*	10.63	*12*	0.855	*4*	3.01	*11*
12	−4.2	*12*	4.9	*12*	0.998	*10*	9.39	*9*	0.814	*14*	2.96	*13*
13	0.5	*1*	0.8	*1*	1.000	*1*	4.84	*8*	0.844	*8*	4.26	*2*
14	−9.1	*15*	9.7	*15*	0.998	*11*	14.62	*15*	0.782	*15*	1.50	*15*
15	−1.1	*5*	3.2	*7*	0.996	*12*	2.02	*3*	0.842	*10*	3.66	*7*

MBE, RMSE, R^2, t-statistic, Willmott's index of agreement d, and Muneer's accuracy score AS. For each of these performance indices, the corresponding model ranking is indicated. As could be expected, the models that use detailed atmospheric information (particularly on aerosols and water vapor) perform better than those with little or no such inputs. The main disturbing fact, however, is that the ranking methods disagree widely, particularly for diffuse and global irradiance. This confirms the need for more in-depth investigations on this issue.

7 Conclusions

The primary focus of this chapter has been to emphasize to the newcomer as well as the experienced solar radiation model developer, tester, or user, the nuances of model validation and performance evaluation. Section 2 addressed seven criteria describing typical solar radiation model approaches or types. Sections 3 and 4 described the principles of model validation and uncertainty analysis required for both measured data and uncertainties in model estimates. Sections 5.1 and 5.2 addressed some aspects of qualitative and quantitative model performance. Section 5.3 emphasized seven constituent elements of model validation that must be addressed in any evaluation. These include validation and input data quality, independence, and consistency of temporal and spatial extent, and validation limits. Section 6 discussed evolution and validation of model component parts, the importance of, and difficulties associated with, interpreting independent model validation, as well as demonstrated the practice (and difficulties) of comparing the performance of many models).

As the solar energy industry and scientific research into the detailed energy balance of the Earth continues to grow and evolve, it is critical that computational models be validated and tested as stringently as possible to provide decision makers, and the scientific community in general, with the most accurate, comprehensive, and well documented information possible.

References

Abdelrahman MA, Said SA, Shuaib AN (1988) Comparison between atmospheric turbidity coefficients of desert and temperate climates. Solar Energy 40: 219–225

Alados I, Olmo FJ, Foyo-Moreno I, Alados-Arboledas L (2000) Estimation of photosynthetically active radiation under cloudy conditions. Agric. For. Meteorol. 102: 39–50

Alados-Arboledas L, Olmo FJ, Alados I, Pérez M (2000) Parametric models to estimate photosynthetically active radiation in Spain. Agric. For. Meteorol. 101: 187–201

ASHRAE (2005) Handbook of Fundamentals, SI Edition. American Society of Heating, Refrigerating and Air-Conditioning Engineers, Atlanta, GA

Atwater MA, Ball JT (1978) A numerical solar radiation model based on standard meteorological observations. Solar Energy 21: 163–170

Badescu V (1997) Verification of some very simple clear and cloudy sky models to evaluate global solar irradiance. Solar Energy 61: 251–264

Battles FJ, Olmo FJ, Tovar J, Alados-Arboledas L (2000) Comparison of cloudless sky parameterizations of solar irradiance at various Spanish midlatitude locations. Theor. Appl. Climatol. 66: 81–93

Bellocchi G, Acutis M, Fila G, Donatelli M (2002) An indicator of solar radiation model performance based on a fuzzy expert system. Agron. J. 94: 1222–1223

Bevington PR, Robinson DK (2003) Data reduction and error analysis for the physical sciences. McGraw-Hill

BIPM (1995) Guide to the expression of uncertainty in measurement. ISBN 92-67-10188-9, International Bureau of Weights and Measures (BIPM), International Standards Organization

Bird RE, Hulstrom RL (1981a) A simplified clear sky model for direct and diffuse insolation on horizontal surfaces. SERI TR-642-761 (Available online at http://rredc.nrel.gov/solar/models/clearsky), Solar Energy Research Institute

Bird RE, Hulstrom RL (1981b) Review, evaluation, and improvement of direct irradiance models Trans. ASME, J. Solar Energy Engng. 103: 182–192

Bristow LL, Campbell GS (1984) On the relationship between incoming solar radiation and daily maximum and minimum temperature. Agric. Forest Meteorol. 31: 159–166

Claywell R, Muneer T, Asif M (2005) An efficient method for assessing the quality of large solar irradiance datasets. Trans. ASME, J. Solar Energy Engng. 127: 150–152

Collares-Pereira M, Rabl A (1979) The average distribution of solar radiation—correlations between diffuse and hemispherical and between daily and hourly insolation values. Solar Energy 22: 155–164

Crandall KC, Seabloom RW (1970) Engineering fundamentals in measurement, probability, statistics, and dimensions. McGraw Hill

Davies JA, Schertzer W, Nunez M (1975) Estimating global solar radiation. Bound. Layer Meteor. 9: 33–52

Davies JA, Hay JE (1979) Calculation of the solar radiation incident on a horizontal surface. Proc. First Canadian Solar Radiation Data Workshop, April 17–19, 1979, Canadian Atmospheric Environment Service

Davies JA, McKay DC (1982) Estimating solar irradiance and components. Solar Energy 29: 55–64

Davies JA, Abdel-Wahab M, McKay DC (1984) Estimating solar irradiation on horizontal surfaces. Int. J. Solar Energy 2: 405–424

Davies JA, McKay DC, Luciani G, Abdel-Wahab M (1988) Validation of models for estimating solar radiation on horizontal surfaces. IEA Task IX Final Report, Atmospheric Environment Service, Downsview, Ont.

Davies JA, McKay DC (1989) Evaluation of selected models for estimating solar radiation on horizontal surfaces. Solar Energy 43: 153–168

De Miguel A, Bilbao J, Aguiar R, Kambezidis HD, Negro E (2001) Diffuse solar irradiation model evaluation in the North Mediterranean belt area. Solar Energy 70: 143–153

Fila G, Bellocchi G, Donatelli M, Acutis M (2003) IRENE_DLL: A class library for evaluating numerical estimates. Agron. J. 95: 1330–1333

Fox J (1997) Applied regression analysis, linear models, and related methods. Sage Publ. See also http://www.princeton.edu/~slynch/SOC_504/multicol linearity.pdf

González J-A, Calbó J (1999) Influence of the global radiation variability on the hourly diffuse fraction correlations. Solar Energy 65: 119–131

Gopinathan KK, Soler A (1995) Diffuse radiation models and monthly-average, daily, diffuse data for a wide latitude range. Energy 20: 657–667

Gordon J, ed (2001) Solar energy—The state of the art, ISES position papers. James & James and International Solar Energy Society, pp Pages

Grenier JC, de la Casinière A, Cabot T (1994) A spectral model of Linke's turbidity factor and its experimental implications. Solar Energy 52: 303–314

Gueymard CA (1987) An anisotropic solar irradiance model for tilted surfaces and its comparison with selected engineering algorithms. Solar Energy 38: 367–386. Erratum, Solar Energy, 40, 175 (1988)

Gueymard CA (1989) A two-band model for the calculation of clear sky solar irradiance, illuminance, and photosynthetically active radiation at the Earth's surface. Solar Energy 43: 253–265

Gueymard CA (1993) Critical analysis and performance assessment of clear sky solar irradiance models using theoretical and measured data. Solar Energy 51: 121–138

Gueymard CA, Jindra P, Estrada-Cajigal V (1995) A critical look at recent interpretations of the Ångström approach and its future in global solar radiation prediction. Solar Energy 54: 357–363

Gueymard CA (1998) Turbidity determination from broadband irradiance measurements: A detailed multicoefficient approach. J. Appl. Meteorol. 37: 414–435

Gueymard CA (2000) Prediction and performance assessment of mean hourly global radiation. Solar Energy 68: 285–303

Gueymard CA (2001) Parameterized transmittance model for direct beam and circumsolar spectral irradiance. Solar Energy 71: 325–346

Gueymard CA (2003a) Direct solar transmittance and irradiance predictions with broadband models. Pt 2: Validation with high-quality measurements. Solar Energy 74: 381–395. Corrigendum: Solar Energy 76, 515 (2004)

Gueymard CA (2003b) Direct solar transmittance and irradiance predictions with broadband models. Pt 1: Detailed theoretical performance assessment. Solar Energy 74: 355–379. Corrigendum: Solar Energy 76, 513 (2004)

Gueymard CA, Kambezidis HD (2004) Solar spectral radiation. In: Muneer T (ed) Solar radiation and daylight models. Elsevier, pp 221–301

Gueymard CA (2005a) Importance of atmospheric turbidity and associated uncertainties in solar radiation and luminous efficacy modelling. Energy 30: 1603–1621

Gueymard CA (2005b) Interdisciplinary applications of a versatile spectral solar irradiance model: A review. Energy 30, 1551–1576

Gueymard CA (2008) REST2: High performance solar radiation model for cloudless-sky irradiance, illuminance and photosynthetically active radiation—Validation with a benchmark dataset. Solar Energy (in press)

Gueymard CA, Myers D (2007) Performance assessment of routine solar radiation measurements for improved solar resource and radiative modeling. Proc. Solar 2007 Conf., Cleveland, OH, American Solar Energy Society

Hay JE, McKay DC (1985) Estimating solar irradiance on inclined surfaces: a review and assessment of methodologies. Int. J. Solar Energy 3: 203–240

Hay JE, McKay DC (1986) Calculation of solar irradiances for inclined surfaces: verification of models which use hourly and daily data. Report to International Energy Agency, SHCP Task IX, Atmospheric Environment Service, Canada

Hay JE (1993a) Solar radiation data: validation and quality control. Renew. Energy 3: 349–355

Hay JE (1993b) Calculating solar radiation for inclined surfaces: practical approaches. Renew. Energy 3: 373–380

Hinzpeter H (1950) Über Trübungsbestimmungen in Potsdam in dem Jahren 1946 und 1947. Zeit. Meteorol. 4: 1–8

Houghton HG (1954) On the annual heat balance of the northern hemisphere. J. Meteorol. 11: 1–9

Hoyt DV (1978) A model for the calculation of solar global insolation. Solar Energy 21 27–35

Ianetz A, Kudish AI (1994) Correlations between values of daily horizontal beam and global radiation for Beer Sheva, Israel. Energy 19: 751–764

Ianetz A, Lyubansky V, Setter I, Kriheli B, Evseev EG, Kudish AI (2007) Inter-comparison of different models for estimating clear sky solar global radiation for the Negev region of Israel. Energy Conv. Mngmt. 48: 259–268

Ineichen P, Guisan O, Razafindraibe A (1984) Indice de clarté. CUEPE Rep. No. 20, University of Geneva, Switzerland

Ineichen P, Perez R, Seals R (1987) The importance of correct albedo determination for adequately modeling energy received by tilted surfaces. Solar Energy 39: 301–305

Ineichen P (2006) Comparison of eight clear sky broadband models against 16 independent data banks. Solar Energy 80: 468–478

Iqbal M (1983) An introduction to solar radiation. Academic Press

Iziomon MG, Mayer H (2002) Assessment of some global solar radiation parameterizations. J. Atmos. Solar Terr. Phys. 64: 1631–1643

Jacovides CP, Kontoyiannis H (1995) Statistical procedures for the evaluation of evapotranspiration computing models. Agric. Water Mgmt. 27: 365–371

Jacovides CP, Hadjioannou L, Pashiardis S, Stefanou L (1996) On the diffuse fraction of daily and monthly global radiation for the island of Cyprus. Solar Energy 56: 565–572

Jacovides CP (1998) Reply to comment on "Statistical procedures for the evaluation of evapotranspiration computing models". Agric. Water Mgmt. 37: 95–97

Jeter SM, Balaras CA (1986) A regression model for the beam transmittance of the atmosphere based on data for Shenandoah, Georgia, U.S.A. Solar Energy 37: 7–14

Kambezidis HD, Psiloglou BE, Gueymard C (1994) Measurements and models for total solar irradiance on inclined surface in Athens, Greece. Solar Energy 53: 177–185

Kasten F (1980) A simple parameterization of the pyrheliometric formula for determining the Linke turbidity factor. Meteorol. Rdsch. 33: 124–127

Kasten F, Czeplak G (1980) Solar and terrestrial radiation dependent on the amount and type of cloud. Solar Energy 24: 177–189

Kasten F (1983) Parametrisierung der Globalstrahlung durch Bedeckungsgrad und Trübungsfaktor. Ann. Meteorol. 20: 49–50

Kasten F, Young AT (1989) Revised optical air mass tables and approximation formula. Appl. Opt. 28: 4735–4738

Katz M, Baille A, Mermier M (1982) Atmospheric turbidity in a semi-rural site—Evaluation and comparison of different turbidity coefficients. Solar Energy 28: 323–327

Lacis AL, Hansen JE (1974) A parameterization of absorption of solar radiation in the Earth's atmosphere. J. Atmos. Sci. 31: 118–133

Li Z, Whitlock CH, Charlock TP (1995) Assessment of the global monthly mean surface insolation estimated from satellite measurements using Global Energy Balance Archive data. J. Clim. 8: 315–328

Lopez G, Rubio MA, Battles FJ (2000) Estimation of hourly direct normal from measured global solar irradiance in Spain. Renew. Energy 21: 175–186

Louche A, Maurel M, Simonnot G, Peri G, Iqbal M (1987) Determination of Angström's turbidity coefficient from direct total solar irradiance measurements. Solar Energy 38: 89–96

Loutzenhiser PG, Manz H, Felsmann C, Strachan PA, Frank T, Maxwell GM (2007) Empirical validation of models to compute solar irradiance on inclined surfaces for building energy simulation. Solar Energy 81: 254–267

Ma CCY, Iqbal M (1984) Statistical comparison of solar radiation correlations Monthly average global and diffuse radiation on horizontal surfaces. Solar Energy 33: 143–148

Martinez-Lozano JA, Tena F, Onrubia JE, Rubia JDL (1984) The historical evolution of the Angström formula and its modifications: Review and bibliography. Agr. For. Meteorol. 33: 109–128

Maxwell EL (1987) Quasi-physical model for converting hourly global horizontal to direct normal insolation In: Hayes J, Andrejko DA (eds) Proc. Solar '87 Conf., Portland OR, American Solar Energy Society 35–46

Maxwell EL (1998) METSTAT—The solar radiation model used in the production of the National Solar Radiation Data Base (NSRDB). Solar Energy 62: 263–279

Michalsky JJ, Anderson GP, Barnard J, Delamere J, Gueymard C, Kato S, Kiedron P, McCormiskey A, Richiazzi P (2006) Shortwave radiative closure studies for clear skies during the Atmospheric Radiation Measurement 2003 Aerosol Intensive Observation Period. J. Geophys. Res. 111D: doi:10.1029/ 2005JD006341

Molineaux B, Ineichen P, O'Neill N (1998) Equivalence of pyrheliometric and monochromatic aerosol optical depths at a single key wavelength. Appl. Opt. 37: 7008–7018

Monteith JL (1962) Attenuation of solar radiation: a climatology study. Quart. J. Roy. Meteor. Soc. 88: 508–521

Muneer T, Gul MS, Kubie J (2000) Models for estimating solar radiation and illuminance from meteorological parameters. Trans. ASME J. Sol. Energy Eng. 122: 146–153

Muneer T, Fairooz F (2002) Quality control of solar radiation and sunshine measurements— Lessons learnt from processing worldwide databases. Build. Serv. Eng. Res. Technol. 23: 151–166

Muneer T, ed (2004) Solar radiation and daylight models, 2nd edn. Elsevier

Muneer T, Younes S, Munawwar S (2007) Discourses on solar radiation modeling. Renew. Sustain. Energy Rev. 11: 551–602

Myers RH (1986) Classical and modern regression with applications. PWS Publishers

Notton G, Muselli M, Louche A (1996) Two estimation methods for monthly mean hourly total irradiation on tilted surfaces from monthly mean daily horizontal irradiation from solar radiation data of Ajaccio, Corsica. Solar Energy 57: 141–153

NREL (2007) National Solar Radiation Database 1991–2005 update: User's manual. National Renewable Energy Laboratory, Golden, CO

Olmo FJ, Vida J, Foyo-Moreno I, Tovar J, Alados-Arboledas L (2001) Performance reduction of solar irradiance parametric models due to limitations in required aerosol data: case of the CPCR2 model. Theor. Appl. Climatol. 69: 253–263

Oreskes N, Schrader-Frechette K, Belitz K (1994) Verification, validation, and confirmation of numerical methods in the Earth sciences. Science 263: 641–646

Perez R, Stewart R (1986) Solar irradiance conversion models. Solar Cells 18: 213–222

Perez R, Ineichen P, Seals R, Michalsky J, Stewart R (1990a) Modeling daylight availability and irradiance components from direct and global irradiance. Solar Energy 44: 271–289

Perez R, Seals R, Zelenka A, Ineichen P (1990b) Climatic evaluation of models that predict hourly direct irradiance from hourly global irradiance: Prospects for performance improvements. Solar Energy 44: 99–108

Perez R, Ineichen P, Maxwell EL, Seals R, Zelenka A (1992) Dynamic global-to-direct irradiance conversion models. ASHRAE Trans. 98 (1): 354–369

Perez R, Seals R, Zelenka A (1997) Comparing satellite remote sensing and ground network measurements for the production of site/time specific irradiance data. Solar Energy 60: 89–96

Perez R, Kniecik M, Zelenka A, Renne D, George R (2001) Determination of the effective accuracy of satellite-derived global, direct, and diffuse irradiance in the Central United States. In: Campbell-Howe R (ed) Proc. Solar 2001 Conf., Washington D.C., American Solar Energy Society, Boulder, CO

Perez R, Ineichen P, Moore K, Kmiecik M, Chain C, George R, Vignola F (2002) A new operational model for satellite derived irradiances, description and validation. Solar Energy 73: 307–317

Power HC (2001) Estimating clear-sky beam irradiation from sunshine duration. Solar Energy 71: 217–224

Psiloglou BE, Balaras CA, Santamouris M, Asimakopoulos DN (1996) Evaluation of different radiation and albedo models for the prediction of solar radiation incident on tilted surfaces, for four European locations. Trans. ASME, J. Solar Engng. 118: 183–189

Reindl DT, Beckman WA, Duffie JA (1990) Evaluation of hourly tilted surface radiation models. Solar Energy 45: 9–17

Remund J, Wald L, Lefèvre M, Ranchin T, Page J (2003) Worldwide Linke turbidity information. Proc. ISES Conf., Gothenburg, Sweden, International Solar Energy Society

Rigollier C, Bauer O, Wald L (2000) On the clear sky model of ESRA—European Solar Radiation Atlas—with respect to the Heliosat method. Solar Energy 68: 33–48

Santamouris M, Mihalakakou G, Psiloglou B, Eftaxias G, Asimakopoulos DN (1999) Modeling the global solar radiation on the Earth's surface using atmospheric deterministic and intelligent data-driven techniques. J. Clim. 12: 3105–3116

Scharmer K, Greif J, ed (2000) The European Solar Radiation Atlas, Vol. 2. Presses de l'Ecole des Mines de Paris

Skartveit A, Olseth JA, Capelak G, Rommel M (1996) On the estimation of atmospheric radiation from surface meteorological data. Solar Energy 56: 349–359

Soler A (1990) Statistical comparison for 77 European stations of 7 sunshine-based models. Solar Energy 45: 365–370

Stone RJ (1993) Improved statistical procedure for the evaluation of solar radiation estimation models. Solar Energy 51: 289–291

Stone RJ (1994) A nonparametric statistical procedure for ranking the overall performance of solar radiation models at multiple locations. Energy 19: 765–769

Suckling PW, Hay JE (1976) Modelling direct, diffuse, and total solar radiation for cloudless days. Atmosphere 14: 298–308

Suckling PW, Hay JE (1977) A cloud layer-sunshine model for estimating direct, diffuse and total solar radiation. Atmosphere 15: 194–207

Taylor KE (2001) Summarizing multiple aspects of model performance in a single diagram. J. Geophys. Res. 106D7: 7183–7192

Thornton PE, Running SW (1999) An improved algorithm for estimating incident daily solar radiation from measurements of temperature, humidity, and precipitation. Agric. Forest Meteorol. 93: 211–228

Wang Q, Tenhunen J, Schmidt M, Kolcun O, Droesler M (2006) A model to estimate global radiation in complex terrain. Bound. Layer Meteorol. 119: 409–429

Watt D (1978) On the nature and distribution of solar radiation. U.S. DOE Report HCP/T2552-01, U.S. Department of Energy

Wilcox W, Anderberg M, George R, Marion W, Myers D, Renné D, Lott N, Whitehurst T, Beckman W, Gueymard C, Perez R, Stackhouse P, Vignola F (2007) Completing production of the updated National Solar Radiation Database for the United States. Proc. Solar 2007 Conf., Cleveland, OH, American Solar Energy Society

Willmott CJ (1981) On the validation of models. Phys. Geogr. 2: 184–194

Willmott CJ (1982a) Some comments on the evaluation of model performance. Bull. Amer. Meteorol. Soc. 63: 1309–1313

Willmott CJ (1982b) On the climatic optimization of the tilt and azimuth of flat-plate solar collectors. Solar Energy 28: 205–216

Willmott CJ, Ackleson SG, Davi RE, Feddema JJ, Klink KM, Legates DR, O'Donnell J, Rowe CM (1985) Statistics for the evaluation and comparison of models. J. Geophys. Res. 90C: 8995–9005

Yang K, Huang GW, Tamai N (2001) A hybrid model for estimating global solar radiation. Solar Energy 70: 13–22

Yang K, Koike T (2005) A general model to estimate hourly and daily solar radiation for hydrological studies. Water Resour. Res. 41: doi:10.1029/ 2005WR003976

Index

Printing: Krips bv, Meppel, The Netherlands
Binding: Stürtz, Würzburg, Germany